Birkhäuser

Thomas Sonar

The History of the Priority Di∫pute between Newton an*d* Leibniz

Mathematics in History and Culture

With an Epilogue by Eberhard Knobloch

Birkhäuser

Thomas Sonar
Technische Universität Braunschweig
Braunschweig
Germany

Translated by Thomas Sonar, Braunschweig, Germany; Keith William Morton, Oxford, UK; Patricia Morton, Oxford, UK

Editor: Project Group "History of Mathematics" of Hildesheim University H.W. Alten, K.-J. Förster, K.-H. Schlote, H. Wesemüller-Kock

ISBN 978-3-030-10226-5 ISBN 978-3-319-72563-5 (eBook)
https://doi.org/10.1007/978-3-319-72563-5

Originally published in German in the series "Vom Zählstein zum Computer" under the title: "Die Geschichte des Prioritätsstreits zwischen Leibniz und Newton. Geschichte - Kulturen – Menschen" (ISBN 978-3-662-48861-4) Springer-Verlag Berlin Heidelberg 2016 ISBN 978-3-662-48862-1 (ebook)

Printed on acid-free paper

This book is published under the imprint Birkhäuser, www.birkhauser-science.com by the registered company Springer International Publishing AG part of Springer Nature
The registered company address is: Gewerbestrasse 11, 6330 Cham, Switzerland

The two warriors immortalised in the University Museum of Oxford [Photos Sonar]

Dedicated to
Keith William (Bill) Morton,
mentor, teacher, and friend in Oxford,
and to his wonderful wife
Patricia (Pat) Morton.

'The world, surely, has not another place like Oxford:
it is a despair to see such a place and ever have to leave it.'

NATHANIEL HAWTHORNE, 1856.
[Morris 1978, p. x]

About the Author

Thomas Sonar was born 1958 in Sehnde next to Hannover. After studying Mechanical Engineering at the University of Applied Sciences ('Fachhochschule') Hannover he became a laboratory engineer in the Laboratory for Control Theory of the same University for a short time, and founded an engineering office. He then studied mathematics at the University of Hannover (now Leibniz University), after which he worked from 1987 until 1989 at the German Aerospace Establishment DLR (then DFVLR) in Brunswick for the orbital glider project HERMES as a scientific assistant. Next he went to the University of Stuttgart to work as a PhD student under Prof Dr Wolfgang Wendland while spending some time studying under Prof Keith William Morton, at the Oxford Computing Laboratory. His PhD thesis was defended in 1991 and Thomas Sonar went to Göttingen to work as a mathematician ('Hausmathematiker') at the Institute for Theoretical Fluid Mechanics of the DLR; there he developed and coded the first version of the TAU-code for the numerical computation of compressible fluid fields, which is now widely used. In 1995 the postdoctoral lecture qualification for mathematics was obtained from the TU (then TH) Darmstadt on the basis of a habilitation treatise. From 1996 until 1999 Thomas Sonar was full professor of Applied Mathematics at the University of Hamburg and is professor for Technical and Industrial Mathematics at the Technical University of Brunswick since 1999 where he is currently the head of a work group on partial differential equations. In 2003 he declined an offer of a professorship at the Technical University of Kaiserslautern connected with a leading position in the Fraunhofer Institute for Industrial Mathematics

ITWM. In the same year Sonar founded the centre of continuing education for mathematics teachers ('Mathelok') at the TU Brunswick which stays active with regular events for pupils also.

Early in his career Thomas Sonar developed an interest in the history of mathematics, publishing in particular on the history of navigation and of logarithms in early modern England, and conducted the widely noticed exhibitions in the 'Gauss year' 2005 and in the 'Euler year' 2007 in Brunswick. Further publications concern Euler's analysis, his mechanics and fluid mechanics, the history of mathematical tables, William Gilbert's magnetic theory, the history of ballistics, the mathematician Richard Dedekind, and the death of Gottfried Wilhelm Leibniz. In 2001 Sonar published a book on Henry Briggs' early mathematical works after intense research in Merton College, Oxford. In 2011 his book *3000 Jahre Analysis* (3000 years of analysis) was published in this series and in December 2014 he edited the correspondence of Richard Dedekind and Heinrich Weber. Altogether Thomas Sonar has published approximately 150 articles and 14 books – partly together with colleagues. He has established a regular lecture on the history of mathematics at the TU Brunswick and has for many years held a lectureship on this topic at the University of Hamburg. Many of his publications also concern the presentation of mathematics and the history of mathematics to a wider public and the improvement of the teaching of mathematics at secondary schools.

Thomas Sonar is member of the Gesellschaft für Bildung und Wissen e.V. (Society for Education and Knowledge), the Braunschweiger Wissenschaftliche Gesellschaft (Brunswick Scientific Society), a corresponding member of the Academy of Sciences in Hamburg, and an honorary member of the Mathematische Gesellschaft in Hamburg (Mathematical Society in Hamburg).

Preface of the Author

Scientific priority - In science, priority is the credit given to the individual or group of individuals who first made the discovery or propose the theory. [...] Priority becomes a difficult issue usually in the context of priority disputes, where the priority for a given theory, understanding, or discovery comes into question.

WIKIPEDIA
en.wikipedia.org/wiki/Scientific_priority
(lastly retrieved on 3rd July 2017)

Quarrels on the priority of scientific methods, theories or inventions occured often in history, and still occur; almost none was so doggedly fought as the famous priority dispute between Isaac Newton (4th January 1643 – 31st March 1727)[1] and Gottfried Wilhelm Leibniz (1st July, 1646 – 14th November, 1716) on the invention of the differential and integral calculus, and none has had such a tremendous impact. Hal Hellman has included this quarrel in his book on the great feuds in mathematics [Hellman 2006] and he described only ten feuds in all. We can therefore be certain we are dealing here with a real scandal.

The controversy grew quickly to an international incident: England against the continent! On Leibniz's side were the pugnacious brothers Jacob and John Bernoulli and all continental mathematicians who understood the new calculus (small in number, of course). Behind Newton stood his English backers and the mathematicians of the Royal Society. Both groups were not really large by comparison with the impact of the quarrel! The huge effect of the priority dispute was the separation of English mathematics from the mathematics on the continent well into the 20. century! The English stood behind Newton and used his cumbersome dot notation $\dot{x}, \ddot{x}$, and so forth for the derivatives of a

[1]Protestant England had not yet introduced the Gregorian calendar in Newton's days since it was thought a papist invention. The English were hence 10 days behind until the year 1700 and 11 days from the 28th February, 1700. In the Julian calendar Newton was born on the 25th December 1642 – he was a Christmas child – and died on 20th March, 1726. Until the year 1752 the New Year began in England at the 25th of March. Therefore, many authors give dates between the 1st of January 1727 and the 24th of March 1727 in the form 1726/27, but in order not to confuse my affectionate readership I have converted all dates to the Gregorian format. Only at a few places I have clearly indicated a Julian date.

function $x(t)$ while the continental mathematicians recognised the superiority of Leibniz's calculus and used the differential quotient dy/dx instead of $\dot{y}/\dot{x}$, which easily led to intuitive calculation schemes.

Thus, mathematics on the continent developed with quantum jumps while the English apparently did not add much to it. In the year 1755, one generation after the two adversaries, the first of two volumes of the opus *Institutiones Calculi Differentialis* of the great Leonhard Euler (1707–1783) appeared, who was a pupil of John Bernoulli. In chapter four Euler remarked in section 116 [Blanton 2000, S.64f.]:

> '*... but there is no doubt that we have won the prize from the English when it is a question of notation. For differentials, which they call fluxions, they use dots above the letters. Thus, $\dot{y}$ signifies the first fluxion of y, $\ddot{y}$ is the second fluxion, the third fluxion has three dots, and so forth. This notation, since it is arbitrary, cannot be critized if the number of dots is small, so that the number can be recognized at a glance. On the other hand, if many dots are required, much confusion and even more inconvenience may be the result. For example, the tenth differential, or fluxion, is very inconveniently represented with ten dots, while our notation, $d^{10}y$, is very easily understood. There are cases where differentials of even much higher order, or even those of indefinite order, must be represented, and for this the English mode is completely inapt.*'

However, Euler hastened to insist in the following section 117 that also the English quite liked the mathematical literature of the Leibnizians from the continent and that continental mathematicians for their part were reading also English books on the calculus of fluxions.

Indeed it would be clearly missing the point to indicate a sudden shortfall of English mathematics as compared to the mathematics on the continent after Newton's death. In a very detailed investigation Niccolò Guicciardini [Guicciardini 1989] has followed the development of Newton's calculus of fluxions from 1700 up to 1800. In effect English mathematicians could very well keep up until the mid 18th century, but then the roaring development of multidimensional differential and integral calculus started on the continent and this progress was barely absorbed in England – England fell behind.

The difference between modern continental mathematics and English mathematics *à la* Newton was so large at the beginning of the 19th century that something *had* to be done. In 1803 mathematics professor Robert Woodhouse (1773–1827) from Cambridge published his book *Principles of Analytical Calculation* in which he used Leibniz's differential and integral calculus, but Woodhouse apparently did not emphasize or enforce these notations and techniques in the syllabus of his university [Hyman 1987, p. 40]. In the year 1810 young Charles Babbage (1791–1871) came to Cambridge to take up his studies in mathematics. He was the son of a wealthy father and later became the inventor of the program-controlled computer. The young man was recommended a French textbook by Sylvestre François Lacroix (1765–1843) which treated Leibniz's calculus and which Babbage bought for an enormous price in England. He knew therefore the Leibnizian calculus was superior to that of his tutors, and hence was dissatisfied with the contents of study in Cambridge. He quickly found like-minded fellow students: together with George Peacock (1791–1858) and John Herschel (1792–1871) he founded the 'Analytical Society' with the purpose of propagating the heresy of 'pure d-ism against the Dot-age of the University' [Guicciardini 1989, p. 135]. They wanted the Leibnizian d-notation instead of Newton's dots! Ironically 'dot-age' sounds very similar to 'dotage'. Honi soit qui mal y pense?

In this book we want to follow the history of this fascinating priority dispute, illuminate and discuss the pre-history of it, the actual quarrel, and the aftermath. This is not the first attempt to do so and different authors have tried to increase the sales figures of their books by choosing a dramatic title in which the word 'war' played a significant role. So did the historian of science Alfred Rupert Hall (1920–2009) who presented the recommendable opus *Philosophers at war* [Hall 1980][2] and quite recently the less recommendable, since lurid, *The Calculus Wars* [Bardi 2006] by the US-American biophysicist and science journalist Jason Socrates Bardi. Both books are comletely free of mathematics. Beyond doubt it was a war: and on the English side even a war of extermination with the aim to sweep Leibniz from the stage of history. I have decided nevertheless to use a less spectacular title[3] and follow therebye the small brochure [Fleckenstein 1956] by the historian of mathematics and astronomy Joachim Otto Fleckenstein (1914–1980). The booklet was published in a second edition in 1977 but significant new findings were not incorporated at that time.

We begin by discussing what it is all about. I ask to be excused by the mathematically educated readership for explaining the basics of the differential and integral calculus, but I felt the responsibility that also the purely historically inclined should have a chance to grasp the proper meaning of this invention before they start wondering about the pungency of the priority dispute. Besides, the first chapter is to show that this book is *not* free of mathematics. As I have argued already in the preface of my book [Sonar 2011] it is impossible to grasp the history of mathematical analysis without looking at the mathematics involved. But have no fear! The mathematical level presented here is not beyond the level at the end of secondary school, and I have taken pains to explain everything as clear as possible so that even mathematics haters should

[2] Also recommendable as a survey is his shorter work [Hall 2002].

[3] The rolling S and the italic d in the title are not printing errors but signify the two important operators – integral and differential – of Leibniz's infinitesimal mathematics.

be able to follow. In case of emergency just skip the mathematics, although then you are going to miss something.

In the second chapter we introduce the political and cultural developments in France, the Netherlands and England around the time of the priority dispute. We then turn to the 'giants' on whose shoulders Newton had claimed to stand. In the mind of Newton these giants are John Wallis (1616–1703) and Isaac Barrow (1630–1677); for Leibniz we have to consider Christiaan Huygens (1629–1695) and Blaise Pascal (1623–1662). If the reader asks himself why we do not discuss the political and cultural developments in Germany I have to answer that 'Germany' did not exist in these days. We would have to deal with the 'German lands' each reigned by a local ruler. Besides, Leibniz was born two years before the Peace of Westfalia was signed and the German lands were devastated from the Thirty Years' War. From our point of view there is thus not much to be reported.

In the third chapter we cast an eye on the early developments of both of our combatants. As with every conflict, this one began to ripen slowly, thrived on misunderstandings and jealousies, and gradually built up. In the centre are the two adversaries – Isaac Newton and Gottfried Wilhelm Leibniz, up to the year 1672. This is the year in which Leibniz arrives in Paris and his astounding mathematical developments begin. Until 1672 Leibniz is a mathematical nobody, although he feels his strength growing. Newton has already made important mathematical discoveries but they are still unpublished. First conflicts with Robert Hooke (1635–1703) on the theory of coulours begin to haunt Newton and make him even more cautious and vulnerabel as he already is.

It is in the fourth chapter where the conflicts start to develop into a cold war – for the time being. The 'annus mirabilis' for Leibniz is the year 1673 in which he launches his new differential and integral calculus while in Paris. It is not yet a calculus, the d and the integral sign $\int$ are still missing, but by 1676, the year in which Leibniz leaves Paris, the new analysis is a true 'calculus'. Here the climax of the cold war which is soon to become hot, lies in the years 1675 and 1676.

In the year 1676 Leibniz and Newton exchange letters via the secretary of the Royal Society, Henry Oldenburg (about 1618–1677). We analyse the two letters known as *Epistola prior* and *Epistola posterior* in the fifth chapter under the heading 'No Trace of the Cold War', since I couldn't follow some earlier authors who already found here traces of Newton's suspicions. But read it for yourself.

In the sixth chapter we focus on Newton's arguably most important work, the *Principia*, the origin of modern physics. Our two warriors have completely different vita: Leibniz gets employed in the province but unfolds a hustle of outrageous breadth in many areas, Newton lives in Cambridge (from the point of view of Oxford certainly also a province) the life of a hermit who does not want to be disturbed. We look thoroughly into the evolutionary history of the *Principia* but also into the aftermath. It is here that Leibniz lays a fuse for conflicts yet to come, in that he reacts to the *Principia* with three of his own publications and states that he did not see the *Principia* beforehand.

The war turns hot with chapter seven. Newton undergoes several crises; in particular a break in the relationship with the young Swiss scholar Fatio de Duilier, whom we shall disrespectfully call 'Newton's monkey', therebye following Frank E. Manuel who called him 'the ape of Newton' [Manuel 1968, p. 191]. This monkey is the first to accuse Leibniz of plagiarism, but for the time being only within the private correspondence with Christiaan Huygens, and Huygens is wise enough not to inform his pupil Leibniz. John Wallis now pushes Newton to eventually publish his mathematical works of long ago; then a quarrel develops between the Royal Astronomer John Flamsteed and Newton, and at last Newton leaves Cambridge to occupy a position at the Royal Mint in London. We now see a Newton other than the withdrawn hermit in Cambridge. Newton is self-confident now and gets his leading position straight after the publication of the *Principia.* Meanwhile Leibniz has found highbrow advocates for his calculus, namely the Bernoullis and the Marquis de l'Hospital who, with great support by John Bernoulli, publishes the first textbook on the Leibnizian differential calculus in the year 1696 [Bradley et al. 2015]. John Wallis, very advanced in years, publishes his mathematical œvre in three volumes and takes Newton's side. In the third volume the *Epistolae* which Newton wrote for Leibniz in the year 1676 are published for the first time, as are some letters from Leibniz in order to show that the priority of the invention of infinitesimal calculus belongs entirely to Newton. Then a public attack of Fatio on Leibniz takes place which Leibniz is able to counter, also with the help of John Bernoulli. Fatio has to admit defeat but the war is clearly declared in 1699. Here is also the right place to critically iluminate Leibniz's behaviour in the Republic of Letters of his day.

With chapter eight the war breaks out with full force, reopened by another of Newton's monkeys, John Keill. As in the case of Fatio, Leibniz looks for help at the Royal Society but this time other circumstances prevail: Newton is president of the Royal Society. Officially a committee was constituted to decide the case, but Newton is behind the committee. The result is the *Commercium epistolicum* in which Leibniz is charged for plagiarising Newton's mathematics. Leibniz and John Bernoulli defend themselves, a flysheet appears – the *Charta volans* – and the warriors on both sides of the English Channel entrench themselves in their intellectual dugouts. Leibniz's death would have given enough reason to end the war, but it continued.

The ninth chapter concerns the war after the death of Leibniz. John Bernoulli doggedly fights farther against the English, but tries to stay on good terms with Newton at the same time. The war only ends with the death of Newton in the year 1727.

We cannot report the whole story of the priority dispute without committing ourselves to its aftermath. The tenth chapter is devoted to the early disputes over the calculus which had surfaced already in Newton's and Leibniz's lifetimes. In the case of Leibniz it is in particular Bernard Nieuwentijt, who massively attacks the calculus on the grounds of the use of higher order differentials, but he can be parried. For Newton's calculus of fluxions it is the bishop George Berkeley who initiates a chain reaction with his critical writings which turn out to be very effective indeed.

What meaning has the war for English analysis? We answer this question in chapter eleven. In the 19th century Newton's calculus of fluxions and fluents was at a dead end. However, driven by English national pride Newton biographies appear which border on hero worship. Only the mathematician Augustus de Morgan breaks this worship and verifies that Newton is not the radiant winner as the English want to see him. By and by England's long journey to accept Leibniz's analysis begins but again a dead end due to the algebraic analysis of Lagrange is reached. Only with the publication of books like *Calculus made easy* by Silvanus P. Thompson with which we have opened this book has England arrived at the mathematics of Leibniz, the Bernoullis, and Euler.

I close our journey through the history of the priority dispute with some personal afterthoughts which are necessary in my opinion. The intensive research on the war over the priority of the invention of the differential and integral calculus has in fact changed some prejudices and assessments which I have adhered to for a long time.

A word concerning the indices: In the text the reader may find names of persons who do not appear in the index of names. This is intentionally so since there are many persons who did not contribute to the history of the priority dispute but deserved to appear in the text; for example this happens in the case of the assistents of Flamsteed. Book titles in the index follow the true titles as they appear on books or journal articles. Hence from time to time all words of a title may be written in lowercase letters (except the first word and names) while other publications appear with upper case letters in nearly every word in the title.

In the realms of the history of mathematics the study of sources plays a central role. Since a large number of documents is available in the case of Leibniz and Newton – letters in particular – it was my concern to let the persons involved 'speak' for themselves as often as possible. Only one who has read the contents and grasped the tone of the correspondence in person can eventually detect swings or changes in mood or can come to other assessments than the ones handed down in the literature. Inserted in the quotations within squared brackets are remarks by the translator. I wish my readership new and deep insights in one of the most famous and significant conflicts the world history of science has ever seen.

Acknowledgements

It was Mr Clemens Heine from Springer-Verlag who initiated this book in an email dating from the beginning of the year 2014. He was looking for a competent author who could treat this topic and without hesitation I proposed Prof Dr Eberhard Knobloch since he is *the* international expert on Leibniz. To my great pleasure Professor Knobloch instead proposed me and promised to critically read and correct the manuscript what he already did in case of my book on the history of analysis [Sonar 2011].

My exceptional gratitude is thus due to Mr Heine and to Professor Knobloch who has accompanied this project with his expert knowledge from start to finish. In particular he has translated Latin, French, and Italian book titles and quotations (nearly in real-time!), has pointed out new literature to me, and has painstakingly read the entire manuscript. Eventually he has even contributed an epilogue. For all he did I cannot thank him enough. Dear Eberhard, I have learned a lot from you and owe you very much!

A further reader of the German manuscript was Karl-Heinz Schlote from the project group 'History of Mathematics' at the University of Hildesheim. I am very grateful that he read the entire manuscript twice with sharp eyes and has given important hints concerning the style. If I have not followed his advice in rare cases then this is only due to my pig-headedness. Franz Lemmermeyer has read parts of the German manuscript for which I am very grateful. Dr Kühn from Leipzig has whipped my bibliography into shape and my friend Prof Dr Klaus-Jürgen Förster from the project group in Hildesheim has read the manuscript during his summer vacation on the beach. Additionally he gave moral support more than once. The last finishing touches on the German manuscript were provided by Prof Dr Alten from Hildesheim. I have to thank them all.

Clemens Heine gave me the choice to publish the German book either as an independent book with Springer or in the series 'Vom Zählstein zum Computer' (From the Pebble to the Computer) of the project group in Hildesheim in which my former book *3000 Jahre Analysis* appeared. Since the cooperation with the colleagues and friends of the project group in Hildesheim concerning my former book was outstanding I decided to publish again in their series. I have no regrets! This book was made possible by the project group in Hildesheim around my friend Klaus-Jürgen Förster and our doyen Heinz-Wilhelm Alten whom I have to thank as much as Heiko Wesemüller-Kock, who in his usual manner has provided photos and figures of high quality. I also have to thank Mrs Anne Gottwald for obtaining the difficult and elaborate authorisation of licenses.

Although working for many years with a distinguished word processor like LaTeX, in a project as large as this one I needed from time to time a TeX-Wizard. At my side was the excellent PhD student Jakob Schönborn from the University of Hildesheim who competently advised me even on weekends, public holidays, and around Christmas (!). May his girlfriend forgive me.

I am grateful to my secretary Mrs Jessica Tietz for her tireless ordering, retrieving, and bringing back books which I could only get hold of by borrowing because buying them was either impossible or prohibitively expensive.

I have to thank Dr Charlotte Wahl from the Leibniz Research Centre Hannover of the Göttingen Academy of Sciences for one of her manuscripts concerning the priority dispute. I also thank Prof Dr Herbert Breger who is the former head of the Leibniz Archives in Hannover, and Prof Dr Wenchao Li from the University of Hannover since they gave me the opportunity to give a talk on the dispute in spring 2015 in the Leibniz house in Hannover. Mrs Anja Fleck from the Gottfried Wilhelm Leibniz Library in Hannover has spent a whole afternoon to show Leibniz manuscripts to me and advised me in choosing some of them for this book. Thank you very much!

This book came into existence while I was the dean of the Carl-Friedrich-Gauß faculty of the Technical University Brunswick and everyone who has experience with this function in our times knows what this means. If I acted as a reasonable and responsible dean, and I can only hope I did, I owe this primarily to my managing director Mrs Imma Braun who steadfastly was my support and my 'boss' in all things administrative. Since we follow the same protestant work ethic I might be allowed to call us both 'war pigs' – she will correctly understand this as a great compliment!

I have to thank my excellent physicians for keeping me running against all odds. In particular I have to thank chief physician Professor Dr med. Max Reinshagen from the Klinikum Braunschweig, Dr med. Jörn Schröder-Richter from the former St. Vincent-Stift in Braunschweig, my GP Dr med. Torsten Prüfer, who sadly passed away in 2017, and Ahmad Rahimi, my internist and angiologist. During the finishing touches of the manuscript my internist Walter Christian was of invaluable help. Their appeals to me to cut back activities and recover fully I have taken seriously every time, but somehow it never does the trick.

I am whole-heartily grateful to my wife Anke for bearing stoically on seeing me proceeding from one book project to the next. The knee-high stacks of books on the floor in the living and dining room for which I have apologised to her already in *3000 Jahre Analysis* have grown in number as well as in height – not to mention the costs for these books. Even though my increased workload has reduced our joint hours further, my love for her has continually increased through 35 years of matrimony. This is also the right place to ensure our children Konstantin, Alexander, Philipp, and Sophie-Charlotte (in order of birth), all true haters of mathematics, that I am very proud of them. Moreover, Sophie-Charlotte may find her namesake here.

This book is dedicated to my thesis advisor and paternal friend in Oxford: Keith William 'Bill' Morton, who also took the pains of correcting my translation. Dear Bill, of all the people who shaped my view on mathematics and its history I owe you the most and can't thank you enough. When Bill's health began to deteriorate in 2016 his wonderful wife Pat took the burden of correcting my English. This book is dedicated also to her.

Thomas Sonar, October 2017

Preface of the Editors

Who does not know the joy of having discovered something new for oneself? In many cases it is quite irrelevant whether others have discovered the same finding before us or at the same time. However, in our commercialised world in which the thought of profitable marketing often comes directly after a happy discovery, it plays an important role in science and economics to secure ones rights to the first finding – the priority – of the idea, discovery, or development by an independent institution. Despite global networking and a permanent exchange of information, even today independent coincidental or nearly concurrent discoveries may occur and can lead to quarrels concerning priority. In former times with their restricted and slow ways of communications and radically different forms of communication as compared to today, the more difficult clarification of priority issues repeatedly got out of control. One of the most well known cases is the priority dispute between Gottfried Wilhelm Leibniz and Isaac Newton concerning the invention of the infinitesimal calculus. The fame of this dispute is not least owed to the fact that not only both of the accepted scientists were involved who, a few years before, had ensured themselves mutual esteem, but that additionally important scientists got involved in the conflict and stood opposite to each other in enemy camps. At the same time the emergence of the infinitesimal calculus, as an important part of the scientific revolution in the 17th century, first of all influenced the further development of mathematics, natural sciences, and technology sustainably; and later on even more sciences were influenced.

In view of the tremendous significance of this scientific discovery, and the involvement of numerous scientists, it is not surprising that the priority dispute was the subject of several studies in the history of science. The individual portrayals differ considerably, owing either to the imprint of national feelings and opinions, or to a one-sided concentration on single aspects or actors in the quarrel, and hence to severely differing choices of sources. Meanwhile many primary as well as secondary sources are well accessible and analysed. The scientific achievements of both of the heroes and their fellow campaigners can be clearly described. The first appearance of the fundamentally new ideas can be dated quite accurately in most cases, in particular when they were written down for the first time. The broader historical context was also examined thoroughly – be it the circumstances of the lives of Newton, Leibniz, and other scientists involved, the intellectual attitudes and philosophical currents of that period, the habits of communication, or the social conflicts.

Thomas Sonar now has turned himself to the task – faithfully to the concerns of the book series 'Vom Zählstein zum Computer' – to compose a well-balanced and clearly legible book for historians of science, mathematicians and other scientists, and for a broader, interested readership. In his own lively style he introduces the reader to continental Europe ravaged by the Thirty Years War and further struggles for power as well as to England, shaken by numerous fights between parliament and monarchy. He brings alive the protagonists Newton and Leibniz as well as scientists as Christiaan Huygens, Johann Bernoulli, John Wallis, and Isaac Barrow with their strenghts and weaknesses, their particularities and their customs in their living environments. It is pointed out in detail how the climate of scientific exchange of ideas changed due to deliberate as well as thoughtless human actions and statements. It is shown that this exchange of ideas indeed breathed the spirit of competition and the ambition for recognition, yet was characterised by mutual appreciation; but then it suddenly turned over into a relentless confrontation and hostility. Thomas Sonar clearly points out that much happened without or at least with not much influence or awareness of the two opponents, but that later on both were not free of unfair actions and contributed to the escalation. In all this controversy the science does not fall by the wayside; and the author confidently imparts the necessary knowledge, be it the Leibnizian calculus, the fluxional calculus of Newton, or physics and astronomy – in particular that concerning the theory of gravitation and the theory of light. This opens the door for a broader view of particular incidents: for example, the inclusion of seemingly separate events in order to understand how the rejection of Newton's theory of gravitation by Leibniz worsened the dispute; or how the quarrels between Newton and Robert Hooke concerning the fundamental ideas of this theory sustainably influenced Newton's later dealings with critical comments.

The present book is the English translation of the German book 'Die Geschichte des Prioritätsstreits zwischen Leibniz und Newton', which constitutes a volume within the series 'Vom Zählstein zum Computer' (From the pebble to the computer). After the book on the history of geometry, '5000 Years of Geometry', Sonar's treatise is the second volume of this series that appeares in an English edition. The project group thanks Thomas Sonar wholeheartedly for finishing the English translation of his book so shortly after the publication of the German edition on the occasion of the 300th anniversary of the death of Gottfried Wilhelm Leibniz in 2016. As is the German original the 'Priority Dispute' is also richly equipped with illustrations for which a special thank is due to the media scientist and associate editor H. Wesemüller-Kock who implemented some necessary changes of the graphic design, and to Mrs A Gottwald for obtaining the difficult and elaborate authorization of licenses. Moreover we thank all the colleagues and friends contributing to the finalisation of the book.

We are grateful to Birkhäuser, in particular to Mrs S. A. Goob, for their cooperation concerning the usual superb layout and the splendid presentation of this English edition of the book.

It is a great pleasure for the project group that the 'Priority Dispute' will now be available to a much broader readership. We hope that the book will impart a lively impression of the development of mathematics not only to historians and expert academics, but to pupils, students, teachers, as well as many other interested readers. This development is not only concerned with abstract theories but at the same time with the people creating them, and as such it is often influenced by everyday characteristics.

Hildesheim, October 2017, for the editors

Karl-Heinz Schlote *Klaus-Jürgen Förster*

Project group 'History of Mathematics' of the University Hildesheim

Advice to the reader

Parentheses contain additional insertions, biographical details, or references to figures.

Squared brackets contain

- omissions and insertions in quotations
- references to the literature within the text
- references to sources in legends of figures

In the figure legends squared brackets mark the author/creator of the particular work. Further specifications appear in common paranthesis.

Figures are numbered following chapters and sections, e.g. Fig. 10.1.4 means the fourth figure in section 10.1 of chapter 10.

The original titles of books and journals appear in italic type, likewise quotations. Further reading or explanations of only shortly described circumstances are marked by references like '(cp. more detailed in...)'.

Literally or textually quoted literature as well as further reading can be found in the bibliography.

Contents

1 An Attunement

T. Sonar, *The History of the Priority Diſpute between Newton and Leibniz*, https://doi.org/10.1007/978-3-319-72563-5_1

1.1 What is it All About?

Already in antiquity Archimedes (about 287–212 BC) succeeded in computing tangents to given curves and areas under curves. However, in his day he was unable to recognise the two operations – computing tangents and areas – are in fact inverse operations: this can be seen from the fundamental theorem of calculus which was first discovered by Newton and Leibniz. Leibniz transformed it from the geometry of Isaac Barrow (1630–1677) into the new language of symbolic algebra, in which form it could display its full power.

But *why*, for God's sake, would one want to compute tangents and areas? This question has been asked certainly by generations of pupils and you are possibly asking it yourself in this moment. I shall try to give an explanation but first try to clarify how one can perform such computations. In this task I revert to a great guide: the book *Calculus made easy* by the English physicist Silvanus Phillips Thompson (1851–1916).

Calculus made easy was published in 1910 and is still in print up to this day! In the year 1998 the well-known author and science journalist Martin Gardner (1914–2010) released the book [Thompson 1998a] with a new introduction and modernised it carefully. In the same year the reprint of the 12th German edition *Analysis leicht gemacht* [Thompson 1998b] appeared. Thompson's mathematics does not match modern precision in all respects but is appropriate for our purposes! If you happen to have experienced a good education in calculus you may skip the following sections, of course.

Fig. 1.1.1. Silvanus P. Thompson [Photo: unknown photographer, before 1900 from: Thompson and Thompson, 1920] and in a lecture for a junvenile audience [Photo: photographer of the "Daily Mirror", 1910, Thompson and Thompson, 1920]

1.2 Slope, Derivative, and Differential Quotient

What is meant if you see a slope of 10 % as a warning on a traffic sign? Do you have to climb a vertical wall if the slope would be 100 %?

Fig. 1.2.1. Beware: 10 % slope

The slope signifies the ratio of the height to climb to the driven route. If a height of 250 m has to be climbed on the next 5 km the slope is

$$s = \frac{250\,\text{m}}{5000\,\text{m}} = 0.05$$

or 5 %. If instead 500 m would have to be climbed on a horizontal route of 5 km then the slope would be 10%, and if on a 5 km route the height of 5 km would have to be overcome, then the slope would be 100 %. Figure 1.2.2 shows the situation whereby Δx designates the horizontal route to drive and Δy is the height.

We can associate an angle α with the slope via the tangent function,

$$\tan\alpha = \frac{\Delta y}{\Delta x},$$

because the **gradient triangle** apparently is a right triangle. In our example of 5 % slope the angle α would be

$$\alpha = \arctan 0.05 \approx 2.86^\circ,$$

10 % slope corresponds to an angle of 5.7°, and 100 % slope leads to $\alpha = \arctan 1 = 45^\circ$.

However, a road in fact does not consist of plane pieces but rather is a function

$$y = f(x)$$

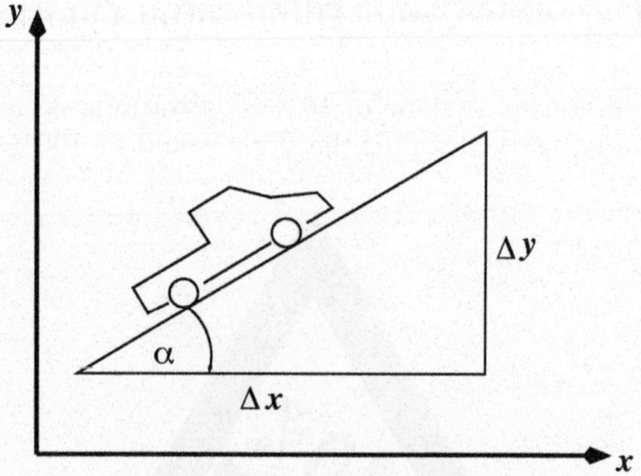

Fig. 1.2.2. Slope and angle

as shown in Figure 1.2.3. As can be seen, the slope $\Delta y/\Delta x$ is no longer the true slope at the point x_0 and the larger we choose Δx the less has our slope to do with the true slope. It is here where the new calculus of Newton and Leibniz comes into play.

We now assume that we take a tiny little step in x-direction. Really very, very tiny. Hence an infinitely small step (or, in the words of Thompson, 'a little bit of x') [Thompson 1998a, p. 42]), and to distinguish this step from Δx we write it in Leibniz's notation as dx. The 'd' can therefore be read as 'a little bit of'.

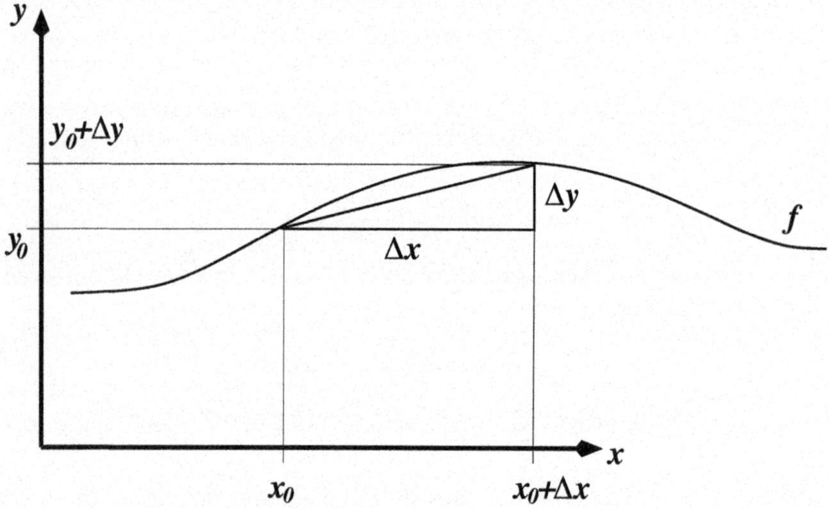

Fig. 1.2.3. Realistic shape of a road

Therefore dx is 'a little bit of x'.

Moving fom x_0 a tiny step dx to the right results in a tiny bit dy in the difference of height. Moving from x_0 to $x_0 + dx$ the function value $y_0 = f(x_0)$ turns into the value $y_0 + dy$. Since we only moved away from x_0 a small bit the **differential quotient**

$$s(x_0) := \frac{dy}{dx} \tag{1.1}$$

is the true slope of the road at the point x_0. This true slope describes the *change* of the road at the point x_0. But how can this quotient be computed from two infinitely small quantities?

We can find a good example in the parabola $y = f(x) = x^2$. If we want to compute the slope at the point $x_0 = 2$ we have to consider the change from $y_0 = f(x_0) = x_0^2$ to $y_0 + dy = f(x_0 + dx)$:

$$y_0 + dy = f(x_0 + dx) = (x_0 + dx)^2 = x_0^2 + 2x_0\,dx + (dx)^2.$$

Now $y_0 = x_0^2$, so that we get

$$dy = 2x_0\,dx + (dx)^2. \tag{1.2}$$

We assumed the quantity dx to be infinitely small. How large then is the square of such a quantity? We may suspect that we can consider $(dx)^2 = 0$ without problems but why then can we *not* already set $dx = 0$ from the beginning? It helps to think of the area of a square with edge length $x + dx$ as in Figure 1.2.4. Obviously, the area is $(x + dx)^2 = x^2 + 2x\,dx + (dx)^2$ and we can observe the following facts: Both of the stripes with area $x \cdot dx$ contribute significantly to the area but the small square with area $(dx)^2$ can surely be neglected.

One speaks of **smallness of higher order**, calls quantities like $x \cdot dx$ **quantities of first order**, quantities like $dx \cdot dx$ **quantities of second order**, and so on, and says that quantities of higher order are negligible in comparison with quantities of first order.

Therefore our equation (1.2) transforms into

$$dy = 2x_0\,dx,$$

and division by dx yields the slope at the point x_0,

$$\frac{dy}{dx} = 2x_0.$$

At the point $x_0 = 2$ the slope of the function $y = x^2$ hence is $dy/dx(2) = 4$.

In this way – and with the **binomial theorem** – we can show that the slope of the family of functions $y = x^n$, n any natural number, is $dy/dx = nx^{n-1}$. This follows from the ansatz

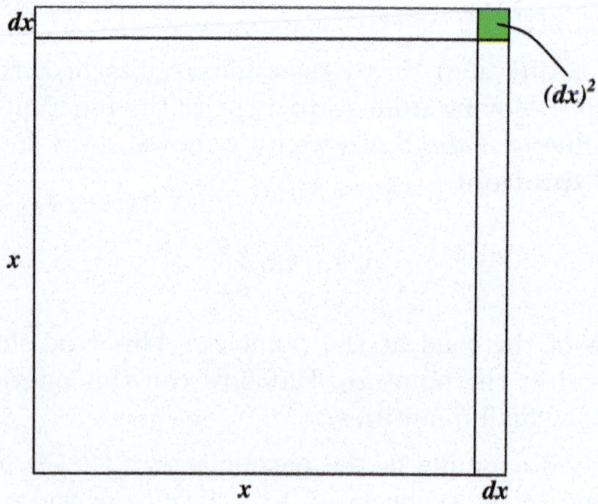

Fig. 1.2.4. Comparing areas in a square

$$y + dy = (x + dx)^n$$

and the binomial expansion of the term in brackets:

$$(x + dx)^n = x^n + \frac{n}{1} x^{n-1}\, dx + \frac{n(n-1)}{1 \cdot 2} x^{n-2}\, (dx)^2 + \ldots$$
$$+ \ldots \frac{n(n-1)(n-2)\cdots 2}{1 \cdot 2 \cdot \cdots \cdot (n-1)} x(dx)^{n-1} + \frac{n(n-1)(n-2)\cdots 2 \cdot 1}{1 \cdot 2 \cdots (n-1) \cdot n} (dx)^n.$$

Certainly you know this expansion for the case $n = 2$ since this is the binomial formula $(a + b)^2 = a^2 + 2ab + b^2$. Since we have already set $(dx)^2 = 0$ we can cross out all terms with even higher powers of dx and get

$$y + dy = x^n + nx^{n-1}\, dx,$$

and since $y = x^n$ the differential quotient results in

$$\frac{dy}{dx} = nx^{n-1}.$$

The binomial theorem is valid not only for natural numbers n as exponents; that is one of the essential findings of Newton. Then the sum is no longer finite but has an infinite number of summands; but that does not need to bother us since we set the higher powers of dx to zero anyway. Hence we can also compute slopes of root functions $y = \sqrt[k]{x}$, since $\sqrt[k]{x} = x^{1/k}$:

$$y + dy = \sqrt[k]{x + dx} = (x + dx)^{1/k} = x^{\frac{1}{k}} + \frac{1}{k} x^{\frac{1}{k}-1} dx + \ldots,$$

(the higher powers of dx are hidden in the dots), and thus it follows that

$$y = x^{\frac{1}{k}} \quad \Rightarrow \quad \frac{dy}{dx} = \frac{1}{k} x^{\frac{1}{k}-1}.$$

With a little bit more effort one obtains the differential quotients of further functions which we present here in a table:

$y = f(x)$	$\frac{dy}{dx}$
x^k	kx^{k-1}
e^x	e^x
$\sin x$	$\cos x$
$\cos x$	$-\sin x$
$\ln x$	$\frac{1}{x}$

(1.3)

Using the methods just described we may easily derive rules for differential quotients like the product and chain rule and rely again on the heuristically motivated method in [Thompson 1998a].

1.2.1 The Product Rule

If a product is given in the form $y = f(x) \cdot g(x)$ then it holds

$$y + dy = (f + df) \cdot (g + dg) = fg + f\,dg + g\,df + df\,dg. \qquad (1.4)$$

Hey now! Why do we not write

$$y + dy = f(x + dx) \cdot g(x + dx)$$

as we did above? Well, it is easier this way[1]! We don't have specific functions for example here but general place holders $f(x)$ and $g(x)$, but we simply write $f(x + dx)$ in the form $f + df$, imagining that the function value of f at the point $x + dx$ is the one of f at x with a little correction df - in so far as it does indeed not matter a bit whether we write $f(x + dx)$ or $f + df$ since $f(x + dx) = f + df$.

Let us now look back to (1.4). Apparently $df \cdot dg$ is a quantity of second order and therefore zero. Moreover, subtracting $y = fg$ yields

$$dy = f\,dg + g\,df,$$

and if we finally divide by dx the **product rule**

[1]If you happen to know Taylor's formula then you know: $f(x+dx) = f(x) + dx \cdot df(x)/dx + \ldots$, where the higher powers of dx are again hidden in the dots. If we neglect these then indeed we get $f(x + dx) = f(x) + df(x)$.

$$y = f(x)g(x) \quad \Rightarrow \quad \frac{dy}{dx} = f\frac{dg}{dx} + g\frac{df}{dx} \tag{1.5}$$

follows. With it we can compute the differential quotient for functions like $y = \underbrace{e^x}_{=f(x)} \cdot \underbrace{(1+x^4)}_{=g(x)}$ simply by using $df/dx = e^x$ and $dg/dx = 4x^3$ to give

$$\frac{dy}{dx} = f\frac{dg}{dx} + g\frac{df}{dx} = e^x \cdot 4x^3 + (1+x^4)\cdot e^x.$$

1.2.2 The Quotient Rule

If instead of a product, a quotient

$$y = \frac{f(x)}{g(x)}$$

of two functions is given, then

$$y + dy = \frac{f + df}{g + dg}$$

leads to the following **polynomial long division**[2]:

$$\begin{array}{l}
f + df : g + dg = \dfrac{f}{g} + \dfrac{df}{g} - \dfrac{f}{g^2}dg \\
\underline{-(f + \dfrac{f}{g}dg)} \\
\qquad df - \dfrac{f}{g}dg \\
\qquad \underline{-(df + \dfrac{df\,dg}{g})} \\
\qquad\qquad -\dfrac{f}{g}dg - \dfrac{df\,dg}{g} \\
\qquad\qquad \underline{-\left(-\dfrac{f}{g}dg - \dfrac{f}{g^2}(dg)^2\right)} \\
\qquad\qquad\qquad -\dfrac{df\,dg}{g} + \dfrac{f}{g^2}(dg)^2.
\end{array}$$

The remainder $-df\,dg/g + f(dg)^2/g^2$ just obtained however is zero since only quantities of second order are involved. Our polynomial long division hence yields

[2]It doesn't matter if you never saw a computation like this before. Just notice that one can divide polynomials exactly in the way one would divide two numbers.

$$y + dy = \frac{f}{g} + \frac{g\,df - f\,dg}{g^2}$$

and subtracting now $y = f/g$ and dividing by dx shows the **quotient rule**:

$$y = \frac{f(x)}{g(x)} \quad \Rightarrow \quad \frac{dy}{dx} = \frac{g\frac{df}{dx} - f\frac{dg}{dx}}{g^2}. \tag{1.6}$$

Instead of the word 'differential quotient' one nowadays uses the word 'derivative' and writes

$$f'(x_0) = \frac{dy}{dx}(x_0)$$

for the slope of a function f at a point x_0. Apparently the slope of functions vanishes at maxima and minima so that by solving the equation

$$\frac{dy}{dx} = 0$$

one can find the points x at which extreme values may be found.

Even **higher derivatives** play an important role. In Leibniz's notation

$$\frac{d^2y}{dx^2} := \frac{d}{dx}\left(\frac{dy}{dx}\right)$$

is the derivative of the derivative, etc. Second derivatives are – amongst other things – useful to decide whether an extremum is a maximum or a minimum. For example, the first derivative of $y = f(x) = x^2$ is

$$f'(x) = 2x$$

and it vanishes at $x = 0$. Hence we surmise an extreme value at $x = 0$. The second derivative

$$f''(x) = 2$$

is positive and that means that we have a minimum at $x = 0$.

1.2.3 The Chain Rule

The true superiority of Leibniz's notation comes to light in the chain rule. This rule is used in cases of compositions of functions like

$$y = f(g(x)).$$

If we call the inner part $u(x) := g(x)$ and the outer part $v(u) := f(u)$ and differentiate both,

$$\frac{du}{dx} = g'(x), \quad \frac{dv}{du} = f'(u) = f'(g(x)),$$

then the complete derivative follows from the product

$$\frac{dy}{dx} = \frac{dv}{dx} = \frac{du}{dx} \cdot \frac{dv}{du},$$

since the du can be formally cancelled:

$$\frac{\cancel{du}}{dx} \cdot \frac{dv}{\cancel{du}} = \frac{dv}{dx},$$

hence we get

$$y = f(g(x))) \quad \Rightarrow \quad \frac{dy}{dx} = g'(x) \cdot f'(g(x)). \tag{1.7}$$

This rule is not confined to two functions in composition. If in the fourfold composition

$$y = f(g(h(k(x))))$$

we use the abbreviations

$$u(x) := k(x), \quad v(u) := h(u), \quad w(v) = g(v), \quad z(w) = f(w),$$

then $y = z(x)$ and from the single differential quotients

$$\frac{du}{dx} = k'(x), \frac{dv}{du} = h'(u), \frac{dw}{dv} = g'(v), \frac{dz}{dw} = f'(w)$$

it follows by multiplication and cancelling

$$\frac{\cancel{du}}{dx} \cdot \frac{\cancel{dv}}{\cancel{du}} \cdot \frac{\cancel{dw}}{\cancel{dv}} \cdot \frac{dz}{\cancel{dw}} = \frac{dz}{dx},$$

and hence

$$\frac{dy}{dx} = \frac{dz}{dx} = k'(x) \cdot h'(k(x)) \cdot g'(h(k(x))) \cdot f'(g(h(k(x)))).$$

Take the function $y = \sin(e^{4x^2+x})$ as an example. We introduce the abbreviations

$$u(x) = 4x^2 + x, \quad v(u) = e^u, \quad w(v) = \sin v$$

and differentiate one by one:

$$\frac{du}{dx} = 8x + 1, \quad \frac{dv}{du} = e^u, \quad \frac{dw}{dv} = \cos v.$$

Therefore it follows

$$\begin{aligned} \frac{dy}{dx} = \frac{dw}{dx} = \frac{du}{dx} \cdot \frac{dv}{du} \cdot \frac{dw}{dv} &= (8x + 1) \cdot e^u \cdot \cos v \\ &= (8x + 1) \cdot e^{4x^2+x} \cdot \cos\left(e^{4x^2+x}\right). \end{aligned}$$

1.2.4 The Rule of Differentiating the Inverse Function

If $y = f(x)$ and if the inverse function $x = f^{-1}(y)$ exists, the **rule of differentiating the inverse function** reads as

$$\frac{dy}{dx} = \frac{1}{\frac{dx}{dy}}. \tag{1.8}$$

To illustrate this rule we look at the logarithmic function

$$y = \ln x$$

for which we know that it is the inverse of $y = e^x$. By applying the e-function we derive from $y = \ln x$ the equation

$$x = e^y,$$

and therefore

$$\frac{dx}{dy} = e^y = e^{\ln x} = x.$$

From the rule of differentiating the inverse function it then follows

$$\frac{dy}{dx} = \frac{d\ln x}{dx} = \frac{1}{\frac{dx}{dy}} = \frac{1}{x}.$$

We have shown that $(\ln x)' = 1/x$ holds true.

1.3 Area, Integral, and Antidifferentiation

We now imagine an interval $[a, b]$ of length $x = b - a$ being divided into infinitely small quantities of length dx. If we sum all these tiny bits the length x has naturally to reappear. Since the pieces of length dx are infinitely small their number is infinitely large so that it may make sense to use a special symbol for the sum. Following Leibniz for this sum we write the **integral**

$$x = \int_a^b dx = b - a$$

which symbolises a sustained 'S' (for sum). The boundary values a and b at the integral sign define it to be a **definite integral**. If we now take a tiny little bit of the length x by applying d we get

$$dx = d\int dx = dx,$$

in other words: **derivative and integral are mutually inverse to one another – the integral turns dx into x (i.e. neutralises the d) and the d neutralises the integral.**

Hence we can compute integrals as the reverse of differentiation. Since for $y = x^n$ the derivative is nx^{n-1}, the integral of $y = nx^{n-1}$ is nothing but

$$x^n = \int nx^{n-1}\, dx.$$

Well, this is not *quite* right, since not only x^n is the **indefinite integral** (one also says: **primitive function**) but also $x^n + c$ for every constant c. If $f(x) = c$ is any arbitrary constant (hence f does not depend on x) then of course

$$y_0 + dy = f(x_0 + dx) = c$$

and since $y_0 = f(x_0) = c$ the quantity dc has to vanish. Thus it holds that

$$d(x^n + c) = dx^n = d \int nx^{n-1}\, dx = nx^{n-1}\, dx.$$

A primitive function is not uniquely defined but only up to arbitrary constants which may be added.

This simple aspect of interpreting integrals as inverses of differentiation, hence interpreting integration as **antidifferentiation**, is sufficient for our purposes! An independent meaning of integration was introduced into mathematics only long after Leibniz and Newton.

But now to a geometrical interpretation. What does $\int f(x)\, dx$ mean geometrically?

In Figure 1.3.1 the interval $[a, b]$ on the axis of abscissae is divided in pieces of length dx. In our figure only four pieces can be seen but we imagine there are infinitely many since dx is infinitely small. If we now consider the product $f(x) \cdot dx$ where always the left value of x is taken on each subinterval. This product is hence the area of one of the rectangles marked green below the curve and summing them all we get the area below the curve f on the whole interval,

$$\int_a^b f(x)\, dx.$$

An integral of a function f on an interval hence is nothing other than the area under the graph of this function.

We have already pointed out the inverse character of integration and differentiation. This important insight is the content of the **fundamental theorem**: If $f : [a, b] \to \mathbb{R}$ is a continuous function then for all $x_0 \in [a, b]$ the primitive function

$$F(x) := \int_{x_0}^{x} f(t)\, dt \tag{1.9}$$

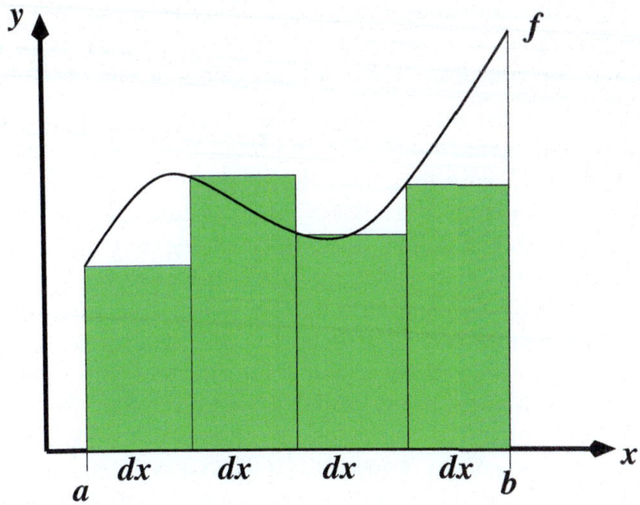

Fig. 1.3.1. Computation of the area below a given curve

is differentiable and it holds $F'(x) = f(x)$ at every point $x \in [a, b]$. For the value of the integral on $[a, b]$ it holds

$$\int_a^b f(x)\,dx = F(b) - F(a). \tag{1.10}$$

1.4 Indivisibles and Infinitesimals

The integration, meaning the computation of the area under a curve, is a good starting point to clarify two perceptions which have played an important role in the history of mathematics: indivisibles and infinitesimals.

The strips of width dx in Figure 1.3.1 are called **infinitesimals**; here they are infinitesimal areas. The width dx of one strip is an infinitely small, but nonzero quantity. If one divides the interval between zero and one into very many subintervals of length dx then one gets infinitesimal line segments. If one divides the interval for example in 10 million parts then $dx = 0.000\,000\,1$, if one chooses 100 million subintervals then $dx = 0.000\,000\,01$, and so on. And here the crucial point can be seen: the more subintervals we require, the smaller dx will be and eventually – if we consider more and more subintervals – dx will be indefinitely small, but never zero!

The infinitesimals correspond to the concept of the ancient philosopher Aristotle (384–322 BC) who, in his book *Physics* [Aristoteles 1995], declared that a straight line can never consist of points but instead of infinitesimals. Aristotle calls the straight line a **continuum** and postulates its arbitrary divisibility. It doesn't matter how often we may divide a continuum, even the smallest parts will always be continua. There are no points in the continuum.

But why can't a straight line be composed from points? This view is indeed possible and was supported by the Greek philosopher Democritus (460 or 456 BC – around 380 BC) who imagined the straight line as being composed of an infinite number of atoms (atomos = undividable). A point, however, has no extension if we follow Euclid's (3. c. BC) axioms and therefore is an **indivisible**. Thinking of the area under the curve in Figure 1.3.1 as being formed by indivisibles we have to imagine infinitely many lines of width zero which 'cover' the area somehow, but that seems impossible since as many lines with width zero as we can think of never can cover anything two-dimensional. The 'sum' of all these infinitely many indivisibles would then be the area. This is hard to swallow but mathematicians like Bonaventura Francesco Cavalieri (1598–1647) and Evangelista Torricelli (1608–1647) – true masters of indivisibles – used this idea to compute the volume of complicated objects and thereby proved the usefulness of the calculus of indivisibles, cp. [Sonar 2011]. The word 'sum' appears here in quotation marks intentionally and its use is dangerous! Cavalieri never spoke of sums but of 'omnes lineae', hence the collection of all lines, cp. [Folkerts/Knobloch/Reich 2001, p. 299].

In the beginning of analysis there was the idea of a calculus of indivisibles and even Leibniz started as a mathematician of indivisibles. With the idea of introducing a dx which can be thought of as small as necessary he became one of the founders of the mathematics of infinitesimals.

Fig. 1.4.1. Democritus (from a Greek banknote), marble sculpture of Aristotle (copy dating from the 1./2. century from a lost bronze sculpture, Musée de Louvre) (Creative Commons Lizenz 2.5 generic), Archimedes (post stamp)

1.5 ... and What is it Good For?

Hopefully we can make clear *how* slopes and areas under curves can be computed, but *why* is it important to be able to do so?

In the times of Leibniz and Newton mechanics was a young science. One generation earlier Galileo Galilei (1564–1642) had derived the laws of free fall. Under the influence of gravitational acceleration $g \approx 9.81\text{m/s}^2$ a body falls a distance of

$$y(t) = \frac{1}{2}gt^2$$

in time t, thereby reaching a velocity (neglecting friction in the air) of

$$v(t) = gt$$

and the acceleration in the gravity field is $a(t) = g$ which, of course, does not depend on the fall time. Using the new differential and integral calculus, mechanics could be completely developed – a program which was carried out by Leonhard Euler (1707–1783) [Szabó 1996]. Velocity is nothing but the temporal change of the distance, i.e.

$$v(t) = y'(t) = \frac{dy}{dt}(t) = gt,$$

and acceleration is just the temporal change of velocity,

$$a(t) = v'(t) = \frac{dv}{dt}(t) = \frac{d^2y}{dt^2}(t) = g.$$

Fig. 1.5.1. Galileo Galilei ([Painting by Justus Sustermans, 1636] National Maritime Museum, Greenwich, London)

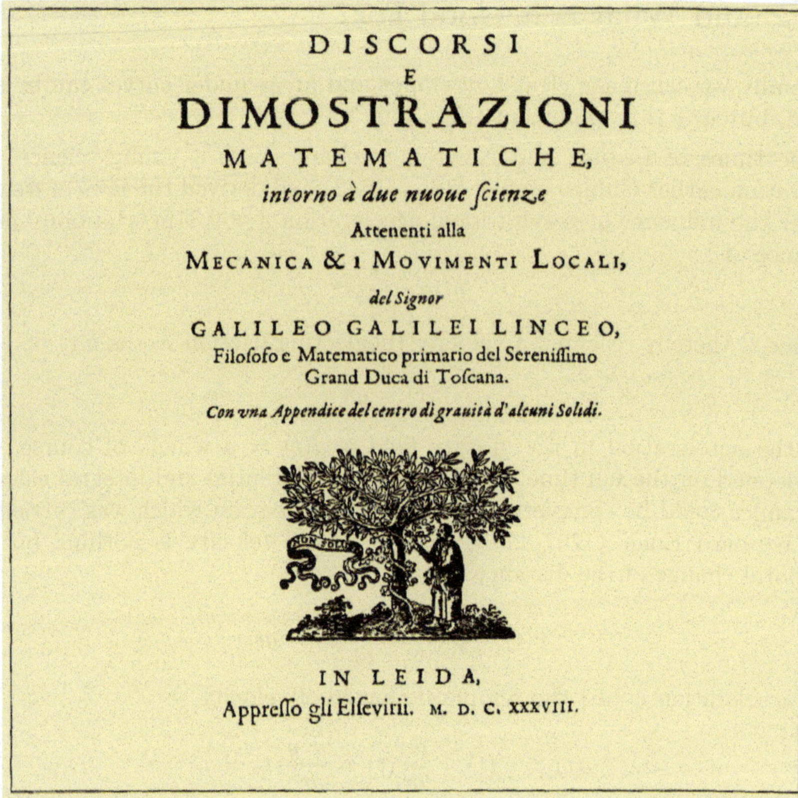
DISCORSI
E
DIMOSTRAZIONI
MATEMATICHE,
intorno à due nuoue ſcienze
Attenenti alla
MECANICA & I MOVIMENTI LOCALI,
del Signor
GALILEO GALILEI LINCEO,
Filoſofo e Matematico primario del Sereniſſimo
Grand Duca di Toſcana.
Con vna Appendice del centro di grauità d'alcuni Solidi.

IN LEIDA,
Appreſſo gli Elſevirii. M. D. C. XXXVIII.

Fig. 1.5.2. Titel page of the *Discorsi* 1638

To rephrase it, we therebye know that the area under the curve v in the velocity-time diagram is nothing but the integral of v:

$$\int_0^t v(s)\,ds,$$

and this quantity has to be the distance travelled since eventually integration is the inverse of differentiation! It therefore holds

$$y(t) = \int_0^t v(s)\,ds.$$

In his book *Discorsi e dimonstrazioni matematiche, intorno à due nuoue scienze attinenti alla meccanica e ai moti locali* ('Discourses and Mathematical Demonstrations Relating to Two New Sciences Concerning Mechanics and the Laws of Free Fall'), published 1638 in Leiden when Galileo was already in the grip of the inquisition in Italy, we find the following Theorem [Galilei 2014, p. 165]:

'The time in which a certain space is traversed by a moveable in uniformly accelerated movement from rest is equal to the time in which the same space would be traversed by the same moveable carried in uniform motion whose degree of speed is one-half the maximum and final degree of speed of the previous, uniformly accelerated, motion.'

This theorem that Galileo proved by geometrical arguments is nothing else but the equality of the integrals

$$y(t_E) = \int_0^{t_E} gt\,dt = \frac{v(t_E)}{2}\int_0^{t_E} dt,$$

where t_E denotes an end time. The equality of the two integrals expresses just the equality of areas in the velocity-time diagram as shown in Figure 1.5.3. The crosshatched rectangular area in yellow is precisely the same area as the one under the straight line $v(t) = gt$ from $t = 0$ to $t = t_E$.

Even this simple example invites us to guess that a science like mechanics can be completely described using differential and integral calculus. Important processes in nature, in engineering sciences, and in financial mathematics lead to differential equations, i.e. equations in which derivatives of a function appear. Equations of this type can only be treated with the help of differential and integral calculus. It may hence appear not surprising that Newton's and Leibniz's differential and integral calculus became a foundation of the technical revolution which began in the 17. century. Bridges, ships, motors, turbines, airplanes, and even smartphones and MP3 players are simply not conceivable without this kind of mathematics. This new mathematics of Newton and Leibniz led to a revolution in nearly all sciences within a very short period of time.

In other words: it was a question of now or never! If one of the wranglers had given in concerning the fight on the priority of the discovery of this new mathematics then his name would be stained forever from suspicion of plagiarism.

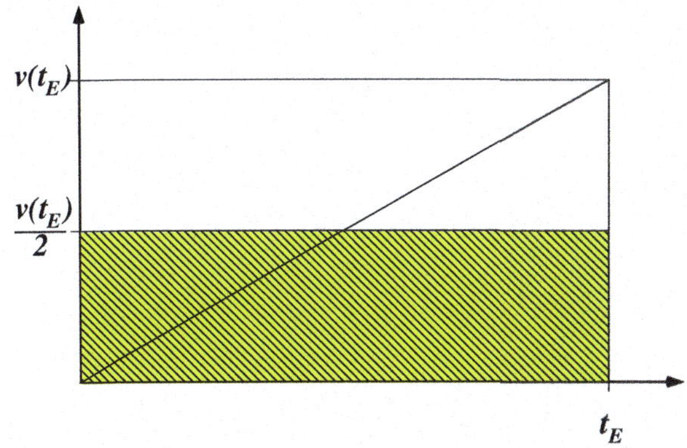

Fig. 1.5.3. Equality of areas in the velocity-time diagram

2 'On ye Shoulders of Giants'

T. Sonar, *The History of the Priority Diſpute between Newton and Leibniz*, https://doi.org/10.1007/978-3-319-72563-5_2

1556–1598	Philip II King of Spain
1558	Elizabeth I Queen of England
1560	Charles IX King of France. The name 'Huguenots' appears for Protestants of the Calvinistic faith
1562	Edict of Saint-Germain: French protestants gain legal acceptance. Massacre of Protestants in Vassy
1562/63	First religious war in France
1568–1648	Eighty Years' War between the Netherlands and Spain ends at the same time as the Thirty Years' War and marks the Birth of the 'Republic of the United Netherlands'
1572	St. Bartholomews's Day massacre, approx. 13 000 Huguenots are killed in France
1577–1580	The Englishman Francis Drake circumnavigates the world for the second time in history
1587	Mary Stuart, Queen of Scotland until 1567, executed
1588	The Spanish Armada is whitewashed by the English
1589	Start of theatrical performances of the works of William Shakespeare in London
1598	The Edict of Nantes ends the religious wars and secures the position of the Huguenots
1603	Elizabeth I dies. The Scottish King James VI becomes King of England as James I
1604	Peace between Spain and England
1605	'Gunpowder Plot' in England, 5. November
1610–1643	Louis XIII King of France
1622	England wages a war against Spain
1624–1642	Cardinal Richelieu becomes chief minister in France
1625	James I dies in England. His son becomes King as Charles I
1629	Charles I dissolves the English Parliament and reigns without it for 11 years. Peace with Spain and France. His opponent and leader of the revolutionary army is Oliver Cromwell
1629	Edict of Alès: any political autonomy of the Huguenots is prohibited
1635	France openly takes the side of the Swedish in favour of the Protestants in the Thirty Years' War
1642	Charles I has to leave London. Civil war begins. Charles I takes winter quarters in Oxford
1648	Second English civil war due to the incursion of the Scottish. Cromwell prevails
1649	Charles I is beheaded. Monarchy in England is abolished
1653	Oliver Cromwell becomes *Lord Protector*
1660	The English Restoration starts with the coronation of Charles II
1661–1715	Louis XIV King of France
1665–1683	Colbert Controller-General of Finances
1665	Plague in London
1672–1679	War of France against Holland
1685	Edict of Nantes abolished. 300 000 Huguenots leave France
1713	Papal Bull 'Unigenitus' against the Jansenists

2.1 Who were the Giants?

In a letter to Robert Hooke dated 15th February 1676 [Turnbull 1959–77, Vol. I, p. 416] Newton writes:

> *'If I have seen further it is by standing on the shoulders of Giants'*

Since the Middle Ages learned men used the parable of the dwarfs standing on the shoulders of giants to honour 'the ancients' and (in the main) to play down their own role in all (false) modesty. This specific understatement surely was also Newton's intention. We should ask anyway: on whose shoulders did Leibniz and Newton actually stand? Which preliminary works were already there for the two men to build upon?

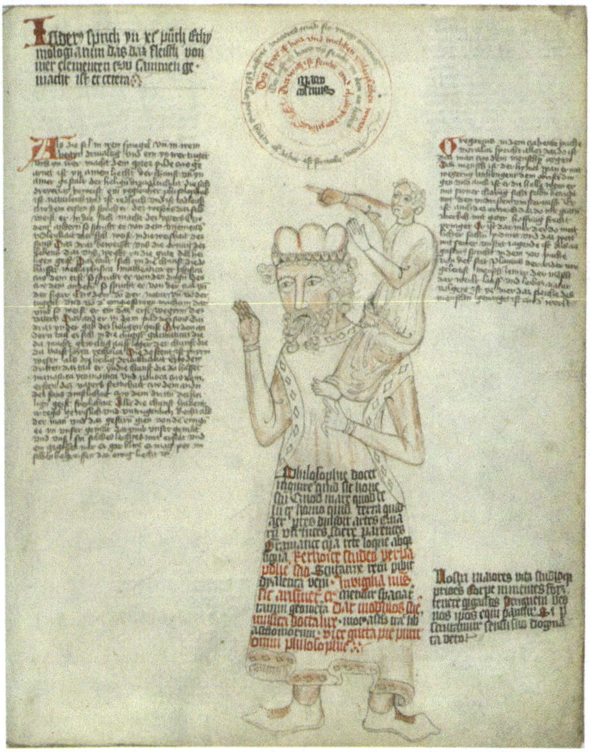

Fig. 2.1.1. Image of people on the shoulders of giants. Precursors of Newton's quote exist since the 12th century (Bernard of Chartres) [encyclopedic manuscript with allegoric and medical drawings, South Germany approx. 1410. Handwritten manuscript from the collection J. Rosenwald. Library of Congress, Washington, Rosenwald 4, Blatt 5r]

In order to be able to answer this question even partly we would actually need a book on its own, but that exists already and I refer the readership gladly to [Sonar 2011]. The idea of a mathematics of infinitesimals was already old in the lifetimes of Newton and Leibniz; its history goes back to Archimedes and even further back. In the generation directly before Newton and Leibniz these ideas became vibrant in a specific manner. Josef Ehrenfried Hofmann has described this situation felicitously [Hofmann 1974, p. 74f.]:

> *'Infinitesimal problems were being hotly pursued simultaneously in France, Italy and England; the improved concept of indivisibles was being used as a guiding principle by Fermat, Pascal and Huygens equally as well as by Torricelli, Ricci, Angeli and Sluse. Gregory got to know it in Italy, and perhaps Barrow had it from Gregory. The characteristic triangle – to take up a particular point – was known already to Fermat, Torricelli, Huygens, Hudde, Heuraet, Wren, Neil, Wallis and Gregory long before it was made public by Barrow. Each of these predecessors had used it, but nobody wanted to expose the jealously guarded secret by which he had found his results. Leibniz learnt of the characteristic triangle not from Barrow's book but, as has been briefly explained earlier on, from reading Pascal's*

We don't want to let ourselves be distracted and confused by the many names and notations, but we should – and have to – inspect four of these giants more closely. In case of Newton: Isaac Barrow (1630–1677), whose successor on the Lucasian Chair Newton became, and John Wallis (1616–1703) whose *Arithmetica Infinitorum* had a groundbreaking effect. In case of Leibniz: Blaise Pascal (1623–1662), whose characteristic triangle meant an initial spark for Leibniz, and Christiaan Huygens (1629–1695) who acted as a teacher in mathematics for Leibniz.

2.2 England in the 17th Century

On 24th March 1603 the English Queen Elizabeth I died aged 69, and thereby England's 'Golden Age' came to an end. Elizabeth had led the underdeveloped, agriculturally shaped England into a glorious time of sea power. She succeeded in anchoring the Protestant religion of her father Henry VIII in the country and, after the whitewash of the Spanish Armada in the year 1588 with which Spain had tried to occupy England, sciences and arts flourished. Not for nothing is it the age of Shakespeare and of the theatre! Elizabeth, as 'Virgin Queen' or 'Gloriana' or 'Good Queen Bess' imprinted in the memory of the people, never got married and left no heirs. With her the Tudor dynasty ceased, which had reigned the country since 1485. The role of the rulers now fell to the Stuarts and they brought major uproar and upheavals.

Fig. 2.2.1. Charles I of England ([Painting by Daniel Mytens, about 1623], Private Collection Photo © Philip Mould Ltd., London/Bridgeman Images) and Queen Elizabeth I of England [Painting by Nicholas Hilliard, after 1575, detail]

In the year 1587 Elizabeth I had the Catholic Queen of Scotland, Mary Stuart, executed. Her son, however, was educated in the Protestant faith and in 1586 a strategic alliance between England and Scotland was forged in the treaty of Berwick, so that Mary Stuart's son was eventually brought to the throne as James I [Haan/Niedhart 2002, p. 149f.]. James was a weak king internally as well as outwardly. Inside Parliament the influence of the lower gentry increased dramatically; at the same time England saw a kind of religious revivalism through the Puritans who insisted on a complete 'cleansing' of the Christian faith and from worship of Roman-Catholic elements. Therefore a strong opposition against the just established Church of England, naturally strongly characterised by Catholic rites, came to life. Many Puritans did not see a future in England any more and emigrated to the English colonies in North America.

In the year 1640 one counted already about 25,000 settlers who crossed the Atlantic in ships like the famous *Mayflower* as pilgrims.

Again England suffered religious strains. With his son Charles I a monarch showing strong Catholic inclination came to the throne after the death of James I in 1625. Additionally, Charles thought of himself as an absolute monarch following the idea of the divine rights of kings. To make things even worse he married Catholic Henriette Marie de Bourbon, the daughter of the King of France, before his coronation on 13 th June 1625.

Fig. 2.2.2. The *Mayflower* 1620 in the harbour of Plymouth ([Painting of William Halsall 1882] Pilgrim Hall Museum, Plymouth, Massachusetts, USA)

When in need of money for a war with Spain, Parliament, now manned with many Puritans, granted him the revenues of tonnage and poundage only for one year and not, as was usual, for all of his reign. Charles fumed. He dissolved Parliament without further ado! But in 1628 he needed it again to raise new money. Parliament presented him with the 'Petition of Rights' with which it wanted to secure itself against the despotism of the king. Charles subscribed to the petition but he never thought of committing to the rules: for 11 years Parliament will not be called in again.

The Scots hadn't introduced the constitution of the Church of England; instead the church followed a Presbyterian constitution, hence the Calvinist reform. Charles and his archbishop of Canterbury, William Laud, wanted to change this. The Scots reacted with a rebellion and marched in to England. Charles again needed a war chest and Parliament was called in on 13th April 1640; but already on 5th May 1640 he dissolved it again. This short time period was hence called 'Short Parliament'. Now Charles had crossed a fictitious border; turmoil started within the nation. Parliament got into action again on 3rd November 1640 and stayed active until 1660; hence the name 'Long Parliament' is used for this period.

Under the Puritan John Pym an impeachment against Charles's closest confidant Thomas Wentworth, who had made himself unpopular due to his hard hand in Ireland, was envisaged. Wentworth had even given an army by the king which he was to use against his unruly subjects in his own country. Called back to England to advise the king he became the first Earl of Stafford. The House of Commons in the Long Parliament did not want to declare war on the Scots and accused Wentworth of high treason, since he had advised the

Fig. 2.2.3. Threefold portrait of Charles I ([Anthonis van Dyck, 1635/36] Gallery Royal Collection, Windsor Castle, London)

king to turn the Irish army against his own people in England. Although the charge collapsed, Wentworth was sentenced to death. The king tried to prevent the execution but even in the House of Lords some members voted to condemn Wentworth. Charles now began to fear for his own life and that of his family members and eventually approved the execution. Wentworth, first Earl of Stafford, was beheaded on the 12th May 1641.

In October 1641 a rebellion in Ireland broke out. For the most part, being Catholics the people of Ireland feared an invasion of Protestant English forces, and arose against England. This, in turn, led to violent reactions of the English whereby thousands of Irish people died. The House of Commons refused to provide the King with additional troops and released the 'Grand Remonstrance' in November 1641, a notice of appeal against the King and his government. In London riots flared up against the King who had to flee with his family. In mid-1642 both parties, the royalist and the parliamentary, started to arm themselves: the English civil war began.

The military leader on the side of the Parliamentary party became Oliver Cromwell (1599–1658), a Puritan who was a member of the House of Commons. In a very short time he founded the 'New Model Army' which inflicted heavy casualties to the royal troops. The king retreated to Oxford but after the battle of Naseby on 14th June 1645, Cromwell's troops besieged Oxford and Charles could only escape disguised as a servant. He fled into the arms of the Scots but was extradited after tenacious negotiations and payment of a considerable sum of money.

Charles would escape his English imprisonment again but was finally arrested and brought before court in 1648. On the 30th of January 1649 the former King of England, Charles I, was beheaded.

Meanwhile Cromwell had led an invasion of Ireland where he had engaged in exceptional brutality. In May 1650 he turned against Scotland which had proclaimed Charles's son King Charles II. In the battle of Dunbar on 3rd September 1650 Cromwell's army won a bloody victory over the Scots. Charles's II attempt to take London with his remaining Scottish forces was baulked by Cromwell in the battle of Worcester on 3rd September 1651 and Charles II had to flee to exile on the continent. He had to stay there until 1660.

Fig. 2.2.4. The trial against the Earl of Stafford in the House of Commons [Wenzel Hollar, Historical Print, 17th c. Thomas Fischer Rare Book Library Wenzel Hollar, Digital Collection Plate No. P551]

Fig. 2.2.5. Oliver Cromwell ([Painting: Circle of Adriaen Hanneman, 17th c.] hampel-auctions, Munich) and Charles II of England ([Painting of Peter Lely (also: Pieter van der Faes), about 1675] Euston Hall, Suffolk; Belton House, Lincolnshire)

In England a Rump Parliament reigned from 1651 to 1653 but without a king groups began to form which fought against each other. Cromwell brought this to an end on the 20th April 1653 when he occupied Parliament and dissolved it by force. A short period of the 'Parliament of Saints' (or 'Barebone's Parliament') followed which consisted of men chosen on grounds of their religious conviction. Actually these men were supposed to prepare the elections for a new Parliament but religious radicals succeeded in scaring most of the members of the Parliament of Saints so much that this transitional Parliament was finally dissolved on 12th December 1653. Incidentally, Cromwell was not a member of the Parliament of Saints. Cromwell was appointed 'Lord Protector' on the 16th December 1653 and would rule like a monarch until his death in 1658.

Through Cromwell's militant puritanism hard times dawned for London and its inhabitants. Public music was forbidden, as were theatre and dance performances. The 'alehouses' suffered badly from Puritanic observation [Haan/Niedhart 2002, p. 186]. Cromwell was offered the crown in 1657 but after a fairly lengthy reflection he declined the offer. When he died he was buried solemnly in Westminster Abbey and his son Richard followed him as Lord Protector. Richard, however, could not win majorities in Parliament and was soon forced to abdicate, in May 1659. Thereupon the leader of the New Model Army, the English Governor in Scotland, moved his troops to London and reinstated Long Parliament which made Charles II king in 1660: the monarchy was finally restored.

Concerning the years between 1640 and 1660 the philosopher Thomas Hobbes (1588–1679) wrote they were the peak of this time, since the struggles for power gave him an overview of all kinds of injustices and follies the world

Fig. 2.2.6. Thomas Hobbes ([Engraving: W. Humphrys, 1839] Wellcome Images/ Wellcome Trust, London]) and Richard Cromwell, successor of Oliver Cromwell as Lord Protector [Engraving by William Bond, 1820]

would ever witness [Haan/Niedhart 2002, p. 167]. Against the background of the civil war, also Hobbes's famous dictum: 'man is a wolf to man' has to be understood. Hobbes's principal work actually is *Leviathan* published in English in the year 1651. In this book every man is but a miserable warrior, since in the 'state of nature' everything is a matter of very survival and one's own advantage. Hence the state has to care for the necessary organisation and safety: people hand over all power to the state, the Leviathan, and pay the price in the loss of their free will. Hobbes also emerged through mathematical works and was convinced he had solved the problem of the quadrature of the circle [Sonar 2011, p. 39ff.]. He fought literary battles with John Wallis who thought justly that Hobbes was not more than an amateur mathematician with mediocre mathematical knowledge, cp. [Jesseph 1999].

For the decade from 1660 there is arguably no better source than the 'secret' diaries of Samuel Pepys (1633–1703). Pepys, the child of a dressmaker, came to fame and honour through the protection of his cousin Edward Montagu, for whom he functioned as a private secretary, and he became clerk in the Treasury in the year 1658. The Treasury was led by George Downing. Since Montagu and Downing were instrumental in the transformation from Republic to Monarchy, Pepys's career went on ever upwards. After becoming scribe in the Naval Office, his rise began there. Already in 1665 Pepys became a member of the Royal Society and developed an interest in mathematics: and from 1684 to 1686 he acted as president of the Royal Society, so that Newton's great work *Philosophiae Naturalis Principia Mathematica*, published in 1687, bears his imprimatur.

The relations with the Netherlands had declined after 1660 and unpleasant incidents took place at sea. Being sea and trading powers, both parties had reasons to look at each other with mistrust. On the 4th March 1665 Charles II declared war. Pepys noted in his diary on the 8th June (jul.) [Pepys 2003, p. 494]:

Fig. 2.2.7. ,
Samuel Pepys [Painting by John Hayls, 1666,
Photo: Thomas Glyn, 2003]

> *'I to my Lord Treasurer's by appointment of Sir Tho. Ingram's, to meet the goldsmiths – where I met with the great news, at last newly come, brought by Bab May from the Duke of Yorke, that we have totally routed the Dutch. That the Duke himself, the Prince, my Lord Sandwich, and Mr. Coventry are all well. Which did put me into such joy, that I forgot almost all other thoughts. The sum of the news is:*
>
> *Victory over the Dutch. June. 3. 1665.'*

However, this was only a first victory in one battle; ultimately 'routing the Dutch' the English could not achieve.

Of particular historical importance are Pepys's depictions of two further incidents: the plague which broke out in 1665 and the Great Fire of London in 1666. On 10th June (jul.) Pepys noted succinctly [Pepys 2003, S. 495]:

> *'In the evening home to supper, and there to my great trouble hear that the plague is come into the City* [...]*'*

With the diary we are able to follow the spread of the epidemic. On 20th July (jul.) Pepys writes:

> *'So walked to Redriffe, where I hear the sickness is, and indeed is scattered almost everywhere – there dying 1089 of the plague this week.'*,

on 31th July (jul.) we read:

Fig. 2.2.8. Map of London after the Great Fire 1666, the destroyed areas are marked in white [Wenzel Hollar]

> ' [...] *only under some difficulty because of the plague, which grows mightily upon us, the last week seeing about 1700 or 1800 dying of the plague.'*,

and on 12th August (jul.)

> *'The people die so, that now it seems they are fain to carry the dead to be buried by daylight, the nights not sufficing to do it in'.*

Even for Isaac Newton in Cambridge this plague epidemic, which remained not restricted to London, will turn out to be fateful.

After a hot and dry summer in 1666 a great fire developed from a smouldering fireplace of a royal baker in Pudding Lane on the early Sunday morning of the 2nd September, which will destroy four-fifths of London.

Samuel Pepys notes in his diary [Pepys 2003, S. 329]:

> *'Some of our maids sitting up late last night to get things ready against our feast to-day, Jane called us up about three in the morning, to tell us of a great fire they saw in the City.'*

In the palace of Whitehall Pepys succeeded in informing Charles II about the fire [Pepys 2003, p. 330]. He suggested knocking down rows of houses in order to keep the fire from spreading, but the house-owners refused to abandon their houses. Hence the fire raged on until 5th September and approximately 400 streets were devastated. The medieval city centre of London was gone.

On issues of religion things calmed down in the 1660s. The explosive mixture of Independents, Levellers, Baptists, Presbyterians, Quakers, and whatever other new religious trends had surfaced [Hill 1991], all had to be controlled somehow. In addition there were the Catholics still waiting for England's re-Catholisation. In the *Act of Uniformity* all clergymen had to submit to the *Common Prayer Book* of the Church of England in 1662. In 1673 and 1678 laws followed making every civil and official office inaccessible for non-members of the Church of England [Haan/Niedhart 2002, p. 192]. Suspicion against the crown still existed since Charles II still ogled Catholicism.

In the year 1673 Parliament extracted the Test Act from Charles II. Every state official had to swear an oath and sign it on paper that he would reject the transubstantiation – the Catholic doctrine of change of bread and wine.

Meanwhile France was allied with the Netherlands and again there was a lack of money on the English side to carry on with further war efforts. Denmark utilised the situation and declared war on England. Since the Dutch were (justly!) worried about their alliance with France the Treaty of Breda was agreed in 1667, leaving England with heavy war depts, but finally ending the second Anglo-Dutch war. Charles II concluded a secret contract with the French 'Sun King' Louis XIV: Louis promised extensive payments, but Charles had to promise to convert to Catholicism. Incidentally, both kings agreed on a war of aggression against the Netherlands. Although the superior strength of the Anglo-French troops was overwhelming, the Dutch could withstand it due to their military leader William III of Orange-Nassau, who became Stadtholder of the Netherlands in 1672. Eventually the Anglo-French navy succumbed and England withdrew from the confrontation through the Peace of Westminster in February 1674.

A legislative proposal to prevent the Catholic brother of the king, James, failed. As a consequence political England split into the *Whigs* and the *Tories*. The Tories stood firmly on the ground of an hereditary monarchy and backed the king while the Whigs stood for the supremacy of Parliament and wanted to prevent a Catholic heir to the throne. To escape the resulting pressure Charles II moved Parliament to Oxford. Outside London the Whigs found less support and so Charles could prevail. When additionally a conspiracy against the King was uncovered (the Rye House Plot) Charles had every reason to actively fight the Whigs.

Then Charles II died in February 1685 and his Catholic brother followed him on the throne as James II. In new elections Parliament became dominated by Tories, but even this Parliament was not easily manipulated. It required

Fig. 2.2.9. William III of Orange-Nassau ([Painting of Willem Wissing about the end of the 17th c.] Rijksmuseum Amsterdam online catalogue) and James II of England [Painting of John Riley, from 1660]

from the King a monopoly of the Church of England and strict compliance to the Test Act. But the King had other plans: starting in 1685 he expanded his standing army to about 20000 soldiers [Haan/Niedhart 2002, p. 197] and wanted to fill the officer's posts with Catholics. To do so, exceptions from the Test Act were necessary and when Parliament opposed this development, James II dissolved Parliament without further ado. He tried to give two-thirds of the positions of Justices of the Peace to Catholics and wanted a general dispensation from the Test Act. That was too much, of course: Whigs and Tories were separated by seemingly insurmountable differences, but on the refusal of the Royal action they agreed.

The political elite sent a document to William of Orange-Nassau on 30th June 1688 requesting his immediate intervention in England. William was married to the daughter of James II and himself a Protestant. The document certified that nearly all of the people of England were dissatisfied with the reign of James; in a military intervention one hoped that large parts of the army would help, as they seemed mistrustful of the papist religion. 'There must be no more time lost', the senders wrote - four Whigs and three Tories [Haan/Niedhart 2002, p. 198].

At the beginning of November William landed with his troops in Cornwall and it turned out that indeed all crucial forces in England were on his side. It now began what is called the *Glorious Revolution* because not a drop of blood was shed. James II was disposed of as king by a newly elected Parliament. This Parliament enacted a *Declaration of Rights* killing off any form of absolutism. William of Orange-Nassau and his wife Mary approved the declaration and on 13th February 1689 they were made King William III and Queen Mary. The topic of revolution was thus removed from the table in England, and the country moved into politically stable times.

When William III died at the beginning of the year 1702 his wife's sister Anne became Queen of England: and under Queen Anne England became involved in the War of the Spanish Succession. From 1707 onwards Anne is also Queen of 'Great Britain', since England and Scotland had joined to form the Realm of Great Britain. On the streets Anne will soon be called 'Good Queen Anne'. She died in August 1714 and with her the House Stuart came to an end. Her successor to the English throne is a Hanoverian from Germany, the employer of Leibniz, Georg Ludwig of Hanover, who was enthroned in 1714 as George I. The last ruler of the House of Hanover would be Queen Victoria who died in 1901 and gives way to the House Saxe-Coburg-Gotha which is the House of Windsor today.

2.3 John Wallis

In the 17th century there is hardly any personality as fascinating as John Wallis. In the civil war he was active on the Parliament's side, became professor of mathematics through Cromwell's patronage, developed into an outstanding and creative mathematician and became a founding member of the Royal Society. Aged 80 he looked back on his life and wrote an autobiography, 'Memorials of my Life' [Scriba 1970], so satisfying a wish of his old friend Dr. Thomas Smith.

Fig. 2.2.10. Queen Anne of England, coloured engraving 1707 (Royal Atlas Amsterdam)

Fig. 2.3.1. John Wallis [Pastel of HWK after a painting of Godfrey Kneller, 1701] and the old school building of Felsted School [© Colin Smith, 2003]

John Wallis was born on 23th November 1616 in Ashfort, East Kent. His father was a clergyman in this small market town who had already died in 1622. John was the eldest son and went to school in Ashfort but was moved three years later to a school in Lay Green in the parish of Tenderden due to an outbreak of plague [Scott 1981, p. 2ff.]. John found accommodation by a Mr. Finch and enjoyed the teaching of a Scottish tutor, Mr. James Movat. In his autobiography Wallis is enthusiastic about him. He writes [Scriba 1970, p. 24]:

> *'For it was always my affectation even from a child, in all pieces of Learning or Knowledge, not merely to learn by rote, which is soon forgotten, but to know the grounds or reasons of what I learn; to inform my Judgement, as well as furnish my Memory; and thereby, make a better Impression on both.'*

In the year 1630 Movat closed his school to work as a private tutor abroad. Young Wallis must have been close to his heart because he wanted to take Wallis with him. This was scuppered, however, by Wallis's mother. Hence Wallis came to Felsted to visit the school of Martin Holbeach. Under Holbeach the school curriculum changed in the direction of the great schools of St. Paul's and Westminster.

A good education in Latin and Greek – syntax, grammar, and use of the languages – was offered; in higher classes Hebrew came along to prepare for theological studies at the two universities, but also an elementary introduction to French was offered. Apart from languages, logic was also offered – an indispensable preparation for the universities of those times [Scriba 1970, p. 25ff.]. Mathematics was unknown to Wallis and he acquired a basic knowledge only by Christmas time 1631[1]. At this time a younger brother had to learn elementary arithmetic in order to become a merchant. Wallis caught fire and started learning *Common Arithmeticke.* He was exceptionally successful, mastering arithmetic in less than two weeks, and the love of mathematics will never leave him again. This love is not satisfied in Felsted, however, not even when Wallis moves to Emmanuel College to start his studies. One could not study mathematics as a topic in those times: one went to Oxford or Cambridge to study philosophy, geography, or astronomy, and in any case in order to become a theologian. Wallis was doing very well in his studies but missed mathematics a lot. Somewhat frustrated, he writes [Scriba 1970, p. 27] that of the two hundred students in his College not two (if any) know more mathematics than himself. In Oxford and Cambridge modern developments were simply overlooked until approximately 1610: one studied the writings of Galen of Pergamon and those of Aristotle, while in London the mathematics of navigation on sea was already a topic. About 50 years later – to be precise, in 1663 – Henry Lucas donated the Lucasian Chair of Mathematics to the University of Cambridge.

We have to handle Wallis's critique of the nonexistent mathematical education in Cambridge with a little care, however. He could have advanced in mathematics so far as to be able to read and understand the writings of René Descartes (1596–1650), of Blaise Pascal (1623–1662) and of Pierre de Fermat (1607–1665), to name a few key figures, only in Cambridge.

In *Hilary Term*[2] 1636/37 Wallis completes his studies for the *Bachelor of Arts* degree, and four years later he is a *Master.* His genius must already then have attracted attention, since we know of an attempt to help Wallis to a fellowship in Emmanuel College. This attempt failed, however, since the statues allowed exactly only one fellow from each county, and the county of Kent was already represented [Scott 1981, p. 5].

The Master of Emmanuel College, one Dr. Oldsworth, even tried to create a new fellowship position specifically for Wallis but then Wallis became a fellow

[1]Scott notes in [Scott 1981, p. 3] the year as being 1630, but in [Scriba 1970, p. 26] 1631 is mentioned twice. Even if we take into account that the new year in England began on 25th March (Annunciation), at least until 1752, the Christmas date in the English or even in the modern calendar can only lie either in 1630 or in 1631.

[2]In Oxford and Cambridge an academic year consisted (and still consists) of three trimesters, called *terms.* Today the terms in Cambridge are called *Michaelmas Term, Lent Term* and *Easter Term,* while in Oxford one still speaks of *Michaelmas Term, Hilary Term* and *Trinity Term.*

Fig. 2.3.2. Three great French mathematicians: René Descartes ([Painting after Frans Hals, 2. half of 17. c.] Louvre Museum Paris), Blaise Pascal [Copy of a painting of François II Quesnel, 1691] and Pierre de Fermat [unknown artist, 17. c.]

of Queens College but did not stay there for long. Already as a child Wallis had seen himself as a clergyman and besides mathematics his actual aim in life was becoming a divine of the Church of England. In the year 1640 he was consecrated by Dr. Walter Curll and became a private chaplain to Sir Richard Darley in Yorkshire shortly afterwards. One year later he left Yorkshire and became private chaplain to Lady Vere, the widow of Lord Horatio Vere, first Baron Vere of Tilbury in Essex.

After the city of Chichester in the south of England fell to the parliamentary party of Cromwell after a siege of troops under William Waller in the civil war on 29th December 1642, an encoded letter of the royal party came into the hands of Wallis through Waller [Scriba 1970, p. 37f.]. In a remarkably short span of time Wallis succeeded in decoding the message and hence recommended himself as a cryptologist to the parliamentary party. Yet before his thirtieth birthday he was awarded the income of the parish of St. Gabriel's Fenchurch in London's Fenchurch Street for the services he had rendered. In 1644 he even advanced to the post of secretary of Westminster Assembly. Financially secured he could think of getting married in the same year. As his wife he chose Susanna Glyde, daughter of John und Rachel Glyde of Northiam in Sussex[3], with whom he had several children of which only one son and two daughters survived infancy.

Wallis was now very involved in his spiritual activities and through these activities had also contact with enthusiasts of sciences in London. In comparison to the Continent England had a late start. Already in 1560 a scientific society with the name *Academia Secretorum Naturæ* was founded in Naples, in 1603 the *Academia dei Lincei* followed in Rome. Everywhere in Europe the

[3]Scott [Scott 1981, p. 7] specifies Northiam in Northamptonshire. In Northamptonshire there is no place with this name however, and Wallis himself writes of 'Northjam in Sussex' [Scriba 1970, p. 39].

Fig. 2.3.3. John Wallis, 1931 [John Wallis by Giovanni Battista Cipriani] (© National Portrait Gallery London)

universities were behind the times. This is the reason why extramural institutions gained so much importance. In England the universities of Oxford and Cambridge were not in good shape either. When the founder of the London Stock Exchange, Sir Thomas Gresham, died in 1579, the *Gresham College*

Fig. 2.3.4. Francis Bacon ([Painting by Frans Pourbus, 1617] Royal Bath Museum, Palace on the Water); John Wilkins [Painting by Mary Beale, about 1670]

could be founded following his last will. In Gresham College seven professors were employed who had to teach (in the English language!) on theology, astronomy, geometry, physics, law, rhetorics and music on consecutive days of the week. Hence a new spirit came to London. The writings of philosopher Francis Bacon (1561–1626), in particular his *Novum Organum* of 1620, provided an atmosphere of delving into the realms of science. The Baconian ideal of propagation of knowledge through observation gained ground [Maurer 2002, p. 208].

With Bacon the era of *New Learning* started in England. It is the awakening of the new sciences which were free of any form of Aristotelianism. 'Knowledge itself is power' is one of the pointed key sentences of Bacon.

Since the 19th century Francis Bacon is called the 'foster father of empiricism' [Krohn 2006, p. 11] and his enormous impact on the development of sciences and philosophy in England is indeed astonishing. Modern science which ultimately can be found in the works of Newton arose philosophically from the ideas of Bacon [Klein 1987, p. 1]. With Bacon not only empiricism but also the inductive method to gain knowledge came into modern England, and thereby this country experienced a boost of first rank towards modernization on all social levels, especially as far as ideas to understand nature were concerned.

Men who were fascinated by these new ideas met regularly in London and among them Wallis was a driving force. They discussed physical theories, conducted experiments and formed hypotheses. One time they met at the home of one Dr. Goddard, whose assistant could grind lenses for telescopes; another time they gathered in Gresham College. We are aware of the initial period of this group, which was later to transform into the famous *Royal Society*, from a document published by Wallis in 1678 [Scott 1981, p. 9f.]:

'I take its first ground and foundation to have been in London about the year 1645, (if not sooner) when the same Doctor Wilkins (then Chaplain to the Prince Elector Palatine[4] in London) ... with myself and some others met weekly.'

John Wilkins (1614–1672) later became Bishop of Chester and first secretary of the Royal Society, but first of all he went to Oxford in 1648 to become Warden of Wadham College. In Oxford he continued the London tradition and gathered a circle of men around him whose interests were in the sciences. A year later, in 1649, Wallis was appointed to the Savilian Chair of Geometry in Oxford. This appointment had nothing to do with Wallis's large mathematical knowledge (which he did not have then) but was a reward from the parliamentary party for his services rendered as a cryptologist in the civil war. His predecessor in the Savilian Chair was Peter Turner (1586–1652) who had fought on the side of the royalists and was therefore removed from office in 1648 [Scott 1981, p. 14]. In the year 1651 the group in Oxford became the *Philosophical Society of Oxford* while in London another group continued to gather in Gresham College. At the instigation of Wilkins, Wallis and others, the two groups merged and in 1662 the Royal Society was founded with the help and support of the king.

The early period of the Royal Society was not easy; many 'enthusiasts' who had no traces of scientific education damaged the reputation of the young Royal Society so much that the writer, gardener and author of diaries John Evelyn (1620–1706) refused twice to act as president of the society. Eventually even the King laughed about his Royal Society as we are told by Samuel Pepys in his diary of 1664 [Scott 1981, p. 11]: The King

'mightily laughed at Gresham College, for spending time only on the weighing of air, and doing nothing else since they sat.'

Among the twelve founding members (Wallis does not count as such) were such illustrious figures as Robert Boyle (1627–1692) who, together with Robert Hooke (1635–1703), improved the air pump and after whom a gas law (Boyle-Mariotte) is named, Christopher Wren (1632–1723), a magnificent architect who, together with Hooke, rebuilt London after the Great Fire in 1666, the Irish nobleman William Brouncker (1620–1684) who found the continued fraction representation of $\pi/4$, and Robert Moray (1608 or 09–1673), a Scottish polymath and freemason.

Already in 1647 the elementary book on algebra, *Arithmeticae in numeris et speciebus institutio: quae tum logisticae, tum analyticae, atque totius mathematicae quasi clavis est* (Introduction to the Arithmetic of Calculating with

[4]This refers to Prince Rupert of the Rhine (1619–1682), Duke of Cumberland, who showed a strong interest in the natural sciences.

Fig. 2.3.5. Three founding members of the Royal Society: Robert Boyle (Chemical Heritage Foundation [Photo: William Brown]), Christopher Wren [John Closterman, about 1690] and William Brouncker [Painting by Peter Lely, approx. 1674]

Numbers and Letters: Which as it were is the Key to Arithmetic as to Analysis and to all of Mathematics) by William Oughtred (1573–1660) fell into the hands of Wallis; the title is usually abbreviated as *Clavis mathematicae* (The Key to Mathematics). Through the *Clavis mathematicae* Wallis's interest in mathematics was ignited again. He estimated this book highly well into his old age, although its contents were already outdated in 1647 due to the works of Thomas Harriot (1560–1621)[5]. Holding the Savilian Chair of Geometry (i.e. of mathematics) since 1649, Wallis now developed into a mathematician of the first rank.

In his book *Arithmetica Infinitorum* [Stedall 2004], dedicated to William Oughtred and published for the first time in 1655, Wallis succeeded in finding the transition from the purely geometrically motivated methods of indivisibles of Cavalieri [Andersen 1985] to a certain algebraic calculus, so that the computation of areas under curves could be accomplished by pure calculation. This unique achievement was possible because Wallis combined two developments, analytical geometry and the method of indivisibles. René Descartes's *La Géométrie* work, in which analytical geometry was born, could be studied in Europe since 1649 through Frans van Schooten's Latin translation *Geometria*. The book *Exercitationes geometricae sex* by Bonaventura Cavalieri dating from 1647 and being the 'Bible' of the method of indivisibles was not available in England, but that was good luck! Cavalieri wrote incomprehensibly and the ideas were presented obscurely, but Wallis could fall back on *Opera geometrica* by Evangelista Torricelli who presented the ideas of Cavalieri's methods much more clearly, more intuitively and more lucidly than Cavalieri himself. The book by Torricelli appeared in print already in 1644: and Wallis succeeded in combining the ideas in those two books to give something completely new!

[5]The history of English algebra and Wallis's excitement concerning Oughtred is beautifully described by Jacqueline Stedall in [Stedall 2002].

(a) View on University College in High Street, Oxford, with a plaque for Boyle and Hooke [Photo P. Öffner]

(b) Plaque for Boyle und Hooke [Photo P. Öffner]

Fig. 2.3.6. Reminder on Robert Boyle and Robert Hooke in Oxford

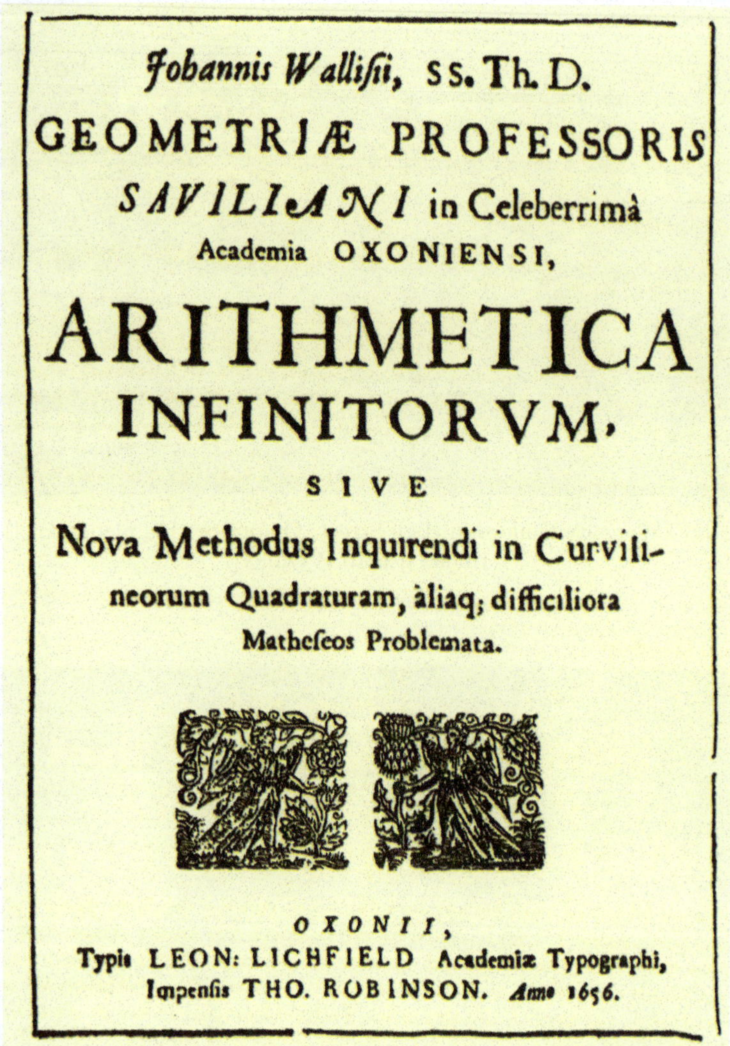

Johannis Wallisii, SS. Th. D.
GEOMETRIÆ PROFESSORIS
SAVILIANI in Celeberrimâ
Academia OXONIENSI,
ARITHMETICA
INFINITORVM,
SIVE
Nova Methodus Inquirendi in Curvili-
neorum Quadraturam, àliaq; difficiliora
Matheseos Problemata.

OXONII,
Typis LEON: LICHFIELD Academiæ Typographi,
Impensis THO. ROBINSON. Anno 1656.

Fig. 2.3.7. Title page of *Arithmetica Infinitorum* by John Wallis 1656

In Proposition 19 [Stedall 2004, p. 26] Wallis showed that the ratio

$$\frac{0^2 + 1^2 + 2^2 + \ldots + n^2}{n^2 + n^2 + n^2 + \ldots + n^2}$$

approaches the value $1/3$ for ever-increasing values of n. In other words: if n grows unboundedly the value of the ratio will be $1/3$.

Wallis 'proves' this proposition with the help of 'induction', where this does not mean complete induction as we know it today. The notion of 'induction' is rather used in the sense of Francis Bacon: If it seems clear that the ratio will approach 1/3 for some values of n, then this is true for all n. Accordingly Wallis computes for $n = 1, 2, 3, 4, 5, 6$:

$$\frac{0+1}{1+1} = \frac{3}{6} = \frac{1}{3} + \frac{1}{6}, \quad \frac{0+1+4}{4+4+4} = \frac{5}{12} = \frac{1}{3} + \frac{1}{12},$$
$$\frac{0+1+4+9}{9+9+9+9} = \frac{7}{18} = \frac{1}{3} + \frac{1}{18},$$
$$\frac{0+1+4+9+16}{16+16+16+16+16} = \frac{3}{8} = \frac{9}{24} = \frac{1}{3} + \frac{1}{24},$$
$$\frac{0+1+4+9+16+25}{25+25+25+25+25+25} = \frac{11}{30} = \frac{1}{3} + \frac{1}{30},$$
$$\frac{0+1+4+9+16+25+36}{36+36+36+36+36+36+36} = \frac{13}{36} = \frac{1}{3} + \frac{1}{36},$$

and concludes that the deviation from 1/3 decreases consistently. Hence it is clear to him that the limit will be 1/3.

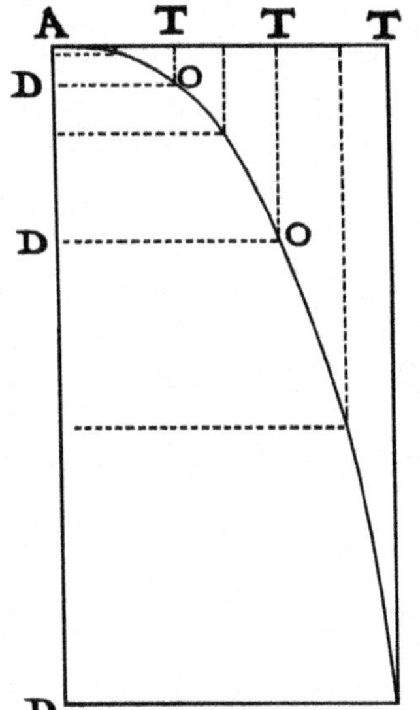

Fig. 2.3.8. The quadrature of the parabola after Wallis

He now turns to the parabola $y = x^2$, cp. figure 2.3.8, and wants to compute the area under the curve between $x = 0$ and $x = 1$. Attention!: The parabola is shown upside down; the abscissa (x-axis) points horizontally to the right, the ordinate (y-axis) points downwards. We know from our first chapter that this area is given by the definite integral

$$A = \int_0^1 x^2\,dx$$

and the Fundamental Theorem of Calculus (which Wallis did not know!) yields $A = (1/3)x^3|_{x=0}^1 = 1/3$. Wallis, however, proceeded as follows: He knows the area of the rectangle $ATOD$ and wants to compute the ratio of the unknown area ATO to $ATOD$. Following Cavalieri's method of indivisibles he would have to 'sum' the indivisibles TO. Every TO has but the length $(OD)^2$ since these are the function values of the parabola $y = x^2$. It now emerges that Wallis completely ignores the difference between indivisibles and infinitesimals, since he assumes that between the lines denoted by TO there is a small gap of size $a > 0$. In order to compute the ratio of the areas Wallis has to know the value of

$$\frac{0^2 + a^2 + (2a)^2 + (3a)^2 + \ldots + (na)^2}{(na)^2 + (na)^2 + (na)^2 + \ldots + (na)^2},$$

where the gap a is ever-decreasing if n grows. Since

$$\begin{aligned}\frac{0^2 + a^2 + (2a)^2 + (3a)^2 + \ldots + (na)^2}{(na)^2 + (na)^2 + (na)^2 + \ldots + (na)^2} &= \frac{a^2(0^2 + 1^2 + 2^2 + 3^2 + \ldots + n^2)}{a^2(n^2 + n^2 + n^2 + \ldots + n^2)}\\ &= \frac{0^2 + 1^2 + 2^2 + 3^2 + \ldots + n^2}{n^2 + n^2 + n^2 + \ldots + n^2},\end{aligned}$$

it follows from Proposition 19 that this term approaches $1/3$ as n increases indefinitely. Hence

$$\frac{\text{area }(ATO)}{\text{area }(ATOD)} = \frac{1}{3}$$

is shown and since the area of the rectangle $ATOD$ is just 1 the area under the parabola is $1/3$.

This result was not new! Other mathematicians knew this long before Wallis, but the actual breakthrough here is the *technique*: Wallis has deduced it himself from complicated geometrical arguments; an algebraic relation (Proposition 19) is merely necessary to compute the area.

Let us look again at figure 2.3.8. If the area of the gusset ATO is $1/3$ then the amount $2/3$ only remains for the complementary area AOD. This area, described by the indivisibles DO, is but the area under the inverse function of $y = x^2$ and this is $y = \sqrt{x}$. Now Wallis calculates with this part of the area AOD as he did before with ATO and gains the relation

$$\frac{\sqrt{0} + \sqrt{1} + \sqrt{2} + \ldots + \sqrt{n}}{\sqrt{n} + \sqrt{n} + \sqrt{n} + \ldots + \sqrt{n}} \overset{n\text{ 'very large'}}{=} \frac{\text{area }ATOD - \text{area }ATO}{\text{area }ATOD} = \frac{2}{3}.$$

Wallis had already shown that for arbitrary k

$$\frac{0^k+1^k+2^k+\ldots+n^k}{n^k+n^k+n^k+\ldots+n^k} \overset{n \text{ 'very large'}}{=} \frac{1}{1+k}$$

holds. For $k=2$ we regain Proposition 19. Under the impression of the term

$$\frac{\sqrt{0}+\sqrt{1}+\sqrt{2}+\ldots+\sqrt{n}}{\sqrt{n}+\sqrt{n}+\sqrt{n}+\ldots+\sqrt{n}} \overset{n \text{ 'very large'}}{=} \frac{2}{3}=\frac{1}{\frac{3}{2}}=\frac{1}{1+\frac{1}{2}}$$

he deduces that $\sqrt{x}$ has to be the same as $x^{\frac{1}{2}}$! Analogously he then moves on to show that $\sqrt[3]{x}=x^{\frac{1}{3}}$ has to hold, and that $x^0=1$.

Therebye Wallis has proved the relation

$$\int_0^1 x^{\frac{p}{q}}\,dx=\frac{q}{p+q},$$

a result which also was not new, though its derivation was.

A further real gem in Wallis's *Arithmetica Infinitorum* is the nowadays famous product

$$\frac{\pi}{4}=\frac{2\cdot 4\cdot 4\cdot 6\cdot 6\cdot 8\cdot 8\cdot 10\cdot 10\cdots}{1\cdot 3\cdot 3\cdot 5\cdot 5\cdot 7\cdot 7\cdot 9\cdot 9\cdots}.$$

With a hint to the author one also finds the continued fraction representation of William Brouncker,

$$\frac{4}{\pi}=1+\cfrac{1^2}{2+\cfrac{3^2}{2+\cfrac{5^2}{2+\cfrac{7^2}{2+\cfrac{9^2}{\ddots}}}}}.$$

Brouncker was so impressed by Wallis's product that he worked to find this beautiful expansion in 1655.

John Wallis was at all times concerned with questions of priority. He suspected the French, Descartes in particular, to have pretended results of Thomas Harriot as being his own. In case of the priority dispute between Newton and Leibniz he clearly took Newton's side and became a true ally of Newton.

2.4 Isaac Barrow

Isaac Barrow was born in 1630 into the family of the wealthy cloth merchant Thomas Barrow in London. Isaac's great-grandfather was the Cambridge-educated physician Philip Barrough who gained recognition in 1583 with his book *Methode of physicke*. Philip's brother – another Isaac – was a fellow of Trinity College in Cambridge and teacher of the future Chancellor of the Exchequer Robert Cecil. Two sons of Philip, Samuel and yet another Isaac, were matriculated at Trinity College and this Isaac became Justice of the Peace. His sons were Thomas, the father of Isaac Barrow, and (yet another) Isaac who became Bishop of St. Asaph in North Wales [Feingold 1990a, p. 1]. Our Isaac Barrow hence came from a Cambridgeshire family with a strong Cambridge-academic background; only his father Thomas was an exception. The harshness of his father drove Thomas away from home, persuaded him to take a trade apprenticeship in London and not to take up a scientific training. In the year 1624 he had worked his way up to become the linen trader for King Charles I and he got married. It seems that the only offspring of this marriage was the son Isaac. The mother died about 1634 and the widowed father gave his four year old son away in the care of his grandfather Isaac, the Justice of the Peace. Within the next two years Thomas got married again and retrieved his son.

Driven out of his family home by his own father, Thomas tried with all his power to allow his son a scientific career which was barred him. Thomas payed

Fig. 2.4.1. Isaac Barrow, statue by Matthew Noble, Chapel of Trinity College, Cambridge [Photo: Andrew Dunn, 2004] and as engraving of William Hall, the younger (after a portrait by Whood)

Fig. 2.4.2. Peterhouse in the year 1815 [Painting by Rudolph Ackermann, 1815]

double fees to Robert Brooke who was the headmaster of Charterhouse, a school known for its solid and excellent education, for giving special attention to Isaac. Unfortunately, Brooke did not measure up to this task and Isaac developed into a ruffian and bully.

His father Thomas was desperate; so much so, in fact, that he *'often solemnly wished, that if it pleased God to take away any of his children it might be his son Isaac'* [Feingold 1990a, p. 2][6]. Hence after two or three years Isaac was taken out of Charterhouse School and was given into the custody of schoolmaster Martin Holbeach at Felsted School in Essex. Holbeach sympathised with the puritans and was involved in the education of Oliver Cromwell's four sons at his school. He did not attach great importance to a purely religious education but Felsted School was known for its discipline and the solid classical education it offered. Already ten years before Barrow John Wallis was a pupil of Holbeach, and as he wrote in his autobiography many pupils of Felsted School went to University.

Isaac must have stayed approximately four years in Felsted. Roughly in the middle of this period, in 1642, Isaac's father Thomas got bankrupted by the Irish Revolution of 1641. Holbeach reacted and took Isaac up to his house since

[6]Feingold gives John Aubrey's *Brief Lives* as a reference but I can't find it in [Aubrey 1982]. It seems more likely that it originates in [Rouse Ball 1960, p. 309].

Fig. 2.4.3. Trinity College in the year 1690
[David Loggan, Cantabrigia, Cambridge 1690]

he couldn't pay for accomodation any longer. Later Holbeach took care that Isaac became the 'little Tutor' of Thomas Fairfax, who later became fourth Viscount Fairfax of Emely in Ireland, so that he could earn some pocket money. Then Isaac was offered a stipend for Peterhouse in Cambridge. We can only suspect that his uncle (also with the name Isaac), who was fellow of Peterhouse, was behind it.

At that time Isaac's uncle lost his seat in Parliament and moved to Oxford to live with Isaac's father. Our Isaac hence remained in Felsted with Fairfax; the stipend was given to another student. To make problems worse, Fairfax fell in love with a girl having a dowry of 'only' 1000 pounds and married her without permit of his father. Thereupon he lost all financial support and Isaac thus his pocket money. Together with Fairfax and his wife he moved to London where he remained without income. Holbeach intervened, found Isaac, and offered him to become his only heir. Isaac declined the offer. Instead, he moved to Trinity College, Cambridge, with an old schoolfriend who promised to support him. Hence, Isaac Barrow became 'sizar' in Cambridge on 10th March 1646, i.e. a poor student who had to serve wealthier ones.

The victory of the republicans had a massive impact on Trinity College. In the year 1645 the puritan clergyman Thomas Hill (died 1653) was appointed the new Master. Hill played a particular role in Barrow's career.

Hill had begun as a sharp and dogmatic Presbyterian on the republican side, but with the years he saw himself as protector of the scholars in his College who showed objectionable (in his sense) religious or political views and his sense of mission declined considerably. When the attacks of the Independents in the Universities in the years 1648 and 1649 became particularly violent, Hill maintained teaching and research activities and persuaded London's mayor,

John Wollaston, to establish a mathematical professorship at Trinity College. If Hill was important in Isaac Barrow's career, so was his tutor James Duport (1601–1679), a classics scholar and presumably the only tutor in Cambridge who openly showed sympathies for royalists and Anglicans. Duport kept his protecting hands on the sons of royalists and Isaac Barrow was one such son.

Barrow started to study mathematics seriously in 1648 or 1649 since in November 1648 John Smith was appointed the mathematical professorship donated by John Wollaston. We do not know exactly whether Barrow really attended Smith's lectures, but most likely Smith lectured on Descartes's geometry. We know that because Smith corresponded with Wallis whom he asked to explain some results in Descartes's *Geometria* [Feingold 1990a, p. 19]. Ralph Cudworth (1617–1688), a Cambridge Platonist, a philosopher and theologian, had lectured on the new mathematics of Descartes in Cambridge already in the year 1642. About his lectures we know nothing, but the register in his library lists *La Géométrie* by René Descartes as also the Latin translation of van Schooten, works by Henry Briggs, Henry Gellibrand, Thomas Harriot, François Viéte, Bonaventura Cavalieri, Evangelista Torricelli, John Wallis and Galileo Galilei, and there are clear signs that he must have been an eminent mathematician [Feingold 1990a, p. 20]. Surely Cudworth's lectures must have stimulated further studies by his pupils.

This new teaching had always to fight against two problems: there were only a few interested students who were talented enough to benefit, and it was difficult to find donors for mathematical studies. Barrow did not let this disturb him. He studied mathematics on his own and with all his strength. When Gilbert Clerk reported to Isaac Newton in 1687 that he and Barrow tried for 40 years to establish mathematics at the University of Cambridge[7], he was certainly not exaggerating [Feingold 1990a, p. 21]. Cartesian philosophy and mathematics had fully arrived in Cambridge about 1660, in fact. Isaac Barrow was elected a Fellow of Trinity College in 1649 due to his achievements, and he was the leading figure of the young royalists in College. It is no wonder that despite the protection of Thomas Hill problems occurred. On 5th November 1651 Barrow used the Memorial Day of the Gunpowder Plot to express his attachment with the House of Stuart and only the repeated intervention of Thomas Hill spared Barrow's expulsion from College. This royalist attitude caused Barrow to get in contact with a clergyman Henry Hammond (1605–1660) who defended and kept up the role of the Church of England even in the days of the civil war. It was through Hammond that Barrow managed to educate children of royalists. Furthermore he was involved in some kind of underground work in 1651/52, conveying letters, news, and books between some royalists.

[7] *'contributed nearly 40 years since, as much or more than any to others, (to speak modestly) in diebus illis, to bring these things into place in ye university.'*

In the year 1656[8] Isaac Barrow's first mathematical work, an edition of Euclid's book *The Elements*, was published. *The Elements* originated about 300 BC and became *the* source of inspiration in the west for classical geometry after Adelard of Bath translated it from the Arabic for the first time in the 12th century. Barrow's first book was a result of his experiences of mathematical teaching and became a bestseller immediately.

Since about 1654 Barrow had tried to get a position as lecturer at the university. Friends tried to help but Barrow's well-known royalist attitudes spoiled all opportunities. In the time of the most intense attacks of religious extremists he therefore decided to undertake a tour on the Continent. He applied for one of three travel bursaries of Trinity College and got a three-year stipend, conditional on reporting regularly. On 4th May 1655 – even before his edition of *The Elements* appeared in print – he received his pass and in July he left England with his friend Thomas Allen of Gonville and Caius College.

Their first stop was Paris where Isaac Barrow reunited with his father whom he gave financial assistance. Isaac stayed in Paris for 10 months and reported home that Trinity College and Cambridge had nothing to fear in comparison with the schools in Paris. At the same time as Isaac Barrow, the great Dutch Christiaan Huygens, the later teacher of Leibniz, was in Paris. Barrow knew and admired the writings of Huygens. Whether the two men met is not known, but Christiaan's father Constantijn maintained as a diplomat very good relationships with England and Christiaan was elected a member of the Royal Society in 1663. In a letter Barrow mentioned the mathematician Gilles de Roberval (1602–1675), cp. [Sonar 2011, p. 273ff.], but in this case we do not know whether the two men met. In any case Huygens did meet Roberval at this time.

Some day in February 1656 Barrow moved on to Florence where he went to stay more than eight months, mainly due to the raging plague in Italy. In Florence Barrow visited the library Medicea Laurenziana of the Medici and studied old coins. He improved his tricky financial situation through purchasing coins for the enhancement of the collections of two merchants, one being the generous James Stock. At that time he befriended the Florence mathematician Carlo Renaldini (1615–1698). Florence in these days was a melting pot for mathematicians of the first rank. The last pupil of Galileo Galilei, Vincenzo Viviani (1622–1703), was, like Barrow an eager student of ancient Greek geometry. In the second half of the 1650s he worked on the reconstruction of the fifth book of Apollonios of Perga's *Konika*, an ancient book on conic sections for which Barrow also showed great interest. Other famous scientists found themselves in Florence. What did Barrow take with him from his sojourn? We don't know exactly, but later on Barrow showed an outstanding

[8]The title page shows the (English, Julian) date 1655, compare [Folkerts/Knobloch/Reich 2001, p. 63], but the book was published at the end of the Julian year 1655, hence at the beginning or the Gregorian year 1656, cp. [Feingold 1990a, p. 40].

Fig. 2.4.4. Vicenzo Viviani, Painting by Domenico Tempesti, about 1690 and Christiaan Huygens ([Painting by Bernard Vaillant] Museum Hofwijck, Voorborg, it very much resembles the painting by Caspar Netscher of 1671, Fig. 2.7.6)

knowledge of Torricelli's and Cavalieri's method of indivisibles so that we may assume that he learned these techniques in Florence.

In November 1656 Barrow boarded a ship to Turkey in order to get to Constantinople. The way to Rome was obstructed by the plague. Travelling via Venice was impossible as well, and hence he seemed to have chosen a way to the east. It was simply the only way for Barrow to travel on. During the sea voyage the ship was attacked by Algerian pirates. It is typical of Barrow that he fought side by side with the crew. Later asked why he hadn't left fighting to the crew he answered: *'Holy Liberty is dearer than my vital breath'* and added that the prospect of enslavement by infidels was worse than death [Feingold 1990a, p. 50]. The damaged ship was repaired on the Isle of Milos in the Aegean and the casualties were cared for. Then they sailed further to Smyrna, now Izmir, in Turkey. Barrow stayed seven months enjoying the hospitality of the English Consul Spencer Bretton and befriending with the Chaplain of the English colony, Robert Winchester.

In summer 1657 Barrow eventually arrived in Constantinople where he was accommodated by the English ambassador Sir Thomas Bendish and where he got generous financial support from the merchant Jonathan Dawes. Barrow stayed one and a half years in Constantinople. In May 1658 the time granted for his travels expired but Barrow got permission to extend the time limit. He got news concerning the death of his benefactor James Stock in December 1658 and decided to go back to England. He certainly bought coins for Stock also in Constantinople. He improved his language skills in Greek, too, and satisfied his interests in the Eastern Church and its rites. Not for nothing was the Holy Chrysostomos later his preferred Church Father. The return journey was not easy, however. The ship had barely anchored in Venice when it caught fire and Barrow's belongings were destroyed. From Venice he traveled on an overland route through the German lands and the Netherlands and arrived in Cambridge in September 1659.

Meanwhile, John Arrowsmith, Master of Trinity College, had passed away. John Wilkins became the new Master on 3rd September with whom Barrow became closely befriended. Not only the statutes of Trinity College demanded Barrow's ordination but Barrow saw himself as a theologian. Hence he was ordained and got a further Bachelor in Theology in 1661.

Then the restoration took place. Charles II was crowned King of England and Barrow as a royalist was overjoyed for the time being. John Wilkins supported his election as Regius Professor for Greek; in 1662 he additionally accepted the professorship of geometry at Gresham College in London. He also accepted a deputy professorship in astronomy because only very few students were interested in Greek lessons. In Gresham College Barrow occupied the same rooms in which the inaugural meeting of the Royal Society took place on 28th November 1660. On 17th September 1662 Barrow himself became a member of this society, probably through recommendation of John Wilkins. He was not very active in the Royal Society, however.

One month prior to his death in July 1663 the politician and clergyman Henry Lucas specified in his last will and testament that a professorship for mathematics should be established at Cambridge University, financed from his assets. Isaac Barrow was chosen the first professor for the Lucasian Chair. We know that Barrow was a devoted teacher who cared much for his students [Feingold 1990a, p. 65f.]. His most famous student was Isaac Newton, one of our wranglers. In the year 1683 Barrow's mathematical lectures stemming from the 1660s were published under the title *Lectiones Mathematicae* [Barrow 1973]. His optical investigations *Lectiones Opticae et Geometricae* were published already in 1669. In the preface 'Letter to the Reader' Barrow emphasised the active involvement of his pupil Isaac Newton who not only took on the proof-reading but also contributed some of his own ideas [Shapiro 1990, p. 111f.]. It seems almost certain that Newton indeed attended Barrow's lectures on optics in 1667. At that time Newton's own research was advanced so far that he could appreciate Barrow's mathematical approach.

The man behind Barrow's publications was John Collins (1625–1683), also a member of the Royal Society since 1667. The priest Marine Mersenne (1588–1648) worked in Paris as a mathematician, but his great achievements were the communication activities with numerous other French mathematicians. Mersenne acted like a spider in its web and got and spread mathematical results from French mathematicians. In much the same sense Collins was acting in London and hence can be seen as Mersenne's equivalent in England. He is therefore often called the 'English Mersenne'. After the plague and after the Great Fire of London 1666 in particular, publishing scientific books and papers was not easy; yet Collins was a skilful and imaginative publisher.

But Barrow did not stay in Cambridge. He handed over the Lucasian Chair to his pupil Isaac Newton in the year 1669. Many modern historians have attributed this to Barrow's wish for further career success, but Feingold's detailed analysis in [Feingold 1990a, p. 79ff.] reveals another motive. Barrow was a theologian in the first place who over the years got a feeling that he was neglecting his true calling. He was a serious believer and hence departed from his secular tasks; he even accepted loss of wages. Friends tried without success to help him to a well-paid position within the Church of England; they then turned to look for a professorship at a university or the position as a College's Master. It says much in favour of Barrow that even great men like the Archbishop of Canterbury acted on his behalf. Finally he became Master of Trinity College on 15th February 1673. Barrow passed away in London on 4th May 1677, following a fever. He was 46 years of age, a bachelor, and was buried three days later in Westminster Abbey.

Fig. 2.4.5. Westminster Abbey, Main entrance [Photo: Σπάρταχος 2013]

Earlier authors have seen Barrow as the actual inventor of the differential and integral calculus; in particular J.M. Child [Child 1916] claimed to have found the fundamental theorem of calculus already in Barrow's works. Child wrote in the preface of [Child 1916]:

'Isaac Barrow was the first inventor of the Infinitesimal Calculus; Newton got the main idea of it from Barrow by personal communication; and Leibniz also was in some measure indebted to Barrow's work, obtaining confirmation of his own original ideas, and suggestions for their further development, from the copy of Barrow's book that he purchased in 1673.'

We can't leave it like this; the text was written from a view of a mathematician in the year 1916 and Child had interpreted Barrow in this modern view. Today one thinks differently, as can be read in a text by Mahoney in [Mahoney 1990, p. 180f.]:

'At the same time that recent work provides a richer understanding of Barrow's mathematical work taken on its own terms, it also removes him from a major role either in the development of the calculus or in Newton's early mathematical thinking.'

Presumably Mahoney's position is also too extreme; at least it is unfair. A large influence of Barrow on Newton's early mathematical thinking can certainly be attested as is confirmed by Niccolò Guicciardini [Guicciardini 2009, p. 4]:

'Barrow's defence of geometry as a model of reasoning and his idea that, since geometrical magnitudes are generated by motion, a causal relationship can be captured in such mechanically-based geometry must have impressed the young scholar. These typical Barrovian ideas remained the backbone of Newton's views about mathematics.'

It is very clear that Barrow was bonded to classical geometry, his mathematical love, and that he did not discover the differential and integral calculus. Let's see for ourselves.

Barrow's mathematics is rooted in considering the movement of a point moving with time-dependent velocity $v(t)$. The velocity curve in figure 2.4.6 shows $y = f(x)$ plotted downwards. The upper curve $z = A(x)$ shows the increase of area under the curve $y = f(x)$ from 0 to x. This is the curve we are interested in. Let the point D be given by the coordinates $(x_0, 0)$ on the x-axis; for the point T

$$DT = \frac{DF}{DE} = \frac{A(x_0)}{f(x_0)}$$

is supposed to be valid.

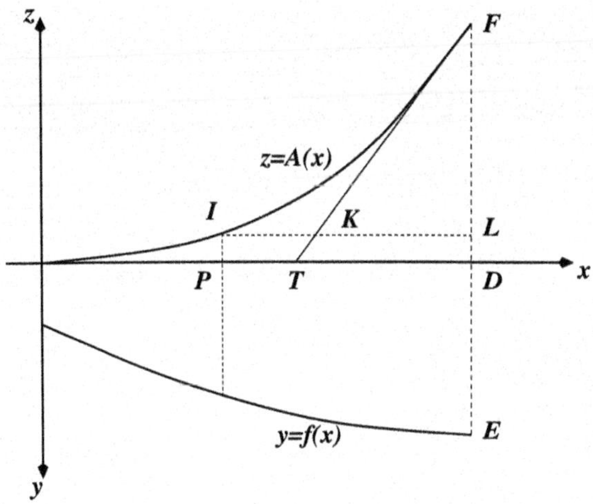

Fig. 2.4.6. Barrow's way towards the fundamental theorem

Barrow now shows that the straight line TF touches the curve $z = A(x)$ only in the point x_0 in this construction. For this purpose he remarks that the slope of TF is given by

$$\frac{DF}{DT} = \frac{A(x_0)}{\frac{A(x_0)}{f(x_0)}} = f(x_0).$$

If Barrow had written here that TF is the tangent to $A(x)$ at the point x_0 and that in fact

$$\frac{dA}{dx}(x_0) = f(x_0)$$

holds he indeed would have had discovered the fundamental theorem of calculus. But that he did not write; instead he stayed close to classical geometry. If I is the point on the curve $z = A(x)$ at $x = x_1 < x_0$ and K the intersection point of TF with IL – the parallel of the x-axis, then Barrow showed that K is always to the right of I. To do so he starts with

$$\frac{LF}{LK} = \frac{DF}{DT} = DE$$

and gets $LF = LK \cdot DE$ from this. But on the other hand it holds that

$$LF = DF - PI = A(x_0) - A(x_1) < \underbrace{DP \cdot DE}_{\text{area of rectangle}}$$

and hence it follows that

$$LF = LK \cdot DE < DP \cdot DE \quad \Longrightarrow \quad LK < DP = LI,$$

which was to be proven. The case $x_1 > x_0$ is completely analogous.

Fig. 2.4.7. James Gregory [unknown painter] and a replica of his telescope ([Photo: Sage Ross 2009] Replica about 1735, Putnam Gallery, Harvard Science Center)

Edwards [Edwards 1979, p. 140f.] has pointed out that a similar result can be found a little earlier in the works of the young Scottish mathematician James Gregory (1638–1675). This is perhaps not surprising since Gregory was an outstanding mathematical talent who found some results of the differential and integral calculus simultaneously with his contemporaries. He invented a particular reflecting telescope, was a member of the Royal Society since 1668, professor of mathematics at the University St. Andrews, and from 1674 at the University of Edinburgh. He died after suffering a stroke when only 36 years of age. Since he published rarely his results stayed unknown for a fairly long time.

Let us finally look at what Barrow had to say concerning the computation of tangents. It is here where he used a technique going back to Fermat. He considers functions of the form $f(x,y)=0$. If o_1 and o_2 denote infinitely small quantities then certainly

$$f(x+o_1, y+o_2) = f(x,y) = 0.$$

Let us consider with Edwards [Edwards 1979, p. 133] the example of the Folium of Descartes

$$f(x,y) = x^3 + y^3 - 3xy.$$

Since $f(x+o_1, y+o_2) = f(x,y) = 0$ it follows

$$(x+o_1)^3 + (y+o_2)^3 - 3(x+o_1)(y+o_2) = x^3 + y^3 - 3xy = 0.$$

Multiplying out and tidying, this is

$$3x^2 o_1 + 3x o_1^2 + o_1^3 + 3y^2 o_2 + 3y o_2^2 + o_2^3 - 3x o_2 - 3y o_1 - 3 o_1 o_2 = 0.$$

Since o_1 and o_2 are supposed to be infinitely small Barrow sets all higher powers to zero ('for these terms have no value'). Of course, also the product $o_1 o_2$ has to be set to zero. It remains

$$3x^2 o_1 + 3y^2 o_2 - 3x o_2 - 3y o_1 = 0,$$

which gives the slope of the tangent as

$$\frac{o_2}{o_1} = \frac{y - x^2}{y^2 - x}.$$

2.5 France and the Netherlands in the 17th Century

2.5.1 France on its Way to Absolutism

The old mocker Voltaire reports in his *Letters concerning the English Nation*, Letter XIV, published in 1733 for the first time: 'On Des Cartes and Sir Isaac Newton' [Voltaire 2011, p. 61]:

> *'A Frenchman who arrives in London, will find Philosophy, like every Thing else, very much changed there. He had left the World a plenum, and he now finds it a vacuum. At Paris the Universe is seen, composed of Vortices of subtile Matter; but nothing like it is seen in London. In France, 'tis the Pressure of the Moon that causes the Tides; but in England 'tis the Sea that gravitates towards the Moon; so that when you think that the Moon should make it Flood with us, those Gentlemen fancy it should be Ebb, which, very unluckily, cannot be proved. For to be able to do this, 'tis necessary the Moon and the Tides should have been enquired into, at the very instant of the Creation.'*

The discrepancy which Voltaire pointedly describes here is the discrepancy between the physics of René Descartes on the French side, and Isaac Newton's physics on the English side; but in the 17th century the countries France and England were very different too and probably still are today.

On the 14th May 1610 the French King Henry IV was stabbed to death by a religious fanatic. His oldest son, the nearly nine years old Louis XIII, became the new king; due to his young age the King's mother Maria de' Medici took over the reign and kept it until 1617. Unfortunately, Maria was anything but a wise politician [Haupt et al. 2008, p. 172]. Henry IV had succeeded in curbing the French high nobility, but now Italian favourites won influence at court and the high nobility revolted again. The Protestants in France had to live again in fear since the policy of tolerance and reconciliation of Henry was at stake. At a meeting of the 'Estates General' in the year 1614 old and new conflicts broke open. The Italian favourite of Maria, Concino Concini, succeeded in renewing

Fig. 2.5.1. Young Louis XIII ([Painting by Frans Porbus 1611] Louvre-Lens, Pas-de-Calais [Photo: Jean-Pol Grandmont]); his mother: Maria de Medici ([Painting by Piertro Fachetti, about 1594] Palazzo Lancelotti, Rom)

the central power and was able to beat down the uprising (the 'Fronde') of the noblemen, but the young heir to the throne was not on his mother's side. Maria's regency ended in 1617; she was even forced to leave court. Concini was assassinated with the knowledge of the king.

The young king's trusted friend was Charles d'Albert de Luynes who took action against the Protestants in France, since Louis pursued a radically Catholic policy. As a consequence of this policy a man was called in the Royal Council in the year 1624 who became immortal through Alexandre Dumas's *The Three Musketeers*: Armand-Jean du Plessis, Cardinal-Duke of Richelieu and of Fronsac (1585–1642).

Richelieu was descended from an old noble family and became Cardinal in 1622. Initially on the side of the King's mother, he succeeded by and by in blackening de Luynes in the King's mind who did not have much self-confidence. On 20th November 1630, the 'Day of the Dupes'[9], Richelieu could get rid of the last resistance fighters in the Royal Council – the so-called devotees. The way was free now for his policy which turned out to be decisive for France.

For Richelieu two main goals were paramount: the elimination of the threat against the inner unity by the Protestants and the nobility, and a foreign

[9]It is not quite clear whether it was November 10, 11, or 12 in the Julian calendar.

Fig. 2.5.2. Triple portrait of Cardinal Richelieu ([Painting by Philippe de Champaigne about 1640] National Gallery, London)

policy clearly directed against the Habsburg dynasty to keep Spain's role small in Europe. Between 1626 and 1629 Louis and Richelieu took military action against the rebellious Huguenots in the south-west of France. After a siege of one year the Protestant fortress La Rochelle fell in 1628. Yet a year later the king issued an edict of mercy for the Huguenots, but any military and political organisation of the Protestants was prevented. The Protestants could thereby continue to exist in France, but as a political party they were meaningless. The rebellious nobility could not be combatted in a military strike; Richelieu betted on exemplary measures against them. If the nobility violated the law – for example by violation of edicts against duels – they were uncompromisingly persecuted and sometimes even their castles were pulled down [Haupt et al. 2008, p. 177]. In consequence of this policy the monarchy became the only powerful entity in France.

In foreign affairs Richelieu tried to push back the influence of the Habsburg dynasty and of Spain. Although a catholic cardinal Richelieu did not hesitate to make agreements with protestants powers in Europe if need be. France paid supporting benefits (subsidies) to protestant Sweden for instance, and positioned itself openly as an ally on the Swedish side in 1635. This policy of Richelieu in the Thirty Years' War ultimately established France's supremacy in Europe. Since Richelieu died in 1642 he could not take notice of the Peace of Westphalia in 1648 which brought many advantages to France. However, the tax burden grew unbearably high for the French people due to the open intervention of France in the Thirty Years' War. Following [Haupt et al. 2008, p. 178] the direct taxes increased threefold between 1624 and 1661. Such a tax burden the country couldn't bear; the economic performance was too low. Taxes were vigorously pressed from the people; the national debt increased at the same time. Many popular uprisings in these days document the consequences of this tax policy. Royal tax collectors were attacked. With respect to religious beliefs things calmed down a bit since the resolutions of the Council of Trent, which took place from 1545 to 1563, were introduced in France. Important were the abolition of accumulation of offices in priesthood, a reform of the practice of indulgence, and the introduction of theological colleges (seminars). After the numerous religious wars large parts of the elites now turned towards an inner piety. Into this climate came a catholic reform movement from the Netherlands to France, Jansenism, which got its name from the bishop of Ypres, Cornelius Jansen (1585–1638). Jansenists wanted to return to the primordial Christian faith; in particular they built on Augustine of Hippo's treatise of grace. Man in his sin has no influence but is completely and utterly at God's mercy. Jansenism attracted many French, including parts of the elite and of the clergy. Blaise Pascal and Antoine Arnauld were well-known Jansenists and the monastery of Port Royal des Champs close to Versailles became a centre of Jansenism. The Abbess Angélique Arnauld, sister of Antoine Arnauld, acted as the Principal of this monastery. Quickly conflicts developed between the Jansenists and the mighty Jesuits; France certainly became no more tolerant, on the contrary! Many great thinkers who could no longer live in the intellectual narrowness of Catholic France left their country; René Descartes for example, who sought and found exile in the Netherlands.

Richelieu passed away in 1642; Louis XIII followed him in 1643. Louis's son, Louis XIV, was six years old when his father died.

Again a female regent took the affairs of state in her hands, this time the mother, Anne of Austria, who ruled with the help of Richelieu's successor, the Italian cardinal Jules Mazarin (1602–1661). Again a time of insecurity began - again there were riots first in Paris and then in the provinces. When Mazarin died in 1661 Louis XIV at last took over the government. He had experienced the riots himself and started immediately to counteract the loss of authority of the monarchy, emphasising the royal omnipotence very clearly. Owing to this self-confidence of Louis, who was a strong character, we today have the notion of 'absolutism'.

Louis decided to rule without a 'first minister' like Richelieu or Mazarin, but instead with a small circle of advisors. Of them Jean-Baptiste Colbert (1619–1683) became famous and was responsible for administration, economy and finances. The basic principles of his French economic model are called 'Colbertism' today. Colbert reduced the direct taxes and increased the indirect ones; he was also responsible for regulatory measures in the financial system.

No monarch before Louis XIV put himself as well as his country so much in the limelight; none before him encouraged artists and architects to put him

Fig. 2.5.3. Louis XIV, the 'Sun King' ([Painting by Hyacinthe Rigaud 1701] Louvre Museum, Louis XIV Collection)

so effectively on the scene. The royal residence in Versailles, having devoured enormous assets, is an eloquent testimony of his self-assessment.

In terms of foreign policy he was 'in search of glory' [Haupt et al. 2008, p. 201]; a standing army was built up and diplomacy was perfected. Louis XIV had married a daughter of the King of Spain, Maria Theresa of Spain, in 1660. When his father-in-law (who also was his uncle) died in 1665 Louis claimed the inheritance law of his wife. To this end he waged the War of Devolution between 1667 and 1668 in which Louis claimed the Spanish Netherlands and

Fig. 2.5.4. Versailles at the end of the reign of Louis XIV
[Painter: Pierre-Denis Martin le Jeune, 1722]

put his new troops to test. In the Treaty of Aix-la-Chapelle of 2nd May 1668, Spain had to permanently transfer some of its territory to France. In a further war 1672–1679 against the Netherlands, which was actually directed against Spain, more and more European powers took the side of the Netherlands. It cost the French enormous efforts to appear as winners over Spain in the Treaties of Nijmegen (1678 peace with the Netherlands, 1679 with Sweden and the Holy Roman Empire), although Louis had to return his Dutch conquests. At least Louis was able to incorporate Burgundy. Not a single shift in the power structure of Europe was now possible without Louis XIV.

Louis found a serious opponent in Pope Innocent XI concerning the right of use of the incomes from unmanned dioceses. Most French clergymen were hostile to the Pope, 'Gallican', and they supported their king, thereby intensifying the conflict. This led to Louis's excommunication in the end. However, a change of the royal policy in the 1690s brought France closer to the Vatican again. The protestants could not derive benefit from this conflict: on the contrary; the Edict of Nantes, warranting religious freedom to the protestants since 1598, was revoked by Louis on the 18th of October 1685 in Fontainebleau. This was without doubt a clear signal in the direction of the pope with whom Louis wanted to come to good terms again. Pastors of the protestant faith had now to leave France; churches of the huguenots were destroyed and at least

300 000 huguenots – among them many excellent craftsmen and technicians – eventually had to leave France and went to the Netherlands, to Switzerland, and to the German countries.

But Louis was not only fighting the protestants, but also turned against the Jansenists. Louis refused to accept a religious niche in which asceticism and morality were outstanding values and in which the consistent devotion to the grace of god stood in the focus – especially since the Jansenist doctrine was directed against the Jesuits so that large parts of the French elite sympathised with it. In 1668 Pope Clement IX could still achieve a partial settlement between Louis and the Jansenists, but the actual problems remained and were carried far into the 18th century after Louis's death.

After the Treaties of Nijmegen, Louis carried out the so-called 'reunions', i.e. the incorporation of territories belonging to France in the French legal opinion. These reunions – like the incorporation of the imperial city of Strasbourg, aroused resistance in the remainder of Europe, and the suspension of the Edict of Nantes led to the founding of the Grand Alliance (League of Augsburg) between protestant and catholic imperial estates, with the Emperor and Spain, to oppose France. When Louis remained unimpressed and invaded Southern Germany on the basis of legally problematic hereditary titles of his sister-in-law in 1688, Louis's arch-enemy William of Orange, now King William III, waged war against France together with the Netherlands, further German imperial cities and even the Swedish, and the Grand Alliance. This war ended with the Treaty of Ryswick in the year 1697. France had to give up some of the reunions, even the Palatinate, but the Alsace could be retained.

However, Louis kept with his political power line. When the Spanish throne became vacant after the death of Charles II of Spain in 1700 he stood up for his grandson and heir to the throne Philip of Anjou. Again Louis had to face a mighty alliance. The war between France and England was now also fought in the North American colonies. The year 1713 saw the Treaty of Utrecht with which the War of the Spanish Succession found its end. Louis XIV died on 1st September 1715 in Versailles. His country was prepared to enter the 18th century as a mighty nation.

2.5.2 The Netherlands and the Persistent Conflict with Spain and England

The territory which was called 'the Netherlands' consisted of 17 provinces in 1550. These provinces were organised in an Imperial Circle which belonged to the German Empire [Wielenga 2012, p. 19]. Today the Netherlands are often called 'Holland', but Holland is but one county of the 17 provinces. When the Holy Roman Emperor Charles V resigned in 1555 the Netherlands fell to his son Philip II who inherited the Spanish royal crown at the same time.

Fig. 2.5.5. Father and son: Charles V ([Painting by Lambat Sustris, 1548] attributed initially to Tizian, later corrected; Alte Pinakothek, Munich) and Philip II of Spain ([Painting by Tizian 1551] Prado Museum Madrid)

Formally the Netherlands were now under catholic reign. We don't want to follow the history of the combat operations that followed in the second half of the 16th century and refer the reader to history books like [Wielenga 2012]. In any case the northern parts of the Netherlands succeeded in founding an independent republic of seven provinces which waged war with Spain for approximately 80 years, only interrupted by a ceasefire of twelve years from 1609 to 1621. Only with the Peace of Westphalia 1648, which ended the catastrophe of the Thirty Years' War, was the Dutch Republic accepted by Spain. The territories in the south which remained with Spain were called Spanish Netherlands.

The Dutch Republic quickly established itself as a mighty trading and sea power, playing an important role on the European stage. The decades after 1650 are therefore called the 'Dutch Golden Age' [Wielenga 2012, p. 110ff.]. The rise of the Netherlands attracted a large number of foreign workers who fuelled the economic growth further. The textile industry in Leiden benefited on an enormous scale from labourers coming from the German lands, Scandinavia, Poland, England, and France; on the ships of the Dutch East India Company VOC (Vereinigde Oost-Indische Compagnie), founded in 1609, 40% of the mariners were of foreign nationality [Wielenga 2012, p. 113].

Fig. 2.5.6. Replica of the Batavia, sailing ship of the Dutch East India Company [Photo: ADZee 2007], Transom of the replica [own photo 2004]

Protestants who were religiously harassed poured into the country which had distinguished itself by its tolerant social climate through all of the 17th century. A famous 'refugee' was René Descartes who, despite his catholicism, could live significantly quieter in Holland than in France since he openly displayed his freethinking. The religious freedom was responsible for the peaceful coexistence of jews, protestants, and catholics. The dominant religion was Calvinism, but the calvinists had divided into remonstrants and contraremonstrants due to a quarrel which had developed over the question of the doctrine of predestination.

Humanist ideas found no resistance in a country with religious freedom. Desiderius Erasmus Roterodamus (1466–1536), a Dutchman, eventually became the figurehead of this intellectual current. In particular, humanism had initialised a movement for education, and hence many scholars – Descartes serves as an example – were attracted. The arts flourished in these days, too – the museums of the Netherlands being witnesses today. But even tolerance had limits. When the jewish philosopher Baruch Spinoza published in 1670 in Amsterdam his work *Tractatus theologico-politicus*, in which he advocated freedom of religion, tolerance, and a state which guaranties all freedom, he had to publish it anonymously. He went too far and the *Tractatus* was indeed banned in 1674.

In the Dutch towns three social strata could be found. Vagrants, beggars, and casual labourers comprised between 10 and 20 % of the urban population; not much more were wageworkers – mariners, craftsmen, soldiers. Together with the middle class of self-employed persons these classes represented approximately 90 % of the overall urban population [Wielenga 2012, p. 115]. The remaining 10 % comprised the bourgeoisie: wealthy traders, entrepreneurs,

Fig. 2.5.7. René Descartes ([Painting by Jan B. Weenix about 1648] Centraal Museum Utrecht)

civil servants, and the upper classes with not more than 1 % in proportion. At the head of society were the aristocrats, the ruling class. It was not easy to keep the different provinces together and to control the conflicts between the citizens and the nobility. Conflicts were solved by compromises in general, which entailed further compromises. The clamp which kept everything together was nothing but the threat from outside; if this threat weakened the inner conflicts increased. Soon after 1650 a myth developed, stemming from the dominating calvinists, that the Dutch Republic was a chosen country, a second Israel. This led to a strong self-esteem. This republic saw itself in the tradition of the ancient world, and this tradition was further developed. The large wealth due to the trade additionally ensured calm in the country. The mighty Jan de Witt (1625–1672) became the leading Dutch statesman as the so-called Grand Pensionary for nearly 20 years. He put the destination of the republic in the words, that [Wielenga 2012, p. 129]

> *'the concern of the state lies utterly in peace and quiet, so that the trade may not be hindered.*[10] *'*

[10]'das Interesse dieses Staates ganz und gar in Ruhe und Frieden besteht, so dass der Handel nicht behindert wird.'

Although England and France were confederates and allies in the fight against Spain, this situation changed in the 1660s, when France developed into a potential enemy. The conflict broke in 1672 when an attack against the Dutch Republic was led by English and French troops, supported by the dioceses of Münster and Cologne. The 'catastrophic year' 1672 thrust the House of Orange back to the throne. How could that happen? As a seapower England followed the same goals as the Netherlands and both nations quickly became competitors on the oceans. To get rid of the competitor, England under Cromwell proclaimed the 'Navigation Act' in 1651. According to the Navigation Act foreign countries could only deliver goods to England which actually originated in its home country. This the Dutch could not accept, having grown to No. 1 in world trade. A year later the First Anglo-Dutch War broke out which found its end in 1654. The Dutch had a 'snowball's chance in Hell' in this naval war so that they had to accept the English monopoly of trade from the colonies in the Treaty of Westminster. The Navigation Act was not withdrawn [Ploetz 2008, p. 1055].

Jan de Witt thereupon tried a seesaw policy between France and England which did not result in success but in the Second Anglo-Dutch War (1665–1667). This war brought a tremendous victory for the Dutch, since Admiral Michiel Adrianszoon de Ruyter sailed up the River Thames and succeeded in wiping out the fleet that lay at Chatham. The Dutch wisely did not humiliate the English in the Treaty of Bredam, since the French had invaded the Spanish Netherlands in 1667 and wanted to expand their territory northwards under their king Louis XIV. There was, however, a defence agreement between France and the Dutch Republic since 1662. The Dutch therefore possibly needed the English under these circumstances and thus a three partner coalition was concluded in 1668 with the two partners being England and Sweden. This coalition was betrayed by the English, however, who made public the secret clause which was directed against France. De Witt couldn't believe that an Anglo-French alliance was possible but exactly that happened in the catastrophic year 1672.

The House of Stuart was again on the throne in England; in the Netherlands the member of the House of Orange-Nassau, William III, was Stadtholder (Figure 2.2.9). Jan de Witt and his regents were made responsible for the military collapse of 1672; the member of the House of Orange seemed to offer a saving hand [Wielenga 2012, p. 141]. In August 1672 Jan de Witt and his brother were lynched by a mob in Den Haag; the Stadtholder of the Netherlands now was William III of Orange-Nassau.

William was a wise politician. He cooperated with all nations which also felt threatened by the French expansionism. As of 1673 a four partner coalition existed in which the Dutch acted together with Spain, the Habsburg emperor, and Lorraine, who forced Louis XIV to fight at several fronts simultaneously so that the Dutch Republic could leave its defensive position. The land-based troops always were neglected in favour of the navy; William changed that. The

Fig. 2.5.8. Admiral Michiel de Ruyter ([Painting of Ferdinand Boel 1667] National Maritime Museum Greenwich, London) and Johann (Jan) de Witt [Portrait by Adriaen Hannemann, presumably after 1650]

army became an effective power under his influence and already in 1674 Louis XIV had lost all of his Dutch conquests except the town of Maastricht. All conquered territories fell back to the Netherlands in the Treatise of Nijmegen of 1678. Nevertheless, William couldn't catch the people's hearts. He was moody and reserved and probably did not want to please [Wielenga 2012, p. 149]. In the year 1677 the 26-year-old married his 15-year-old cousin Mary II Stuart, daughter of the English heir apparent James II. In 1685 James II was crowned King of England and hence William's father-in-law sat on the throne. Nevertheless, the Netherlands and England did not come closer since James followed a pro-catholic policy. In the year of James's coronation Louis XIV revoked the Edict of Nantes by which protestants were accepted in France. Therefore a flow of French protestants, the huguenots, poured over Europe; to the Dutch Republic came approximately 35000 [Wielenga 2012, p. 150]. Anti-French sentiments grew in a short time and increased further when Louis revoked a trade clause from the Treaty of Nijmegen in 1687, thereby increasing the trade custom. Yet there was hope that after James's death his protestant daughter, wife of William, would ascend the throne; this hope dwindled with the birth of a male heir in England. For this reason an Anglo-French alliance became ever more likely and that had to be prevented under all circumstances. A letter from the highest circles in England was written on 30th June 1688, asking William to intervene in England, and it came at just the right time: William chased his father-in-law James II from his throne, became himself King of England, Scotland and Ireland in 1689, and commenced the Glorious Revolution in England. Until William's death the Dutch Republic and England walked on the same trails. From 1689 they fought successfully together in the Nine Years' War (1688–1697) against France. Finally, in the Treaty of Ryswick Louis XIV had to acknowledge William III as the English king. France had occupied the principality of Orange; now William got it back and Louis had to renounce all claims in the Southern Netherlands which now had to be cleared. The adverse trade tariff decree was also withdrawn.

William was preparing yet another war front, the War of the Spanish Succession, which began in 1702 and lasted until 1713, but he did not live long enough to see the outbreak of this war. At the end of this war the Dutch Republic was severely weakened, and memories of the disaster in 1672 awoke, when in the War of the Austrian Succession a part of the province of Zeeland was occupied by France in 1742. The 'Dutch Golden Age' was over once and for all.

2.6 Blaise Pascal

Blaise Pascal is a 'fascinating, but elusive personality'[11] [Loeffel 1987, p. 7]. He was born on 19th June 1623 in Clermont, today Clermont-Ferrand in the Auvergne in the heart of France. The family was inconspicuous and belonged to the class of the lower provincial nobility [Attali 2007, p. 21]. His father Étienne Pascal was attracted more by mathematics than by his duties in jurisdiction and administration. He was an extraordinarily cultured and educated man who was also capable of his own mathematical work: thus is the limaçon of Pascal, defined in polar coordinates as

$$r = a \cos \varphi + b,$$

named after him. The father educated his children not as was customary in these days; he is strongly influenced by the pedagogical ideas of Michel de Montaigne (1533–1592), author of the magnificent *Essays*.

It was certainly clear to the father that at least two of his children, Blaise and his sister Jacqueline, were most notably gifted. Blaise's sister Gilberte was slower on the uptake. Étienne came to the University of Paris Sorbonne in 1608 where he got to know some unconventional thinkers. He studied law and was able to buy the office of a District Administrator for the county of lower Auvergne in Clermont after his father had passed away. He privately studied mathematics, married Antoinette Begon in 1616, and bought the more prominent office of the second president at the tax court of Montferrand in 1624 [Attali 2007, p. 25].

From the start little Blaise suffered from nervous spasms and he kept in feeble health all his life. In the summer of 1626 his mother died and left the 38-year-old father with three children. Étienne did not marry again which was very unusual in those days; he instead looked after the education of his children who did not go to school. The highest aim of education was learning for the

[11] 'faszinierende, aber schwer fassbare Persönlichkeit'

sake of the pleasure of understanding. So Blaise and his two siblings learned Latin, history, and geography without pressure, but the abstract realm of mathematics the father would keep from them. The father set no great store by sporting activities or love of nature. When Étienne was hindered by a conspiracy to buy the office of the president of the tax court he sold his office to one of his brothers and settled in Paris to increase his knowledge in the circles of educated men, so that he could teach Blaise.

There were approximately 15 'Academies' in Paris at the beginning of the 1630s; Étienne felt attracted to the *Academia Parisiensis* which was led by the Minim friar and 'secretary of Europe' [Attali 2007, p. 43], Marine Mersenne (1588–1648), cp. figure 2.7.2. In the circle around Mersenne he was quickly welcomed; in particular he was befriended by the mathematician Gilles Personne de Roberval, who had just got the mathematical chair at the Collège de France.

One evening coming back from a meeting with Mersenne Étienne saw his 11-year-old son prove the theorem on the sum of the angles in a triangle while sitting on the floor. Since Blaise did not know the correct terms he called a straight line 'rod' and a circle 'round'. The father recognised that he couldn't keep his son any longer from studying books on mathematics. He asked Mersenne to allow his son as a quiet listener at the meetings of the academy and his wish was fulfilled. There, in the year 1637, Blaise witnessed a discussion on the quarrel between Descartes and Fermat concerning the priority of the discovery of analytical geometry. Blaise got to know the mathematics of Descartes and, although Descartes always pretended his amity with Pascal, he will treat him later with jealousy and disdain [Attali 2007, p. 47]. At the meetings with Mersenne Pascal also attended discussions on Descartes's *Discours de la méthode*.

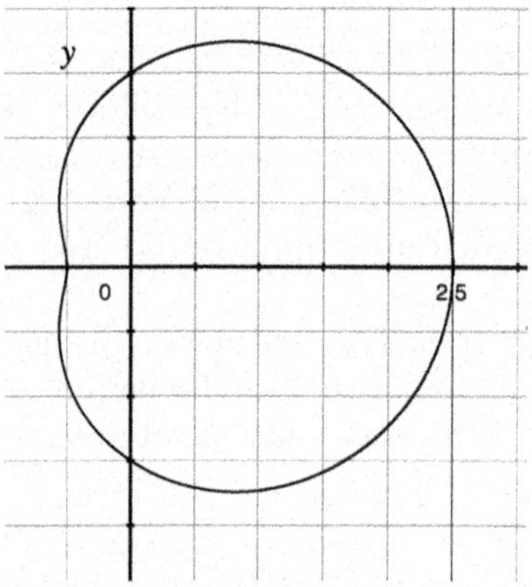

(a) $a < b < 2a$

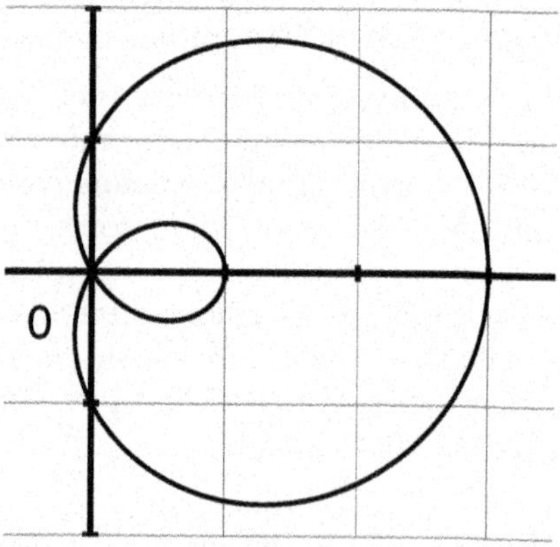

(b) $b < a$

Fig. 2.6.1. Two limaçons corresponding to different ranges of values of the parameters a und b

Fig. 2.6.2. Michel de Montaigne [Painting by Thomas de Leu, about 1578] and Blaise Pascal ([Painting by an unknown artist, about 1690] Palace of Versailles)

From 1635 France was involved in the Thirty Years' War and thereby the fiscal pressure on its citizens became larger. Étienne's income rested nearly completely on interest rates from bonds which he had purchased. As interest rates dropped further and eventually ceased to be paid he protested together with 400 other independent gentlemen in front of the seat of the minister of justice in 1638. Étienne was seen as the ringleader and had to flee Paris for the Auvergne to escape imprisonment. He left his children in Paris with their governess (who, in fact, was also Étienne's mistress); Blaise retreated completely to mathematics, while Jacqueline fell ill with smallpox, so that her father returned secretly to Paris to be at the side of his fatally ill daughter. Like a miracle the daughter survived the smallpox, only marked by a few scars in her face. In April 1639 the highly gifted Jacqueline played Cassandra in a play at court and Cardinal Richelieu was enraptured by the 13-year-old and her talent. Jacqueline took advantage of the Cardinal's favour and asked him to bring her father from exile, and the powerful man agreed. Étienne was received by Richelieu in May 1639 who congratulated him on his daughter [Attali 2007, p. 56ff.].

Meanwhile, 16-year-old Blaise had written a *Traité des coniques* (Treatise on Conic Sections) which was based on works by Desargues but nobody – except Blaise – could understand it yet. It was the birth of projective geometry. Leibniz will hold a copy of this manuscript in his hands some time later and will express his excitement. Mersenne and the members of his academy are also enthusiastic when the 16-year-old recites his results.

As of January 1640 Étienne was appointed 'Acting Member of his Majesty for taxes and tax collection' in Rouen, the capital of Normandy, by Richelieu. He can now live like a lord with his family. On 13th June 1641 Gilberte married her cousin Florin Périer who helped her father as a financial officer and lived

Fig. 2.6.3. A Pascaline of the year 1652 (Musée des arts et métiers) [Photo: David Monniaux 2005]

in her father's house. Blaise, however, was ill. He suffered severe head and stomach pains; then dreadful toothache developed. Gilberte later reported that he never lived a single day without pains after his 18th birthday. No physician could do anything and Attali suspects a psychosomatic illness in connection with the marriage of his older sister [Attali 2007, p. 69f.].

To help his father with the computation of taxes the 18-year-old Blaise constructed and built a calculating machine – the 'Pascaline'. The machine was a two operations machine, i.e. it could only add and subtract, but more was perhaps not needed by a tax officer. The first machine was completed in 1645; in 1649 a patent was granted to Blaise, but it took until 1652 before the final model was available [Schmidt-Biggemann 1999, p. 13f.].

In those years Jansenism became established in the religious landscape of France. Its doctrines, being close to Augustin's doctrine of grace, proved incompatible with the Catholic church and the Jesuits took up the fight against the Jansenists. The Cistercian nunnery Port Royal des Champs south-west of Versailles became the centre of Jansenist learning.

When Étienne Pascal fell on an icy road and broke a leg in the year 1646, he was cared for by two Jansenist physicians. Under their influence Étienne and Jacqueline began to read Jansenist writings and found inner peace in them. Blaise was also bitten by the bug of Jansenist ideas, cp. page 60; one talks of his 'first conversion' [Loeffel 1987, p. 17]. In a typical overeagerness and captured by religious zeal Blaise even sued a theologian of Rouen for theological errors; the theologian had to revoke.

In the year 1646 Blaise repeated the barometer experiments of Torricelli and in 1647 published a book on the vacuum. On 23rd and 24th September there came a meeting with René Descartes but the two men had nothing to say to

Fig. 2.6.4. Cornelius Jansen [Painting by E. d'Ypres]; Port Royal des Champs about 1710 [Painting by an unknown artist]

each other, their views being much too different. Pascal's treatise *Generatio conisectionum* (Generation of Conic Sections) appeared in 1648; in the same year his father moved back to Paris and so his children got in close contact to the nunnery of Port Royal. These are the days of the 'Fronde', the uprising of the nobility, and the family had to retreat briefly to Clermont. After their return to Paris Étienne Pascal died in September 1651. Against the resistance of her brother Jacqueline became a nun in Port Royal, which was led by the abbess Angélique Arnauld in 1653.

Blaise resided in Clermont between October 1652 and May 1653. He was in the company of a beautiful, educated lady [Loeffel 1987, p. 21] and worked on problems of probability theory in a correspondence with Fermat. Pascal's triangle of the binomial coefficients is an offspring of this correspondence. In 1654 Blaise's health further deteriorated. In the night from 23rd November to 24th November 1654 a mystico-religious experience or spiritual enlightenment overtook him; this is the 'second conversion'. He felt an inner fire in which he sensed the presence of god. Then he became unconscious and awoke again at dawn [Attali 2007, p. 192]. In the morning he committed a strange text to paper, the *Mémorial*, and kept this text sewed into his coat ever since [Béguin 1998, p. 111f.] [Attali 2007, p. 194]. Strange as this incident may seem to us: at the age of 31 Blaise Pascal has abjured all worldly things; he has entered a religious life.

From 7th January to 21st January 1655 he lived in the nunnery of Port Royal in the 'Granges', an estate reserved for pious men. He caused a spike belt be made for him and wore it on the naked flesh to remind him on his religious duties [Schmidt-Biggemann 1999, p. 22], but time and again he also appeared as a public person.

Fig. 2.6.5. Jacqueline Pascal, sister of Blaise Pascal, as nun in Port Royal des Champs [unknown painter, probably 1652–1661] and Antoine Arnauld [Copperplate by Louis Simonneau after Philippe de Champagne], both belonging to the Jansenists

Meanwhile the pressure grew on the Jansenists. On 14th January 1656 the greatest defender and leader of the Jansenists, Antoine Arnauld, brother of abbess Angélique Arnauld, was condemned by the Sorbonne in Paris.

Now Pascal no longer held back. He chose the pseudonym 'Louis de Montalte' and published the polemic and satirical *Lettres à un Provincial* from 23rd January 1656. These are 18 letters of a fictive Paris traveller named Montalte in which Pascal made fun of the catholic church and of the Jesuits, and taught them theology. The specific objective of ridicule was the hair-splitting reasoning of the Jesuits, the 'Casuistry'. The *Lettres à un Provincial* were forbidden after the fifth letter but they were not without influence, even if the Jesuits prevailed ultimately.

Even in the days after the 'night of fire' Pascal dealt repeatedly with mathematics; his most famous treatise from these days being certainly the *Lettres de A. Dettonville.* Actually it was a prize competition (*Première lettre circulaire relative à la cycloide*) in which all notable mathematicians (among them Huygens, Wren, Wallis, Carcavi) were asked in June 1658 to solve problems concerning the cycloid which were included in the letter, within a certain amount of time. Sixty Spanish dubloons were exposed as a prize but Pascal decided that not even one of the solutions sent in was worth a reward. Somehow arrogantly Pascal replied to the senders under the pseudonym Amos Dettonville in a series of letters which he then published collectively under the title *Histoire de la roulette* (History of the Cyloid). The pseudonym Amos Dettonville is an anagram of Pascal's pseudonym Louis de Montalte (and hence a pseudonym of a pseudonym). Anagrams are word plays in that the order of the letters of a word or a sentence is changed. Anagrams were very

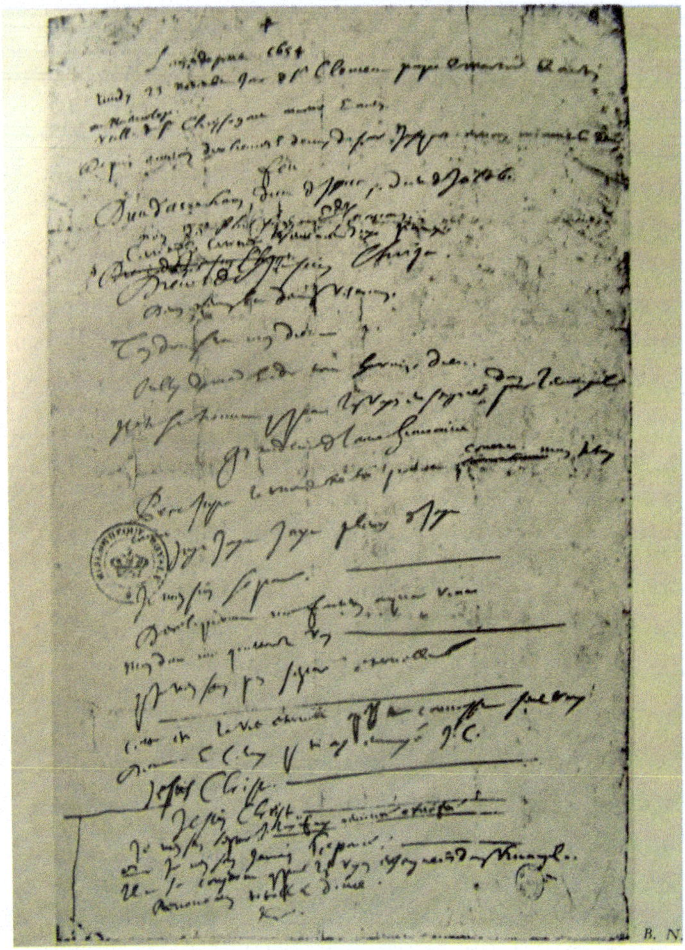

Fig. 2.6.6. Pascals *Mèmorial* (Bibliothèque Nationale de France)

popular in the 17th century; we'll see anagrams also in the priority dispute between Newton and Leibniz. Figure 2.6.7 shows the changes of letters to get from 'Louis de Montalte' to 'Amos Dettonville' and back.

Pascal's mathematics is hardly intelligible for us today, because he abstained completely from mathematical symbols. He explained everything in his perfect French so that Bourbaki writes about him [Bourbaki 1971, p. 222f.]:

> *'Pascal's language is exceedingly clear and precise and although one cannot understand why he abstained from algebraic terminology – not only that of Descartes, but also the one introduced by Viète – one can only admire the tour de force he accomplished, which his mastery of the language alone permitted.'*

(Pascals Sprache ist ganz besonders klar und präzise, und wenn man

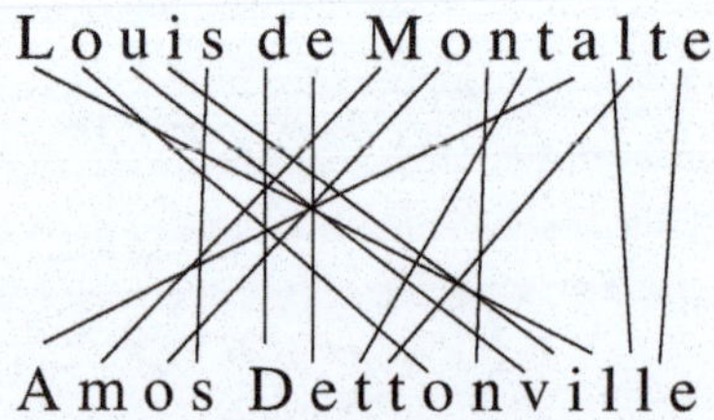

Fig. 2.6.7. Anagram of Amos Dettonville

> auch nicht begreift, warum er sich den Gebrauch algebraischer Bezeichnungsweisen – nicht nur der von Descartes, sondern auch der von Vieta eingeführten – versagt hat, so kann man doch nur die Gewalttour, die er vollbringt und zu der ihn allein seine Beherrschung der Sprache befähigt, bewundern.)

Meanwhile the religious quarrels increased. Jacqueline died in 1661; one reason was probably the pressure on Port Royal and on the Jansenists which she couldn't resist any longer. In a letter to Fermat dating from 10th August 1660, Pascal finally dismissed mathematics [Béguin 1998, p. 104f.]:

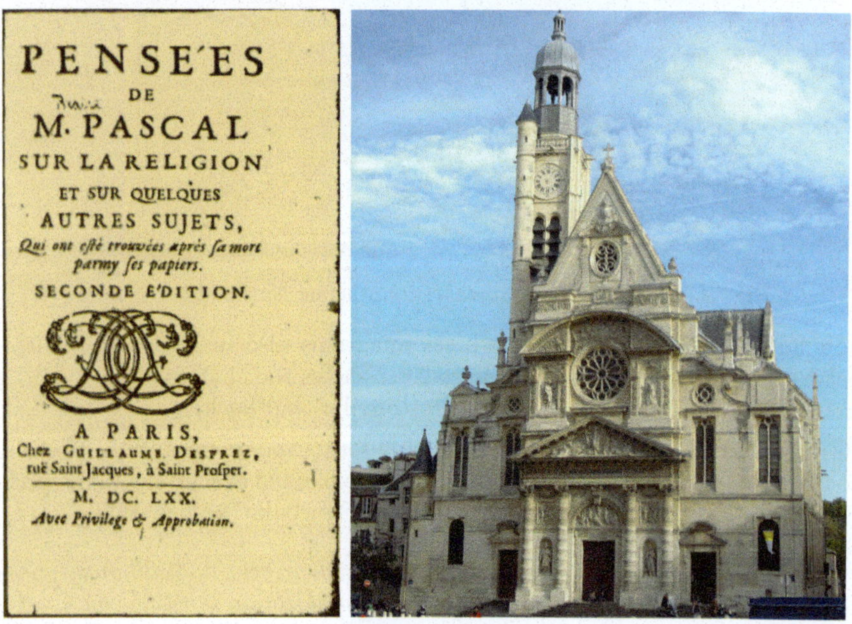

Fig. 2.6.8. Title page of the *Pensées* and west façade of Saint-Étienne-du-Mont in Paris [Photo: Pline 2008]

'Then, to tell you an open secret about mathematics: I take it to be the highest school of the mind; but at the same time I recognise it as so useless that I make little difference between a man being only a mathematician and a skillful craftsman ... In my case there is added the fact that I am absorbed in studies that are so far from the spirit of mathematics, that I can hardly remember that such spirit exists.'

(Denn, um Ihnen ein offenes Wort über Mathematik zu sagen: Ich halte sie zwar für die höchste Schule des Geistes; gleichzeitig aber erkannte ich sie als so nutzlos, daß ich wenig Unterschied mache zwischen einem Manne, der nur ein Mathematiker ist, und einem geschickten Handwerker ... Bei mir kommt aber jetzt noch hinzu, daß ich in Studien vertieft bin, die so weit vom Geist der Mathematik entfernt sind, daß ich mich kaum mehr daran erinnere, daß es einen solchen gibt.)

The 'studies' mentioned above are the studies concerning a long planned apology of Christianity which was published posthumously under the title *Pensées* (Thoughts). Today the *Pensées* belong to the standard literature of the philosophy of religion [Pascal 1997].

Pascal, already terminally ill, got a royal patent in January 1662 for the foundation of the charitable transport company 'Carosses à cinq sols' – the first bus line ever in Paris. At the beginning of August 1662 he wrote his last will and testament, then he received the last rites and passed away on 19th August 1662, aged 39. He was buried behind the choir of the parish church of Saint-Étienne-du-Mont in Paris.

It is important for the mathematics of Pascal that he modified the indivisibles of Torricelli and Cavalieri to allow for 'infinitesimal rectangles' [Loeffel 1987, p. 99f.]. Further mathematical achievements of Pascal can be found in [Loeffel 1987] and [Sonar 2011]. We are going to present here a result from Pascal's paper *Traité des sinus du quart de cercle* (Treatise on the Ordinates of the Quarter Circle), from which Leibniz later got the idea of the general characteristic triangle in a flash of inspiration.

At the quarter circle, Pascal considered the tangent at the point D as in figure 2.6.9 and with it formed the triangle $EE'K$ which Leibniz will later call the characteristic triangle. The triangles ODI and EKE' are similar triangles, i.e. they can be mapped onto each other by shifts, dilations and reflections. It follows from this similarity, that

$$DI : OD = EK : EE',$$

i.e. the line segment DI is to OD as EK to EE'. With the notation in the figure this is the same as

$$y : r = \Delta x : \Delta s,$$

hence

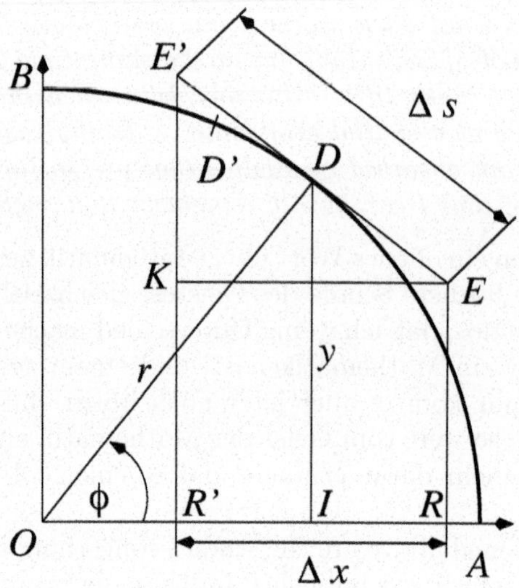

Fig. 2.6.9. The characteristic triangle at the quarter circle

$$y \cdot \Delta s = r \cdot \Delta x. \tag{2.1}$$

Now Pascal sums the 'ordinates', i.e. he wants to compute the sum

$$\sum DI \cdot \mathrm{arc}(DD').$$

Let the values of the abscissa corresponding to the ends of the arc DD' be named x_1 and x_2, respectively. One can find this arc everywhere on the circumference of the circle and not only at the point D, and if one assumes the arc being 'indefinitely small' (in the terminology of our first chapter) and hence accepts infinitely many summands, the arc can certainly be replaced by the line segment EE'. With arbitrary accuracy it then holds

$$\sum DI \cdot \mathrm{arc}(DD') = \sum DI \cdot EE'.$$

Using the notations in the figure 2.6.9 we can also write

$$\sum DI \cdot \mathrm{arc}(DD') = \sum DI \cdot EE' = \sum y \cdot \Delta s.$$

The sum therefore runs over the entire arc length s of the arc DD' – in fact from $s := s_2$, corresponding to the value of the abscissa x_2, up to $s := s_1$ which corresponds to $x_1 < x_2$, thereby traversing the arc in mathematically positive direction (i.e. anti-clockwise). Now we use (2.1) and get

$$\sum_{s_2}^{s_1} y \cdot \Delta s = -\sum_{x_2}^{x_1} \frac{r \cdot \Delta x}{\Delta s} \Delta s = -\sum_{x_2}^{x_1} r \cdot \Delta x = r(x_2 - x_1), \tag{2.2}$$

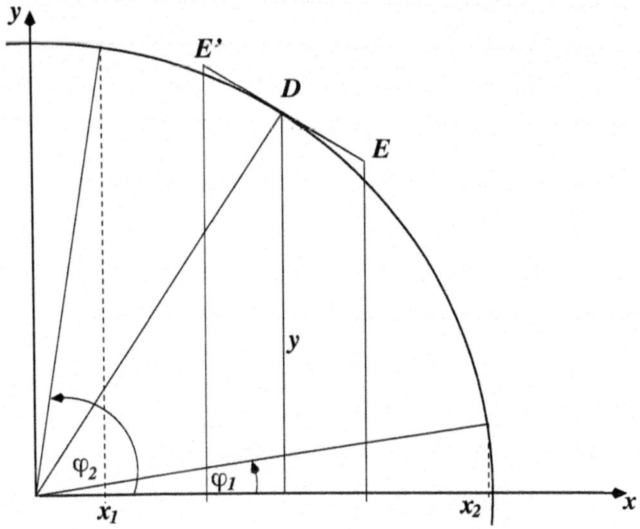

Fig. 2.6.10. The position of x_1, x_2 and the corresponding angle

where the negative sign results in traversing the abscissa backwards, i.e. from right (x_2) to left ($x_1 < x_2$).

What did Pascal achieve by this means? To see that we again help ourselves with the notation of Leibniz. The arc length relates to radius and angle, namely in the form

$$s = r \cdot \varphi.$$

Hence the differentials are

$$ds = r \cdot d\varphi.$$

Furthermore we note that apparently it holds

$$x = r \cdot \cos \varphi, \quad y = r \cdot \sin \varphi.$$

Therefore we can write the points x_1, x_2 as

$$x_1 = r \cdot \cos \varphi_2, \quad x_2 = r \cdot \cos \varphi_1$$

and (2.2) results in

$$\int_{x_1}^{x_2} y \cdot ds = \int_{\varphi_1}^{\varphi_2} \underbrace{r \sin \varphi}_{=y} \cdot \underbrace{r d\varphi}_{=ds} = r(x_2 - x_1) = r^2(\cos \varphi_1 - \cos \varphi_2).$$

Now we can divide by r^2 to get

$$\int_{\varphi_1}^{\varphi_2} \sin \varphi \, d\varphi = \cos \varphi_1 - \cos \varphi_2 = -\cos \varphi \Big|_{\varphi_1}^{\varphi_2};$$

hence Pascal has computed the definite integral of the sine function.

2.7 Christiaan Huygens

To consider Christiaan (Christian) Huygens (1629–1695) only as the teacher of Gottfried Wilhelm Leibniz would be a fatal misunderstanding. Although Leibniz would certainly have arrived at mathematical peak performances without Huygens, on the other hand we would do gross injustice to a Dutch genius if we ignored his influence. Physics can't be imagined without the name of Huygens today – e.g. Huygens's principle in the theory of wave optics, the theory of collision and a theory of light. Furthermore Huygens was the first to construct a pendulum clock with rate of high precision, and he invented the cycloid pendulum with an even higher rate-precision. Moreover, he discovered the rings of Saturn and Saturn's moon Titan, with the help of a telescope the lenses of which he had ground himself.

Despite his universal impact, Huygens was scarcely noted in the history of science for a long time. The first scientific biography which preserved his name appeared in 2005 for the first time [Andriesse 2005], even if older works like [Bell 1947] are still worth reading.

Christiaan Huygens was an offspring of a famous family. Already the grandfather, also a Christiaan, acted as a secretary to William I of Orange who was called 'William the Silent', from 1578. His second son Constantijn (1596–1687) got an excellent school education and showed a mathematical talent early on.

Since the 16th year of his life Constantijn had to wear thick glasses due to an extreme myopia. Besides, he suffered from a posterior vitreous detachment so that all objects in his line of sight seemed to rotate permanently.

Fig. 2.7.1. Christiaan Huygens ([Part of a pastel by Bernard Vaillant, 1686] Huygensmuseum Hofwijck, Vorburg) and Constantijn Huygens ([Part of a painting by Jan Lievens, about 1628] Rijksmuseum Amsterdam)

Fig. 2.7.2. René Descartes ([Painting by Frans Hals 1648] Louvre Museum Paris) and Marin Mersenne

Eventually he became a well-known diplomat but also left behind 75 555 (!) poems – in Dutch, Latin, French, Italian, Greek, German, and Spanish [Andriesse 2005, p. 13]. Christiaan's father, the Constantijn mentioned, can yet only be called a genius. Remarkable and important in the life of Christiaan is Constantijn's friendship with René Descartes who resided in the Netherlands since 1629. The great French philosopher and mathematician visited the Huygens family freely and little Christiaan, only a 4-year old in 1633, took confidence in the man 37 years old. In the year 1637 Descartes's published his opus *Discours de la méthode pour bien conduire sa raison et chercher la verité dans les sciences* (Discourse on the Method of Rightly Conducting One's Reason and of Seeking Truth in the Sciences). In the *Discours* there is an appendix containing the embryonic state of analytical geometry, the treatment of geometric problems with the help of algebraic methods.

Up to his 16th birthday Christiaan was educated through private tuition, even in Latin poetry, in singing and in playing the lute. At the age of 16, in the year 1645, Christiaan and his brother, another Constantijn, enrolled at the University of Leiden. They studied mathematics and law and Christiaan quickly became the best student of Professor Frans van Schooten who in 1649 translated Descartes's analytical geometry into Latin and commented on it. This edition of Descartes's analytical geometry saw a very wide distribution in Europe. In these days mechanics was still a part of mathematics and it is exactly this area in which Christiaan Huygens will shine brightly.

After two years studying in Leiden Christiaan went to the University of Breda which was famous in these days. Breda will be the exile of Charles II of England and the Treaty of Breda will bring an end to the first Anglo-Dutch

War in 1667. The English mathematican John Pell (1611–1685) was teaching at Breda and hence Christiaan had an excellent teacher in mathematics again. After his studies Christiaan travelled to Denmark, Frisia and Rome. He started a correspondence with Father Marin Mersenne in Paris who was a friend of Descartes's and correspondence partner of Christiaan's father Constantijn; he provided the basis for communication between the mathematicians in all of France and beyond.

Mersenne quickly grasped the great talent of the young Dutchman and wrote him approving letters. Through Mersenne Huygens learned of the probability theory of Fermat and Pascal. He also learned from Mersenne the critique of Descartes's theories concerning animals as automata and of the universe filled with vortices of subtle matter. Hence the young admirer of Descartes's philosophy turned between 1648 and 1657 into one of his critics. In his opus *Principia philosophiae*, published 1644 in Amsterdam, Descartes had explained his vortex theory in which the planets moved in vortices of subtle matter [Aiton 1972a]. Cartesian physics was not granted a long life; it was simply too fuzzy. When the *Philosophiae naturalis principia mathematica* Newton's appeared in 1687 and swept away Cartesian physics, it was already discredited; only the vortex theory persisted and appeared again in the work of Kant [Kant 2005].

Even before 1656 Huygens had explained his own theory of motion in his writing *De motu corporum ex percussione*, but it was published only posthumously. His first paper *Cylometria* was published already in 1651. Here 25-year-old Huygens reported on an error in a work of Gregory de St. Vincent who had claimed to have solved the problem of the quadrature of the circle. Already this paper made Huygens known but the more extensive paper *De circuli magnitudine inventa* of 1654 catapulted him in the top range of European mathematicians. Huygens now examined problems which the Ancients had treated – Archimedes, Nicomedes and other Greeks, and which he solved in an algebraic manner.

Already before Mersenne's death in 1648 Huygens planned to visit Paris with his father but the uprising of the nobility foiled such a journey. Huygens came to Paris eventually only in 1655 where besides Roberval he met more intellectual giants of his time. He also got involved in an erotic relationship with the idolised beauty Marie Perriquet from the circle around Pascal [Andriesse 2005, p. 135f.]. In the meantime he built his own telescope which he continually improved. Barely back from Paris he discovered the rings of Saturn with his telescope. On 5th March 1656 he wrote the anagram

> aaaaaaa ccccc d eeeee g h iiiii llll mm nnnnnnnnn ooooo pp q rr s ttttt uuuuu,

meaning

> Annulo cingitur tenui plano nusquam cohaerente ad eclipticam inclinato,

i.e.

> [The planet] *is surrounded by a thin and flat ring which is nowhere connected with it and which is inclined with respect to the ecliptic.*

The popularity of anagrams owed much to a missing copyright law which time and again led to priority disputes. An anagram did not reveal much about a discovery but could be called upon to prove priority.

Before travelling to Paris Huygens had already discovered a moon of Saturn in 1655 and published a small book *Systema Saturnium* in 1656 in The Hague which was send to Paris and to Prince Leopoldo de' Medici to whom it was dedicated. Evidently the book made no impression on other European astronomers. No one seems to have understood the importance of Huygens's discovery.

Between 1655 and 1660 Huygens worked on an accurate pendulum clock. Galileo Galilei had already dealt with such clocks falling back on drawings by Leonardo da Vinci (1452–1519). However, Huygens went a step further and invented the cycloid pendulum. If a mass oscillates between two cycloid-shaped metal blocks then the mass is moving on a tautochrone, i.e. every oscillation takes exactly the same time regardless of the starting point of the oscillation.

Since the thread on which the mass is mounted has to cling against the cycloid-shaped blocks there is increased friction, however. Moreover, Huygens saw that even circular shaped blocks allowed for time-keeping which was not much less accurate [Bell 1947, p. 39]. The development history of the cycloid pendulum is closely linked to Blaise Pascal's price problems which he published under his pseudonym Amos Dettonville. Huygens learned about these price problems in 1658 and he was able to solve the simpler ones. He started a correspondence with Pascal who expressed much praise of the pendulum clock, but the close cooperation he had hoped for failed to materialise. Huygens had explained the theory of his pendulum clock in the book *Horologium*; in his *Horologium oscillatorium sive de motu pendulorum*, published 1673, he demonstrated the complete theory of the cycloid pendulum as well.

Starting in 1655 the problem of rectification, i.e. the calculation of arc length of curves, became topical. It was known from antiquity that the arc length (circumference) of a circle with radius r is $2\pi r$, where the nature of the number π was not known to the ancients except as an approximate value [Sonar 2011, p. 84f.]. Arc length of other bent curves were outside the means of the ancients. Although Thomas Harriot (1560–1621) had rectified the rhumb line (loxodrome) in 1614, i.e. the curved line intersecting all latitudes on earth under the same angle [Sonar 2011, p. 340ff.], he was a mystery-monger and did not publish his result. Now, in 1655, the unfortunate amateur mathematician Thomas Hobbes had attacked the eminent mathematician John Wallis and had accused him of incompetence. To show where Wallis had failed Hobbes

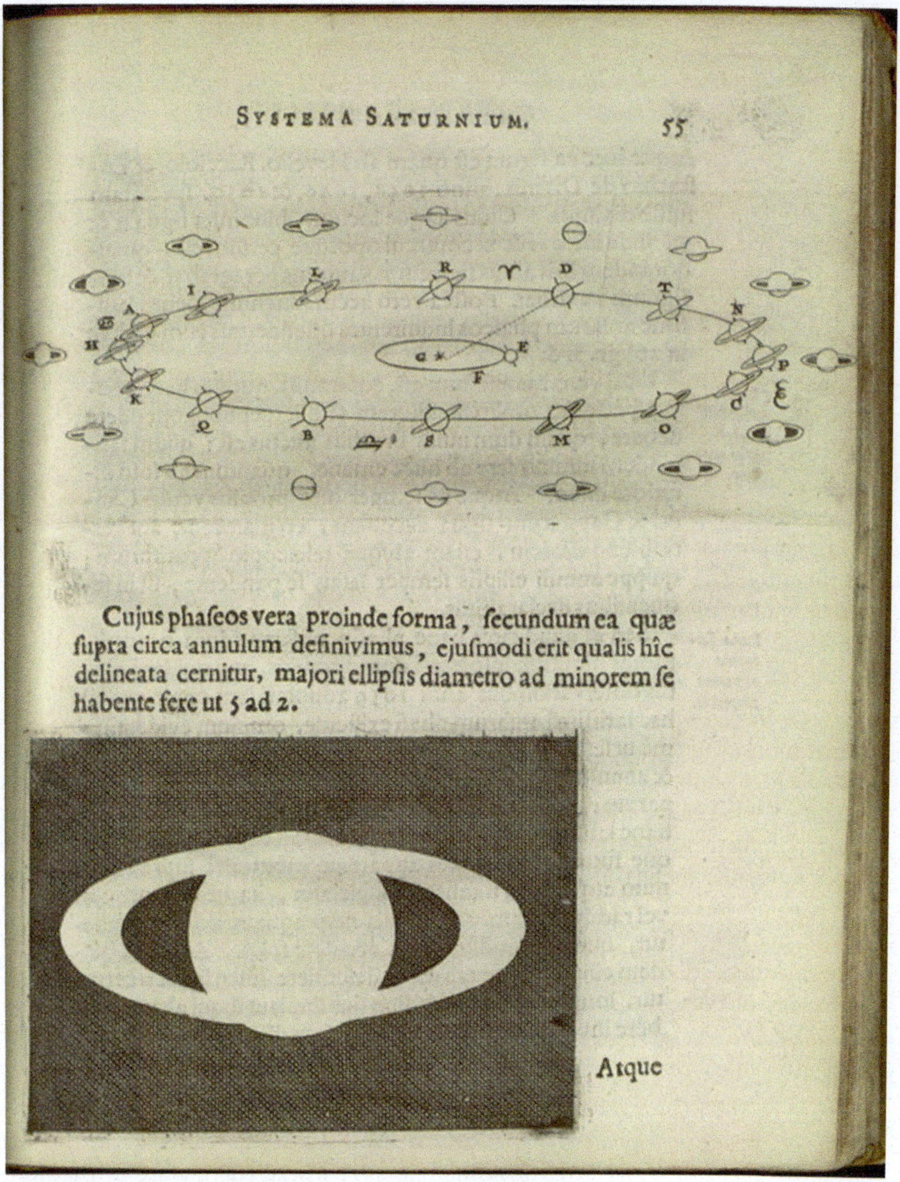
SYSTEMA SATURNIUM. 55

Cujus phaseos vera proinde forma, secundum ea quæ supra circa annulum definivimus, ejusmodi erit qualis hîc delineata cernitur, majori ellipsis diametro ad minorem se habente fere ut 5 ad 2.

Atque

Fig. 2.7.3. A page of Christiaan Huygens's *Systema Saturnium* of 1656

published a rectification of the circle and the parabola, and also a squaring of the circle. His proofs were ridiculously incorrect, however; the conflict between Wallis and Hobbes is extensively examined in [Jesseph 1999]. While Wallis was involved in a conflict, so was Huygens. Already in 1651 Huygens had refuted the 'proof' of the quadrature of the circle of Grégoire de Saint-Vincent in an appendix to his paper *Theoremata de quadratura*. This led to violent attacks of some followers of Grégoire against Huygens [Yoder 1988, p. 117]. But in 1657 there occurred a priority dispute on rectification which revealed a dark side of Huygens.

Huygens was able to prove in 1657 that the arc length of a parabola could be expressed in terms of the area under the hyperbola. Also in the summer of 1657 William Neile succeeded in rectifying the semi-cubical parabola with the help of Wallis's methods [Sonar 2011, p. 343ff.]. A year later Wren supplied the rectification of the cycloid. These results were not immediately published but passed on orally. The world got to know about them through Pascal's *Histoire de la roulette* in 1658. Late in 1659 the problem of rectification was so pressing that Wallis published the results of Neile and Wren in an open letter to Huygens to secure the priority of his English colleagues. Huygens had at that time written to Wallis that there was a general method of rectification by Hendrik van Heuraet (1634–about 1660) in the new edition of the *Geometria* by van Schooten. In particular, he wrote that van Heuraet had succeeded in rectifying the semi-cubical parabola $y^2 - a^2x^2 = 0$, $a > 0$. Now Huygens had derived yet another result in 1657, namely the proof that the surface area of a paraboloid is directly proportional to the area of the circle at the base of the paraboloid. He had communicated these two results to van Schooten and de Sluse, but he had not published them. He obviously felt betrayed of his priority by van Heuraet and started to press hard on van Heuraet to acknowledge Huygens's priority. The apparently good-natured van Heuraet gave in eventually and wrote to Huygens [Yoder 1988, p. 121]:

> *'If you had only known my character, it would not have been necessary to exert so much effort against me, who by no means shall seek to rob*

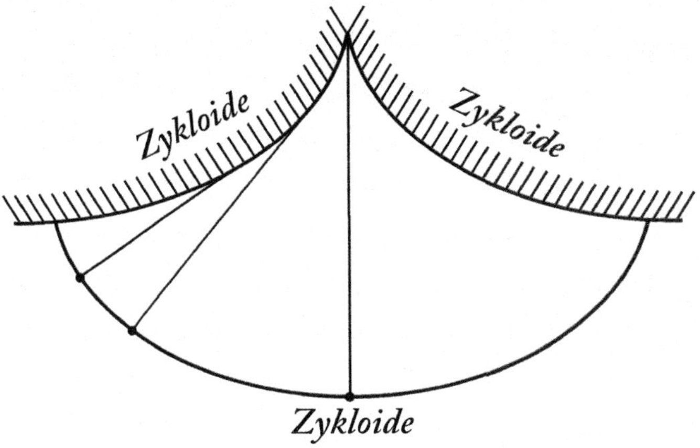

Fig. 2.7.4. Schematical drawing of a cycloid pendulum

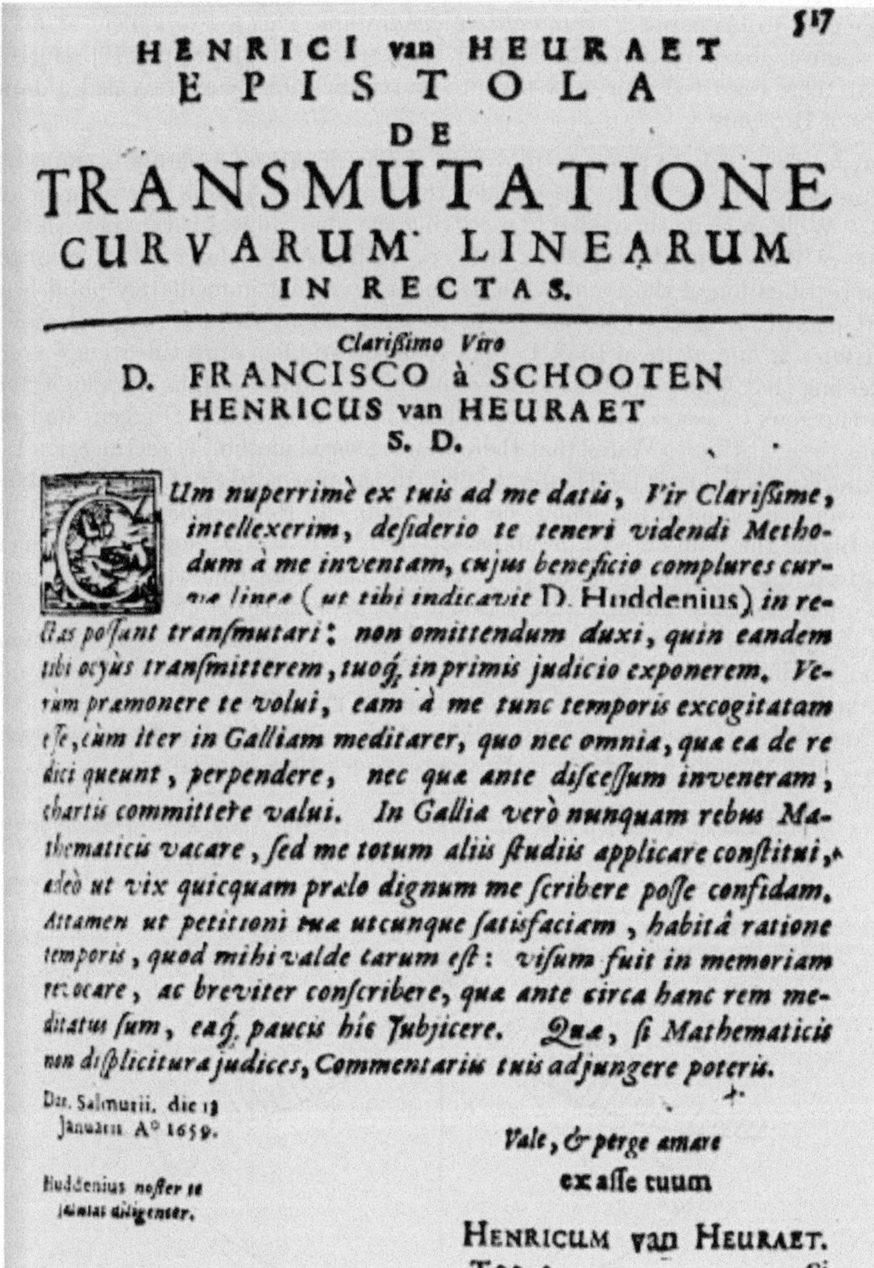

517

HENRICI van HEURAET
EPISTOLA
DE
TRANSMUTATIONE
CURVARUM LINEARUM
IN RECTAS.

Clarissimo Viro
D. FRANCISCO à SCHOOTEN
HENRICUS van HEURAET
S. D.

Cum nuperrimè ex tuis ad me datis, Vir Clarissime, intellexerim, desiderio te teneri videndi Methodum à me inventam, cujus beneficio complures curvæ lineæ (ut tibi indicavit D. Huddenius) *in rectas possunt transmutari: non omittendum duxi, quin eandem tibi ocyùs transmitterem, tuoq; inprimis judicio exponerem. Verùm præmonere te volui, eam à me tunc temporis excogitatam esse, cùm iter in Galliam meditarer, quo nec omnia, quæ ea de re dici queunt, perpendere, nec quæ ante discessum inveneram, chartis committere valui. In Gallia verò nunquam rebus Mathematicis vacare, sed me totum aliis studiis applicare constitui, adeò ut vix quicquam prælo dignum me scribere posse confidam. Attamen ut petitioni tuæ utcunque satisfaciam, habitâ ratione temporis, quod mihi valde carum est: visum fuit in memoriam revocare, ac breviter conscribere, quæ ante circa hanc rem meditatus sum, eaq; paucis hic subjicere. Quæ, si Mathematicis non displicitura judices, Commentariis tuis adjungere poteris.*

Dat. Salmurii, die 13 Januarii A° 1659.

Vale, & perge amare
ex asse tuum
HENRICUM van HEURAET.

Huddenius *noster te salutat diligenter.*

Ttt 3 Si

Fig. 2.7.5. Van Heuraet's treatise on the rectification in van Schootens edition of the *Geometria* of 1659 (gallica.bnf.fr)

Fig. 2.7.6. Christiaan Huygens [unknown artist from Practical Physics, Milikan and Gale, 1920] and Christiaan Huygens ([Painting by Caspar Netscher 1671, detail] Museum Boerhaave, Leiden)

> *you of the pleasure and honour of the aforesaid invention, even if the same might have been found by me long ago.'*

Christiaan Huygens moved in the learned circles of Paris as confidently as he had in those of London. In the year 1663 he was elected a member of the Royal Society. Before that he had already got to know the big players in England. He met John Wallis, Henry Oldenbourg, Robert Boyle and many others and felt the scientific activities in these circles to be considerably more pronounced than those in Paris [Bell 1947, p. 45]. On a diplomatic mission with his father in Paris, Huygens presented one of his pendulum clocks to King Louis XIV. The French scientists now wanted to also found an academy, and this idea was supported by Jean-Baptiste Colbert, minister of finance under Louis XIV. The time came in 1666: the Académie Royale des Sciences was founded and Christiaan Huygens was chosen its first Director. It is nearly inconceivable for us today: a Dutch Protestant became the Director of the French Academy of Sciences! It becomes understandable, however, if we firstly consider that Huygens was the unchallenged number one in European science, and that he also had excellent personal contacts with Colbert.

The year 1672 marks one of the starting points in our history of the priority dispute. It was the year in which Gottfried Wilhelm Leibniz came to Paris. At this time Leibniz was an amateur mathematician at best, but under the influence of Huygens he will develop into a mathematician of first rank with incredible speed. Already in 1674 Leibniz presented a first work on his new infinitesimal calculus to Huygens [Knobloch 1993, p. 9f.]. Huygens always was and stayed a 'traditional' mathematician who used the geometrical methods of the ancients with great skill. He couldn't warm to the mathematics of his pupil and that of Newton. The year 1672 also brings a letter of Oldenbourg (Oldenburg), then secretary of the Royal Society, to Huygens, which contained Newton's work concerning the theory of light from the *Philosophical Transactions.* Huygens criticised this theory unjustifiably and Newton was deeply disappointed by him. But more of this later!

Inspired by Newton's work concerning a theory of light, Huygens started working on his own theory which was published in 1690 in Leiden under the title *Traité de la lumiére.* In this publication which profited from Huygens's work on the theory of collision, Huygens propagated a wave theory of light which he contrasted to the corpuscular theory of Newton.

Huygens had always suffered from phases of illness. In 1670 he lost his memory nearly completely for some time due to an illness, but he recovered again. From 1680 onwards his health deteriorated further and periods of illness alternated with periods in which he felt better. His father died of old age in 1687 and his brother Constantijn followed William of Orange to London in 1688, who became king of England as William III. In the year 1689 Huygens stayed in London for several months. He met Royal Astronomer Flamsteed in Greenwich, attended a meeting of the Royal Society in Gresham College, and together with the Swiss mathematician Fatio de Duillier, who went on to play an inglorious role in the priority dispute, met Newton in person [Bell 1947, p. 84]. Although Huygens had criticised Newton's theory of light and was not in accord with Newton's theory of gravitation as developed in the *Principia Mathematica,* Newton spoke and wrote highly of Huygens. In the same way Huygens admired Newton. Huygens is the only one whom Newton recognized as 'summus' (the eminently great).

After his return to The Hague, Huygens still participated in the solution of mathematical problems stemming from the circle around Leibniz, but in March 1695 he felt so sick that he made his last will and testament. He died on the 8th of July in The Hague and was buried in the Grote Kerk, the town church of The Hague.

Fig. 2.7.7. The Grote Kerk in Den Haag in which the tomb of Christiaan Huygens is located [Photo: Michielverbeek 2010]

The Analysis of the 'Giants'

1631	John Wallis learns *Common Arithmeticke* as a pupil
1639	Blaise Pascal writes in Paris a work on conic sections. He is a 16-year-old and his work will later enthuse Leibniz
1644	Three years before Cavalieri Evangelista Torricelli publishes the foundations of the calculus of indivisibles, the *Opera geometrica*
1645	A group of scientifically interested men meet in Gresham College, London; among them are Wallis and John Wilkins
1646	John Barrow enrolls at the University of Cambridge as a 'subsizar'
1647	Bonaventura Cavalieri publishes the 'bible' of the calculus of indivisibles, *Exercitationes geometricae sex*
1648	Wilkins goes to Oxford as Warden of Wadham College and founds a group of men following the London model in Gresham College
1649	Wallis is appointed the Savilian Chair of Geometry in Oxford. The chair is the reward for his decoding work for the republican side in the civil war
1649	Frans van Schooten publishes a Latin edition of René Descartes's *La géometrie* with several comments and remarks. This edition serves to publicise Descartes's work widely
1651	The Oxford group around Wilkins becomes the *Philosophical Society of Oxford*
1655	Wallis publishes his book *Arithmetica infinitorum*, in which he turns the geometrical methods into an arithmetic form
1656	Christiaan Huygens discoveres the rings of Saturn
1656	Barrow publishes an edition of Euclid's *Elements*. On a long journey which leads him up to Constantinople he gets acquainted with Cavalieri's calculus of indivisibles probably in Florence
1658	Pascal publishes the *Lettres de A. Dettonville*, a collection of prize problems concerning the cycloid
1659	Pascal publishes the *Traité des sinus des quarts de cercle*, in which Leibniz later finds the idea of his 'characteristic triangle'
1660	Foundation meeting for what later becomes the *Royal Society* in Gresham College, London
1662	The London and Oxford groups become the *Royal Society* under the patronage of the King of England. Among the founding members are Robert Boyle, Christopher Wren and William Brouncker. Barrow becomes the first professor for mathematics on the Lucasian Chair in Cambridge
1663	Huygens becomes a member of the *Royal Society*
1666	Establishment of the *Académie Royale des Sciences* in Paris, the first director of which is Huygens
1667	John Collins becomes member of the *Royal Society* and develops into the 'English Mersenne'
1669	Barrow publishes the *Lectiones Opticae et Geometricae*, for which his pupil Newton did some work. Barrow renounces the Lucasian Chair for the benefit of his pupil Isaac Newton
1673	Huygens publishes *Horologium oscillatorium*
1683	Barrow's *Lectiones Mathematicae* are published
1690	Huygens publishes the *Traité de la lumiére*

3 The Warriors Grow Up

T. Sonar, *The History of the Priority Diſpute between Newton and Leibniz*, https://doi.org/10.1007/978-3-319-72563-5_3

1618–1648	Thirty Years' War
1620	The pilgrim fathers, Puritans from England, land in North America at Cape Cod
1633	Galilei has to renounce the Copernican system of the world before the Inquisition
1643–1715	Louis XIV King of France
1643	Isaac Newton is born on 4th January in Woolsthorpe close to Grantham (Gregorian calendar)
1644	Blaise Pascal builds the first mechanical computing machine which is still preserved today
1646	Gottfried Wilhelm Leibniz is born in Leipzig on 1st July
1649	King Charles I of England is beheaded. The Commonwealth is introduced by Cromwell
1660–1685	Charles II King of England
1658–1705	Leopold I Emperor of the Holy Roman Empire
1658	The first Confederation of the Rhine (Confederated States of the Rhine) forms against the emperor
1662	Foundation of the Royal Society in London
1665–1667	Second Anglo-Dutch War
1665	Outbreak of the Plague in London
1666	The Great Fire wipes out large parts of London. Reconstruction under Christopher Wren and Robert Hooke Foundation of the Académie Royale des Sciences in Paris
1672	Leibniz invents the stepped drum as an element of mechanical computing machines
1672–1678	Louis XIV wages a war of conquest against the Netherlands
1683–1699	Turkish war
1685–1688	Catholic James II becomes King of England
1688	Protestants invite William of Orange who moved in London at the end of December. James flees to France
1688–1713	Frederick III prince-elector of Brandenburg. From 1701 as Frederick I King of Prussia
1689–1725	Peter the Great Tsar of Russia
1702–1714	Queen Anne Queen of England
1702–1713	English participation in the War of the Spanish Succession
1703	Foundation of Saint Petersburg
1705–1711	Joseph I Holy Roman Emperor
1709	Ehrenfried Walter von Tschirnhaus and Johann Friedrich Böttger invent the European white hard-paste porcelain
1711–1740	Charles VI Holy Roman Emperor
1714–1727	Georg Ludwig, Prince-Elector of Hanover, becomes the English King George I
1725	Opening of the Academy of Sciences in Saint Petersburg

3.1 The Physicist: Isaac Newton

'No great mathematician is so difficult to study as Newton.' wrote Clifford Truesdell at the beginning of the 1960s [Truesdell 1984, S. 269]. This is no longer true today since with [Whiteside 1967–81] we now have Newton's mathematical writings in a selection of eight thick and large-sized volumes in English language. Newton's correspondence is available in [Turnbull 1959–77] in eight just as voluminous volumes in the English language, and with [Westfall 2006] there exists a definitive biography upholding all scientific standards. In the internet 'The Newton Project' (http://www.newtonproject.sussex.ac.uk) completes the information on Newton and specialist literature is available in large numbers also.

3.1.1 Childhood and Youth

The Gregorian calendar is already valid on the Continent when the English still stay away from this 'papist' reform of the calendar and stay with the Julian calendar. On this calendar Isaac Newton is a Christmas child, born on Christmas day 1642. Following our modern calendar it is 4th January 1643, however. The much too small baby Isaac is born after only seven months of pregnancy in Woolsthorpe-by-Colsterworth in Lincolnshire. The pipsqueak who fits into a quart pot has barely a chance of survival and hence the two midwives walk to North-Witham to fetch strengthening medicine in a dull mood. In any case that is the story of his birth as told by the old Newton [Westfall 2006, p. 49].

Isaac is a half-orphan already at the time of his birth; his father with the name Isaac died during the pregnancy of the mother Hannah, née Ayscough. The place of birth is Woolsthorpe Manor, a modest estate on which the father worked as a yeoman, a free farmer with some property, and fed his family.

However, the father was in a better position than a common yeoman since he was a landowner. The income from Woolsthorpe Manor was approximately £ 30; not an enormous sum, but one from which a small family could live well. The marriage of Isaac Newton with Hannah Ayscough brought another upswing to the Newton family since Hannah brought real property worth approximately £ 50 a year into the marriage. The life in Woolsthorpe Manor hence was quite pleasant; poverty did not threaten, not even after the death of the father.

Little Isaac had to fight for his life approximately for a week, because he was christened only on 11th January – fairly long after his birth. We don't know anything about his first three years, but then something drastic happened: his mother married again in 1646. The new husband, Barnabas Smith, was a clergyman and priest of North Witham, the next village to the south of Woolsthorpe Manor. He was born in 1582 and hence already 63 years old when he

Fig. 3.1.1. Woolsthorpe Manor, the place where Isaac Newton was born

married Hannah Ayscough. Shortly beforehand he had to bury his first wife, and lost no time to complete his household with a new wife. He procreated further three children with Hannah Ayscough Newton Smith before he passed away, aged 71 [Westfall 2006, p. 51].

After Barnabas Smith's death Newton inherited the library of the clergyman, consisting of between 200 and 300 theological books. Anyway, they found their way onto the bookshelves of Newton and can be seen as the nucleus of Newton's later theological library. Furthermore, Smith was rich; his yearly income was approximately £ 500. At the time of the marriage a piece of land was nominated especially for Isaac. Later further land from the heritage of Barnabas Smith came along through his mother. Hence, the Newton family was very well off after Hannah's marriage to the considerably older man. However, there is also a dark side of this marriage which most likely acted devastatingly on Newton's personality. Smith didn't want to be burdened by a three-year old toddler from the first marriage of his wife, and hence Isaac was left behind to live with grandmother Ayscough in Woolsthorpe. In return Smith got her house thoroughly renovated – he could afford it.

For a three-year old child the separation from his mother must have been a huge blow – a world collapsed around the little boy. Does this radical experience perhaps explain the unpleasant personality of Newton? As one of the successors Newton's on the Lucasian Chair in Cambridge, Stephen Hawking, remarked, was his predecessor not at all sociable [Hawking 1988, p. 191]:

'Isaac Newton was not a pleasant man. His relations with other academics were notorious, with most of his later life spent embroiled in heated disputes.'

Undeniably the adult Newton was not a pleasure to be dealt with. Up to his death he practically had no friends but a lot of quarrels, not only with Leibniz but also with Robert Hooke, John Flamsteed, and the forgers in London, who all had to suffer from Newton – we will return to that. Does that all stem from Newton's childhood trauma?

A 'forensic psychoanalysis' would be informative, and such can be found indeed in Frank Edward Manuel's book [Manuel 1968]! Manuel diagnosed Newton as having a serious psychosis, and attributed this psychosis to the separation from his mother and the corresponding rejection he experienced. Long after Barnabas Smith's death the 20 years old Newton wrote down his 'sins' and in 'sin no. 13' we read about his murder fantasies [Manuel 1968, S. 26]:

'Threatening my father and mother Smith, to burn them and the house over them.'

Stepfather Smith passed away in August 1653 and Newton's mother returned with three little children to Woolsthorpe. Sometime within the following two years, probably after his 12th birthday, Isaac was sent to the grammar school in Grantham. There the lad learned extensive Latin, some Greek, but no – or exceptionally obsolete, mathematics.

Studying the Bible was also part of his education and we may presume that together with the theological books of his stepfather these Bible studies laid the foundations for Newton's excessive, strange theological research activities.

In Grantham Newton was accommodated with the pharmacy family of Mr. Clark. Three stepchildren of the pharmacist also lived there and Newton apparently had some problems with them. He had lived too long in isolation with his grandmother to be an 'ordinary boy'. When his intellectual superiority was revealed at school the other boys started hating him, and so he spent his time rather in the society of girls. He started tinkering with furnitures for the dolls of the pharmacist's stepdaughter, who was some years younger than himself, and for her friends. But most of the time he stayed alone. When one of the stepsons of the pharmacist punched him in the stomach on the way to school, Newton struck back until he could press the boy with his nose to the wall of the church while holding him at the ears. But Newton was not satisfied; the opponent had to be properly wiped out intellectually also, and so Newton developed from a rather bad pupil to the undisputed number one.

He is a great inventor already now. He bought tools from the money his mother gave him and showed great skills in the production of mechanical clocks. He made models of windmills which were driven by a mouse in a horse-capstan.

He built storm lamps from foldable paper, and not only used them to illuminate his way to school but also let one fly mounted on a kite during the night

Fig. 3.1.2. Newton's 'signature' on a windowsill at King's School in Grantham [Photo: Fritzbruno 2011]

to scare the residents. As Westfall [Westfall 2006, p. 60] points out rightly, it was very lucky that Grantham did not burn down completely. When a large storm blew over England the day before Cromwell died, Isaac conducted experiments with the wind. He jumped up with and against the wind and tried to calculate the power of the wind by the distances he was blown along. He observed the course of the sun and constructed sundials on the walls of Clark's house – inside and outside, and he is so successful that neighbours as well the Clarks read the time off 'Isaac's dials'. Since he lived in the house of a pharmacist he is also interested in chemicals and chemistry. Later he will spend more time with chemical experiments and investigations than with mathematics and physics. He also was a marvellous draughtsman. The walls in his room were covered with charcoal drawings of birds, men, ships, and plants. And on every board in his room, as on every table he sat at at school, he leaves his name 'Isaac Newton'.

At the end of 1659 the nearly 17-years-old lad left school. He ought to take over the farm in Woolsthorpe but he turned out to be totally unsuitable! Sheep ran away while he was reading a book instead of watching the flock, a pig ravaged a cornfield and a horse led by him unfastened itself and ran home on its own – Isaac later following absentminded with the empty reins in his hands. From the point of view of his mother Hannah Isaac must have turned out to be an absolute disaster and great disappointment. But help was near! The mother's brother, William Ayscough, tried to persuade his sister to allow Isaac to go back to school in order to visit a university later on. Schoolmaster

Fig. 3.1.3. A horse-capstan in which the horizontal rotation of the horses is transformed into the haul of a rope (Wieliczka Salt Mine, Poland [Photo: Rj1979, 2007])

Mr Stokes also tried to persuade Isaac's mother. He offered to waive school fees for Isaac and to accommodate him without costs in his own home. Mother Hannah eventually gave her consent and hence Isaac went back to school in Grantham in the year 1660 to be prepared for study at a university. The nine months which Isaac spent at home must have been as dreadful for him as for the other members of the family. In his list of sins we find the following for the year 1662 [Westfall 2006, p. 64f.]: 'Refusing to go to the close at my mothers command', 'Peevishness with my mother', 'Punching my sister', 'Falling out with the servants', 'Calling Dorothy Rose a jade', and so on.

3.1.2 The Lonely Student

In June 1661 Isaac Newton enrolled at the University of Cambridge. He became a student of Trinity College, bought a lock for his writing desk, an inkwell, a notebook, a pound of candles, and a chamber pot [Westfall 2006, p. 66], and is hence well prepared. The choice of Trinity College was probably influenced by his uncle William Ayscough who was a former student there.

Newton did not come to Cambridge as a son of a rich father. Such spoiled young men stayed for some years in Cambridge or Oxford, pursued their hobbies (dog racing, horses, girls, booze), did not attend lectures, and often left the university without any degree or with one bought by their fathers, if they became convinced that the seriousness of life should start for their offsprings. Poor students had to serve the richer ones to finance their studies at the university. Such poor students were called 'sizars' and 'subsizars' and Newton became one of them. Subsizars had to follow the same rules as sizars but had to pay tuition fees and care for their nourishment themselves. At that time Newton's mother commanded a significant annual income of approximately £ 700; nevertheless Newton had to serve as a subsizar and obtained only £ 10 a year from Hannah. The only acceptable explanation lies in the disturbed relationship between Isaac and his mother. His mother must have been so very disappointed by his uselessness as an heir to the farm that she was not willing to give her 'failure' more financial help than absolutely necessary.

If Newton already had problems with his classmates at the school in Grantham, then this continued in Cambridge. Only with his roommate John Wickins who came to Cambridge in 1663 could he establish such a thing as a friendly relationship. Newton must have made a solitary and afflicted appearance as a student in Cambridge. Additionally a kind of religious crisis developed during the summer of 1662, without which his list of sins can't be explained properly. Newton must have lived under permanent puritan pressure; he always felt guilty, regardless of whether he experienced 'wet dreams' or drank too much wine [Westfall 2006, p. 77f.].

Fig. 3.1.4. Newton's rooms in Trinity College were located on the first floor just right of the tower, compare fig. 2.4.3 [David Loggan, Cantabrigia illustrata, 1690]

Newton learned Aristotelian logic, Aristotelian ethics and Aristotelian physics. Yet before the traditional curriculum was completed he discovered Galileo and started reading. He read works on history, discovered astrology, astronomy, Descartes, Robert Boyle, and other contemporary authors. He compiled a notebook under the title *Quaestiones quaedam Philosophicae* (Some questions of philosophy) in which he included questions concerning 'natural philosophy', i.e. physics, and in which he noted extracts from his reading. Newton seemed to have begun the *Quaestiones* in the year 1664 and with it he became an experimenter. His interests were not restricted to mechanics but also covered optics and light. In his experiments he is ruthless with himself. He looked into the sun with unprotected eyes and pushed a hatpin behind his eyeball to press the retina in order to examine the development of colours in the human eye.

Undoubtedly the *Quaestiones* formed the nucleus of both of Newton's great works *Philosophiae naturalis principia mathematica* and *Opticks*. We do not find chemistry in the *Quaestiones* although this will occupy a large part of Newton's work in later years. But Newton is not only the experimentalist in physics. He viewed himself as a philosopher who wants to understand the nature of things completely [Westfall 2006, p. 96] and he discovers mathematics! In the year 1699, more than thirty years after the events described, Newton remembers as follows [Westfall 2006, p. 98]:

> *'July 4th 1699. ... I find that in ye year 1664 ... I bought Schooten's Miscellanies & Cartes's Geometry (having read this Geometry & Oughtred's Clavis more than half a year before) & borrowed Wallis's works & by consequence made these Annotations out of Schooten & Wallis in winter between the years 1664 & 1665. At wch time I found the method of Infinite series. And in summer 1665 being forced from Cambridge by the Plague I computed ye area of ye Hyperbola at Boothby in Lincolnshire to two & fifty figures by the same method.'*

The 'Miscellanies' by van Schooten relate to the fifth book of *De organica conicarum sectionum in plano descriptione* (On an organic description of conic sections in the plane), a very readable book of 1646 on geometry by Frans van Schooten to which the author added *Exercitationum mathematicarum libri quinque* (Five books of mathematical exercises) in 1657. Newton made excerpts and commented on them [Whiteside 1967–81, Vol. I, p. 21f.][1].

The citation above is as far as possible identical to the content of a letter from Newton to John Wallis stemming from the second half of the year 1692 [Turnbull 1959–77, Vol. VII, S. 394]:

> *'The plague was in Cambridge in both ye years 1665 & 1666 but it was in 1666 yt I was absent from Cambridge & therefore I have set down*

[1]As nearly the complete mathematical work of Newton the early 'Annotations' are carefully translated, edited and commented in the complete edition [Whiteside 1967–81]; here in the first volume on pages 25 to 142.

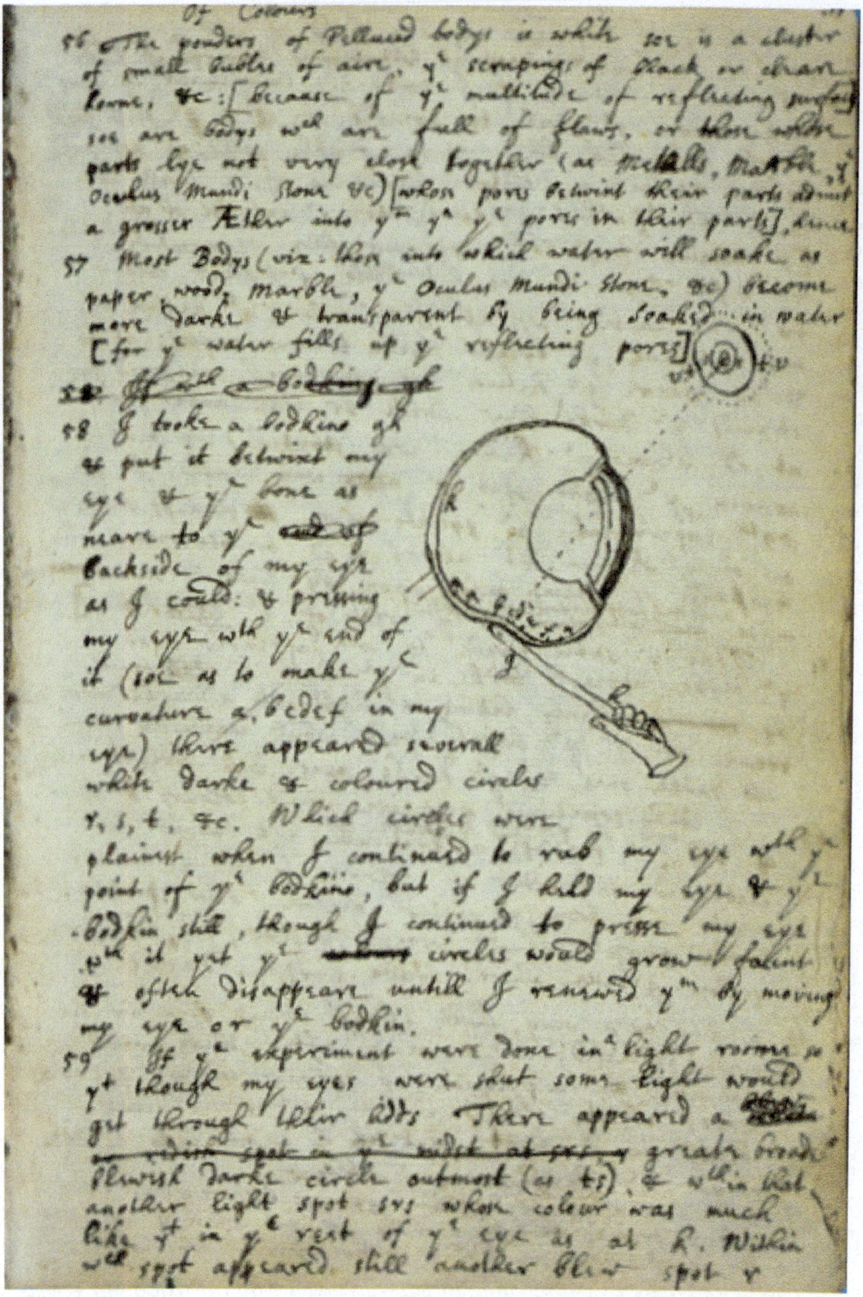

Of Colours

56 The powders of Pellucid bodys is white soe is a cluster of small bubbles of aire, ye scrapings of black or charcoal horne, &c: [because of ye multitude of reflecting surfaces] soe are Bodys wch are full of flaws, or those whose parts lye not very close together (as Metalls, Marble, ye Oculus Mundi Stone &c) [whose pores betwixt their parts admit a grosser Æther into ym yn ye pores in their parts], hence

57 Most Bodys (viz: those into which water will soake as paper, wood, marble, ye Oculus Mundi Stone, &c) become more darke & transparent by being soaked in water [for ye water fills up ye reflecting pores]

~~58 If with a bodkine gh~~

58 I tooke a bodkine gh & put it betwixt my eye & ye bone as neare to ye backside of my eye as I could: & pressing my eye wth ye end of it (soe as to make ye curvature a, bcdef in my eye) there appeared severall white darke & coloured circles r, s, t, &c. Which circles were plainest when I continued to rub my eye wth ye point of ye bodkine, but if I held my eye & ye bodkin still, though I continued to presse my eye wth it yet ye ~~[illegible]~~ circles would grow faint & often disappeare untill I renewed ym by moving my eye or ye bodkin.

59 If ye experiment were done in a light roome so yt though my eyes were shut some light would get through their lidds There appeared a ~~[illegible]~~ greate broade bluish darke circle outmost (as ts), & wthin that another light spot srs whose colour was much like yt in ye rest of ye eye as at k. Within wch spot appeared still another blue spot r

Fig. 3.1.5. Newton's experiment with the hatpin behind the eyeball (reproduced by kind permission of the Syndics of the Cambridge University Library Ms. Add. 3995 p. 15)

> *an amendmt of ye year. I wrote to you lately that I found ye method of converging series in the winter between ye years 1665 & 1666. For that was ye earliest mention of it I could find then amongst my papers. But meeting since wth the notes wch in ye year 1664 upon my first reading of Vieta's works Schooten's Miscelanies & your Arithmetica Infinitorum, I took out of those books & finding among these notes my deduction of the series for the circle out of yours in your Arithmetica Infinitorum: I collect yt it was in ye year 1664 that I deduced these series out of yours. There is also among these notes Mercators series for squaring the Hyperbola found by ye same method wth some others.'*

The Mercator mentioned is Nicolaus Mercator, born in Holstein about 1620, who taught mathematics in London from 1658 to 1682 and became a member of the Royal Society in 1666 [Hofmann 1949a]. In the year 1668[2] he published *Logarithmotechnia* [Mercator 1975] in which he succeeded in squaring the hyperbola. He could prove that the area under a strip of the hyperbola corresponds to the logarithm and gave the infinite series:

$$\int_0^x \frac{d\xi}{1+\xi} = x - \frac{1}{2}x^2 + \frac{1}{3}x^3 - \frac{1}{4}x^4 \pm \ldots = \ln(1+x).$$

Newton had found this results earlier but did not publish it, compare [Hofmann 1939].

We have to note here: Newton was a mathematical autodidact! The Lucasian Chair for mathematics was established in 1663 and Isaac Barrow, its first holder, began his lectures on 24th March 1664. Barrow was *not* Newton's tutor, but Newton attended Barrow's lectures and Barrow certainly had fuelled Newton's mathematical interests as we already have pointed out.

Newton's own studies led to diminishing activities in his study of the actual curriculum, and hence it got more and more difficult for him to gain one of the 62 stipends, leading to a fellowship at Trinity College. Such a fellowship would have resulted in financial security for Newton; and hence he threw himself back into the curriculum and took his examinations literally at the very last moment. Following John Conduitt who married Newton's niece Catherine Barton, and whom we have to thank for an early biography of Newton, Newton's tutor Pulleyn recognised the genius of his pupil and referred him to Barrow [Westfall 2006, p. 102]:

> *'When he stood to be scholar of the house, his tutor sent him to Dr Barrow then Mathematics Professor to be examined; the D^r examined him in Euclid wch Sr I. had neglected & knew little or nothing of, & never asked him about Descartes's Geometry wch he was master of*

[2]The first edition was dated 15. VIII. 67, but is lost. That this edition existed is deduced by Hofmann [Hofmann 1949a, p. 56] by the date 1667 on the edition of 1668, compare figure 3.1.6.

LOGARITHMO-TECHNIA:
SIVE
Methodus conſtruendi
LOGARITHMOS
Nova, accurata, & facilis;
SCRIPTO
Antehàc Communicata, Anno Sc. 1667.
Nonis *Auguſti*: Cui nunc accedit.
Vera Quadratura Hyperbolæ,
&
Inventio *Summæ* Logarithmorum.

AUCTORE *NICOLAO MERCATORE*
Holſato, è Societate Regia.

HUIC ETIAM JUNGITUR
MICHAELIS ANGELI RICCII Exercitatio
Geometrica de Maximis & Minimis; hîc ob Argumenti
præſtantiam & Exemplarium raritatem recuſa.

LONDINI,
Typis *Guilielmi Godbid*, & Impenſis *Moſis Pitt* Bibliopolæ, in
vico vulgò vocato *Little Britain*. Anno M. DC. LXVIII.

Fig. 3.1.6. Title page of Mercator's *Logarithmotechnia* 1668
(Thomash Collection Images)

Sr I. was too modest to mention it himself & Dr Barrow could not imagine that any one could have read that book without being first master of Euclid, so that Dr Barrow conceived then but an indifferent opinion of him; however he was made a scholar of the house.'

But how could Newton still yet win a stipend at the beginning of 1664? Had he wrongly assessed the impression he made on Barrow when he in his old age told Conduitt the story of his examination? Did the responsible people in College sense Newton's genius despite his meagre performance? In the case of such a reclusive young man this is not very likely. It is more likely that Newton had influential advocates. One possible advocate certainly was Barrow who took care that Newton was chosen as his successor in the Lucasian Chair in 1669. Even more likely as an advocate was Humphrey (Humfrey) Babington (ca. 1615–1691), however, a clergyman of the Church of England who had studied at Trinity College. Newton had employed 'Mr Babington's Woman' for some time to care for him, and he stayed in Boothby, close to Woolsthorpe, during the time of the plague, where Babington worked as a clergyman. Additionally, Babington was the brother of the pharmacist's wife Mrs Clark with whom Newton lived in his first period at the school in Grantham. At the time of Newton's election, Babington was on his way to becoming a 'senior fellow' of Trinity College, one of only eight fellows at the top of the college [Westfall 2006, p. 102f.]. Until Babington's death Newton was befriended by him, so that the clergyman was very likely an advocate.

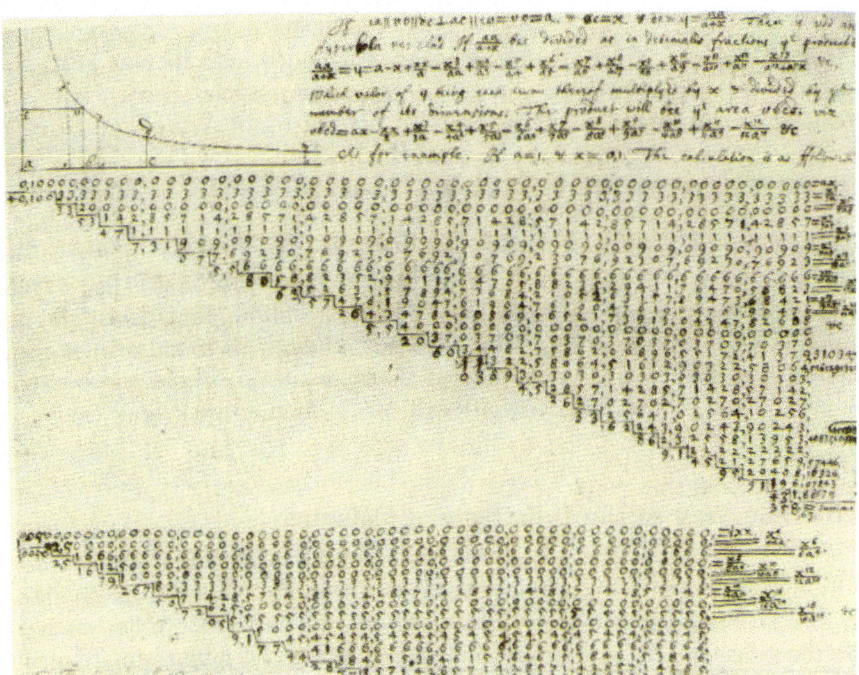

Fig. 3.1.7. Newton's squaring of the hyperbola, part of a manuscript page of 1665 (reproduced by kind permission of the Syndics of the Cambridge University Library)

Fig. 3.1.8. Isaac Newton ([Painting: Godfrey Kneller 1689] Farleigh House, Portsmouth)

With the stipend, Newton's life as a subsizar ended, and he now got money from the college for at least the next four years. Associated with it was the prospect to become a lifelong fellow after his Master of Arts exam. Now that the pressure on him was released he threw himself in his own work and thereby often forgot his meals. His cat grew fat, as Newton later told his niece, because it ate the meals left over by Newton. He also forgot to sleep as his roommate Wickins reported. Even as an old man, will Newton not be found in his bed until a problem he is working on is solved. If the old Newton is called to lunch or dinner by his servant, this call comes at least half an hour before the meal is ready. When Newton then finally left his study, if he found a book on his way, then it could well have happened that this meal got cold and had to be removed. Today we would certainly call such a man a 'workaholic'.

3.1.3 The Way to the Infinitesimal Calculus

Why did we call Newton a physicist in the heading? Well, firstly he showed an inclination towards experiments early on; and furthermore he was led by the physical idea of 'motion' on his way towards the infinitesimal calculus. Starting from the reading of Wallis's *Arithmetica Infinitorum*, Newton had squared the quarter circle and had seen on his way how the squaring of any curves

$$(1+x^2)^{0/2}, \quad (1+x^2)^{1/2}, \quad (1+x^2)^{2/2}, \quad \text{etc.}$$

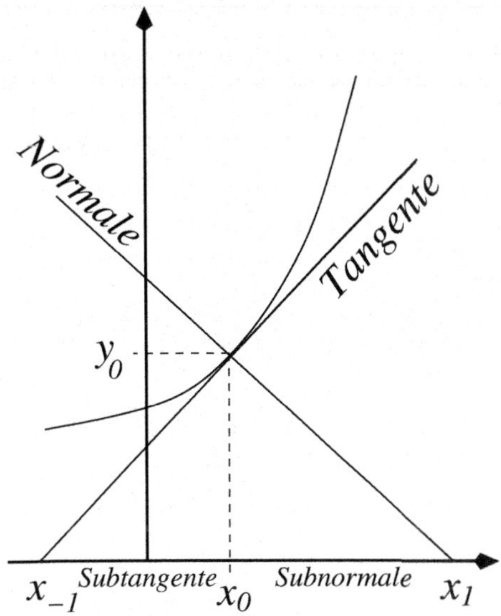

Fig. 3.1.9. Subtangent and subnormal

could be accomplished. Fifty years later, in 1715, at the peak of the priority dispute, Newton will write in the anonymously published *Account of the Commercium Epistolicum* [Hall 1980, p. 263-314] that his quadrature method (method of squaring) rests on three pillars: the squaring of $x^{m/n}$, the addition and subtraction of areas, and the reduction of functions to infinite series.

In an investigation of the subnormal in the autumn of 1664 [Whiteside 1967–81, p. 213ff.] (or in summer 1665, the dating is uncertain, cp. [Westfall 2006, p.123]), Newton's breakthrough to a general theory of quadrature is documented. Considering the tangent and the normal at a point (x_0, y_0) of a curve as in figure 3.1.9 the subnormal is defined as the projection of the normal onto the abscissa, hence the line segment $[x_0, x_1]$, while the subtangent is defined as the projection of the tangent onto the abscissa, hence $[x_{-1}, x_0]$.

René Descartes had presented an algebraic method to find the tangent at a curve, and Newton now starts to exploit it, cp. [Sonar 2011, p. 247f.].

To compute the tangent at a point $(x_0, f(x_0))$ of a function, Descartes thought of a circle with radius r and centre v on the abscissa, cp. figure 3.1.10. Such a circle K satisfies the equation

$$y^2 + (x - v)^2 = r^2.$$

If the radius is large enough there are at least two points of intersection with the function, hence we can insert $y = f(x)$ and get

$$(f(x))^2 + (x - v)^2 = r^2.$$

What characterises a tangent to f at the point $(x_0, f(x_0))$? If there exist exactly two points of intersection of K with f then the straight line marked by r in figure 3.1.10 is certainly not orthogonal to f, but if there is exactly one point of intersection, then the tangent to f is just the tangent to K at the point of contact and the straight line marked with r is orthogonal to f. For Descartes the function f is always polynomial, hence f^2 is a polynomial. In the case of exactly one point of intersection, the polynomial $(f(x))^2+(y-v)^2 = r^2$ has to have a double root and hence can be written in linear factors as

$$(f(x))^2 + (x - v)^2 - r^2 = (x - x_0)^2 \sum_k c_k x^k$$

where the factor $\sum_k c_k x^k$ is a polynomial of lower degree. We now insert the equation for f, multiply out, and compare the coefficients corresponding to equal powers. Doing so, the slope of the normal to f at x_0 emerges as

$$-\frac{f(x_0)}{v - x_0};$$

and from the geometrical relations as shown in the right-hand part of figure 3.1.10 it then follows that the slope of the tangent is given as

$$\frac{v - x_0}{f(x_0)}.$$

The length of the subnormal follows as $v - x_0$ in passing.

Newton studied this method early in the autumn of 1664. In spring 1665 he devised how to come from the quadrature of the functions $y = x^n$ (the quadrature is $\frac{1}{n+1}x^{n+1}$) to quadratures of more general functions [Whiteside 1967–81, p. 225]:

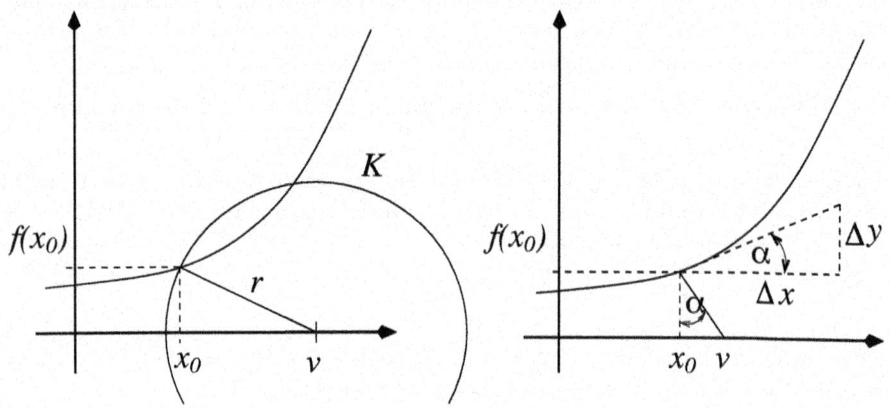

Fig. 3.1.10. Descartes's circle method

'wch shewes y^e nature of another crooked line y^t may be squared.'

We follow [Westfall 2006, S.123ff.] and consider a function defined as

$$y^2 = rx;$$

Newton writes *'The line cdf is a parabola'*.

Using the notations of figure 3.1.11 let

$$ac = \frac{r}{4},$$
$$ap = dt = a.$$

The subnormal apparently is the line segment an or eg. From Descartes's method Newton knows that the subnormal is of constant length along the curve and that this length is $r/2$, hence

$$an = eg = \frac{r}{2}.$$

At $x = c = 0$ we have $y = 0$, at $x = c + \frac{r}{4}$ (which is the point a) we have $y = ad = \sqrt{r(c + r/4)} = r/2$, so that it follows

$$4ac = 2ad = r.$$

Since the subnormal's length is always $r/2$ we arrive at the chain of equations

$$4ac = 2ad = r = 2an = 2eg.$$

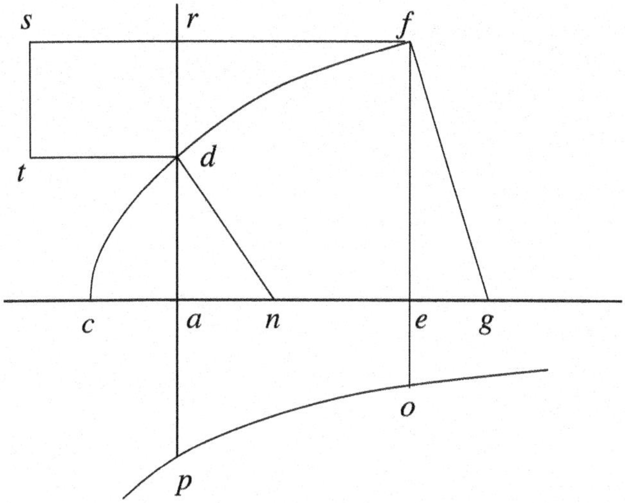

Fig. 3.1.11. Newtons way to the fundamental theorem of calculus

Newton denotes ce by x, then $y = ef$. The quotient of interest

$$\frac{eg}{ef}$$

then is

$$\frac{eg}{ef} = \frac{\frac{r}{2}}{y} = \frac{r}{2\sqrt{rx}} =: \frac{eo}{ap} = \frac{eo}{a}.$$

We now introduce a new variable z designed to define the new curve po, i.e. $z = eo$. It then follows that

$$\frac{eg}{ef} = \frac{r}{2\sqrt{rx}} = \frac{z}{a}$$

and

$$z = \frac{ar}{2} \frac{1}{\sqrt{rx}}.$$

Newton writes [Whiteside 1967–81, p. 299]:

> '*wch shews y^e^ nature of y^e^ crooked line po. now if dt = ap. y^n^ drst= eoap. for supposeing eo moves uniformely from ap & rs moves from dt wth motion decreaseing in y^e^ same proportiō y^t^ y^e^ line eo doth shorten.*'

Indeed[3]

$$\frac{z}{a} = \frac{dy}{dx},$$

hence

$$z\,dx = a\,dy,$$

and if we introduce a 'variable of time' t to describe Newton's 'motion', then

$$a\frac{dy}{dt} = z\frac{dx}{dt}.$$

Hence

$$\text{Area}(eopa) = \int_{r/4}^{x} z\,dx = \int_{r/2}^{y} a\,dy = \text{Area}(rstd).$$

Let us note: The new curve $z(x)$ constructed by Newton by means of an argument of motion[4] is the derivative of the initial curve up to a factor a. Newton established a connection between the area under the curve of the derivative and the given curve itself and thus stood here at the door leading

[3] It nearly hurts at this point that one explains Newton's thoughts with the help of Leibniz's notation, but the reader looking at Newton's text [Whiteside 1967–81, p. 299] with his own eyes and trying to comprehend will be very thankful for this technique applied even by Whiteside.

[4] Here he follows Barrow.

to the fundamental theorem of calculus. Starting from his technique of the subnormal he had developed a general differential calculus!

In rapid succession Newton applied his new knowledge to other functions. He wrote a manuscript entitled *A Method whereby to square those crooked lines* w^{ch} *may bee squared*, in which he clearly showed his understanding of the fundamental theorem. Based on the functions $y = 3x^2/a$ and $z = x^3/a$ he showed the area under y being proportional to differences of z. Up to one further manuscript this is the only point where Newton delivers something like a proof of the fundamental theorem.

Until Newton the quadrature of curves was something static: Wallis had yet summed areas of quadrangles. But Newton had arrived at a *kinetic* method in that he had imagined points moving along curves with certain velocities. We can now also understand the word 'fluxion' which Newton used for his derivatives since he imagined changes of functions as a fluent process. Derivatives for Newton *are* velocities and it is no wonder that he titled a manuscript dated 13th November 1665 *To find* y^e *velocitys of bodys by* y^e *lines they describe.*

After several attempts in which he time and again returned to mathematics, [Westfall 2006, p. 134f.] the famous 'October tract' *The October 1666 tract on fluxions* in [Whiteside 1967–81, p. 400ff.] arised, actually titled by Newton as *To resolve Problems by Motion these following Propositions are sufficient.* Here Newton summarised his results in a virtuoso manner and the publication of this tract would have brought all European mathematicians to admire the not yet 24 years old Newton [Westfall 2006, p. 137], but Newton did not publish anything! Even Barrow knew in 1666 only that Newton existed; he still could guess nothing about Newton's way to the first mathematican in Europe.

In the 'October tract' we also find a rigorous derivation of the fundamental theorem of calculus – with an argument of motion, of course! Newton did not yet use his later dot notation for his 'fluxions' $\dot{x}$ und $\dot{y}$, but variables $\mathcal{X}$, $\cdot\mathcal{X}$ und $\mathcal{X}\cdot$, the modern meaning being

$$\mathcal{X} = f(x, y) = f(x, y(x)) = 0, \quad \cdot\mathcal{X} = x\frac{\partial f}{\partial x}, \quad \mathcal{X}\cdot = y\frac{\partial f}{\partial y}.$$

Since $\frac{d\mathcal{X}}{dx} = 0$ we get:

$$\frac{\partial f}{\partial x} + \frac{\partial f}{\partial y}\frac{dy}{dx} = 0.$$

Again we use Leibniz's notation to explain Newton's mathematics.

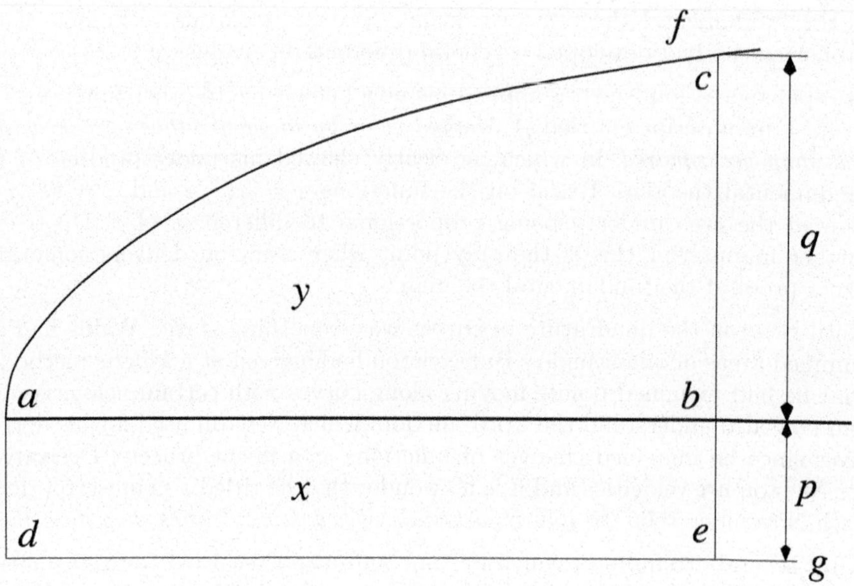

Fig. 3.1.12. The fundamental theorem of calculus

Our notations and figure 3.1.12 are consistent with the corresponding figure in the October tract. We think of the area y under the curve f as being generated by the variable line segment bc moving from a to the right. At the same time ad moves with velocity $\dot{x} = 1$ and hence forms the rectangle $abde$. Newton argues as follows [Whiteside 1967–81, p. 427]:

> *'Now supposing y^e line cbe by parallel motion from ad to describe y^e two superficies [=areas] ae $= x$, & abc $= y$; The velocity wth wch they increase will bee, as be to bc: y^t is, y^e motion by wch x increaseth being be $= p = 1$, y^e motion by wch y increaseth will be bc $= q$. which therefore may bee found by prop: 7th. viz: $\frac{-\mathcal{X}y}{\mathcal{X}x} = q =$ bc.'*

With the notations introduced above thus

$$q = \frac{-\cdot\mathcal{X}y}{\mathcal{X}x} = \frac{dy}{dx}$$

holds and the area under the curve f is

$$\int q\,dx = y,$$

hence

$$\int \frac{dy}{dx}\,dx = \int dy = y,$$

which is the fundamental theorem of calculus. Incidentally, constants of integration do never occur in the work of Newton. He always assumed that the functions considered pass through the origin.

He called the quantities x and y 'fluents', i.e. quantities which flow, and their (temporal) change $\dot{x}$ and $\dot{y}$ are the fluxions.

3.1.4 The 'anni mirabiles'

We have described how Newton succeeded in a complete penetration of a new differential and integral calculus in the years 1664 to 1666[5]. The years from 1664 to 1666 were 'magic years' for Newton with regard to yet more developments.

We may think of Newton's time as a student in Cambridge as being quite strange. While in his spare time he made breathtaking mathematical discoveries – he discovered the binomial theorem in the winter of 1664/65 and laid the foundation for a calculus of fluxions, he had to face medieval disputations at the university in order to get his BA in the year 1665. From this time stems the only preserved letter of mother Hannah to her son Isaac. Unfortunately, due to the crumpled paper some letters are missing in the left upper corner [Westfall 2006, p. 141], [Turnbull 1959–77, Vol. I., p. 2]:

> *'Isack*
> *received your leter and I perceive you*
> *letter from mee with your cloth but*
> *none to you your sisters present Thai*
> *love to you with my motherly lov*
> *you and prayers to god for you I*
> *your loving mother*
> *hanah*
> *wollstrup may the 6. 1665'*

Already from the orthography (even if we consider 'old' English) we can see that mother Hannah could barely write and that she probably did not write often. Was there a lack of understanding on the side of an uneducated mother for the ingenious son whose intellectual world she *could not* understand?

In the summer of 1665 the plague broke out in England for the last time – in Cambridge as well. On 11th September all public meetings were prohibited, on 20th October the senate of the university prevented preachings in St Mary and all activities in the public schools [Westfall 2006, p. 141]. The university was as good as deserted during the eight months to come; only in the spring of 1667 operations as usual could be continued. Many students departed with their

[5] Although he did not work with differentials we use this term here for simplicity.

Fig. 3.1.13. Removal of plague victims 1665

tutors to continue their studies but Newton was already a BA and independent and hence departed to his mother in Woolsthorpe probably before 17th August 1665. He is back in Trinity College on 30th March 1666 and leaves again probably in June. He finally returned to Cambridge early in May 1667. Nearly 50 years later he writes retrospectively [Westfall 2006, p. 143]:

> *'In the beginning of the year 1665 I found the Method of approximating series & the Rule for reducing any dignity of any Binomial into such a series. The same year in May I found the method of Tangents of Gregory & Slusius, & in November had the direct method of fluxions & the next year in January had the Theory of Colours & in May following I had entrance into ye inverse method of fluxions. And the same year I began to think of gravity extending to ye orb of the Moon & (having found out how to estimate the force with w^ch [a] globe revolving*

> *within a sphere presses the surface of the sphere) from Keplers rule of the periodical times of the Planets being in sesquialterate proportion of their distances from the centre of their Orbs, I deduced that the forces wch keep the Planets in their Orbs must [be] reciprocally as the squares of their distances from the centers about wch they revolve: & therebye compared the force requisite to keep the Moon in her Orb with the force of gravity at the surface of the earth, & found them answer pretty nearly. All this was in the two plague years of 1665-1666. For in those days I was in the prime of my age for invention & minded Mathematicks & Philosophy more then at any time since.'*

Newton has described here everything with which he should be busy with after 1666 until his death: The theory of light, the theory of gravitation (or somehow more general: celestial mechanics), and the calculus of fluxions and fluents. The actual 'magic year' is indeed 1666 or rather the time which Newton stayed in Woolsthorpe. How was it possible to arrive at such ingenious ideas in a period like this? Was it the 'mother's womb' in which Newton withdrew and which caused psychological reactions as suspected by Manuel [Manuel 1968, p. 80]? Or was it the rural tranquility of Woolsthorpe which gave Newton leisure for far-reaching ideas? We do not know. On the other hand, 1666 was no less magical for Newton than 1665 and 1664; hence he would have arrived at his achievements probably even if he had stayed at Cambridge.

Following the October tract 1666 Newton turned away completely from mathematics to study mechanics intensively. He started already in January 1665. He is interested in particular in circular motion and the impact of a body onto another one. He was not convinced of the discussions he found in Galileo's *Dialogue Concerning the Two Chief World Systems* [Galilei 1982] and in Descartes' *Pinciples of Philosophy* [Descartes 2005]. He thought about the notion of force, emphasised mass inertia as a central property and thus laid the foundation of his dynamics, and prepared his intellectual world for his magnus opus *Philosophiae naturalis principia mathematica* which will be published more than 20 years later. In his earlier investigations concerning mechanics Newton was quite close to the works of Christiaan Huygens without knowing it. Both men will later meet and show mutual appreciation of each other.

In the time of the 'anni mirabiles' falls the famous story of the falling apple. In 1666 Newton thought about the nature of gravitation when living with his mother. As we are told by Conduitt [Westfall 2006, p. 154] he saw an apple fall from a tree and this event ought to have brought Newton the idea that gravitation not only pulls the apple down to earth but also the moon, and that the moon exerts a force on the earth.

This story became so famous that it even found its way into a German book for children by the illustrator Petra Weigandt and the writer Peter Tille [Weigandt/Tille 1988]. Whether this event really took place we don't know; it is highly unlikely, since the law of gravitation certainly did not stem from a single flash of inspiration.

Fig. 3.1.14. Tribute to Newton via stamps (Great Britain 1987)

A further area of research to which Newton turned was the theory of colours and also here he is very close to the investigations of Huygens. In the year 1664 Robert Boyle had published *Experiments and Considerations Touching Colors*; a year later Robert Hooke published his experiments with the microscope under the title *Micrographia* [Hooke 2007].

Boyle as well as Hooke had tried to explain colours but Newton was not satisfied. Hooke's statements concerning colours in particular must have made Newton mad. Hooke and Newton will soon be joined in a hostility which will last 40 years. For Hooke the colour blue was an oblique and disturbed impulse of light the weak part of which reaches the retina first and the strong part of which follows[6].

Early in the year 1666 Newton ground optical lenses; he probably owned already a prism. Back in Cambridge he directs sunlight from a cutout in his shutter through a prism and saw a spectrum of colours.

This 'Experimentum crucis' reveals to him that light consists of all colours and that colours do not result, as in Hooke, from different velocities in different media.

[6] *'That Blue is an impression on the Retina of an oblique and confus'd pulse of light, whose weakest part precedes, and whose strongest follows. And, that Red is an impression on the Retina of an oblique and confus'd pulse of light, whose strongest part precedes, and whose weakest follows.'* [Hooke 2007, p. 64].

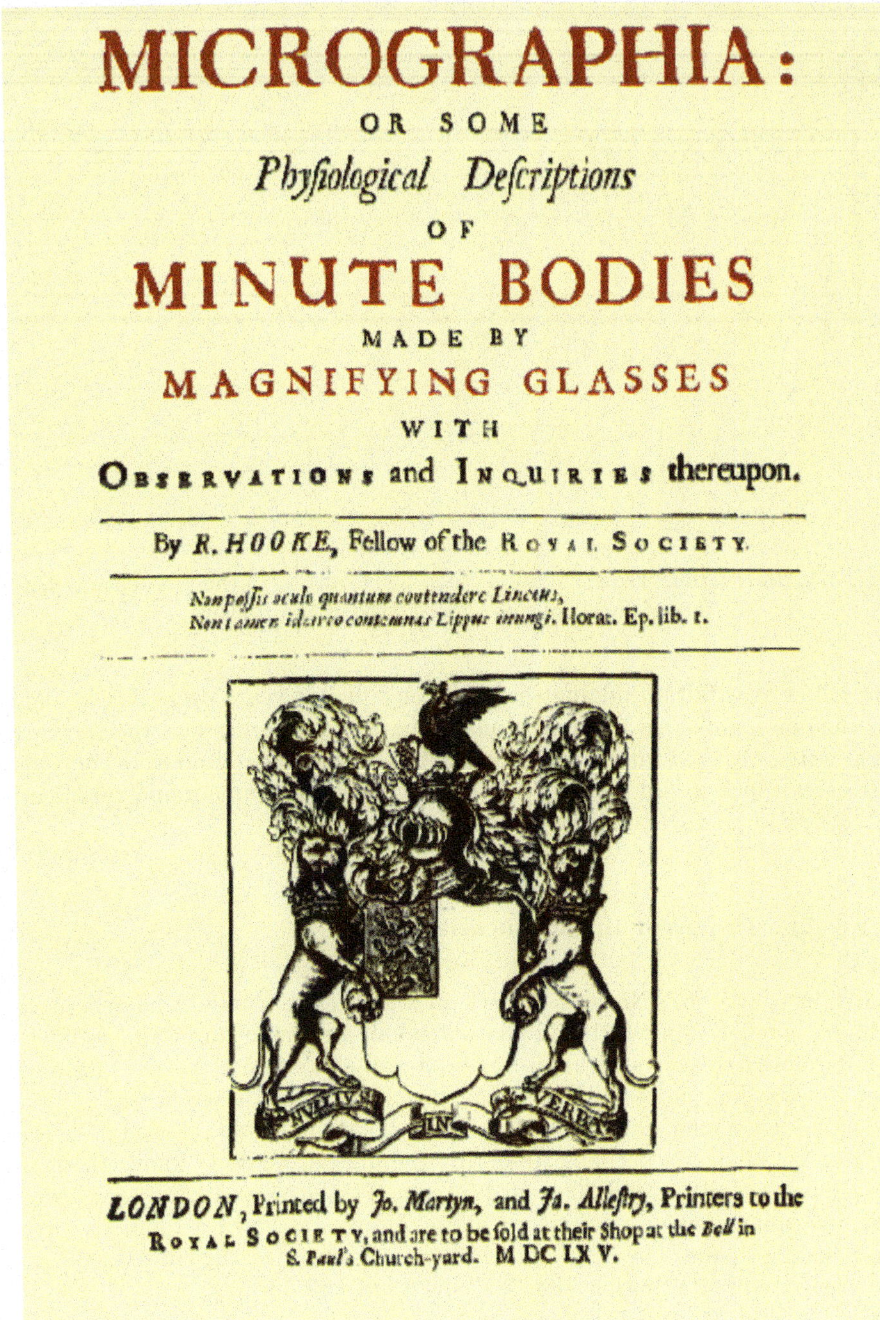

MICROGRAPHIA:
OR SOME
Physiological Descriptions
OF
MINUTE BODIES
MADE BY
MAGNIFYING GLASSES
WITH
OBSERVATIONS and INQUIRIES thereupon.

By *R. HOOKE*, Fellow of the ROYAL SOCIETY.

Non possis oculo quantum contendere Linceus,
Non tamen idcirco contemnas Lippus inungi. Horat. Ep. lib. 1.

LONDON, Printed by *Jo. Martyn*, and *Ja. Allestry*, Printers to the ROYAL SOCIETY, and are to be sold at their Shop at the *Bell* in S. *Paul's* Church-yard. M DC LX V.

Fig. 3.1.15. Title page of *Micrographia* of 1665

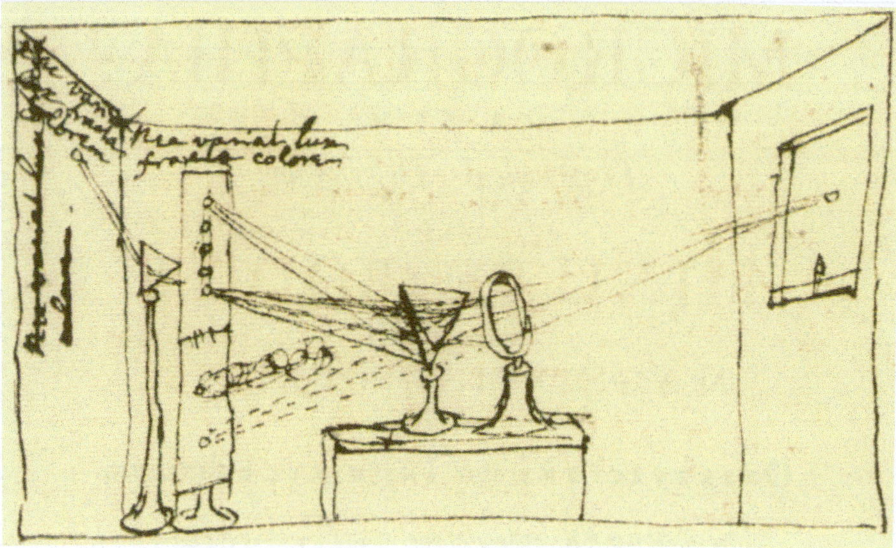

Fig. 3.1.16. Newtons 'Experimentum crucis' [© Courtesy of the Warden and Scholars of New College, Oxford/Bridgeman Images]

We follow Westfall in judging that Newton's 'magic years' were not so magic on the one hand – his theories in mathematics, mechanics and the theory of light were only in a state of germination, but on the other hand were enormous successes which would have served to make Newton's name immortal already in 1666, had he only published them promptly.

3.1.5 The Professor in the Lucasian Chair

Early in May 1667, Newton is back in Cambridge, but will he succeed in winning a fellowship which will guarantee him a durable stay? All candidates for the nine free positions had to pass an oral examination lasting four days, but Newton had neglected the university curriculum. He celebrated his BA, bought a gown, the traditional cloth of a scholar, and – was elected. Whether he had an advocate, perhaps Humpfrey Babington, we don't know; there are no official signs [Westfall 2006, p. 177]. On 12th October 1667 Isaac Newton became 'minor fellow' of Trinity College. He was assigned a room and lived on a small stipend. On 11th April 1668 he was awarded the title Master and he was promoted a 'major fellow'; a new gown was due.

We have to imagine Newton's life in Cambridge still as a very lonely one. Nevertheless – or just because of his reclusiveness? – he won great respect with time. Perhaps his absent-mindedness even caused awe. We know about only three more intimate friends: John Wickins with whom he shared a room until

Wickins's leave in 1683, Humphrey Babington, and finally Isaac Barrow. In 1669 Newton took up (al)chemical studies but also turned again to optics and mathematics. The year before *Logarithmotechnia* by Nicolaus Mercator, cp. page 101, was published, in which the area under the hyperbola was shown to be the logarithm – a result which Newton arrived at earlier and independently. During the first months of the year 1669 John Collins had sent the book to Isaac Barrow who put it in Newton's hands. John Collins (1625–1683) was an English mathematician who became a kind of English Mersenne. He was a royal official who corresponded with all of the important mathematicians in his country and with some abroad. He edited the writings of Isaac Barrow and John Wallis and became a member of the Royal Society in 1667. At the end of July 1669, Collins received answer from Barrow [Westfall 2006, p. 202], who reported about one of his friends:

> *'that hath a very excellent genius to those things, brought me the other day some papers, wherein he hath set down methods of calculating the dimensions of magnitudes like that of Mr Mercator concerning the hyperbola, but very general ...'*

Ten days later Collins received from Barrow a work with the title *De analysi per aequationes numero terminorum infinitas* (On the analysis by means of equations of an infinite number of terms), and at the end of August Collins learned the name of the author, too [Westfall 2006, p. 202]:

> *His name is Mr Newton; a fellow of our College, & very young (being but the second yeest[7] Master of Arts) but of an extraordinary genius & proficiency in these things.*

We do only know from these letters that Newton and Barrow had a personal relationship indeed for quite some time.

Mercator's book had shown Newton that others were on his tracks. Hence he wrote down *De analysi* hastily compiled from his old records and quickly handed it to Barrow. Then, however, Newton must have been beset by doubts – perhaps scared by possible critique about his work – and Barrow was not allowed to send the work to Collins. Only when Collins showed enthusiastic interest was Barrow allowed to send the work to London and to reveal Newton's name.

It was the October tract from which *De analysi* was composed, but also expanded. The infinite series of e^x, $\sin x$ and $\cos x$, which appear here for the first time in the history of European mathematics, is one example of Newton's work beyond the October tract [Westfall 2006, p. 205].

Collins, who was himself rather a mediocre mathematician, recognised Newton's genius immediately from the accomplishments in *De analysi.* Before he

[7] The meaning of the word 'yeest' is not clear but 'the second year' seemed to Westfall the most likely translation.

ANALYSIS
Per Quantitatum
SERIES, FLUXIONES,
AC
DIFFERENTIAS:
CUM
Enumeratione Linearum
TERTII ORDINIS.

LONDINI:
Ex Officina PEARSONIANA. *Anno* M.DCC.XI.

DE ANALYSI
Per Æquationes Numero Terminorum
INFINITAS.

Methodum generalem, quam de Curvarum quantitate per Infinitam terminorum Seriem mensuranda, olim excogitaveram, in sequentibus breviter explicatam potius quam accuratè demonstratam habes.

BASI *AB* Curvæ alicujus *AD*, sit Applicata *BD* perpendicularis: Et vocetur *AB* = *x*, *BD* = *y*, & sint *a*, *b*, *c*, &c. Quantitates datæ, & *m*, *n*, Numeri Integri. Deinde,

Curvarum Simplicium Quadratura.
REGULA I.
Si $ax^{\frac{m}{n}} = y$; *Erit* $\frac{an}{m+n}x^{\frac{m+n}{n}}$ = *Areæ ABD.*
Res Exemplo patebit.

1. Si x^2 (= $1x^{\frac{2}{1}}$) = y, hoc est, $a = 1 = n$, & $m = 2$; Erit $\frac{1}{3}x^3$ = ABD.
2. Si

Fig. 3.1.17. Title 'Analysis' and *De Analysi* in the printed version of 1711

returned the work as he had promised, he made a copy, showed it to others, and corresponded on it with James Gregory, René de Sluse, and other mathematicians in France and Italy. Newton's anonymity was shattered. Collins and Barrow now had the idea to publish *De analysi* in an appendix to Barrow's book on optics, but Newton could not be convinced. He withheld the publication, and thereby showed a pattern which will become typical for him for the first time.

When Newton handed *De analysi* to Barrow his teacher was just thinking to hand over the Lucasian Chair, cp. page 53. If Barrow had not previously been fully convinced of Newton's genius, he was by now at the latest. Barrow withdrew in favour of Newton and hence Isaac Newton became the new professor of mathematics on the Lucasian Chair in the year 1669. His first lecture was not concerned with the contents of *De analysi*, however, but with optics. He conducted experiments furthermore, bought prisms and worked on a theory of colours. His manuscripts of these days will lead to the publication of his book *Opticks* [Newton 1979] in 1704 which will also play a large role in the priority dispute yet to come. He has now left Christiaan Huygens behind in his optical investigations [Westfall 2006, p. 217], but put them aside in 1670 for the time being to turn again to mathematics. In 1672 and 1690 he will come back to optics. Concerning Isaac Barrow's *Lectiones XVIII* on optics he made two improvements for which Barrow thanked him in the preface, and Newton also contributed to the *Lectiones geometricae* of 1670. Barrow encouraged Newton to revise the *Algebra* by Gerard Kinckhuysen (1625–1666) which Collins had earlier translated from Danish. In late November or early December 1669 Newton travelled to London and there met Collins who will correspond with Newton from then on. At first Newton was impressed by Collins's attention, but when Collins tried to urge Newton to publish his computations regarding a problem in interest rates Newton retreated and wrote on 28th February 1670 [Westfall 2006, p. 222]:

> *'soe it bee w*th*out my name to it. For I see not what there is desirable in publick esteeme, were I able to acquire & maintaine it. It would perhaps increase my acquaintance, y*^{e} *thing w*^{ch} *I chiefly study to decline.'*

Newton also refused his authorship of *Observations on Kinckhuysen.* The revised book on algebra by Kinckhuysen was eventually published with an addition 'enriched by another author'. A silence of nine months followed between Newton and Collins, but in mid-1671 Collins wrote a light-hearted letter to Newton noting that the Kinckhuysen book would sell better if Newton's name would appear on it. Newton responded with impolite rejection and asked Collins not to send him books any more. He was working on a new mathematical manuscript which later will be known as *Tractatus de methodis serierum et fluxionum* (Treatise on the Method of Series and Fluxions). It is an explanation of Newton's methods drawn from *De analysi* and the October tract in which he thoughtfully considers the foundations of his calculus of fluxions. He tries to move away from a static understanding of infinitesimals and introduces the notion of 'last ratios', his attempt to characterise a limit. Newton started working on *De methodis* in winter 1670/71 but he did not finish it. In May 1672 he writes to Collins that the manuscript grew longer than he had expected. Only in 1736 will John Colson publish an English translation in London under the title *The Method of Fluxions and Infinite Series; With its Application to the Geometry of Curve-Lines*; in Latin it will be published as late as 1779, the editor being Samuel Horsley. Since then the work is frequently cited as *Methodus fluxionum et serierum infinitarum.* Certainly Collins would have done everything to send this manuscript off to the press. Instead, Collins had to confess in a letter of 29th October 1675 to Gregory [Turnbull 1959–77, Vol. I, p. 356] that he had heard nothing from Newton for 11 or 12 months and that he had probably got stuck in chemical studies.

In *De methodis* [Whiteside 1967–81, Vol. III, p. 32-372] Newton took up the idea of area-preserving transformations in 'Problem 8'. He considered two functions $v = f(x)$ and $y = g(z)$ defining the curves *FDH* and *GEI*, cp. figure 3.1.18.

As we already know from Newton, he imagines the areas s and t as being produced by the motion of the line segments *BD* and *EC*, respectively. The temporal change of the area s hence is the product of length v with the velocity $\dot{x}$ in a horizontal direction. Something analogous holds for the temporal change of t; hence it follows

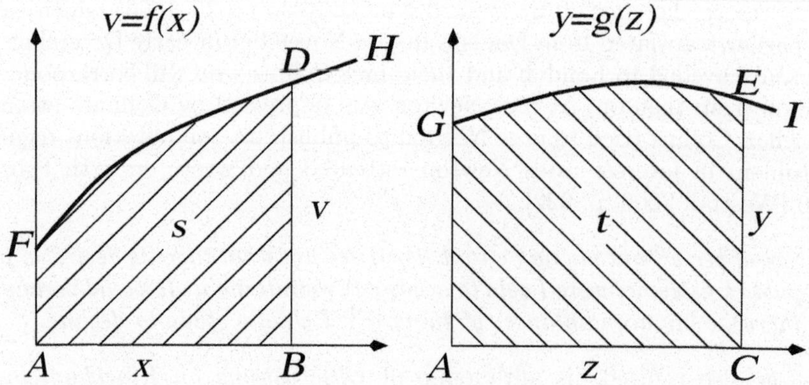

Fig. 3.1.18. Regarding area-preserving transformations

$$\frac{\dot{s}}{\dot{t}} = \frac{v\dot{x}}{y\dot{z}}.$$

If we set $\dot{x} = 1$ then $\dot{s} = v$ follows. Since $\dot{t} = y\dot{z}$ it further follows that $y = \dot{t}/\dot{z}$. Demanding equal areas, i.e. $s = t$, then $\dot{s} = \dot{t} = v$ follows and thereby

$$y = \frac{v}{\dot{z}}. \tag{3.1}$$

Assuming now a functional relationship

$$z = \phi(x), \quad x = \psi(z)$$

between x und z, (3.1) is a definition of that function $y = g(z)$, for which equal areas can be achieved, namely

$$y = \frac{v}{\dot{z}} = \frac{f(x)}{\phi'(x)\dot{x}} = \frac{f(\psi(z))}{\phi'(\psi(z))\dot{x}} \stackrel{\dot{x}=1}{=} \frac{f(\psi(z))}{\phi'(\psi(z))}.$$

From our functional relationship it follows $x = \psi(z) = \psi(\phi(x))$ and thereby $\dot{x} = 1 = \psi'(z)\phi'(x)$, hence $\phi'(x) = \phi'(\psi(z)) = (\psi'(z))^{-1}$. Thus $y = g(z)$ has been found:

$$y = f(\psi(z))\psi'(z).$$

Hence, Newton has found the rule of integration by substitution

$$\int f(x)\,dx = \int f(\psi(z))\psi'(z)\,dz$$

which we give here in its modern form. With Newton this result appears in his own, equivalent symbolism.

From a letter from the year 1669 [Turnbull 1959–77, Vol. I., p. 3f.] we learn for the first time about a reflector telescope which Newton had made with his own hands. At a meeting with Collins at the end of 1669 Newton reports on his telescope and he probably presented it in Cambridge, because Collins heard it from a fellow of Trinity College and John Flamsteed (1646-1719), the later 'Astronomer Royal', heard about it from London and Cambridge.

The Royal Society now became aware of Newton. They wanted to see the telescope. When it finally arrived it caused a stir and was a great sensation. In January 1672 Henry Oldenburg, secretary of the Royal Society, wrote a letter to Newton [Turnbull 1959–77, Vol. I., p. 73] announcing a letter to Huygens in Paris in order to secure Newton's invention, against possible claims of foreigners, in the name of the Royal Society. In Paris, Huygens was excited and on 21st January 1672 Newton became a member of the Royal Society. On 28th January Newton wrote:

> *'I am purposing them, to be considered of & examined, an accompt of a Philosophicall discovery wch induced mee to the making of the said Telescope, & wch I doubt not but will prove much more gratefull then the communication of that instrument, being in my Judgment the oddest if not the most considerable detection wch hath hitherto beene made in the operations of Nature.'*

Newton talked about his theory of colours and light, but after one-and-a-half weeks he still did not send off anything. He did not want to make anything public, actually; he hesitated, but the pressure from the Royal Society was high. Finally, on 16th February 1672, Newton submitted his theory of colours to the Royal Society. He will have reason to regret this shortly afterwards.

3.1.6 Till Death: The Fight with Robert Hooke

Of the numerous persons in his life whom Newton counted as personal enemies, Robert Hooke clearly stood out. Hooke, born on 28th July 1635 in Freshwater on the coast of the Isle of Wight, was long neglected in the history of science, perhaps because the light of his worst adversary Isaac Newton shone too brightly. The 20th century saw a greater interest in Hooke from among the members of the Royal Society, and on the occasion of the 300th year of his death in 2003 two readable biographies were published, [Inwood 2002] and [Jardine 2003]. With [Purrington 2009], [Chapman 2004] and others; we now also have modern assessments of his scientific achievements.

More than any other name, Hooke's is linked with the first decades of the Royal Society. Since he was an excellent engineer and draughtsman, a position as organist and scholar at Christ Church College in Oxford brought him in 1653 in contact with the group around John Wilkins which founded the Royal Society in 1660. Already in 1655, aged 20, Robert Hooke became assistant to the anatomist and chemist Thomas Willis, learned astronomy from Seth Ward

Fig. 3.1.19. Replica of Newton's reflecting telescope of 1672 with a mirror of 15 cm in diameter (Whipple Museum of the History of Science, Cambridge) [Photo: Andrew Dunn 2004]

(1617–1689), and became an assistant to Robert Boyle in 1658 with whom he worked at Boyle's air pump, cp. figure 2.3.6. Whether he ever finished his BA we don't know, but the title MA was awarded in 1663 [Purrington 2009, p. 2].

Christopher Wren and Robert Boyle became faithful friends of Hooke; and he was acquainted with all of the other eminent scientists. In 1662 Hooke went to London; he was 27 years of age and already a promising natural philosopher who had demonstrated his brilliance. Now Hooke became a professional scientist in the service of the Royal Society – perhaps the first of his kind. One of his responsibilities was the presentation of an experiment at the weekly meetings. Thus, he conducted hundreds of experiments in the 30 years of his active employment. Without Robert Hooke the Royal Society and their meetings can hardly be imagined. Through his activities and his numerous talks with scientists of the first rank he also became a scientist of first rank himself, with an independent mind.

After the Great Fire in 1666, it was Wren and Hooke who rebuilt London; but today we cannot find a sign of Hooke's life in London. Even on 'The Monument', erected to remember the fire by himself and Christopher Wren, his name cannot be found. Wren and Hooke built 51 churches in London after the fire; half of them stem from Hooke alone [Purrington 2009, p. 7]. In elasticity theory 'Hooke's law' is named after him; he built a spring mounted clock simultaneously with Huygens (but compare [Hall 1993a] who is sceptical about this), he guessed the theory of gravitation [Hall 1993b], discovered the great red spot on Jupiter, he described a cell for the first time, and his *Micrographia* [Hooke 2007] shows him as a keen observer and a magnificent draughtsman.

In former times one often assumed belligerence and cantankerous behaviour in Hooke. Although Hooke certainly had some rough edges and was able to fight for his discoveries and inventions, the image of a squabbler does not fit a man who had so many good friends and who was a welcome guest in London's coffee houses. In the quarrel with Newton, which Hooke certainly saw as a purely scientific struggle, he became Newton's mortal enemy.

The manuscript on the theory of colours, sent in 1672 by Newton in the form of a letter to Henry Oldenburg, secretary of the Royal Society, was criticised by Robert Hooke within two weeks.

All other responses were extremely positive; even Huygens very much praised the work, although he later commented on it quite critically. The work was printed on 29th February 1672 in the *Philosophical Transactions*, the journal of the Royal Society, together with a description of the reflector telescope. Newton had risen to be an international figure. Even an unknown young man with the name Gottfried Wilhelm Leibniz, who corresponded with the Royal Society since 1670, informed Oldenburg in 1673 that he had seen the work. John Flamsteed, soon to be the first Astronomer Royal, commented on the work with little insight [Westfall 2006, p. 240]. Robert Hooke, since the publication of his *Micrographia* the recognised expert on optics in England, responded with rejection. In a letter to Newton he hinted that he had performed all of Newton's experiments earlier, but that Newton had drawn incorrect conclusions from these experiments. Newton did not yet show anything, though he must have been boiling inside, since Oldenburg had to wait no less than three months for an answer to Hooke's critique. Newton had planned to answer in the form of a larger work, but this was never sent off. Had he not seen what happened when he made his works public? He was criticised, and that he could not stand. In addition Oldenburg asked him to leave out the names of Hooke and Ignace-Gaston Pardies, a Parisian scientist who had also criticised Newton's theory of colours, in his written reply. After three months Newton then sent only extracts of what he had prepared to Oldenburg [Turnbull 1959–77, Vol. I., p. 171ff.]. Instead of skipping Hooke's name Newton mentioned him more than 27 times [Westfall 2006, p. 246]. The tone was aggressive; Newton instructed Hooke how he could manufacture better optical instruments and wrote that Hooke's theory of colours was not only

Fig. 3.1.20. Hooke's signature

Fig. 3.1.21. The old Gresham College as Robert Hooke knew it [Popular Science Monthly volume 81, 1912]

'insufficient', but also 'unintelligible' [Turnbull 1959–77, Vol. I., p. 176]. The letter was one long insult; the laboriously suppressed hatred against Hooke can be clearly felt, and the wrath of the writer is evidently recognisable. In a covering letter to Oldenburg [Turnbull 1959–77, Vol. I., p. 193] Newton remarked complacently that Hooke surely would not find anything objectionable in his answer.

To top it all, Huygens started to criticise Newton's theory and Newton threatened Oldenburg to leave the Royal Society. To Huygens Newton remained always polite, although determined.

Yet the relationship between Newton and Hooke was not completely ruined. At the end of 1679 a correspondence on celestial mechanics began which lasted for six weeks. Inspired by Hooke in this correspondence Newton found the key to his theory of gravitation [Purrington 2009, p. 177]. When Hooke in 1684 saw Newton's *De motu corporum in gyrum* and shortly after the *Principia*, presented 1686 before the Royal Society, he was thunderstruck. Here he saw his ideas mathematically developed and perfected. Already at the presentation of the reflecting telescope Hooke had bitterly remarked that he had already presented to the Royal Society how systems of mirrors may help to shorten telescopes, what he saw as the actual invention of the reflecting telescope. Already then the allegation of plagiarism against Newton came up; but now Hooke felt betrayed and cheated out of the fruits of his labour. The last, final, ugly split between the two men was executed. In all fairness we have to remark that Hooke was not able to understand Newton's theories, however much help he might have been given in the correspondence [Purrington 2009, p. 229]. When Hooke died in 1703, Newton became President of the Royal Society and lived on for another 24 years. That was long enough to wipe out the

Fig. 3.1.22. Henry Oldenburg and a memorial stone for Robert Hooke in London [Photo: Rita Geer 2009]

memory of the uncomfortable, but magnificent scientist Robert Hooke. But did Newton really actively wipe out this memory?

When the German Zacharias Conrad von Uffenbach visited the Royal Society in the year 1710 he reported about a portrait of Robert Hooke. No portrait of Hooke, however, has come down to us and this missing portrait has opened the way to speculations of Newton having ordered the destruction of the portrait when he became President of the Royal Society. Obviously, this would fit Newton's character. In [Jardine 2003] Lisa Jardine has presented a portrait which she assumed to represent Hooke. Today we know that Jardine was wrong [Chapman 2004, p. 262ff.]. Still we do not have any portrait of Hooke. Did Newton do it after all? As Chapman pointed out there is even no portrait in the luxury edition of *Posthumous Works* of Hooke's friend Richard Waller which also contains the first biography of Hooke. Hence it is most likely that a portrait of Hooke never existed. This is the point of view shared officially by the Royal Society; von Uffenbach certainly saw a portrait, but not that of Hooke. Newton is acquitted.

Nevertheless we cannot reject out-of-hand the thought that Newton did everything after the death of Hooke to wipe out any reminiscence of his arch-enemy. It is indeed strange that barely any reminiscence of him has survived. Almost all of the buildings he designed were demolished or were destroyed by German bombs in the second World War. His mortal remains were mixed with others in the 19th century and transferred to a mass grave in North London, and the IRA finally ruined the last memory of Hooke in that a painted window in the church St Helen Bishopsgate in which he was buried, inserted in his honour, was destroyed by a bomb on the 10th April 1992 [Purrington 2009, p. 248].

3.2 The Lawyer: Gottfried Wilhelm Leibniz

In comparison with Newton's mathematical works [Whiteside 1967–81] and his correspondence [Turnbull 1959–77], the situation is not as comfortable in the case of Leibniz. Initiated at the beginning of the 20th century, the Leibniz tradition is carried on by the Berlin-Brandenburg Academy of Sciences and Humanities and the Göttingen Academy of Sciences and Humanities after the German reunification, thereby involving research centres in Berlin, Hanover, Münster, and Potsdam. The Gottfried-Wilhelm-Leibniz-Gesellschaft in Hanover also promotes the Leibniz edition. Containing eight series all writings and letters are, or will be, published in transcribed form (even on the internet), but regrettably up to now – in contrast to what the English did concerning Newton's writings – there are only a few translations of the texts. Hence, one is confronted with French or Latin language, cp. [Leibniz 2008b]. Two notable exceptions are [Leibniz 2011], published in 2011, and [Knobloch/Schulenburg 2000] published in 2000, containing translations of the main texts of Leibniz concerning the mathematics of finance and insurance. In the first-mentioned publication all of the mathematical journal articles of Leibniz are now available in German language. In case of the mathematical writings of Leibniz this new translation replaces the one from the 19th century [Leibniz/Newton 1998].

Slightly better is the situation concerning Leibniz biographies. Still from the 19th century stems the two-volume biography by Guhrauer [Guhrauer 1966] which still contains interesting details; in recent times the biography by Aiton [Aiton 1991] has to be emphasised, cp. the book review of the original English edition by Herbert Breger [Breger 1987]. Even younger is the recommendable biography by Maria Rosa Antognazza [Antognazza 2009]. Indispensable for a detailed description of Leibniz's vita is still the chronology [Müller/Krönert 1969] which does not seem to have been translated into any other language.

3.2.1 Childhood and Youth

The Thirty Years' War is going to rage for two years more; the country is already bled to death, when on Sunday 1st July 1646 at a quarter to seven pm a son is born to the actuary and professor of moral philosophy Friedrich Leibniz (Leibnütz) and his wife Katharina in Leipzig. Two days later this son is christened with the name Gottfried Wilhelm. His father noted proudly in the family chronicle [Müller/Krönert 1969, p. 3]:

> *'21st June* [jul.] *on Sunday 1646 was born to the world my son Gottfried Wilhelm, after six o'clock in the evening a quarter to 7 o'clock, in Aquarius.'*

Fig. 3.2.1. The father: Friedrich Leibniz (Leibnütz), Portrait of an unknown painter, mid 17th century (© Custody of the University Leipzig)

> (21. Junii [julianisch] am Sontag 1646 Ist mein Sohn Gottfried Wilhelm, post sextam vespertinam 1/4 uff 7 uhr abents zur welt gebohren, im Wassermann.)

It was not the first marriage of the father; in those days the mortality of women in childbed was so high that second and third marriages of men were no rarity. He married in 1625 for the first time; the son Johann and the daughter Anna Rosina were the offsprings of this relationship. The second wife died childless in 1643 and hence Friedrich Leibniz married the daughter of a Leipzig attorney, Katharina Schmuck, a wise, pious and friendly woman [Aiton 1991, p. 26].

Already on the occasion of the christening, the father believed he saw signs for a special vocation of his son, when the little one lifted up his head and opened his eyes as if asking for the baptismal water. In 1648 the little boy fell off a table when his mother was in church and his father lay ill in bed. The aunt who should have cared about the child was as horrified as the father, but Gottfried Wilhelm lay at a much larger distance from the table than he would have been able to jump, and he laughed unharmed at his father and aunt [Müller/Krönert 1969, p. 3].

On 11th August 1648 Gottfried Wilhelm's sister Anna Catharina is born. She will marry the preacher Simon Löffler who will bring a son Friedrich Simon from a first marriage into this relation, who eventually will become Gottfried Wilhelm Leibniz's universal heir.

Fig. 3.2.2. The old Nikolai school in Leipzig
[Photo: Appaloosa 2009]

The father died on 15th September 1652 when Gottfried Wilhelm was six years old. His education was taken over by his mother. The father had early on successfully tried to awaken the love of literature in his son. In retrospect Leibniz wrote [Müller/Krönert 1969, p. 4]:

> *'The history and poetry as also the knowledge of literature I have loved more than games even as a boy'*
>
> (Die Histori und poësin auch notitiam rei literariae habe ich als noch ein Knabe anstatt des Spiels geliebt.)

In July 1653 Leibniz became a pupil of the Nikolai school which he attended until Easter 1661. Since his father was professor at the University Leipzig Leibniz was already matriculated at the time of his start at school. The teaching of Latin began which Leibniz was able to shorten considerably. He found an illustrated edition of Livius and the history book *Thesaurus* by Sethus Calvisius. Since he possessed German texts on history he could easily understand the *Thesaurus*, but Livius delayed him. Later he remembered [Müller/Krönert 1969, p. 4]:

> *'I get caught quite often in Livius; for the world of the ancient and their linguistic peculiarities were unknown to me ... I understood barely a line, to be honest. But because it was an old edition with woodcuts I viewed them eagerly, read here and there the words below them, not*

> *being distressed by the darker passages, and what I did not understand I skipped. When I did that several times and browsed through the whole book and started again after some time, I could understand much more. Thereupon highly pleased I continued without a dictionary until most of it became clear to me and I could penetrate deeper into the meaning.'*
>
> (In dem Livius dagegen blieb ich öfter stecken; denn da mir die Welt der Alten und ihre sprachlichen Eigentümlichkeiten unbekannt waren, ... verstand ich, ehrlich gesagt, kaum eine Zeile. Weil es aber eine alte Ausgabe mit Holzschnitten war, so betrachtete ich diese eifrig, las hier und da die darunterstehenden Worte, um die dunklen Stellen wenig bekümmert, und das, was ich gar nicht verstand, übersprang ich. Als ich dies öfter getan, das ganze Buch durchgeblättert hatte und nach einiger Zeit die Sache von vorn begann, verstand ich viel mehr davon. Darüber hoch erfreut, fuhr ich so ohne irgendein Wörterbuch fort, bis mir das meiste ebenso klar war, und ich immer tiefer in den Sinn eindrang.)

In this self-educated manner he learned the first-class Latin, in which he later wrote his scientific treatises. The capabilities of his young pupil could not escape the attention of the teacher and so he encouraged his mother and aunts to allow him access to works not actually suited to lads of his age. A friend of the family, an educated young man, witnessed a conversation and thereupon started to question young Leibniz who could impress him with his knowledge. He caused the mother to promise the opening of the library of his father to him, which he was not allowed access to before. Hence the eight year old Leibniz got access to the Latin classics and the writing of the church fathers in 1654. Following his own account he was able to understand Latin without difficulty at age 12 and also the basics of Greek [Aiton 1991, p. 27].

The thirteen year old boy presented his own Latin poem in 300 hexameters at Pentecost 1659, but another passion of the pupil, logic, started developing. In higher grades Aristotelian syllogistic was taught and Leibniz became not only perfect in the applications of this logic, but even recognised the limits of syllogistic logic [Müller/Krönert 1969, p. 5]:

> *'Not only could I easily apply the rules to specific cases, which I alone was able to do among all my classmates to the astonishment of my teachers, but I also had doubts and carried already then new ideas which I wrote down in order not to forget them. What I wrote down at fourteen years of age I have read again afterwards and felt an extraordinary pleasure thereupon.'*
>
> (Ich konnte nicht nur leicht die Regeln auf konkrete Fälle anwenden, was ich zur Verwunderung meiner Lehrer allein unter meinen Schulgenossen fertigbrachte, sondern ich hatte auch meine Zweifel und trug mich schon damals mit neuen Ideen, die ich aufschrieb,

um sie nicht zu vergessen. Was ich damals mit vierzehn Jahren niederschrieb, habe ich lange nachher wiedergelesen und ein außerordentliches Vergnügen daran gehabt.)

These intensive studies in logic can be seen as the nucleus of research to come, also in mathematics: Leibniz looked for an 'alphabet of human thoughts', a universal language, called 'characteristica universalis', in which all contexts could be discovered and proved. The wise notations dx and $\int$ of differential and integral calculus and the resulting easy manageability which we could admire in chapter 1 are offsprings of the thoughtful quest for such a language.

Leibniz is now also occupied with medieval scholasticism, metaphysics and theology. In theology, Protestant and Catholic controversy theologists ('Kontroverstheologen' in German)[8] stood in his focus, and although Leibniz always followed the Lutheran faith he will later advocate a reunion of the confessions – without success.

3.2.2 The Student

Leibniz changed to Leipzig University at Easter 1661. We do know his university professors [Müller/Krönert 1969, p. 6] and hence know that he attended an introductory course of Euclidean geometry by Johann Kühn (1619–1676). It is barely imaginable that other students except Leibniz were able to follow the lectures. Later Leibniz writes about the deplorable standard of the course and remarks that if he had grown up in Paris like Pascal he would have been able to enrich sciences much earlier [Aiton 1991, p. 29]. His professor of philosophy, Jakob Thomasius (1622–1684), however, he estimated highly. Thomasius became the supervisor of his Bachelor thesis *Disputatio metaphysica de principio individui* (Metaphysical Treatise on the Principle of Individuality), published in 1663. Thomasius judged his pupil [Müller/Krönert 1969, p. 6]:

> *'The highly learned, young Gottfried Wilhelm Leibniz enjoyed lines of thought in which he could exercise his wits and his zeal. Despite his youth he is up to the most complicated and extensive controversies.'*
>
> (Der hochgebildete, junge Gottfried Wilhelm Leibniz hatte Freude an Gedankengängen, in denen er seinen Verstand und seinen Eifer üben konnte. Trotz seiner Jugend ist er den schwierigsten und weitläufigsten Kontroversen gewachsen.)

The Bachelor thesis already contains the nucleus of the later theory of monades [Leibniz 2005]; in fact the notion of a monade, entering philosophy through Leibniz, goes back to his teacher Thomasius [Aiton 1991, p. 30].

[8]Today we would speak of 'oecumene theologists'.

Fig. 3.2.3. Seal of the University of Leipzig, Erhard Weigel

On 29th June 1663 Leibniz enrolled at the University of Jena. Well into the 20th century it will remain customary for students to leave their university and study at another for one or more semesters, and Leibniz chose Jena probably because Erhard Weigel (1625–1699) taught mathematics there. Weigel certainly was no first-rate mathematician but he was also a moral philosopher and an excellent thinker concerning natural law. With the help of mathematical methods he examined propositions of scholastic philosophy and mocked them, thereby freeing philosophy finally from scholastic disputations. It is out of the question that Leibniz learned a lot from Weigel. At this time Leibniz also became a member of the academic society 'Societas quaerentium' in which Weigel weekly gathered professors and students alike, to exchange opinions and to discuss about books. Similar gatherings also took place in Leipzig and Leibniz attended these too [Aiton 1991, p. 33].

Beginning with the winter semester in October 1663, Leibniz is back in Leipzig and starts his juridical professional studies. Because of his knowledge in history and philosophy the study of law was no problem for him. A friend who was a trainee lawyer at the supreme court in Leipzig showed him by means of examples how judgement was passed in the juridical praxis. Although Leibniz was certainly attracted by the post of a judge, he shied away from the intrigues of the lawyers. At the start of the year 1664, on 7th February, he became Master of Philosophy with the thesis *Specimen quaestionum philosophicarum ex jure collectarum* which was published in December. Then his mother died on 16th February *'following a catarrh which tied up her respiratory tracts'*[9] [Müller/Krönert 1969, p. 7]. Aiton writes [Aiton 1991, p. 27]:

[9] 'infolge eines Katharrs, der ihr die Atemwege zuschnürte.'

> *'Her virtue and piety shaped Leibniz from an early age. Hence he much resembled his mother in his admirable moral attitude.'*
>
> (Ihre Tugend und Frömmigkeit prägten Leibniz von Kind an. So glich er in seiner bewundernswerten sittlichen Haltung der Mutter in vielem.)

This citation comes from an English expert on Leibniz who is not under suspicion to transfigure Leibniz on nationalistic grounds. The older Leibniz certainly had his rough edges, but in contrast to Newton Leibniz seemed to have posessed an evenly and friendly temper.

At the beginning of 1665, Leibniz became a member of the 'Societas conferentium' in Leipzig and accepted the office of treasurer twice. On 28th September he acquired the academic degree of a 'juris utriusque baccalaurius'; he became baccalaureate of both laws, church as well as constitutional law. Leibniz now turned to his habilitation thesis for the philosophical faculty. 'Habilitation thesis' must not be understood in the modern sense, but it actually was another thesis following the Master thesis, which with one could apply for positions at a university [Leonhardi 1799, p. 568f.]. His thesis *Disputatio arithmetica de complexionibus*, with which he habilitated on 17th March 1666, can be seen as a preparatory study of his *Dissertatio de arte combinatoria* which appeared 1666. The *Disputatio arithmetica de complexionibus* is distinguished by great originality and was received positively by the scholars of the time. Later, Leibniz will call this thesis an 'early work' and regret that he published it before having studied mathematics [Aiton 1991, p. 35]. Just as a reminder: We have called the years 1664-1666 the 'anni mirabiles', the magic years, of Newton, in which he accomplished unimaginable progress in mathematics. At this time, Leibniz is still far away from the modern mathematics of his time!

In the *Dissertatio de arte combinatoria* Leibniz further developed the idea of an 'alphabet of human thoughts' of his schooldays. Following Leibniz the 'ars combinatoria' is a logic of discoveries which is different from the syllogistic logic of Aristotle. Leibniz is influenced by the 'Great Art' of Raimundus Lullus (about 1232–1316), who had derived a table of categories consisting of six rows and nine absolute attributes, nine relations, nine questions, nine virtues, and nine vices. He then derived relations between them with the help of a mechanical tool. Leibniz's construction is essentially identical to Pascal's triangle, but in a completely new arrangement. It allows for the translation of, say, geometrical notations into symbols, so that geometric definitions result from different combinations. This is the first big step towards the 'characteristica universalis', the great dream of Leibniz, in which one could think and carry out proofs by means of mathematical algorithms. In an appendix Leibniz gave a cosmological proof of God in the form of an Euclidean proof, and most likely it was Weigel's lecture in Jena which inspired him [Aiton 1991, p. 41].

For a young scholar well-known in Leipzig and beyond, the PhD promotion would have been the next step. This was not possible in Leipzig, however. Assessors of the faculty for law were convened in the order of their appointment to their Doctor title. Hence, the assessors had no interest whatsoever in seeing a whippersnapper like Leibniz overtake them. Using the argument of Leibniz being too young, the promotion to a Doctor of both laws was declined.

3.2.3 The Young Doctor Utriusque Iuris

'When I noticed the intrigue of my competitors I changed my plans and decided to go on a journey and study mathematics. For I was of the opinion that a young man be not allowed to dwell at one and the same place like he were nailed on to it, and already for a long time I was aspiring to gaining larger fame in the sciences and to getting to know the world.'

(Als ich die Intrige meiner Konkurrenten bemerkte, änderte ich meine Pläne und beschloß, auf Reisen zu gehen und Mathematik zu studieren. Denn ich war der Meinung, ein junger Mann dürfe nicht wie angenagelt an ein und demselben Ort verweilen, und schon lange brannte ich vor Begierde, größeren Ruhm in den Wissenschaften zu erwerben und die Welt kennenzulernen.)

This is what Leibniz [Müller/Krönert 1969, p. 9] wrote before his farewell from Leipzig at the end of September, 1666. On 4th October he enrolled at the University of Altdorf close to Nuremberg in the law faculty.

The University of Altdorf had its flowering period from after the Thirty Years' War well into the first quarter of the 18th century. When Nuremberg and hence also Altdorf fell to the Kingdom of Bavaria in 1806 came the last hours of the university: on 24th September 1809 the University of Altdorf was dissolved by King Maximilian I Joseph. Altdorf had become famous through its student Albrecht von Waldstein, called 'Wallenstein', who enrolled there on 29th August 1599. As a student he was involved in the murder of an ensign of the Altdorf militia; he also heavily mistreated his own servant with a whip and eventually left the university in 1600 without any degree.

Leibniz had finished his PhD dissertation already in Leipzig and submitted it now at Altdorf University. The title was *De casibus complexis in jure* (On difficult legal cases) and was published already in November 1666. On 15th November Leibniz wrote [Müller/Krönert 1969, p. 9]:

'Shortly after I acquired the degree of a doctor at the Nuremberg university under greatest applause, being 21 years of age.'

(Kurz darauf erwarb ich an der nürnbergischen Universität mit einundzwanzig Jahren den Doktorgrad unter größtem allgemeinen Beifall.)

Fig. 3.2.4. The University of Altdorf in the year 1714

The actual ceremonious promotion to the doctor of both laws took place on 22nd February 1667 with two talks given by Leibniz with great approval. He received congratulation poems from Altdorf, and from Leipzig; also from his teacher Jakob Thomasius. Leibniz was so very successful that it was proposed to keep him in Altdorf as a professor. An offer came from Johann Michael Dilherr who was the head of the educational system of the town of Nuremberg, but Leibniz declined, since [Müller/Krönert 1969, p. 10]:

'... my mind moves in a totally different direction.'

(... mein Geist bewegt sich in einer ganz anderen Richtung.)

On 2nd October 1666 he had already issued a promissory note in the amount of 30 Taler to his sister Anna Katharina Löffler; on 14th February 1667 one follows to a relative of his and asset manager of the family, Christian Freiesleben, in the amount of 30 Taler. When he moved to Nuremberg shortly after his promotion, probably in March 1667, he issued Christian Freiesleben another promissory note on 60 Taler, and on 14th July yet another on 40 Taler. From spring to summer 1667 Leibniz consorted in the house of the protestant theologist Daniel Wülfer (1617–1685).

Wülfer was the president of an alchemist society in Nuremberg and he installed Leibniz as the secretary, who had to document the course of experiments. Guhrauer [Guhrauer 1966, Vol. 1, p. 46] speaks about a society of

Fig. 3.2.5. Daniel Wülfer (town archive/town library Trier, portrait collection signature: Port 4383 [Artist: Peter Troschel (1620–1667)]) and Johann Philipp Schönborn (Elector and archbishop of Mainz. *Beschreibung und Abbildung Aller Königl. und Churfürstl. Ein-Züge, Wahl und Crönungs Acta...* [Merian 1658])

Rosicrucians, but we have no evidence to support this; the older biography by Eckhart [Eckhart 1982] contains no clear attribution of this secret society. While being in the service of this society Leibniz excerpts works of famous alchemists. In contrast to Newton Leibniz is indeed open in regard to alchemistic experiments, but is fundamentally sceptical. He wrote [Müller/Krönert 1969, p. 11]:

> *'I also did not show a lack of curiosity, but kept it within limits due to the necessary critique. I have seen that [Johann Joachim] Becher and other people well-known to me suffered shipwreck, while they believed to be sailing with the favourable wind of their alchemist dreams.'*
>
> (Auch ließ ich es nicht an Neugier fehlen, hielt sie jedoch durch die gebotene Kritik in Grenzen. Ich habe gesehen, wie Becher und andere mir sehr bekannte Leute Schiffbruch erlitten, während sie mit dem günstigen Winde ihrer Alchemistenträume zu segeln glaubten.)

Eventually Leibniz departed. He wanted to go to Holland and even further, but this did not happen. He wrote [Müller/Krönert 1969, p. 11]:

> *'In my 21st year I was promoted to doctor of both laws with great approbation and then went on a journey. But when I came through Mainz, having in mind to go to Holland and further, I became employed by the in those days famous elector Johann Philipp, who kept*

me, gave me a post in his revision council and other graces, and there I started to correspond with the most learned men in Germany and abroad.'[10]

(Im 21. jahr habe ich in utroque jure mit ungemeiner approbation promovirt, und darauff meine reisen angetreten, als ich aber durch Maynz passiret, der meinung nach Holland und weiter zu gehen, bin ich bey dem damahligen berühmten Churfürsten Johann Philipp in Kundschafft kommen, der mich bey sich behalten, mir munus Consiliarii Revisionum conferiret und andere gnaden gethan, und habe alda angefangen mit den gelehrtesten Leuten in und außer Teutschland zu correspondieren.)

3.2.4 Lawyer and Diplomat

Johann Philipp von Schönborn (1605–1673) was elector and archbishop of Mainz from 1647 onwards. He had shown himself willing to compromise with the Protestants.

Baron Johann von Boineburg was the long-serving minister of Schönborn in Mainz before a French intrigue overturned him in 1664. Approximately at the time when Leibniz arrived in Mainz, it came to a reconciliation by reason of the marriage of Boineburg's oldest daughter with the elector's nephew. But Leibniz got to know Boineburg even earlier, since in a letter of Boineburg to the polymath and personal physician to the Swedish Queen Kristina, Hermann Conring (1606–1681), he wrote in April 1668 [Müller/Krönert 1969, q. 13]:

'I know the author [i.e. Leibniz] *very well, he is a doctor of law, twenty-two years old, very educated, a rare philosopher, full of endurance, able in speculative trains of thought, and determined.* [...] *He now lives in Mainz, not without my intercession.'*

(Ich kenne den Autor [d.i. Leibniz] sehr genau, er ist Doktor der Rechte, zweiundzwanzig Jahre alt, sehr gebildet, ein vortrefflicher Philosoph, voll Ausdauer, zu spekulativen Gedankengängen fähig und entschlossen. [...] Er wohnt jetzt in Mainz, nicht ohne meine Fürsprache.)

We don't have any knowledge on the 'when and where' of the first meeting between Leibniz and Boineburg. On his journey, in inns and pubs Leibniz had composed the essay *Nova methodus discendae docendaeque jurisprudentiae* (A new method of teaching and learning in jurisprudence), the first edition of which was published in May 1667 by the printer Johann David Zunner

[10] The original text is in a quite old German with interspersed Latin pieces. I have given a modernised translation which only captures the meaning.

in Frankfurt am Main [Antognazza 2009, p. 82]. Hence, the first meeting of Leibniz with Boineburg was presumably located in Frankfurt. Leibniz handed his work over to the elector personally and thereby came into service on the basis of a weekly fee, in order to help privy councillor Hermann Andreas Lasser with his revision of the Roman law. Here he could really show his juridical knowledge.

Leibniz is also employed by Boineburg and with time a personal friendship developed between the two men. When Boineburg is sought to go on a mission to influence the election of the Polish king for the benefit of the Palatine of Neuburg, Leibniz supported him with a little work which he attributed to an unknown Polish nobleman. Although the writing was published after the election (The palatine was not elected) we can be sure that Boineburg counted on it in his arguments and talks [Aiton 1991, p. 47f.].

Boineburg was a pious man who had converted to Catholicism. Although Leibniz stayed a Lutheran the two men agreed on the idea to reunite the confessions. Leibniz wrote a riposte *Defensio Trinitatis per nova reperta logica* (Defense of Trinity by means of recently found arguments) to a writing from the unitarian Andreas Wissowatius to Boineburg. In this riposte he clearly layed open the logical errors of Wissowatius. In the fight against the enemies of Christendom and religion Boineburg and Leibniz were also united. Boineburg sent Leibniz's essay against Wissowatius to friends, and the theologist Gottlieb Spitzel from Augsburg printed the essay in 1669 without mentioning Leibniz's name under the title *Confessio naturae contra atheistas* (Confession of nature against the atheists). In this writing we already find the idea of Leibniz's important principle of sufficient reason and of the pre-established harmony.

When Boineburg returned from Poland he instructed Leibniz to provide a new edition of the book *Anti-Barbarus seu de veris principiis et vera ratione philosophandi contra Pseudophilosophos* (Anti-barbarian or on the true principles and the right way of thinking in philosophy against the pseudo-philosophers) which was written by the Italian Marius Nizolius and published in 1553. This new edition was published on the occasion of the book fair of 1670. Leibniz had accompanied Boineburg already the year before to Bad Schwalbach where Boineburg followed a course of treatment. There Leibniz met the lawyer Erich Mauritius who pointed him to the publications of Christopher Wren and Christiaan Huygens in the 'Philosophical Transactions' of the Royal Society. Thereby Mauritius gave him the impulse to write down the *Hypothesis physica nova* (A new physical hypotheses), which Leibniz began already in Bad Schwalbach [Aiton 1991, p. 54]. The final version appeared 1671 anonymously, but bears the initials G.G.L.L. on the title page. This final version consisted of two works which supplemented each other, the *Theoria motus concreti* and the *Theoria motus abstracti*, which were dedicated to the Royal Society and the Académie Royale des Sciences in Paris, respectively. In *Hypothesis physica nova* Leibniz supported a continuous motion without breaks. The central notion is that of *conatus* (endeavor). Any moving body is

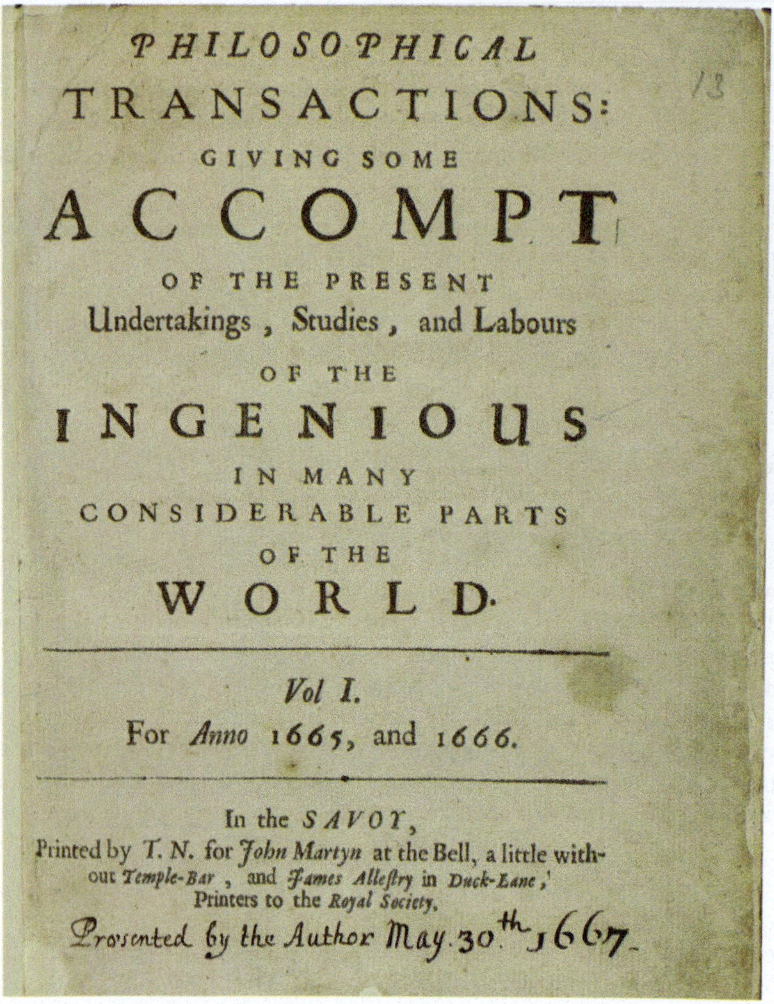

Fig. 3.2.6. Title page of the first volume of the 'Philosophical Transactions', issued by Henry Oldenburg, 1666

anxious to conserve its *conatus*. If a moving body is stopped by other bodies, the *conatus* moves the stopping bodies. A body which moves along a curve has the *conatus* to move along the curve's tangent. In a letter to Heinrich Oldenburg Leibniz attached great importance to the fact that he was the first to discover this principle. Shortly after the publication of the *Hypothesis physica nova* in mid-1671 the Royal Society published a new edition, and Heinrich Oldenburg revealed the author to Christiaan Huygens. Oldenburg wrote [Müller/Krönert 1969, p. 22]:

Fig. 3.2.7. Duke John Frederick [unknown painter] and the young Leibniz (cienciart files wordpress.com)

> *'He does not seem to be an ordinary mind, but he has examined in detail, what the great philosophers of antiquity and of modern times have said about nature, and since he found many open problems he has tried to solve them. I cannot judge to what extent he was successful. Nevertheless I can safely say that his thoughts are worth considering. Among other things he made observations about the laws of motion, as they were formulated by yourself and Mr. Wren.'*
>
> (Er scheint kein gewöhnlicher Geist zu sein, sondern er hat genau geprüft, was die großen Philosophen der Antike und der neueren Zeit über die Natur gesagt haben, und da er viele offene Probleme vorfand, hat er versucht, sie zu lösen. Ich kann nicht beurteilen, inwieweit er Erfolg gehabt hat. Dennoch wage ich zu behaupten, daß seine Gedanken erwägenswert sind. Unter anderem stellt er Überlegungen über die Gesetze der Bewegung an, wie sie von Ihnen und Herrn Wren aufgestellt worden sind.)

At approximately the same time Leibniz got in contact with the Royal librarian in Paris, Pierre de Carcavi, and referred to his calculating machine for the first time. Carcavi and the mathematician Jean Gallois wanted to promote Leibniz to become a member of the French Academy of Sciences, and they suggested his coming to Paris in person. Carcavi asked Leibniz to send his calculating machine to Paris in order to present it to minister Colbert [Aiton 1991, p. 61].

At the end of the year 1669 Leibniz received the first invitation of Duke John Frederick to come and work in Hanover. Leibniz declined; Hanover had become a residential town only in 1636 and was a small, remote one-horse town without any importance. But Duke John Frederick admired Leibniz and Hanover would eventually become his final destiny. For the time being he wanted rather to work with Lasser on the Roman law. In May 1671 Leibniz sent two religious writings as well as a short description of his ideas concerning the *Hypothesis physica nova* to Duke John Frederick in Hanover. A further letter to the Duke followed in October 1671 whom Leibniz might have met at the beginning of the month in Frankfurt am Main in person [Aiton 1991, p. 62]. He reports to the Duke that he wanted to go to Paris, that Colbert wanted to see his calculating machine, and he asks John Frederick for a letter of recommendation.

In August 1670 Leibniz accompanied Boineburg again to Bad Schwalbach and they talked about how peace could be kept in Europe. These talks, put on the agenda due to the hegemony of France under Louis XIV, constitute the impulse towards a daring manoeuvre of distraction, the *Consilium Aegyptiacum*, which will eventually lead Leibniz to Paris.

The Duke of Lorraine felt (justifiably!) threatened by Louis XIV and desired an alliance with the Electorate of Mainz. After consulting Leibniz the desired alliance was declined by Boineburg. A few weeks later the Duke of Lorraine was expelled. Boineburg had private reasons to stay on good terms with the French king, since there were pension and lease payments overdue. In December 1671 an envoy of Louis appeared in Mainz asking for unhindered passage of French ships on the Rhine, since France wanted to attack the Netherlands. To keep Louis from military action against Mainz Leibniz projected a plan for a kind of crusade to Egypt, just the *Consilium Aegyptiacum*, which he planned to present to the French court personally. A vaguely written note made the court in Paris so curious that the French foreign minister asked for further information. In a letter dating 4th March 1672 Boineburg informed the minister that he he would send Leibniz.

Already on the 19th March 1672 Leibniz started in the direction of Paris; twelve days later he arrived there. He was equipped with all the powers of Boineburg to care for all claims concerning pensions and rents, a travel advance, and a letter of recommendation for the minister. Shortly before the journey, Leibniz's sister Anna Catharina Löffler had died.

Activities of the Adolescent Warriors

1661	ISAAC NEWTON enrolls at Trinity College, Cambridge
1661	GOTTFRIED WILHELM LEIBNIZ enrolls at the University of Leipzig
1664-1666	NEWTON's so-called 'anni mirabiles'
1664	NEWTON starts his notebook *Quaestiones quedam Philosophicae*
1664	LEIBNIZ becomes MA
1664-1665	NEWTON intensively studies the writings of VIETA, VAN SCHOOTEN, DESCARTES and WALLIS
1664	NEWTON wins a scholarship at Trinity College
1664	ROBERT BOYLE publishes *Experiments and Considerations Touching Colors*
1664 or 1665	NEWTON achieves a breakthrough concerning a general theory of quadrature
1665	ROBERT HOOKE's *Micrographia* is published in London
1665	NEWTON discovers the method of infinite series and with it computes the area under a hyperbola
1666	NEWTON writes the 'October tract' in which he summarises his results concerning the calculus of fluxions virtuously. He works on a theory of colours and on the theory of gravitation
1666	LEIBNIZ publishes the *Dissertatio de arte combinatoria*
Autumn 1666	LEIBNIZ leaves the University of Leipzig, since promotion is refused due to his youth. He enrolls at the University of Altdorf
1667/68	NICOLAUS MERCATOR publishes the *Logarithmotechnia* containing the quadrature of the hyperbola
1667	NEWTON becomes 'minor fellow' at Trinity College
1667	LEIBNIZ becomes doctor of both laws at the University of Altdorf. He joins the service of the elector of Mainz as a lawyer
1668	NEWTON becomes 'major fellow' at Trinity College
1669	ISAAC BARROW sends NEWTON's manuscript *De analysi per aequationes numero terminorum infinitas* to JOHN COLLINS in London
1669	NEWTON becomes BARROW's successor on the Lucasian Chair
1669	NEWTON invents the reflecting telescope
Winter 1670	NEWTON starts writing the manuscript *De methodis*
1672	NEWTON becomes member of the *Royal Society*. He submits a manuscript on the theory of light and the emergence of colours to the *Royal Society*, whereupon a priority dispute with ROBERT HOOKE breaks out
1672	LEIBNIZ travels to Paris in political mission

4 The Cold War Begins

T. Sonar, *The History of the Priority Diſpute between Newton and Leibniz*, https://doi.org/10.1007/978-3-319-72563-5_4

4.1 The Mathematician: Leibniz in Paris

Our actual story starts only now. It was the end of March 1672 when Leibniz arrived in Paris. He had no inkling of mathematics, he did not know Newton (and *vice versa*!), but he was quite confident concerning his own skills. Within a few years to come a cold war will develop which will turn hot only much later.

At the end of March England had began to wage war on the Netherlands; France entered into the war a week later, so that the secret political mission of Leibniz became invalid – Europe was at war again. Even an audience with the foreign minister was not granted. Leibniz had good contacts to the Jansenist Antoine Arnauld (cp. section 2.6) and he stayed in Paris to push Boineburg's pension claims through, and also to study further. In the first days of November Boineburg informed Leibniz, that his son Philipp Wilhelm

Fig. 4.1.1. At the end of the Eighty Years' War the north of the Netherlands achieved its independence in 1648. The south remained with Spain. England and the Netherlands waged war twice between 1652 and 1668; they were competing sea powers. France under Louis XIV waged war against Spain 1667/68 to win the Spanish Netherlands for themselves. This was prevented by an alliance (England, Sweden, Netherlands). In 1672 French and English troops attacked the Netherlands. England withdraw from this war in 1674. In the peace treaty of 1678 the Netherlands vowed to stay neutral; Louis XIV had extended his power without having finally defeated the Netherlands militarily

[Map: Wesemüller-Kock]

Fig. 4.1.2. Louis XIV crosses the Rhine 1672 at Lobith near Nijmwegen ([Painting by A. Frans van der Meulen, 1690] Rijksmuseum Amsterdam SK-A-3753). The Franco-Dutch War continued for another six years and many European countries got involved.

would come to Paris and that Leibniz should oversee his son's studies. On 16th November young Boineburg arrived, but Leibniz will soon give up his role as a governer. Philipp Wilhelm von Boineburg was accompanied by Melchior Friedrich von Schönborn (1644–1717), nephew of the elector and son-in-law of Boineburg. Then Johann Christian von Boineburg suddenly suffered a stroke and died on 15th December. Leibniz had lost not only a good friend, but also a strong supporter.

Besides Antoine Arnauld, Leibniz kept company with the royal librarian Pierre de Carcavi who had asked to send his calculating machine. Leibniz worked with high pressure on this machine and would present a still not perfectly functioning model at the beginning of 1673[1].

In autumn 1672 Leibniz visited Christiaan Huygens who lived in Paris in a flat in the royal library. Huygens was at the height of his fame. Leibniz has described this visit in the *Historia et origo calculi differentialis* (History and

[1]As Mackensen wrote in [Mackensen 1973, p. 26]: *'Many different mechanics worked on the completion of a safely functioning calculating machine of Leibniz until the death of its inventor, ...'* ('An der Fertigstellung einer sicher funktionierenden Leibnizschen Rechenmaschine arbeiteten jedoch verschiedene Mechaniker noch bis zum Tode des Erfinders, ...')

origin of the differential calculus), written as a reaction to the priority dispute very much later [Gerhardt 1846]. An English translation can be found in [Child 2005]. Huygens recommended to him reading John Wallis's *Arithmetica infinitorum* and Grégoire de Saint-Vincent's *Opus geometricum* which Leibniz immediately borrowed from the library but found to be very cumbersome [Hofmann 1974, p. 15]. Since Leibniz indicated that he found a method for the summation of infinite series [Aiton 1991, p. 70], Huygens presented him with a problem which he himself had solved in 1665 as a test, the summation of the reciprocal triangular numbers. Triangular numbers were already known in Pythagorean antiquity [Sonar 2011].

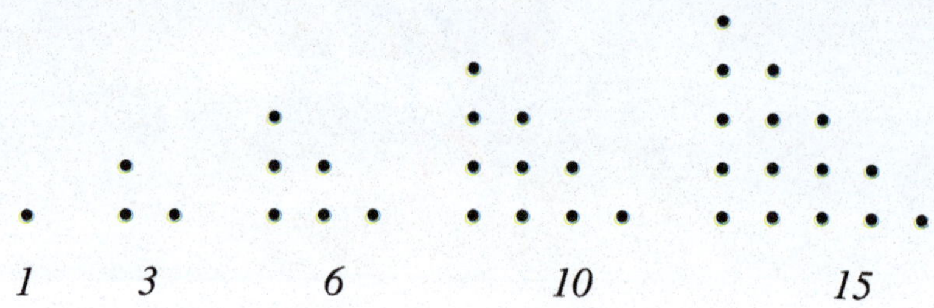

Fig. 4.1.3. The first five triangular numbers

They involve the number of points which result in triangles on an equidistant grid, as shown in figure 4.1.3. The law of construction of triangular numbers apparently is

$$D_n := \frac{n(n+1)}{2}.$$

Huygens's test hence consists of finding the value of the series

$$\frac{1}{1} + \frac{1}{3} + \frac{1}{6} + \frac{1}{10} + \ldots + \frac{1}{D_n} + \ldots$$

if this value exists. In fact, Leibniz had already dealt with sums of differences, i.e. with infinite series of the form

$$d_1 + d_2 + d_3 + d_4 + \ldots,$$

where every summand is given by the difference $d_i := a_i - a_{i+1}$. Since

$$\begin{aligned} d_1 + d_2 + d_3 + d_4 + \ldots + d_{n-1} =& (a_1 - a_2) + (a_2 - a_3) + (a_3 - a_4) + \ldots \\ & \ldots + (a_{n-2} - a_{n-1}) + (a_{n-1} - a_n), \end{aligned}$$

it follows $\sum_{i=1}^{n-1} d_i = a_1 - a_n$. Today we would call this sum a 'telescoping sum'. If the sequence (a_i) is a zero sequence, then the value of the infinite series $\sum_{i=1}^{\infty} d_i$ is just a_1,

$$d_1 + d_2 + d_3 + d_4 + \ldots = a_1. \tag{4.1}$$

The key to the solution of Huygens's problem lies in Pascal's triangle. Pascal's triangle

$n =$	
0	1
1	1 1
2	1 2 1
3	1 3 3 1
4	1 4 6 4 1
5	1 5 10 10 5 1
6	1 6 15 20 15 6 1
7	1 7 21 35 35 21 7 1
8	1 8 28 56 70 56 28 8 1
9	1 9 36 84 126 126 84 36 9 1

contains the coefficients of the binomial expansion $(a+b)^n$. In case $n = 2$ we read off the coefficients 1, 2 and 1, and in fact we know that

$$(a+b)^2 = 1 \cdot a^2 + 2 \cdot ab + 1 \cdot b^2 = a^2 + 2ab + b^2.$$

For $n = 5$ we read off

$$\begin{aligned}(a+b)^5 &= 1 \cdot a^5 + 5 \cdot a^4 b + 10 \cdot a^3 b^2 + 10 \cdot a^2 b^3 + 5 \cdot ab^4 + 1 \cdot b^5 \\ &= a^5 + 5a^4 b + 10a^3 b^2 + 10a^2 b^3 + 5ab^4 + b^5.\end{aligned}$$

Every entry in Pascal's triangle is the sum of the two numbers diagonally above to the left and right. Leibniz now constructed a triangle in which every entry is the difference between two neighbouring numbers, and he called it an 'harmonic triangle':

$n =$	
0	$\frac{1}{1}$
1	$\frac{1}{2}$ $\frac{1}{2}$
2	$\frac{1}{3}$ $\frac{1}{6}$ $\frac{1}{3}$
3	$\frac{1}{4}$ $\frac{1}{12}$ $\frac{1}{12}$ $\frac{1}{4}$
4	$\frac{1}{5}$ $\frac{1}{20}$ $\frac{1}{30}$ $\frac{1}{20}$ $\frac{1}{5}$
5	$\frac{1}{6}$ $\frac{1}{30}$ $\frac{1}{60}$ $\frac{1}{60}$ $\frac{1}{30}$ $\frac{1}{6}$
6	$\frac{1}{7}$ $\frac{1}{42}$ $\frac{1}{105}$ $\frac{1}{140}$ $\frac{1}{105}$ $\frac{1}{42}$ $\frac{1}{7}$

This harmonic triangle is difficult to read off. It is better to choose another arrangement like:

$$
\begin{array}{cccccccc}
\frac{1}{1} & \frac{1}{2} & \frac{1}{3} & \frac{1}{4} & \frac{1}{5} & \frac{1}{6} & \frac{1}{7} & \cdots \\
\frac{1}{2} & \frac{1}{6} & \frac{1}{12} & \frac{1}{20} & \frac{1}{30} & \frac{1}{42} & \cdots & \cdots \\
\frac{1}{3} & \frac{1}{12} & \frac{1}{30} & \frac{1}{60} & \frac{1}{105} & \cdots & \cdots & \cdots \\
\frac{1}{4} & \frac{1}{20} & \frac{1}{60} & \frac{1}{140} & \cdots & \cdots & \cdots & \cdots \\
\frac{1}{5} & \frac{1}{30} & \frac{1}{105} & \cdots & \cdots & \cdots & \cdots & \cdots \\
\frac{1}{6} & \frac{1}{42} & \cdots & \cdots & \cdots & \cdots & \cdots & \cdots \\
\frac{1}{7} & \cdots & \cdots & \cdots & \cdots & \cdots & \cdots & \cdots
\end{array}
$$

Convince yourself that every row is the difference sequence of the row directly above, and that every row is a zero sequence. Hence we can apply Leibniz's result (4.1) and get from the second row

$$\frac{1}{2}+\frac{1}{6}+\frac{1}{12}+\frac{1}{20}+\ldots=1,$$

for the third

$$\frac{1}{3}+\frac{1}{12}+\frac{1}{30}+\frac{1}{60}+\ldots=\frac{1}{2},$$

for the fourth

$$\frac{1}{4}+\frac{1}{20}+\frac{1}{60}+\frac{1}{140}+\ldots=\frac{1}{3},$$

and so on. Leibniz hence managed to sum in this way infinitely many infinite series, and the solution to the problem of Huygens's follows from the sum of the second row by multiplication with a factor 2:

$$\frac{1}{1}+\frac{1}{3}+\frac{1}{6}+\frac{1}{10}+\ldots=2.$$

Already before his supporter Boineburg died, a diplomatic visit in London was planned. Now that Boineburg was dead Leibniz undertook the journey together with Melchior Friedrich von Schönborn. On 21st January 1673 they arrived in Dover; they reached London probably on 24th January.

4.1.1 The First Journey to London

As soon as possible Leibniz visited his countryman Heinrich (Henry) Oldenburg, working as secretary of the Royal Society.

Oldenburg stemmed from Bremen where he was born probably in 1619 [Boas Hall 2002, p. 4]. Oldenburg – he is already the third with name Heinrich; father and grandfather shared the same first name – attended the Paedagogeum in Bremen, a classical Latin school, and changed in May 1633 to the 'Gymnasium Illustre'. Following the reformation new schools to replace the Catholic schools were founded in Protestant lands, to which Bremen belonged. A new school was launched in the rooms of Saint Catherine's monastery already in 1528. In the year 1610 the school was rearranged into the Paedagogeum consisting of six grades, and the further leading Gymnasium Illustre. This was not a school comparable to our secondary schools today, but replaced a university education in the four subjects law, medicine, philosophy, and theology. The Gymnasium Illustre in Bremen was liquidated only in 1810 by Napoleon Bonaparte; afterwards there were discussions about founding a university, but they led to nothing. Hence, the modern University of Bremen, having opened only in the winter semester 1971/72, is the first university following the Gymnasium Illustre in Bremen. Heinrich Oldenburg studied theology, which perhaps is not crucial for our story, but Marie Boas Hall points out in [Boas Hall 2002, p. 5], that Oldenburg was a master of Latin, had studied the seven liberal arts, and had far more than elementary knowledge of astronomy and mathematics. In the year 1641 he started to travel to the north of the Netherlands which was independent of Spain. In Utrecht he visited the university at which the new, radical, philosophical doctrines of René Descartes were taught. Following academic traditions in Holland he called himself Henricus Oldenburg, but he did not follow a university career, although he kept a love of academia throughout his life. From his letter book in which he noted his correspondence we may assume that he worked as a private tutor for some English pupils [Boas Hall 2002, p. 7f.] and that he conducted journeys to France, Italy, Switzerland, and Holland with them. Very likely he also visited England; at least he learned the English, Dutch, Italian and French languages so well that he could use them with great mastery in his later correspondence, and which he also spoke fluently.

He was again in Bremen in the year 1653 and was sent as a diplomat of the senate to England to negotiate with Oliver Cromwell about the neutrality of Bremen in the Dutch-English war. He changed his name to Henry Oldenburg, met John Milton (1608–1675), the great poet of *Paradise Lost*, the philosopher Thomas Hobbes (1588–1679), and numerous other philosophers and natural scientists; and Robert Boyle appointed him teacher of his nephew, with whom he would travel through Europe and visit some meetings of the group around Marin Mersenne in Paris, cp. sections 2.4 and 2.6.

When the Royal Society was founded in 1660, Oldenburg was on the list of members and was elected secretary, an office which he partially shared with John Wilkins. Following the model of Mersenne, he organised a network of correspondents in Europe with the centre in London. He also functioned as an editor of the journal 'Philosophical Transactions' of the Royal Society, which was published for the first time in March 1665.

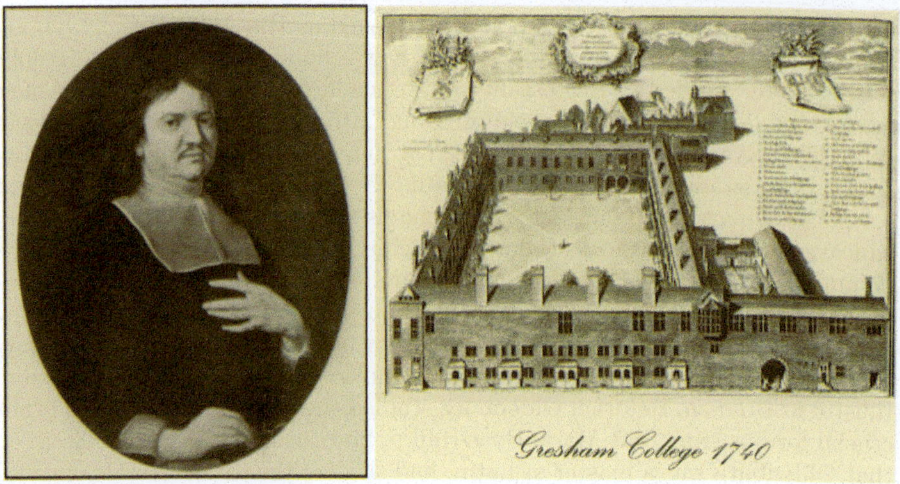

Fig. 4.1.4. Henry Oldenburg and Gresham College, in which the Royal Society was founded in 1660. Early members were Christopher Wren, R. Boyle, W. Brouncker, R. Hooke, S. Pepys, J. Wallis, Henry Oldenburg, and others

An interesting episode in Oldenburg's life is his imprisonment in the year 1667 [McKie 1948]. He had published under his nom de plume 'Grubendol' (an anagram of 'Oldenburg') during the second Anglo-Dutch war and was therefore suspected of being a spy. In 1663 Oldenburg married Dorothy West, a forty-year old woman from the parish of St. Paul Covent Garden [Boas Hall 2002, p. 80f.], of whom we know nothing more than that she died already in 1665 and that she left her husband a young ward named Dora Katharina Dury, being ten years old in 1664. After the death of his first wife, Oldenburg married this ward in 1668 – he being a man about 50, she a girl approximately 14 years old. The couple had two children, Rupert and Sophia [Boas Hall 2002, p. 312ff.]. Heinrich Oldenburg died on 5th September 1677 following a short, serious illness, of which we only know from the diary of Robert Hooke [Robinson/Adams 1955], that Oldenburg became sick on 3rd September and that he died two days later. It seems like a bad joke that we are informed about the illness and death of Oldenburg by Hooke, whose ever more bitterly growing quarrel with Oldenburg overshadowed the last year of the latter's life.

Already in 1670 Leibniz had taken up correspondence with the Royal Society, and Oldenburg had overseen the reprint of *Hypothesis physica nova*. Oldenburg had sent this book to Wallis, Hooke and Boyle for review. We know nothing about Boyle's reaction, Wallis was positive, and Hooke negative [Hofmann 1974, p. 23f.]. Hence Leibniz was not unknown within the Royal Society. He now wanted to present his calculating machine to the Royal Society, and he got his chance to do so during the meeting on 8th February 1673.

Fig. 4.1.5. Replika of the calculating machine of Leibniz (Gottfried Wilhelm Leibniz Library – Niedersächsische Landesbibliothek Hannover (Hanover), Leibniz's Four-Species-Calculating Machine)

Robert Hooke, the experimenter of the Royal Society, examined the model, which did not function correctly, thoroughly. Then he read about reflecting telescopes, thereby attacking Newton's construction. Leibniz was also present during the meeting on 8th February 1673, where Oldenburg read a letter by Sluse on tangents and where Hooke made further remarks on improvements of reflecting telescopes. More important to Leibniz, however, was a meeting with Robert Moray (1608 or 1609–1673), one of the most influential members of the Royal Society. It was Moray who informed Leibniz, that Samuel Morland had also build a calculating machine.

Thereupon Oldenburg arranged a meeting between Leibniz and Morland, on the occasion of which it is revealed that Morland's machine could not automatically multiply and divide, while Leibniz's machine was a complete four-function machine [Hofmann 1974, p. 25].

Leibniz was the guest of Robert Boyle some days after the meeting with Morland and showed his interest in chemical experiments. In the evening Leibniz got to know the mathematician John Pell (1611–1685), who, besides Wallis, was rated as one of the leading English mathematicians.

Pell was 60 years of age; he had liver trouble and was not in a good mood. Additionally, he had no sympathy whatsoever for the young man from the German lands [Hofmann 1974, p. 16]. Leibniz reported on his success concerning the summation and interpolation of difference series, unsuspecting what progress was made by the English in the field of summation of series. To Oldenburg Leibniz wrote [Malcolm/Stedall 2005, p. 221]:

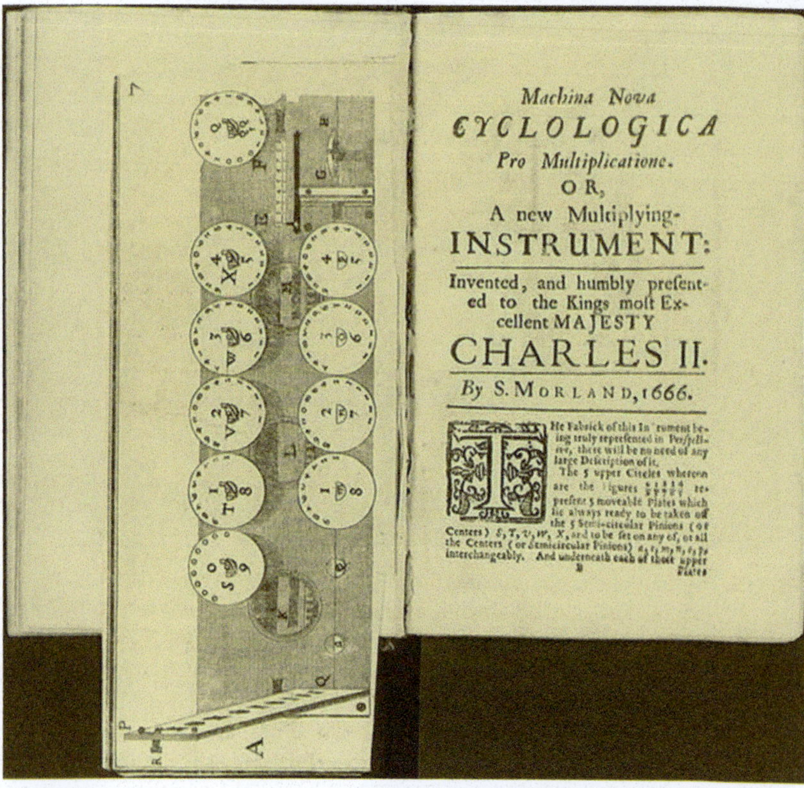

Fig. 4.1.6. Samuel Morland's calculating machine 1666

> *'When I was yesterday at the very illustrious Mr Boyle's, I met the famous Mr Pell, a notable mathematician, and the topic of numbers chanced to come up ...'*

Leibniz was completely surprised when Pell told him that the topic of difference series was already treated thoroughly in the book *De diametris apparentibus solis et lunae* by the French Gabriel Mouton, in which already results by Regnauld could be found. Pell made Leibniz look like a plagiarist and it came to a small scandal. Leibniz wrote further, that

> *'the famous Pell answered that this was already in print ... in the book by the meritorious Mouton, De diametris apparentibus solis et lunae.'*

Leibniz had never seen the book before,

> *'for which reason, picking it up at Mr Oldenburg's ... I ran through it hastily, and found that what Pell had said was perfectly true.'*

Fig. 4.1.7. Sir Samuel Morland ([Painting by P. Lely, 1645] Collection of Mr. James Stunt) and John Pell [Painting by Godfrey Kneller, 17th c.]

Leibniz clearly had failed and it was probably Oldenburg who advised him to leave a written explanation of the scandal at the Royal Society. It can be seen from the hastily written manuscript, how little was the knowledge of the mathematical literature by Leibniz [Hofmann 1974, p. 27].

> In the later indictment *Commercium epistolicum D. Johannis Collins et aliorum de Analysi promota* of the Royal Society from 1712 (date of print) Leibniz is accused of not even knowing the triangle of Pascals *Triangle arithmétique* at this time.

The feud with Pell had ended for the time being, but Leibniz was not invited to the next meeting of the Royal Society on 15th February. On the occasion of this meeting Hooke spoke only poorly about the calculating machine of Leibniz and promised to present his own, simpler model of such a machine. It was only fair that Oldenburg informed Leibniz concerning the negative report of Hooke, referring to Hooke as being a belligerent and cantankerous colleague. He urgently requested that Leibniz should quickly improve his machine. But the machine was sufficient to make Leibniz a foreign member of the Royal Society; on 19th April Leibniz was unanimously elected a member [Aiton 1991, p. 79].

At that time Schönborn wanted to leave London as soon as possible, and the departure came so fast that Oldenburg could not even say goodbye to Leibniz. Instead he wrote a letter and accompanied this with a letter to Huygens, containing the latest number of the Philosophical Transactions with Sluse's work concerning his tangent method [Sonar 2011, p. 281ff.].

4.1.2 The Aftermath of the Pell Affair

The Pell affair concerning difference series had a longer aftermath. Leibniz wanted to know from Pell, what the latter thought about his explanation of the scandal written for the Royal Society. As an answer to his request he only got a short note from Oldenburg. A more detailed answer followed later from John Collins (cp. page 52), who was Oldenburg's consultant in matters mathematical. The mathematical correspondence of the Royal Society with Wallis, Barrow, Gregory, Newton, and others went over Collins's desk. Collins had become irreplaceable within the Royal Society due to his good contacts with the booksellers of London, who published the works of English mathematicians quickly, and due to his contacts with the greatest scientists of England. He was a mediocre mathematician and did not understand much of the new developments, but he drafted letters (often those with mathematical contents) for Oldenburg, who never had learned the modern mathematics of his time. Collins was a nationalist dyed in the wool; he didn't like Descartes and the Cartesians and had reservations against the French in general. From his point of view only the achievements of his countrymen were of value – his scientific horizon was strictly limited [Hofmann 1974, p. 30f.]. Indeed, as Hofmann remarked, Collins's (questionable) conduct was partly responsible for the outbreak of the priority dispute more than a generation later.

Collins had not met Leibniz during the first visit to London of the latter. Leibniz now had to learn in the correspondence with Oldenburg – designed and drafted by Collins – that all series of figurate numbers were already included in Mengoli's *Novae quadraturae arithmeticae* printed in 1650. 'Figurate numbers' are numbers which correspond to certain geometrical figures; think of the triangular numbers. Pietro Mengoli (1626–1686) was an Italian priest

Fig. 4.1.8. Bonaventura Cavalieri, Pietro Mengoli

and mathematician who had studied with Cavalieri and who became his successor at the University of Bologna. At any rate, Leibniz had proved nothing new, only the way of proof was new. But it was not the Pell affair which ultimately led to English reservations about Leibniz, but the presentation of the calculating machine. During the meeting of 15th March 1673 Hooke had presented his own model of a calculating machine and had laid open the complete construction plans of it some weeks later. Thereupon Oldenburg now pressed Leibniz to eventually complete his construction and to lay open the construction plans, but Leibniz was occupied with other duties. Moreover, he already had answered his election as a foreign member of the Royal Society with only some hastily composed words of thank. He even had to be reminded in writing by Oldenburg that he had to confirm his election. All that left not a nice impression in London.

4.1.3 Leibniz Conquers Mathematics

Directly after his return to Paris Leibniz met the mathematician Jacques Ozanam (1640–1718) and discussed with him the solution of equations. Leibniz informed Oldenburg, asked questions about the theory of equations, and received an answer in a letter dated 16th March 1673. Not until one month later, on 16th April 1673, did Leibniz receive a detailed overview on the achievements of the English mathematicians concerning the solution of equations. The letter consisted of three parts, whereof the first two parts are of no interest to us. In the third part Collins gave an overview on the state of the art of mathematics, with a clear focus on English mathematics. Leibniz became aware of the works of Pascal, Mersenne, Descartes, Roberval, and others, and he began to be interested in the beginnings and foundations of the new mathematics. Even some of Newton's results are mentioned in the letter, but their description remained obscure and were certainly not intelligible even to experts.

In the eyes of Collins, Leibniz was a Francophile and he therefore mistrusted him, but he always stayed fair. The two men will meet in the year 1676; Leibniz will have changed then into a mathematician of first rank. Collins will be enchanted by Leibniz's charm.

The year 1673 marks the time of the great discoveries of Leibniz in the realm of mathematics. If the years 1664 to 1666 are the anni mirabiles of Newton, then the years 1672 to 1676 are the anni mirabiles of Leibniz, but in particular the year 1673 is *the* annus mirabilis for him. The elector of Mainz, Johann Philipp von Schönborn, had died on 12th February 1673. His successor was Lothar Friedrich von Metternich-Burscheid. Schönborn's nephew, Melchior Friedrich, left Paris immediately for Mainz, in order to have his office as Lord Marshall (Obermarschall) confirmed. Officially, Leibniz had nothing whatsoever to do in Paris, but the new regent allowed him to stay without having to be afraid of losing his appointment [Hofmann 1974, p. 46]. Philipp Wilhelm von Boineburg still lived in Leibniz's flat in Paris, but he wanted to enjoy the life in Paris to the full and was not interested in science. Hence the anger grew on all sides. Since the financial means of Boineburg's widow were meager she reduced the fee which was negotiated with Leibniz and finally Leibniz was dismissed from his services for the Boineburgs.

Already in spring 1673 the Duke of Brunswick-Lüneburg and the ruler of the Principality of Calenberg, John Frederick had offered Leibniz to come to Hanover for an annual salary of 400 thalers. But Leibniz dithered. The unknown small town of Hanover had become a residential town only recently; the town played no role on an international scale, and Leibniz saw himself more in cities like London or Paris, where the sciences were at home.

The letter which Oldenburg sent to Huygens via Leibniz was a good opportunity for the latter to make a visit to Huygens. Huygens was in the midst of work for the publication of his *Horologium oscillatorium.* He presented one copy to Leibniz, and remarked that this work went back to Archimedes' methods for the calculation of barycentres. Leibniz somewhat naively remarked, that a straight line through the barycentre of a convex figure would always divide the figure into two equal halves, which is simply not true. Huygens must really have felt something for the young lad, since he corrected him and directed his attention towards the corresponding works of Pascal, Gregory, Descartes, Sluse, and others. Retrospectively, Leibniz wrote to Ehrenfried Walther von Tschirnhaus (1651–1708) in 1679 [Child 2005, p. 215]:

> *'Huygens, as soon as he had published his book on the pendulum, gave me a copy of it; and at that time I was quite ignorant of Cartesian algebra and also of the method of indivisibles, indeed I did not know the correct definition of the centre of gravity. For, when by chance I spoke of it to Huygens, I let him know that I thought that a straight line drawn through the centre of gravity always cut a figure into two equal parts; since that clearly happened in the case of a square, or a circle, an ellipse, and other figures that have a centre of magnitude, I imagined that it was the same for all other figures. Huygens laughed when he heard this, and told me that nothing was further from the truth. So I, excited by this stimulus, began to apply myself to the study of the more intricate geometry, although as a matter of fact I had not at that time really studied the Elements. But I found in practice that one could get on without a knowledge of the Elements, if only one was master of a few propositions. Huygens, who thought me a better geometer than I was, gave me to read the letters of Pascal, published under the name of Dettonville; and from these I gathered the method of indivisibles and centres of gravity, that is to say the well-known methods of Cavalieri and Guldinus.'*

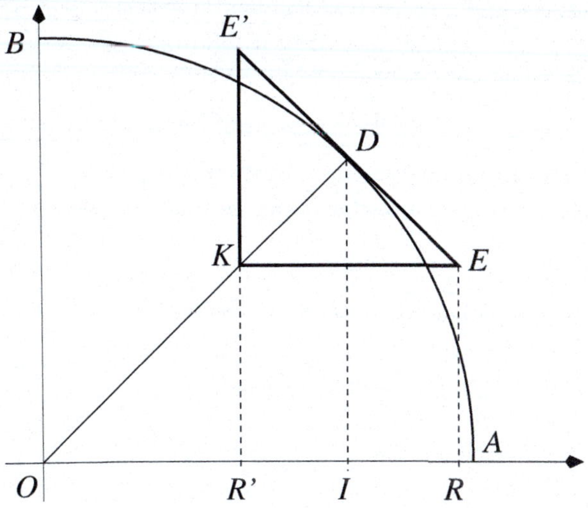

Fig. 4.1.9. The characteristic triangle of Pascal

Leibniz immediately borrowed the recommended books from the royal library and threw himself into a study of mathematics. Eventually, this will lead to the great discoveries of the year 1673.

To begin with, there is Pascal's writing *Traité des sinus du quart de cercle* (Treatise on the Ordinates of the Quarter Circle), cp. page 77f., where we have discussed this work. While reading this work, Leibniz experienced a flash of inspiration which is going to have unpredictable consequences! In his *Historia et origo* he wrote [Child 2005, p. 38][2]:

> *'Later on from one example given by Dettonville, a light suddenly burst upon him, which strange to say Pascal himself had not perceived in it.'*

Let us look again at Pascal's example in a somewhat simplified form in figure 4.1.9, cp. figure 2.6.9. The line segment $E'E$ is tangent to the circular arc ADB. The triangle $E'KE$ is similar to the triangle OID, hence it follows that

$$\frac{OD}{E'E} = \frac{DI}{EK},$$

i.e.

$$DI \cdot E'E = OD \cdot EK = OD \cdot R'R.$$

If we now call $y := DI$, $r = OD$, $\Delta s = E'E$ and $\Delta x = R'R$, then it follows that

$$y\Delta s = r\Delta x.$$

[2]Original [Gerhardt 1846] in Latin.

Understanding Δs and Δx as indivisibles, and looking at the 'entirety' of those lines, then in Leibnizian notation it follows that

$$\int y\,ds = \int r\,dx.$$

After the first rule of Guldin the area of the hemisphere, arising as a body of revolution by revolving the quarter circle around the abscissa, is just

Length of the arc ADB $\times$ Circumference of the circle arising by revolving the barycentre of the arc.

Let us think of the tangent segment $E'E$ as being infinitesimally small (i.e. 'an infinitely small step' in the language of our chapter 1), then, due to symmetry, the barycentre of $E'E$ is D, the circumference of the circle with radius y is $2\pi y$, and the 'length of the arc' here is just ds, hence the area of the infinitesimal frustrum of a cone is given by $2\pi y\,ds$. The sum of these areas of the infinitesimal truncated cones gives the surface of the hemisphere; hence

$$A := \int_{ADB} 2\pi y\,ds = 2\pi r \int_0^r dx = 2\pi r^2.$$

Leibniz's 'light' that 'suddenly burst upon him' when he saw this problem, consisted in the insight that the characteristic triangle could not only be used with a circle, but with arbitrary curves. The characteristic triangle at an arbitrary curve is shown in figure 4.1.10. The two triangles shown are similar; n is the normal to the curve, v the subnormal, cp. figure 3.1.9. Due to the similarity it holds that

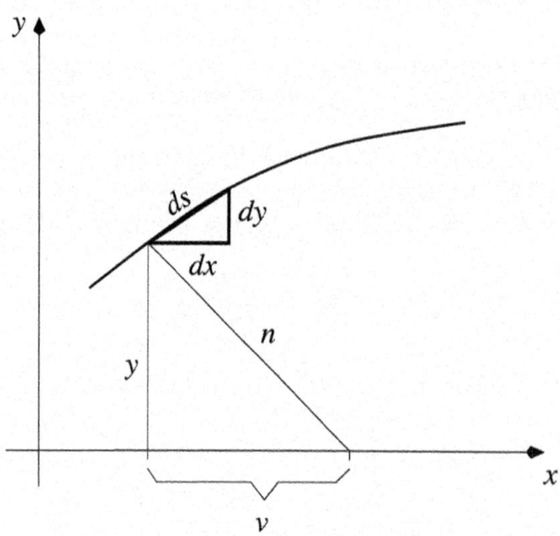

Fig. 4.1.10. The characteristic triangle of Leibniz

$$\frac{ds}{n} = \frac{dx}{y}, \quad \text{or} \quad y\,ds = n\,dx.$$

Summation of the infinitesimals then yields

$$\int y\,ds = \int n\,dx.$$

But be cautious! Leibniz will not find this notation until 1675.

Together with the study of the *Traité des sinus du quart de cercle* Leibniz also resumed the study of van Schooten's edition of the *Geometria* of René Descartes in two volumes. There he learned about the description of curves by means of coordinates and the most important theorems on equations, whereby Leibniz showed greatest interest in the arithmetical and graphical treatment of cubic and bi-quadratic equations. Leibniz also saw there the singular method of Descartes to determine the normal – and thereby the tangent – at curves; the circle method which we presented in figure 3.1.10. In the appendices accompanying the *Geometria* he also found van Heuraet's method of rectification.

The main focus of Pascal's *Lettres de A. Dettonville* was the cycloid. These *Lettres* were published as a challenge to other mathematicians in 1658 (cp. section 2.6). The *Lettres* were answered by John Wallis and Honoré Fabri in 1659, and by Antoine de Lalouvère in 1660. Pascal could not make up his mind to award a prize. What were sought were the areas under the cycloid and the barycentres; furthermore, the volume and barycentre of a body of revolution resulting from revolving the cycloid. Leibniz also studied the answers by Wallis, Fabri, and Lalouvère, but he did not seem to have taken notice of Torricelli's *Opera geometrica* in his time in Paris [Hofmann 1974, p. 51]. Probably Leibniz assumed the habit of drawing the abscissa (x) downwards in his coordinate systems – it becomes the ordinate with Leibniz – and the ordinate (y) to the right. Our modern coordinate system arises from Leibniz's through counter-clockwise rotation of 90°, and we will always use only diagrams in the modern form here.

The most important insight during these months in Paris was the theory of infinitesimals, which went far beyond the method of indivisibles of Cavalieri in that it does not interpret an area as the entirety of lines without breadth, but as a very large number of very small rectangles. Only sometime later Leibniz will read the *Arithmetica infinitorum* by John Wallis and will find there the concept of infinitesimals of equal breadth, cp. section 2.3 and there figure 2.3.8. However, Leibniz did not like Wallis's quadrature of the parabola. His 'interpolation' and the incomplete induction in particular could simply not be accepted by Leibniz. Nobody, not even Leibniz, saw the close connection between Pascal's techniques of quadrature and the ones of Wallis in his *Arithmetica infinitorum* [Hofmann 1974, p. 54].

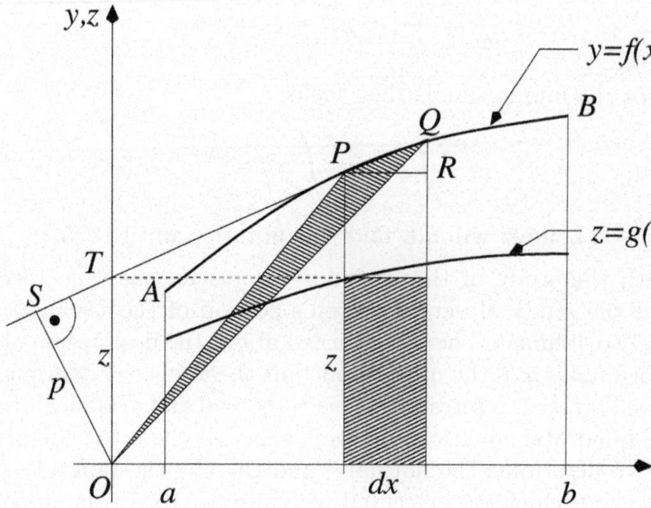

Fig. 4.1.11. Concerning the transmutation theorem

Leibniz excerpted Pascal's *Lettres*, and in doing so the idea seems to have grown slowly in him to build areas not only from infinitesimal rectangles, but also from infinitesimal triangles. This idea led him to the discovery of the transmutation theorem. Let P denote a point with coordinates (x, y) on a curve $y = f(x)$, and let Q be infinitesimally close with coordinates $(x+dx, y+dy)$. It is not difficult to recognise the characteristic triangle as being PRQ. Let us now, however, look at the triangle OQP. Continuing the tangent segment PQ, the tangent intersects the ordinate in the point T, having coordinates $(0, z)$. We can compute z, of course, since a straight line passing through the points $P = (x, y)$ and $(0, z)$ as the tangent does, is given by the equation

$$y = mx + z,$$

where we already know the slope m, since $m = dy/dx$. Hence, we can resolve the linear equation for z and get

$$z = y - \frac{dy}{dx}x. \tag{4.2}$$

Let the line segment OS be the normal to the tangent which intersects the latter in the point S and has length p. We now look for similar triangles and find OTS and PRQ. Then it holds

$$\frac{dx}{p} = \frac{ds}{z}, \quad \text{or} \quad p\,ds = z\,dx.$$

By elementary geometry the area of the infinitesimal triangle OQP is

$$a(OQP) = \frac{1}{2}p\,ds,$$

if we denote the length of the segment PQ by ds. We just saw that $p\,ds = z\,dx$, hence it follows

$$a(OQP) = \frac{1}{2} z\,dx.$$

Summing all these areas results in the area of the figure which is crosshatched in green in figure 4.1.12,

$$a(OBA) = \frac{1}{2} \int_a^b z\,dx.$$

The function z was already computed in (4.2), hence

$$a(OBA) = \frac{1}{2} \int_a^b y - \frac{dy}{dx} x\,dx$$

follows. But we are actually interested in the area under the curve $y = f(x)$ between $x = a$ and $x = b$, i.e. $\int_a^b y\,dx$, and hence the triangle crosshatched in red in figure 4.1.12 has to be added, but then the triangle OaA is too much! The area of the crosshatched red triangle is $\frac{1}{2} b f(b)$, that of triangle OaA is $\frac{1}{2} a f(a)$, and hence we get

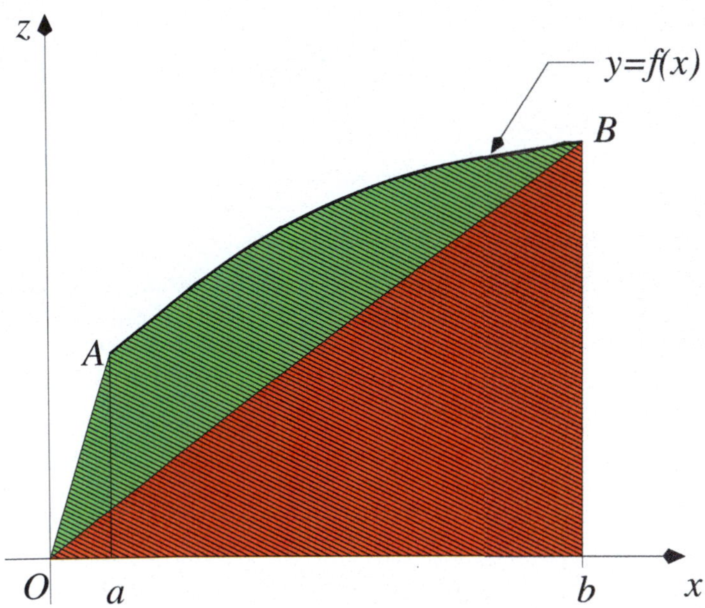

Fig. 4.1.12. Concerning the transmutation theorem

$$\int_a^b y\,dx = \frac{1}{2}bf(b) - \frac{1}{2}af(a) + a(OBA)$$
$$= \frac{1}{2}\,xy\big|_a^b + \frac{1}{2}\int_a^b z\,dx,$$

where we used the usual shorthand notation

$$xy\big|_a^b := ay(a) - by(b) = af(a) - bf(b).$$

Thereby we have deduced the transmutation theorem

$$\boxed{\int_a^b y\,dx = \frac{1}{2}\left(xy\big|_a^b + \int_a^b z\,dx\right).}$$

Inserting the term (4.2) for z yields

$$\int_a^b y\,dx = \frac{1}{2}\left(xy\big|_a^b + \int_a^b y - x\frac{dy}{dx}\,dx\right).$$

If we now pull apart the difference in the last integral and remember that we can simply reduce $\frac{dy}{dx}dx$ to dy (cp. chapter 1), then it follows

$$\int_a^b y\,dx = \frac{1}{2}\,xy\big|_a^b + \frac{1}{2}\int_a^b y\,dx - \frac{1}{2}\int_a^b x\,dy,$$

hence

$$\int_a^b y\,dx = xy\big|_a^b - \int_a^b x\,dy,$$

and that is the formula for integration by parts.

In the days of Leibniz there existed many transformation rules like the transmutation theorem, as Bos [Bos 1980, p. 64] has pointed out. In Barrow's *Lectiones geometricae* one can find a large number of such formulae, for example; however, they are in geometrical disguise and concerning the derivation not comparable with Leibniz's formula.

With the help of the transmutation theorem Leibniz was able to achieve many quadratures which were already known before. Looking at the functions

$$y^q = x^p, \quad q > p > 0,$$

then, since $y = x^{p/q}$, the differential quotient is given by $dy/dx = (p/q)x^{(p-q)/q}$. Multiplying with q and dividing by $y = x^{p/q}$ leads to

$$\frac{q}{y}\frac{dy}{dx} = \frac{p}{x}.$$

With a view to (4.2), it holds

$$z = y - \frac{dy}{dx}x = x^{p/q} - \frac{p}{q}x^{(p-q)/q}x = x^{p/q} - \frac{p}{q}x^{p/q} = \frac{q-p}{q}y.$$

Now it follows from the transmutation theorem

$$\begin{aligned}\int_a^b x^{p/q}\,dx &= \frac{1}{2}\left(xy\big|_a^b + \int_a^b z\,dx\right)\\ &= \frac{1}{2}\,xy\big|_a^b + \frac{1}{2}\frac{q-p}{q}\int_a^b y\,dx\\ &= \frac{1}{2}\,xy\big|_a^b + \frac{1}{2}\frac{q-p}{q}\int_a^b x^{p/q}\,dx.\end{aligned}$$

Moving the last integral to the left-hand side, it follows that

$$\frac{q+p}{2q}\int_a^b x^{p/q}\,dx = \frac{1}{2}\,xy\big|_a^b,$$

and therefore

$$\int_a^b x^{p/q}\,dx = \frac{q}{p+q}\,xy\big|_a^b = \frac{q}{p+q}\,x^{(p+q)/q}\Big|_a^b.$$

The most important application of the transmutation theorem is, however, the arithmetic quadrature of the circle, which Leibniz now started to work on. The problem of the 'quadrature of the circle' dates back to antiquity, and asks for the *construction* of a square with an area coextensive to that of a given circle. The word 'construction' means, that only a (mathematically perfect) compass without any scale of angles, and a (mathematically perfect) ruler without any length scale must be used [Sonar 2011, p. 39f.]. It is clear since the 19th century that this problem is not solvable, (wherein modern algebra plays a large role). In the case of the 'arithmetic quadrature of the circle' we are concerned with the calculation of the circle's area by means of convergent infinite series, the summands of which are but rational numbers, hence fractions [Knobloch 2002, p. 59].

A circle with radius 1 and centre $(1,0)$ can be represented as

$$y^2 + (x-1)^2 = 1.$$

Hence the semicircle shown in figure 4.1.13 is given by

$$y = \sqrt{2x - x^2}.$$

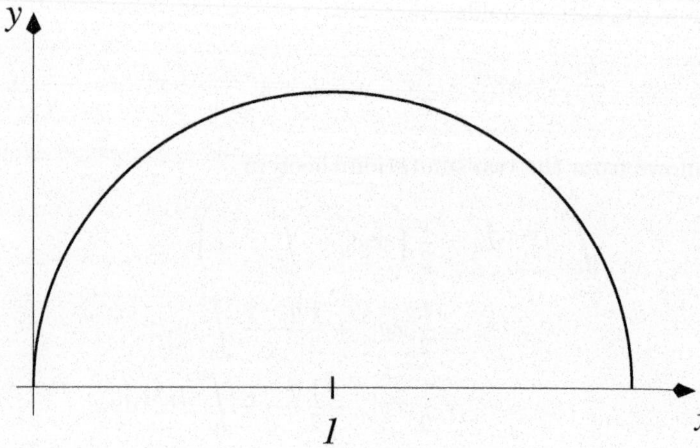

Fig. 4.1.13. Concerning the arithmetic quadrature of the circle

Since $y = (2x - x^2)^{1/2}$ it follows that the differential quotient is given by

$$\frac{dy}{dx} = \frac{1}{2}(2x - x^2)^{-1/2}(2 - 2x) = \frac{1 - x}{\sqrt{2x - x^2}} = \frac{1 - x}{y}.$$

From (4.2) we get

$$z = y - \frac{dy}{dx}x = y - x\frac{1 - x}{y} = \frac{y^2 - x + x^2}{y} = \frac{2x - x^2 - x + x^2}{\sqrt{2x - x^2}}$$
$$= \sqrt{\frac{x}{2 - x}},$$

so that

$$x = \frac{2z^2}{1 + z^2} \tag{4.3}$$

follows. Leibniz now applies his transmutation theorem to the quarter circle:

$$\int_0^1 y\,dx = \frac{1}{2}\left(xy\big|_0^1 + \int_0^1 z\,dx\right)$$
$$= \frac{1}{2}\left(x\sqrt{2x - x^2}\Big|_0^1 + \int_0^1 z\,dx\right). \tag{4.4}$$

Drawing the function $z = \sqrt{x/(2 - x)}$ between $x = 0$ und $x = 1$ yields a diagram as shown in figure 4.1.14. The graph of the function z divides the unit square in the yellow and red crosshatched parts, respectively. The crosshatched area in red is

$$\int_0^1 z\,dx,$$

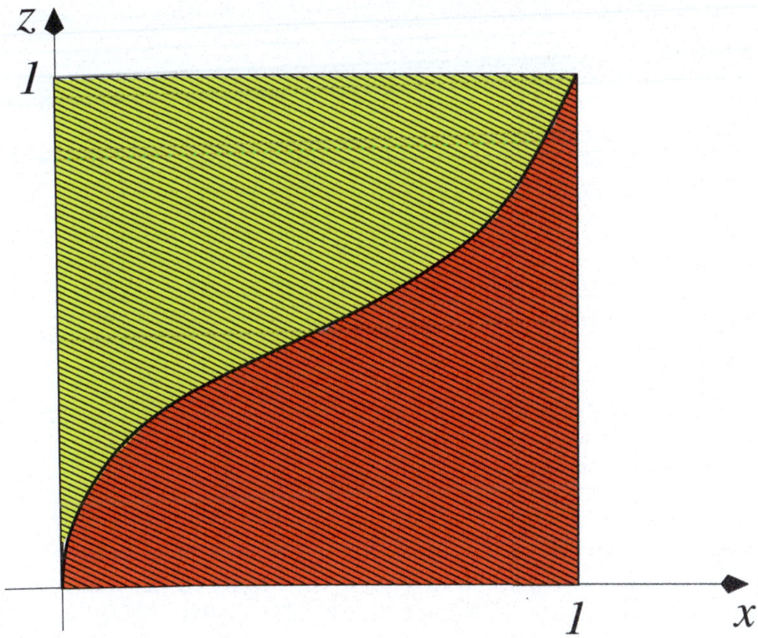

Fig. 4.1.14. The graph of z divides the unit square

the one which is crosshatched in yellow is

$$\int_0^1 x\,dz,$$

and therefore it holds

$$\int_0^1 z\,dx = 1 - \int_0^1 x\,dz.$$

Replacing thus the integral $\int_0^1 z\,dx$ in (4.4) yields

$$\begin{aligned}\int_0^1 y\,dx &= \frac{1}{2}\left(\underbrace{x\sqrt{2x-x^2}\Big|_0^1}_{=1} + \int_0^1 z\,dx\right)\\ &= \frac{1}{2}\left(1 + \left(1 - \int_0^1 x\,dz\right)\right)\\ &\overset{(4.3)}{=} 1 - \int_0^1 \frac{z^2}{1+z^2}\,dz.\end{aligned}$$

The term

$$\frac{1}{1+z^2}$$

is the sum of the geometric series $\sum_{k=0}^{\infty}(-1)^k z^{2k}$, which can be seen from the polynomial division

$$1:(1+z^2) = 1 - z^2 + z^4 - z^6 + z^8 \mp \ldots.$$

Incidentally, Leibniz here follows the techniques from Mercator's *Logarithmotechnia* [Hofmann 1974, p. 60]. Now Leibniz writes

$$\int_0^1 y\,dx = 1 - \int_0^1 \frac{z^2}{1+z^2}\,dz = 1 - \int_0^1 z^2(1 - z^2 + z^4 - z^6 \pm \ldots)\,dz$$

and integrates the infinite series term-wise:

$$\begin{aligned}\int_0^1 y\,dx &= 1 - \left(\frac{1}{3}z^3 - \frac{1}{5}z^5 + \frac{1}{7}z^7 \mp \ldots\right)\Big|_0^1 \\ &= 1 - \frac{1}{3} + \frac{1}{5} - \frac{1}{7} + \frac{1}{9} - \frac{1}{11} \pm \ldots.\end{aligned}$$

Since the days of Greek antiquity we know the area of the quarter circle; it is $\pi/4$, although the ancients didn't know the nature of the number π. There existed good approximations, though [Sonar 2011, p. 84f.]. With his arithmetic quadrature of the circle Leibniz has found the wonderful result

$$\boxed{\frac{\pi}{4} = 1 - \frac{1}{3} + \frac{1}{5} - \frac{1}{7} + \frac{1}{9} - \frac{1}{11} \pm \ldots}. \tag{4.5}$$

When Leibniz published this result in February 1682 on pages 41-46 in the newly founded journal 'Acta Eruditorum' under the title *De vera proportione circuli ad quadrantum circumscriptum in numeris rationalibus expressa* (The True Relationship of the circle to the circumscribed Square Expressed in Rational Numbers) [Leibniz 2011, p. 9–18], he added a citation of Vergil's 8th Eclogue 76 [Vergil 2001]:

> 'Uneven numbers are gods delight'
>
> (Numero deus impare gaudet)

Looking at Leibniz's mathematics of the year 1673, certain connections to Barrow's work may attract attention. Hence it may seem understandable that Child [Child 1916] thought to have found the foundation of Leibniz's mathematics in Barrow's writings. We can't accept this point of view today, however, because it is verifiably wrong. Certainly, Leibniz could have got Barrow's *Lectiones geometricae* from Huygens, but Huygens was seriously ill in 1670 and suffered a complete loss of his memory, cp. section 2.7. His brother Lodewijk went to Paris to take care for him, but following the advice of the attending physician Huygens went back to Den Haag in autumn 1670. He received the *Lectiones geometricae* from England in July 1670 but was so weak that he could not deal with it. In fact, he did not recommend its reading to his pupil. Leibniz had bought Barrow's *Lectiones geometricae* bound together with the *Lectiones opticae* during his first travel to London, whereby the latter part seems to have been of greater interest to him than Barrow's geometry. When he began to study the *Lectiones geometricae* in earnest he had already made his crucial discoveries in the infinitesimal calculus [Hofmann 1974, p. 75f.].

In the summer of 1674 there finally was a calculating machine available which had little in common with the rough wooden model of the year 1672. Leibniz had finally found an outstanding mechanic in Paris, the master craftsman Olivier. But even the machine of Olivier did not function without problems and further machines from later years all had mechanical insufficiencies. Leibniz wrote to Oldenburg [Turnbull 1959–77, Vol. I, p. 313ff.] on the 15th July 1674. He planned to present the machine personally to the Royal Society, but he did not know when he would be able to come to London. He also informed Oldenburg about some important mathematical theorems he had found, and mentioned the arithmetic quadrature of the circle and the quadrature of the cycloid segment (a problem from Dettonville's *Lettres*) in particular. He also pointed out that he already presented his results to the most famous geometers (in Paris), who had confirmed the originality of Leibniz's achievements. The letter was not posted by the ordinary postal service, but Leibniz entrusted the Danish nobleman Walter with the delivery, whom he had befriended in Paris. The letter was not only thought to serve as information for the Royal Society, but also to introduce Walter to Oldenburg. Months passed by without Leibniz getting an answer from London. We know, however [Hofmann 1974, p. 94, footnote 56], that a reply existed, because Oldenburg noted at the end of Leibniz's letter in French: *'received 12th July* [(jul.)] *1674, delivered by Walter, answered on 15th July* [(jul.)]'.

For want of an answer Leibniz wrote a further letter to Oldenburg on the16th October 1674 [Turnbull 1959–77, Vol. I, p. 322ff.]. He again pointed out his arithmetic quadrature of the circle [Turnbull 1959–77, Vol. I, p. 324]:

> *'You know that Viscount Brouncker and the celebrated Nicholas Mercator showed an infinite series of rational numbers equal to the area of an hyperbola. But up to now no one has been able to do this in a circle, although the celebrated Wallis and the illustrious Brouncker have produced rational numbers gradually approaching it. Nobody, however, has given a progression of rational numbers, the sum of which, continued to infinity, is exactly equal to a circle. This at last has come out successfully for me; for I have an extremely simple series of numbers whose sum exactly equals the circumference of a circle, given that the diameter is unity.'*

Leibniz did not know that James Gregory had sent this series (together with some others) already on the 15th February 1671 to Collins [Turnbull 1959–77, p. 61ff.]. Newton's results in *De analysi* of 1669 were already communicated to Leibniz by Oldenburg/Collins in the letter dated 16th April 1673, but their meaning and exact formulation of Newton's method could not be grasped by Leibniz from this letter.

In the later indictment *Commercium epistolicum D. Johannis Collins et aliorum de Analysi promota* of the Royal Society from 1712 (date of print) this report on Leibniz's progress was printed. As a comment it was noted that Collins had sent Newton's series to his friends already years ago. Moreover Gregory had communicated his results years ago within the circle of his acquaintances, while Leibniz had mentioned nothing of this kind during his first visit to London. He had only sent the letter with the $\pi/4$-series when he had learned the result through the letter of Oldenburg. Concerning the results of the English mathematicians these assertions are true. Collins had sent his report on Newton's results concerning series to Sluse in the autumn of 1669. In the spring Collins sent the series of the circular zone to Gregory, who was in possession of his interpolation technique as of 1668, with which he was able to derive such series [Hofmann 1974, p. 98]. To conclude from this, that Leibniz's achievements depended on the preparatory works of the English, is absurd, however. Leibniz had discovered the arithmetic quadrature of the circle already in the autumn of 1673 and in his copy of the *Commercium epistolicum* there is a note in the margin in his handwriting, that the statement concerning himself was wrong and that his series was different from that of Newton.

To avoid the loss of a further letter, Oldenburg sent his reply letter dated 18th December 1674 with Walter, who returned to Paris and had already to carry also a letter to Huygens. In relation to Leibniz's calculating machine Oldenburg remained cautious. Oldenburg wrote, that Leibniz should remember having given a commitment to the Royal Society at that time, and that it would be very desirable if Leibniz would present his machine during a public meeting of the Royal Society. Concerning the theory of quadratures Oldenburg reported that Newton, as also Gregory, had found a general method applicable to all curves and to the circle in particular. Starting from the equation of the curve under consideration one could calculate its arc length, area, barycentre, and the surface and volume of the corresponding body of revolution. Should Leibniz really have succeeded in finding the exact quadrature of the circle, then he should be congratulated, but Gregory would work on a proof of the impossibility of the quadrature of the circle and therefore it might be necessary to look at the proof in more detail.

The reference to Gregory's impossibility proof certainly came from Collins but is irrelevant here, since Gregory wanted to prove that π cannot occur as a root of a polynomial with integer coefficients, and that had nothing to do with Leibniz's series representation of $\pi/4$ [Hofmann 1974, p. 99f.].

4.2 The Priority Quarrels of Huygens

We have already reported on the priority dispute between Hendrik van Heuraet and Huygens concerning the rectification of curves in section 2.7. The good-natured Heuraet had withdrawn without a fight, although he was in the right. We have also given an account of the rejection of Newton's corpuscular theory of light by Huygens in section 3.1.6. Here is now the right place to closely illuminate the priority quarrels and the trouble of Leibniz's teacher Huygens with the colleagues from England.

4.2.1 The Quarrel Concerning the Rectification of Curves

Rectification (from the Latin: straightening) of curves denotes the problem of computing the length of curves. While in the case of the circle this problem was considered theoretically already in antiquity – the circumference of the circle[3] with radius r is $2\pi r$ – it could not be solved in case of more general crooked curves for a long time. In fact, the treatment of problems of rectification was for a long time hindered by philosophical concerns, which we will summarise following Hofmann [Hofmann 1974, p. 101ff.].

The problem of the quadrature of the circle stems from Greek antiquity and has resulted in innumerable failed attempts to construct a coextensive square from a given circle. In the case of the circumference of the circle this meant that it was apparently impossible to construct a straight line having the same length as the crooked curve of the circle. Aristotle had already noted that there could not be a rational relation between the crooked and the straight. Furthermore, Archimedes had said in his work *On the Sphere and the Cylinder I* in the modern translation of Netz [Archimedes 2004, p. 36]:

> '*POSTULATES*
>
> 1. *That among lines which have the same limits, the straight [line] is the smallest.*
> 2. *And, among the other lines (if, being in a plane, they have the same limits): that such [lines] are unequal, when they are both concave in the same direction and either one of them is wholly contained by the other and by the straight [line] having the same limits as itself, or some is contained, and some it has [as] common,* **and the contained is smaller**'.[4]

Let us compare with the classical translation of Heath [Archimedes 2002, p. 3 f.]:

---

[3]The number π was not known to the ancients, but Archimedes had already computed excellent approximations [Sonar 2011, p. 84f.].

[4]My boldface.

Fig. 4.2.1. Christiaan Huygens, honoured with a stamp (The Netherlands 1928) and Nicholas of Cusa, Painting by the 'Master of the Life of the Virgin' ([Johan van Duyren], contemporary donor portrait, Chapel of the St Nicholas Hospital, Bernkastel-Kues)

'ASSUMPTIONS.

1. *Of all lines which have the same extremities the straight line is the least.*
2. *Of other lines in a plane and having the same extremities, [any two] such are unequal whenever both are concave in the same direction and one of them is either wholly included between the other and the straight line which has the same extremities with it, or is partly included by, and is partly common with, the other;* **and that [line] which is included is the lesser [of the two]**'.[5]

A Latin translation of this work of Archimedes by Jakob of Cremona (Beware: *not* Gerard of Cremona) was published including many translation errors about 1460 and was read by Nicholas of Cusa (Cusanus) (1401–1464). Cusanus was a brilliant thinker, but was hampered by his numerous political duties as a cardinal, since he was often traveling on a diplomatic mission [Flasch 2004], [Flasch 2005], [Flasch 2008]. As Eberhard Knobloch has pointed out, a line can be drawn from Cusanus via Galileo Galilei to Leibniz in the case of the treatment of infinity in mathematics [Knobloch 2004].

Following Cusanus, an arc of a circle can be approximated by piecewise straight lines with arbitrary accuracy. An exact rectification, however, could never be possible. This view was quickly extended to other crooked curves. There were other views, though, for example such as in Regiomontanus (1436–1476), but he died before he could publish anything concerning this problem

[5]My boldface.

area. The views of Regiomontanus were received by François Viète (Vieta) (1540–1603), but could not prevail [Hofmann 1974, p. 102]. Yet Descartes believed that no geometric curve could be rectified geometrically – in the case of mechanical curves he thought it possible –, but Fermat, probably influenced by Viète, recognised that the parabola was rectifiable. In the year 1640 Thomas Hobbes had conjectured that the length of a part of an Archimedean spiral could be expressed in terms of a part of equal length of a parabola, and Roberval succeeded in proving it. Torricelli (1608–1647) knew the rectification of the logarithmic spiral which was accomplished for the first time by Thomas Harriot (1560–1621), but was not published [Sonar 2011, p. 337ff.]. Yet Torricelli's result was not published before 1919. The transformation of the spiral into a parabola then appeared again in Gregory's *Geometriae pars universalis*, since Gregory was a pupil of Stefano degli Angeli (1623–1697) in Padua, who himself was a pupil of Cavalieri and adored the works of Torricelli.

As a 21-year-old, Huygens worked through Torricelli's *Opera geometrica* and found a contradiction in the use of Cavalieri's indivisibles, thus causing him to reject techniques employing indivisibles. The problem of rectification became virulent only through an attack of Thomas Hobbes on John Wallis in the *Six Lessons to the Savilian Professors of the Mathematics* of the year 1656. Here Hobbes rectified a parabola, but his derivation was wrong, which was immediately observed by Huygens. Hobbes tried to correct his approach, but failed again, as was exposed by Huygens. A little later, Roberval reported to Huygens about Roberval's discovery, that an arc of a spiral had the same length as a certain piece of a parabola. Huygens replied that the result was correct; he had found it himself, but couldn't prove it. The hunting for the rectification of curves was open and Huygens turned out to be the winner.

Huygens succeeded in rectifying the parabola already on 17th October 1657. Although being cautious concerning a publication he secured his priority through remarks in letters to his friends.

At this time Hendrik van Heuraet and Johan Hudde were pupils of Frans van Schooten and very curious to learn about the mathematical discussions taking place between Huygens and van Schooten.

Van Heuraet wanted to understand the quadrature of the parabola but he had only the result by Huygens, who kept the proof under lock. At the end of 1657 van Heuraet possessed a method of quadrature and soon after he understood the technique of solving problems of rectification. Then he went with Hudde on a 'Grand Tour' so that van Heuraet's rectification was only published in 1659 in the appendix to van Schooten's Latin edition of *Geometria* by Descartes. On the other side of the Channel John Wallis had announced 1659 in his *Tractatus de cycloide*, that William Neile (1637–1670) had found the rectification of the semi-cubic parabola in the summer of 1657. Neile gave a talk on his rectification in Gresham College in the summer of 1658, but the result was not published, probably because nobody really knew Neile [Hofmann 1974, p. 108].

Fig. 4.2.2. Johann Hudde, mayor and mathematician [Part of a painting by Michiel van Musscher 1686]

Meanwhile Christopher Wren had found the arc length of the cycloid as asked for in Pascal's prize contest. Most likely the proof must have emerged before his eyes at the beginning of July 1658; it was sent to Paris in autumn 1658 and was highly praised by Pascal in December. Huygens spoke very positively about Wren's achievement and called it the very first rational rectification of a curve, because he didn't know the works of Descartes and Torricelli on the rectification of the spiral. Only now did Huygens make his rectification of the parabola public, but it was too late. In the eyes of his French correspondents he could not claim priority; he had seen the method of van Heuraet too late, saw the details of Wren's work and Neile's rectification only at the beginning of 1660 when he received Wallis's *Tractatus de cycloide.* Up to this point everything was in good order.

Then Huygens wrote his *Horologium oscillatorium* and explained the history of rectification from his own point of view. He praised Wren, mentioned van Heuraet, but suspected that Neile had not understood rectification as well as van Heuraet. Concerning van Heuraet's discovery, Huygens wrote that his communication on the rectification of the parabola from the year 1657 had given him the impulse. John Wallis received his personal copy of *Horologium oscillatorium* from Huygens on 8th June 1673 when Huygens was in London

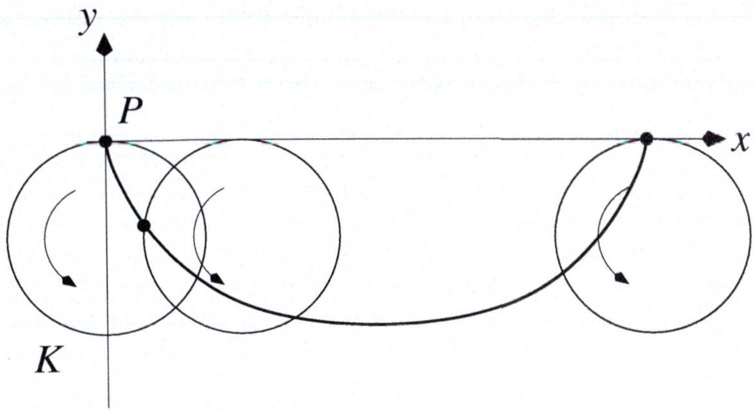

Fig. 4.2.3. Production of a cycloid through unrolling of a circle

during a short trip. Wallis was so deeply hurt in his feelings that he wrote to Huygens already on the next day. Priority was to belong to Neile alone, who understood rectification no less than van Heuraet; only the result was not published. Thereupon Huygens defended himself: if the result would have been published there would not have remained differences. This is the end of the correspondence between Wallis and Huygens. Meanwhile Huygens was involved in a dispute with Newton over the latter's theory of light. When Newton wrote via Oldenburg on 3rd July 1673 to Huygens to defend his theory of light he also mentioned his own work concerning the rectification of curves, being closely related to Huygens's methods. Huygens did not answer this letter and even left further letters of Oldenburg unanswered. Only in May 1674 Huygens wrote again to Oldenburg, but only to declare that he didn't want further inconveniences with fellows of the Royal Society. Indeed, the correspondence was not taken up again before January 1675. At that time the invention of the balance wheel was concerned, for which Huygens claimed priority. Unfortunately, he slipped into his next priority dispute – this time with Robert Hooke.

4.2.2 Turbulent Times: Hooke versus Huygens

We have reported on the relationship between Newton and Hooke already in section 3.1.6. The tireless researcher, tinkerer, and experimenter Robert Hooke made numerous inventions in his lifetime, but often did not care about publication due to lack of time. Hence priority disputes with this pugnacious man were no rarity. Unfortunately it now came hard on Christiaan Huygens who had just got over the trouble concerning the dispute on rectification.

Huygens had already found out in December 1659 that the falling motion of a body in the gravity field of the earth without friction along a cycloid is a tautochrone movement (tauto=the same, chronos=time).

A cycloid is the path of a point P fixed on the boundary (or in the interior) of a circle K which unrolls slip-free on a straight support, as shown in figure 4.2.3. The tautochrone is the curve on which a moving body takes the same time to reach an end-point regardless of the starting point. It was Huygens' great hope that this property of the cycloid might be used to construct a very precise pendulum clock, as we have described already in section 2.7. Since the evolute (i.e. the envelope of the normals) of the cycloid is again a cycloid, the mass of a pendulum moves on a cycloid if the pendulum's thread moves between cycloid cheeks.

For this reason the period of oscillation of a cycloid pendulum is independent of its deflection. In August 1661, Huygens again perfected his pendulum clock by introducing a moveable slide on the pendulum's pole in order to regulate the clock. Another important element of a pendulum clock is the escapement which, in the case of the pendulum, releases the energy which is stored in its chained weights.

The escapement is said to be an invention of Robert Hooke, but that is controversial. The balance wheel is an invention of Christiaan Huygens dating back to the years 1673-74. The invention of the balance wheel was communicated to Monsieurs Cassini, Picard, and Mariott at a meeting of the Parisian Académie des Sciences on 23rd January 1675; Oldenburg also received an informatory letter including an anagram to secure Huygens's priority [Hofmann 1974, p. 118f.]. Also in 1675, Huygens got a patent from Colbert, but thereby the situation got confusing – to say the least!

The Paris clockmaker Isaac II. Thuret[6] suddenly claimed the invention of the balance wheel for himself. Huygens, requiring strict secrecy, had commissioned Thuret to build a first model of a clock regulated by a balance wheel. Now Thuret presented 'his' clock to Colbert. The affair became unpleasant so quickly that Huygens intended to leave Paris, but then a letter of apology arrived from Thuret. Now Abbé Jean de Hautefeuille entered the scene and claimed the invention of the spiral spring – a central component of a balance wheel – for himself. The emerging feud was quickly suppressed by political means. Since Thuret had made the invention of the balance wheel public, great interest began to develop also in England. Huygens hurried to send a description of his invention, which was intended for the 'Journal des Sçavans' and published there, to London. This work of Huygens was read at a meeting of the Royal Society, whereupon Hooke immediately declared to have made the same invention a long time ago. On 25th March 1675 Leibniz published his own invention concerning the regulation of clocks by means of a system of

[6]During three generations, starting with Isaac II., the Thurets have written clockmaker history in France.

Fig. 4.2.4. Portrait of Robert Hooke from the year 2004. Hooke was portrayed here by the artist Rita Greer after descriptions by his friends. On the table there lies a pocket watch, a spring, and a fossil ammonite, things which Hooke dealt with intensely . The starry sky should remind us that Hooke was also a gifted astronomer. Directly in front of Hooke under his hands lies the street map of London, which he decisively shaped after the Great Fire of 1660.

springs in the Journal des Sçavans [Müller/Krönert 1969, p. 37]. Throughout the text he used the notations of his teacher Huygens and referred to the work of the latter, which must have been quite pleasurable for Huygens.

The skilful Hooke built a clock which was presented to the English king in the summer of 1675. Hooke now really came to life: he accused Oldenburg and the entire Royal Society of having revealed his great idea to Huygens. Thereby he put the president of the Royal Society, Lord Brouncker, in an impossible position. The president was neither able to estimate the scope of the invention nor may Hooke's behaviour have pleased him. Brouncker asked the good-natured Oldenburg to commission a clock of Huygens's design di-

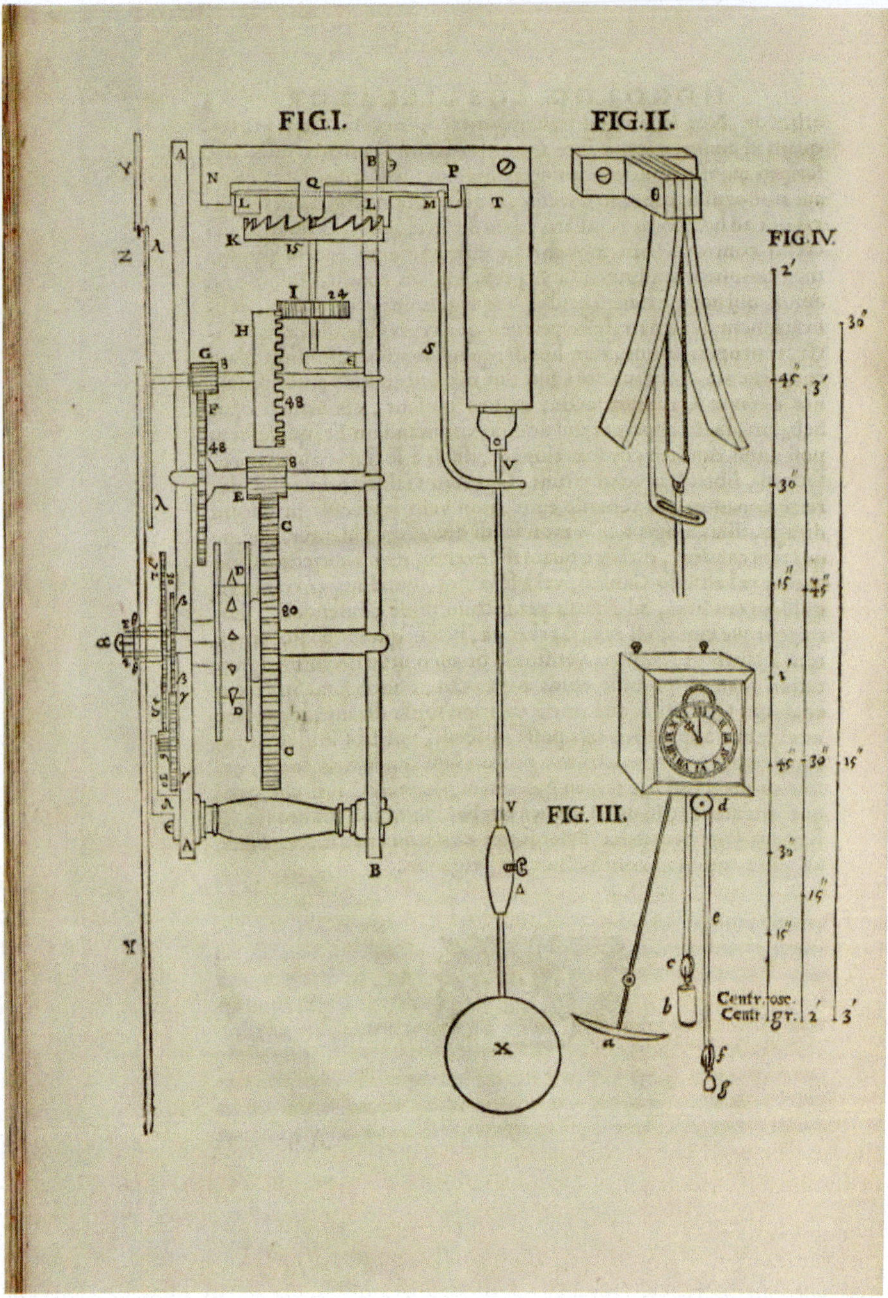

Fig. 4.2.5. The description of Huygens's clock in *Horologium oscillatorium.* In the partial picture headed 'FIG.II.' the cycloid cheeks of the pendulum's mounting can be observed (by courtesy of Sophia Rare Books, Copenhagen)

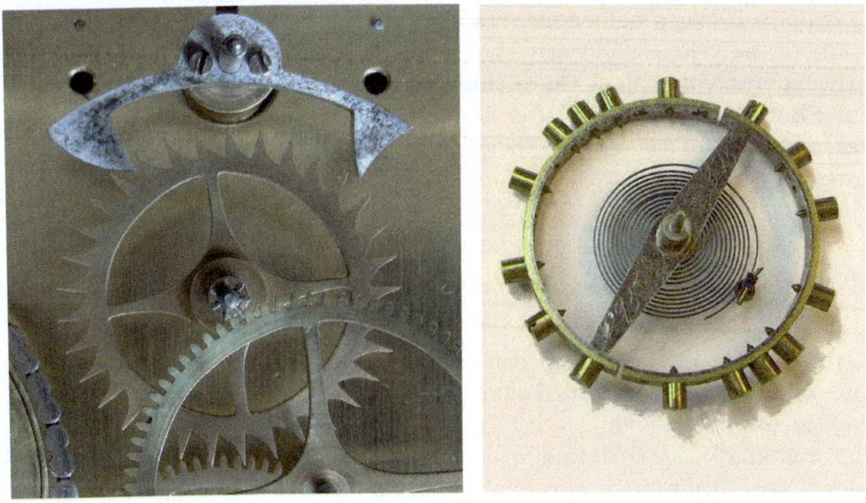

Fig. 4.2.6. An escapement in a real watch [Photo: Timwether 2008], a modern balance wheel with compensating weights [Photo: Chris Burks 2011]

rectly from Huygens in order to compare it to the clock of Hooke. The Italian actor Biancolelli, often engaged in diplomatic missions, served as a courier for the clock, and the complete operating instructions written in Huygens's hand, and brought everything to London. Unfortunately, an unexpected failure occurred, letting the clock stop daily at 12 o'clock. One couldn't fix the failure in England (which says something about the art of clockmaking in England) and Oldenburg urged Huygens to send another clock as a replacement [Hofmann 1974, p. 122]. Meanwhile Hooke's verbal attacks on Huygens grew intemperate. Sweet-tempered Oldenburg had only made some cautious remarks about those attacks to Huygens, but Huygens must have heard from other sources, probably travellers, about the violence of Hooke's accusations. He endeavoured and then started a counter-attack. Thereupon Hooke declared publicly that Oldenburg alone was to blame: he alone had betrayed the secret of the balance wheel to Huygens and hence was nothing but a French spy. Although Hooke had to withdraw this outrageous accusation immediately afterwards, he created an unpleasant and dangerous climate for Oldenburg. Oldenburg was a German and not an Englishman; he accepted the English citizenship only shortly before he died, namely in 1677. He was in prison for a short time already in 1667 because he was accused to be a traitor. His situation hence was anything but enviable as he was attacked by Hooke. Hooke did not even moderate when Huygens's father Constantijn, a well respected diplomat in England, intervened. Several meetings of the Royal Society were necessary to discuss the case. Eventually it was decided to print an official statement for the benefit of Oldenburg in the 'Philosophical Transactions'.

To explain Hooke's behaviour, not exculpating it, we have to note that two letters of the year 1665 clearly show that Huygens was well aware of Hooke's early experiments with clocks at this early time [Jardine 2003, p. 197ff.]. Hooke found both letters in 1677, after Oldenburg's death, when he was elected the new secretary of the Royal Society and could take a look at Oldenburg's correspondence.

4.2.3 Atmospheric Disturbances

It is important in the present context to understand the atmosphere in London, which now was charged and essentially characterised by anti-German resentments. The conflict with Huygens was easily transferred to Leibniz who was known as an intimate of Huygens, since he also had published on clocks. This atmosphere also explains later misunderstandings, reservations, and petty jealousies concerning Leibniz's correspondence with London in the years 1675 and 1676. Leibniz had already sent results on the summation of series to the Royal Society and one now had worries that he would start a dispute on the priority of Newton – after all Newton hadn't published anything on 'his' mathematics. Collins in particular felt a personal responsibility, since he had sent reports on Newton's mathematical results to Gregory, whereof Newton was not pleased. Furthermore Newton was increasingly annoyed about the persistent critique of his theory of colours and decided not to publish anything of his new results either in optics or in the calculus of fluxions. In this state of mind it can only be called favourable that Newton will soon write two letters to Leibniz. It is quite obvious that in the 1670s Newton did not see a direct competitor in Leibniz, nor did he care about questions of priority. There were misunderstandings on both sides, to be sure, and as if they were not enough there was an additional problem of understanding: when Newton spoke about 'Analysis' he meant his theory of series; when Leibniz spoke of 'Analysis' he meant his differential and integral calculus! And even that was not all. The 'mediator' of the correspondence, John Collins, neither had a higher mathematical education, nor was competent enough to assess the correspondence. In contrast to Collins Oldenburg was an excellent diplomat and a warmhearted man, who certainly would have achieved a balance between Newton and Leibniz, but he was advised by Collins in all matters mathematical and thus was lost.

4.3 Times are Changing

On 9th January 1675 Leibniz had the opportunity to present his calculating machine to the Académie des Sciences in Paris [Müller/Krönert 1969, p. 37]. Colbert was apparently impressed, since he ordered three machines; one for the king, one for the royal observatory, and one for himself. In April Leibniz reported that each machine would cost 200 Pistols, meaning the sum of 737 Reichsthalers[7]. He invested large sums of his own money into the manufacture of the machines and hence the search for a well-paid future position became pressing. His hope to get a good position in one of the centres of sciences was not fulfilled; the offer from Hanover alone remained. With a heavy heart Leibniz decided to accept this offer. On 21st January he wrote to Duke John Frederick [Müller/Krönert 1969, p. 37]:

> *'Paris is a place where it is hard to distinguish oneself: one finds there the most experienced men of their times in all areas of sciences, and much work is necessary and a bit of perseverance to establish a good reputation there.* [...] *I still have the same feelings and believe indeed, that a man like me, having no other interests as to make my mark in art and science through spectacular discoveries and to commit the public through valuable works, has only to look for a great prince who possesses insight to assess the value of things, has a generous way of thinking and adjust his actions to the principles of fame, provided that his businesses allow him to lend his ear to beautiful things or give his support.*
>
> (Paris ist ein Ort, wo man sich nur schwer auszeichnen kann: man findet dort in allen Wissenschaftsbereichen die versiertesten Männer der Zeit, und es ist viel Arbeit nötig und ein wenig Beharrlichkeit, um dort seinen Ruf zu begründen. [...] Ich habe noch die gleichen Gefühle, und tatsächlich glaube ich, daß ein Mann wie ich, der kein anderes Interesse besitzt als das, sich durch aufsehenerregende Entdeckungen in der Kunst und Wissenschaft einen Namen zu machen und die Öffentlichkeit durch nützliche Arbeiten zu verpflichten, nur einen großen Fürsten suchen muß, der genügend Einsicht besitzt, den Wert der Dinge beurteilen zu können, eine großzügige Denkungsart hat und seine Handlungen nach den Grundsätzen des Ruhmes ausrichtet, vorausgesetzt, daß ihm seine Geschäfte erlauben, den schönen Dingen sein Ohr oder seine Unterstützung zu leihen.)

But Leibniz did not think of going to Hanover immediately.

[7]One Reichsthaler in Leibniz's days had an equivalence to between 16 and 21 Euro; an information I owe to Prof Dr Eberhard Knobloch. One calculating machine hence devoured between 11800 and 15500 Euro. For the sake of comparison: the Duke of Hanover offered Leibniz an annual income of 400 thalers, hence between 6400 and 8400 Euro.

4.3.1 Leibniz's Letter of 30th March 1675 and its Immediate Consequence

At the beginning of 1675 Leibniz was fully concentrated on his mathematics. In a letter to Oldenburg from 30th March 1675 [Turnbull 1959–77, Vol. I, p. 336ff.], he stated that his arithmetical quadrature of the circle did not contradict a theorem of the impossibility of squaring the circle by Gregory. Then he went on writing [Turnbull 1959–77, Vol. I, p. 337f.]:

> *'You write that your distinguished Newton has a method of expressing all squarings, and the measures of all curves, surfaces and solids generated by revolution, as well as the finding of centres of gravity, by a method of approximations of course, for this is what I infer it to be. Such a method, if it is universal and convenient, deserves to be appraised, and I have no doubt that it will prove worthy of its most brilliant discoverer. You add that some such discovery was known to Gregory also. But since Gregory in his book Geometriae Pars Universalis admits that he did not yet know the method of measuring hyperbolic and elliptic curves, you will show me, if you please, whether either he or Newton has found it since that time, and if they have, whether they have it absolutely, which I can scarcely credit, or from an assumed squaring of the circle or the hyperbola.'*

It was this part of the letter which ultimately led to a contact between Newton and Leibniz by letter.

The letter continued with Leibniz referring to the finite sums of the series $\sum 1/k$, $\sum 1/k^2$, and $\sum 1/k^3$ which were previously communicated to him by Collins. Leibniz wrote that Collins probably had already published these results but he had no knowledge about such a publication. Should Oldenburg – the consent of provided, of course – send him the publication then Leibniz would send him the $\pi/4$-series including the proof to London. Then Leibniz wrote further [Turnbull 1959–77, Vol. I, p. 338]:

> *'Again, the arithmetical squaring of the circle has very far-reaching consequences, and it opens up the way to many new things, as you too will readily judge: accordingly I have no doubt that you will show yourselves accommodating in communicating to me the methods of Newton and Gregory. It is certainly not my practice to let slip opportunities of naming with many expressions of honour the authors of eminent discoveries.'*

In this paragraph Leibniz clearly showed that he went quite far with his methods concerning the arithmetical quadrature of the circle. His offer to name Newton and Gregory in his publications and give them honour concerning certain discoveries was certainly serious, but in the reprint of this letter in the

Commercium epistolicum the whole last part of the letter was intentionally left out. Hence it looked as if Leibniz was only interested in a communication concerning the methods of Newton and Gregory.

The written reply by Oldenburg bears the date 22nd April. Since Oldenburg knew too little mathematics he asked Collins for advice, then translated Collins's English text into Latin, and transferred it to the letter. Oldenburg reported to Leibniz that there were doubts in the Royal Society concerning his proposed clock. It is likely that Oldenburg repeated here the objections of Hooke in somewhat moderate form [Hofmann 1974, p. 131]. Then came the part concerning mathematics which came from Collins and which is significantly more detailed than the comparable part in the letter of 16th April 1673. Leibniz learned here about a new result of Gregory concerning a circle with radius r, namely the representation

$$\pi r = \frac{4r^2}{2d - \frac{1}{3}e - \frac{1}{90}\frac{e^2}{d} - \frac{1}{756}\frac{e^3}{d^2} - \dots}, \tag{4.6}$$

where $d = \sqrt{2}r$ denotes the edge length of the square inscribed in the circle and $e = (\sqrt{2} - 1)r$ is the difference between that edge length and the radius. However, Leibniz couldn't do anything with this communication since the term $-\frac{1}{90}\frac{e^2}{d}$ was missing in Oldenburg's letter! Collins further informed Leibniz, that series like this were communicated to him after Mercator's *Logarithmotechnia* appeared in July 1668, and that he sent the *Logarithmotechnia* to Barrow in Cambridge. From there he received the answer that Newton had found the series of the logarithm some time before the publication of the *Logarithmotechnia* and that he had applied it to the quadrature of all (!) geometrical and mechanical curves[8]. Thereby Collins referred to Newton's *De Analysi* which he had received in the summer of 1669 and had it immediately copied. Collins also gave Newton's series for the sine; if $x = \sin z$ this is

$$z = x + \frac{1}{6}x^3 + \frac{3}{40}x^5 + \frac{5}{112}x^7 + \dots, \text{ or} \tag{4.7}$$

$$x = z - \frac{1}{6}z^3 + \frac{1}{120}z^5 - \frac{1}{5040}z^7 \pm \dots. \tag{4.8}$$

[8]The Greek antiquity knew the classification of curves into three classes: plain, solid, and linear. Plain curves are constructible by means of compass and ruler, solid are those for which the conic sections have to be used, and linear are those which require 'more composite lines' [Mancosu 1996, p. 71]. René Descartes discussed this classification in his *Geometria* [Descartes 1969, p. 19ff.] and introduced the distinction in 'geometrical' and 'mechanical' curves. Curves are called 'geometrical' if they are exact and mathematically precisely describable; 'mechanical' curves are those for which this is not possible. This distinction is not important to us but we have to clarify the concepts [Sonar 2011, p. 244f.]. Please note that one can also construct geometrical curves by means of mechanical devices, and Descartes has even indicated some of them.

A series for the area of a circle and a hyperbola, respectively,

$$2\int_0^a \sqrt{r^2 \mp x^2}\,dx$$

was also given; one for the area $2\int_0^a \sqrt{2rx - x^2}\,dx$ of a segment of the circle was given as well. Collins attributed these series to Gregory, but he 'forgot' to mention that Gregory had this series already in 1668, when no one had heard a word about Newton's theory of series. Even the series of the tangent function and its inverse function was communicated to Leibniz; they also trace back to Gregory who had found them already by repeated differentiation [Hofmann 1974, p. 135].

In the later indictment *Commercium epistolicum D. Johannis Collins et aliorum de Analysi promota* of the Royal Society from 1712 (date of print) it is said that Leibniz had shown his own series for the inverse of the tangent function, the arctangent, to his French friends only after he had received the letter of 22nd April. The comments which Leibniz noted in the margins of his copy of the *Commercium epistolicum* are clear, however: These accusations were unabashed and malicious. In fact, Leibniz had presented his series in the letters of 15th July and of 16th October 1674, and his method of the deduction of the arctangent series, sent to Huygens together with his arithmetical quadrature of the circle in October 1674, was completely different and independent of the one of Gregory.

Collins wrote not much more about Gregory's achievements, but instead he wrote about Newton's, whereby he again exploited the manuscript *De analysi*. The letter also contained material concerning algebra, in particular results of Pell. The theory of algebraically solving equations was a favourite subject of Collins, as was Davenant's problem to find the first four successive terms in a geometric progression, when the sum of their squares and their cubes are given.

Following Hofmann, this 'long, poorly arranged' letter [Hofmann 1974, p. 139] shows many similarities to the letter of 16th April 1673. For the first time Leibniz learned a little more about Newton's and Gregory's research; only results, however, not the methods. Leibniz assumed that Collins had sent the most recent thoughts of the two Englishmen, while Collins in fact had only described problems which were already solved for quite a long time. Leibniz was very much interested in algebraic methods for the solution of equations and Collins's letter may have contributed to Leibniz now starting an intensive occupation with this topic. He discussed this topic with Roberval in Paris, as well as with Ozanam.

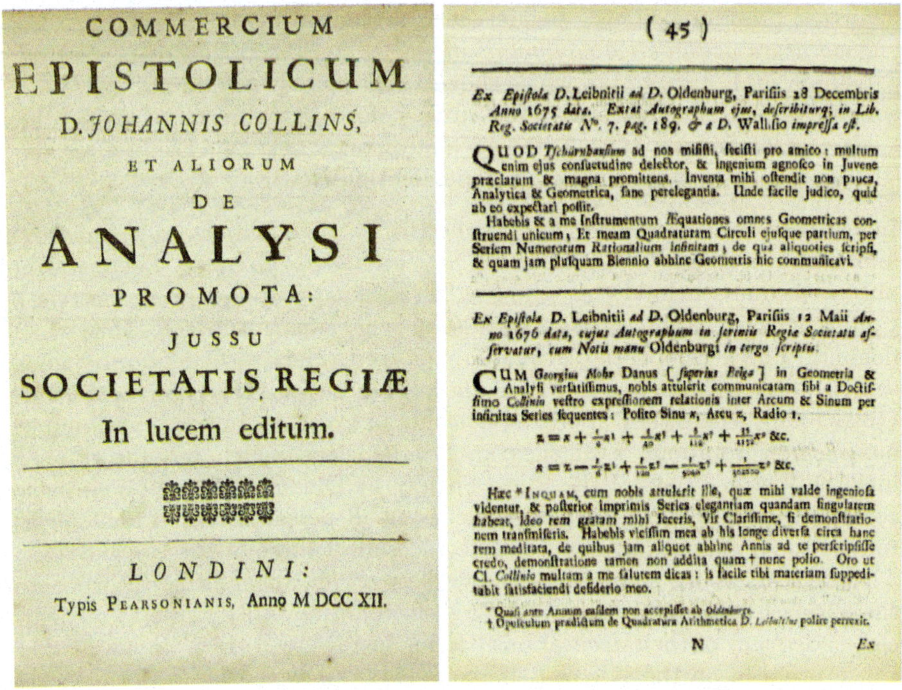
COMMERCIUM
EPISTOLICUM
D. JOHANNIS COLLINS,
ET ALIORUM
DE
ANALYSI
PROMOTA:
JUSSU
SOCIETATIS REGIÆ
In lucem editum.

LONDINI:
Typis PEARSONIANIS, Anno M DCC XII.

(45)

Ex Epistola D. Leibnitii *ad D.* Oldenburg, Parisiis 28 Decembris *Anno 1675 data. Extat Autographum ejus, describiturq; in Lib. Reg. Societatis N°. 7. pag. 189. & a D.* Wallisio *impressa est.*

QUOD *Tschürnbausium* ad nos misisti, fecisti pro amico: multum enim ejus consuetudine delector, & ingenium agnosco in Juvene praeclarum & magna promittens. Inventa mihi ostendit non pauca, Analytica & Geometrica, sane perelegantia. Unde facile judico, quid ab eo expectari possit.

Habebis & a me Instrumentum Æquationes omnes Geometricas construendi unicum, Et meam Quadraturam Circuli ejusque partium, per Seriem Numerorum *Rationalium infinitam*, de qua aliquoties scripsi, & quam jam plusquam Biennio abhinc Geometris hic communicavi.

Ex Epistola D. Leibnitii *ad D.* Oldenburg, Parisiis 12 Maii *Anno 1676 data, cujus Autographum in scriniis Regiae Societatis asservatur, cum Notis manu* Oldenburgi *in tergo scriptis.*

CUM *Georgius Mohr* Danus [*superius Belga*] in Geometria & Analysi versatissimus, nobis attulerit communicatam sibi a Doctissimo *Collinio* vestro expressionem relationis inter Arcum & Sinum per infinitas Series sequentes: Posito Sinu x, Arcu z, Radio 1.

$$z = x + \frac{1}{6}x^3 + \frac{3}{40}x^5 + \frac{5}{112}x^7 + \frac{35}{1152}x^9 \text{ \&c.}$$

$$x = z - \frac{1}{6}z^3 + \frac{1}{120}z^5 - \frac{1}{5040}z^7 + \frac{1}{362880}z^9 \text{ \&c.}$$

Hæc * INQUAM, cum nobis attulerit ille, quæ mihi valde ingeniosa videntur, & posterior imprimis Series elegantiam quandam singularem habeat, ideo rem gratam mihi feceris, Vir Clarissime, si demonstrationem transmiseris. Habebis vicissim mea ab his longe diversa circa hanc rem meditata, de quibus jam aliquot abhinc Annis ad te perscripsisse credo, demonstratione tamen non addita quam † nunc polio. Oro ut Cl. *Collinio* multam a me salutem dicas: is facile tibi materiam suppeditabit satisfaciendi desiderio meo.

* Quasi ante Annum eandem non accepisset ab *Oldenburgo*.
† Opusculum praedictum de Quadratura Arithmetica D. *Leibnitius* polire perrexit.

N *Ex*

Fig. 4.3.1. Title page of the indictment of 1712 and p. 45 showing letters by Leibniz to Oldenburg from 28th December 1675 and 12th May 1676 (PBA Galleries/pbagalleries.com)

On 20th May 1675 Leibniz thanked Oldenburg only briefly for the letter. Then he made a mistake: he thanked him for the series sent to him by Collins and wrote that he had not yet compared these results with his own, some years older results, but as soon as he could do this he would send more about this. This assertion was blatant, since Leibniz couldn't have understood Gregory's first series (4.6) and had better confessed it.

In the later indictment *Commercium epistolicum D. Johannis Collins et aliorum de Analysi promota* of the Royal Society from 1712 (date of print) this answer of Leibniz is incorporated and commented on: Leibniz had never sent the promised comparison with his series, but had instead required information about Newton's methods when he received the series of the sine and the arcsine once again from the Danish mathematician Georg Mohr (1640–1697). Later Leibniz had determined the series of the arctangent of Gregory, without even mentioning the letter of 22nd April 1675 in which all of the details were already presented.

Further proof of Leibniz's plagiarism is served by his introduction to his *Quadratura arithmetica communis* in the Acta Eruditorum of April 1691 [Leibniz 2011, p. 103-113], which showed that Leibniz could not have found his arithmetic quadrature of the circle before 1675 before receiving the letter of 22nd April 1691. Leibniz's analysis was developed only after the letter and hence was completely dependent on Newton's works.

In the year 1712 it seemed clear to the British that Leibniz had plagiarised, and that everything originated from Newton and/or Gregory. There was not only wickedness behind this: the English referred to the drafts of letters by Collins which contained more details on what was ultimately sent by Oldenburg. Furthermore, the letters written by Leibniz to Oldenburg were essentially taken from the third volume of John Wallis's *Opera mathematics* (1693-1699), where they appeared not only in excerpts, but also with numerous misinterpretations. Even the copies of the letters which were incorporated into the 'Letter Book' of the Royal Society by Oldenburg were by no means complete. In particular, Oldenburg left out all personal remarks in the letters and everything that apparently could snub fellows of the Royal Society. Thereby Oldenburg acted out of laudable motives, but had he incorporated the letters in all completeness the English would have been able to recognise the inadequacy of their accusations [Hofmann 1974, p. 141f.]. Moreover, there was a problem on the side of Leibniz in 1712: he owned all copies of letters and drafts which could have proven his independence from the English in all matters mathematical, but the sheer mass of his correspondence and confused notes had outgrown him.

4.3.2 Analysis Becomes Calculus

At the end of the year 1675 Leibniz's analysis transformed into a calculus (meaning 'calculation'). In a manuscript dated 29th October 1675 with the title *Analyseos tetragonisticae pars secunda* (Second part of the quadrating analysis) [Leibniz 2008b, p. 288ff.], [Child 2005, p. 76ff.] he still wrote

$$\frac{\overline{\text{omn}.\ell}^2}{2} = \text{omn}.\,\overline{\overline{\text{omn}.\ell}\cdot\frac{\ell}{a}},$$

where 'omn.', the abbreviation for the Latin 'omnes' (all), stands for the 'entirety of all lines after Cavalieri'. The over-bar denotes parenthesis and the ℓ is the notation later to be replaced by dy. The term a has to be equated with dx, but as already Newton had done, Leibniz also set $a = dx = 1$. In modern (Leibnizian) notation the equation above is simply

$$\frac{1}{2}\left(\int dy\right)^2 = \int\left(\int dx\right)dy.$$

Leibniz [Child 2005, p. 80] then goes on:

'Another theorem of the same kind is:

$$\text{omn. } x\ell = x\,\text{omn. } \ell - \text{omn. omn. } \ell, \tag{4.9}$$

where ℓ is taken to be a term of a progression, and x is the number which expresses the position or order of the ℓ corresponding to it; or x is the ordinal number and ℓ is the ordered thing.'

In modern notation the equation (4.9) reads as

$$\int x\,dy = x \sum dy - \int\int dy = xy - \int y\,dx,$$

whereby it is not always clear whether $\int y\,dy$ or $\int y\,dx$ is meant when Leibniz writes $\int y$ [Edwards 1979, p. 253].

In the manuscript, then, the integral sign appears for the first time in history, cp. figure 4.3.2. Leibniz writes [Leibniz 2008b, p. 292], cp. figure 4.3.3:

'It will be useful to write $\int$ instead of omn. so that $\int \ell$ = omn. ℓ, or the sum of the ℓ. Thereby

$$\frac{\int \overline{\ell}^2}{2} = \int \overline{\int \overline{\ell}\frac{\ell}{a}} \quad \textit{and} \quad \int \overline{x\ell} = x \int \overline{\ell} - \int\int \ell.\text{'} \tag{4.10}$$

(Utile erit scribi $\int$. pro omn. ut $\int \ell$ pro omn. ℓ. id est summa ipsorum ℓ. Itaque fiet $\frac{\overline{\int \ell}^2}{2} \sqcap \int \overline{\int \overline{\ell}\frac{\ell}{a}}$ et $\int \overline{x\ell} \sqcap x \int \overline{\ell} - \int\int \overline{\ell}$.)

The symbol $\int$ is meant to symbolise the 'S' in 'sum' and hence appears as an elongated 'S'. The symbol $\sqcap$ is used by Leibniz to denote the equality sign.

Writing $\ell = dx$ in the first of the two equations (4.10) Leibniz finds again

$$\int x\,dx = \frac{1}{2}x^2.$$

With $\ell = x\,dx$ in the second equation he gets

$$\int x^2\,dx = x \int x\,dx - \int\int x\,dx = x\frac{x^2}{2} - \int \frac{x^2}{2}\,dx,$$

so that

$$\int x^2\,dx = \frac{1}{3}x^3$$

follows.

In the manuscript of 29th October 1675, Leibniz still writes $\ell = y/d$, but only three days later, at 1st November, he writes the manuscript *Analyseos*

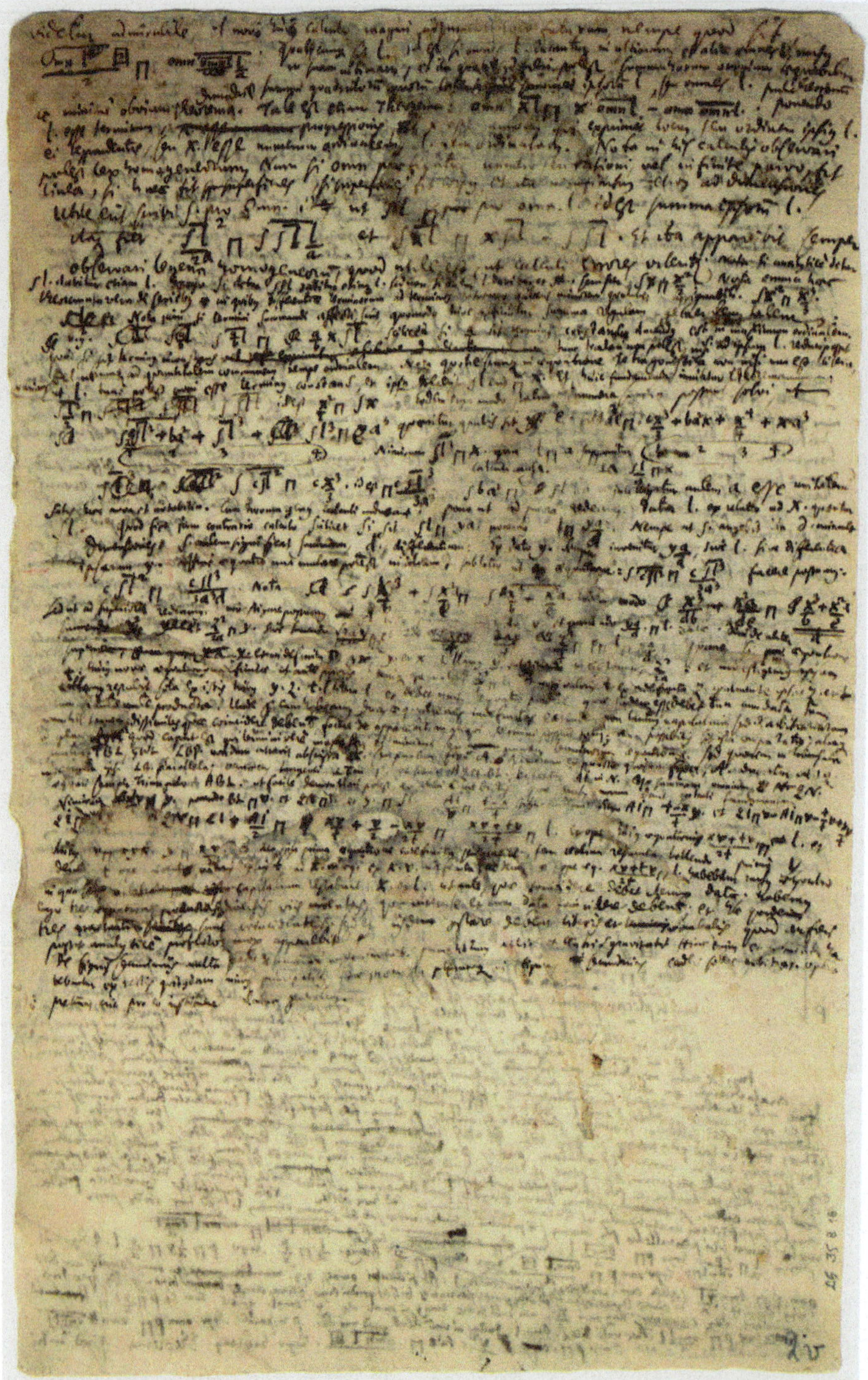

Fig. 4.3.2. A page of a manuscript of 29th October 1675 (Gottfried Wilhelm Leibniz Bibliothek - Niedersächsische Landesbibliothek Hannover (Hanover), Sig. LH XXXV, VIII, 18, Bl. 2v)

Fig. 4.3.3. The line 'Utile erit scribi ...'. Part of the manuscript of 29th October 1675

tetragonisticae pars tertia [Leibniz 2008b, p. 310ff.], [Child 2005, p. 84ff.] and here $\ell = y/d$ is turned into the well-known dy. On 11th November he writes *Methodi tangentium inversae exempla* [Leibniz 2008b, p. 321ff.], [Child 2005, p. 93ff.] and asks whether $dx\,dy$ is the same as $d(xy)$, and whether $d(x/y)$ is the same as dx/dy. He is apparently searching for the product and the quotient rule. He is able to answer both questions negatively, since he remarks that

$$d(x^2) = (x + dx)^2 - x^2 = 2x\,dx + (dx)^2 = 2x\,dx$$

and

$$(dx)(dx) = (x + dx - x)(x + dx - x) = (dx)^2$$

hold. The 'calculus' starts working here! In a manuscript dated 11th July 1677 [Child 2005, p. 128ff.] product and quotient rule are apparent. Leibniz calculates

$$d(xy) = (x + dx)(y + dy) - xy = x\,dy + y\,dx + dx\,dy,$$

and since $dx\,dy = 0$ (the product of two infinitesimals is small of higher order; cp. our discussion in chapter 1!), this yields

$$d(xy) = x\,dy + y\,dx.$$

In the case of the quotient rule he calculates

$$d\frac{y}{x} = \frac{y + dy}{x + dx} - \frac{y}{x} = \frac{x\,dy - y\,dx}{x^2 + x\,dx}$$

and remarks, that $x\,dx$ is incomparably small in comparison to x^2, so that

$$d\frac{y}{x} = \frac{x\,dy - y\,dx}{x^2}$$

follows. Here a point for some criticism can be obviously found. We have argued in chapter 1 that $x\,dx$ is not small of higher order and therefore should contribute something. In fact, Leibniz does not strike off the terms in the denominator of the quotient rule! The crucial point here is that $x\,dx$ is small *in comparison with* x^2, but such an argument clearly calls for criticism and we have to deal with some of the critics later on. In his *Historia et origo* [Child 2005, p. 55f.] Leibniz wrote at the end of his life proudly that the calculus (he really calls it 'calculus') did not need to refer back to geometric figures in order to produce results, and that things which were awe-inspiring earlier, were now merely child's play.

In an undated revised version of the manuscript of 11th July 1677 [Child 2005, p. 136], the role of the characteristic triangle appears explicitly in the calculus. Here also the fundamental theorem of differential and integral calculus is proved. If a curve with ordinate z is given and its area is sought, Leibniz assumes that a curve with ordinate y can be found, such that

$$\frac{dy}{dx} = \frac{z}{a}$$

holds, where a is some constant. Then it follows $z\,dx = a\,dy$ and the sought area is

$$\int z\,dx = a \int dy = ay,$$

where Leibniz assumes as always that the curve with ordinate y passes through the origin [Edwards 1979, p. 258]. A quadrature problem in Leibniz's calculus hence is an inverse tangent problem, since in order to compute the area under the curve with z-ordinate one needs a curve with tangent slope

$$\frac{dy}{dx} = z.$$

Setting $a = 1$ and subtracting the area over $[0, x_0]$ from that over $[0, x_1]$ it follows that

$$\int_{x_0}^{x_1} z\,dx = y(x_1) - y(x_0),$$

and that is the fundamental theorem.

4.3.3 Leibniz Gains a Companion

At the end of September 1675 Leibniz got acquainted with the young German mathematician Ehrenfried Walter von Tschirnhaus (1651–1708).

Tschirnhaus had currently got some fame as the actual inventor of European porcelain. In 1675 he came from England to Paris, carrying a letter of recommendation from Oldenburg to Leibniz. In a letter to Oldenburg dated 18th October 1675 Leibniz wrote [Müller/Krönert 1969, p. 38f.]:

> *'It was intended amicably that you sent Tschirnhaus to us: I find great pleasure in his company and recognise the highly promising and virtuous mind in the young man. He has shown many a thing of analysis and geometry to me, really very beautiful things.'*
>
> (Das war freundschaftlich gedacht, daß Du uns Tschirnhaus zuschicktest: ich finde großes Vergnügen an seinem Umgange und erkenne den vielversprechenden und vortrefflichen Geist in dem jungen Manne. Er hat mir manches aus der Analysis und Geometrie, wirklich sehr schöne Sachen, gezeigt.)

Tschirnhaus had originated from Upper Lusatia and received an education in mathematics and sciences through a private tutor. He went to Leiden to study law but his love was for mathematics and physics. He was also introduced to the philosophy of Descartes and he remained an incorrigible Cartesian for the rest of his life. In 1672 he was involved on the side of Holland in the war against England; he was not involved in fighting, however. Two years later he finished his studies. In The Hague he had contact with the philosopher Baruch de Spinoza who recommended him to Oldenburg in London, where Tschirnhaus made acquaintance with Boyle, Wallis, and Wren [Hofmann 1974, p. 165]. He also became acquainted with Collins and Tschirnhaus wrote to Gregory via Collins.

Tschirnhaus had particularly great algorithmic-algebraic skills with which he even impressed John Pell in London, so that Pell declined to talk to him about scientific questions. Collins counted Tschirnhaus together with Gregory and Newton to be the three most important European algebraists, but Tschirnhaus also aggravated the English with the assertion that all works of Sluse and Barrow would follow from the works of Descartes. In London Tschirnhaus had thought about rules to solve equations of degree up to eight. The meeting with Collins took place on 9th August 1675, shortly before Tschirnhaus's departure to Paris. It is evident from the description of this meeting, which Collins sent to Gregory, that the two men had not met before. Only in the early summer of 1676 did a correspondence on Cartesian methods start between Tschirnhaus and Oldenburg (with Collins in the background), in which

Fig. 4.3.4. Ehrenfried Walter v. Tschirnhaus [Engraving by M. Berningeroth, before 1708, cutout], and Baruch de Spinzoa ([unknown painter, 1665] Herzog August Bibliothek Wolfenbüttel)

Tschirnhaus got information about the infinitesimal methods of the English from Collins, in particular Newton's method of computing tangents.

In the later indictment *Commercium epistolicum D. Johannis Collins et aliorum de Analysi promota* of the Royal Society from 1712 (date of print) this correspondence between Tschirnhaus and Oldenburg/Collins was predated a whole year ahead. If Tschirnhaus should have had this informations already in the summer of 1675, then he could have communicated details of Newton's method to Leibniz. Furthermore Oldenburg's records showed that Gregory's rectification of the circular arc was also shown to Tschirnhaus, so that Leibniz could have got this technique 1675 from Tschirnhaus. This assertion was without any foundation, however, since the document in question was not written before 1676.

Tschirnhaus also had to deliver a letter to Huygens, who received him warmly and questioned him about Spinoza and mutual friends in The Hague. Tschirnhaus could not speak a word of French but tried to learn it quickly in Paris. He gave mathematics lessons in Latin to Colbert's son, since as a scion of a noble family he had no problems to move in the highest circles in Paris.

A deep scientific friendship developed with Leibniz which is documented in many writings [Hofmann 1974, p. 174]. Such writings are printed in Volume VII, 1 of the Academy Edition, No. 23-29, 92, 94. Tschirnhaus reported on his recent results in algebra, thereby remarking that John Kersey's book *The Elements of Mathematical Art, commonly called Algebra* would be worthless, although in England it was seen as an authority in algebra. Both men also talked about Baruch de Spinoza, with whom Leibniz had already communicated on optical questions.

As a philosopher Leibniz took great interest in Spinoza's *Tractatus theologico-politicus* which was published anonymously in 1670, but Spinoza, living in fear of further persecution (the *Tractatus* was prohibited in 1674) like an hermit, had asked Tschirnhaus for absolute secrecy. The two men also talked about rare books in Holland, but Tschirnhaus and Leibniz started talking about mathematics only at the end of November 1675, when Leibniz had already formulated his infinitesimal calculus and the corresponding symbolism.

From London Tschirnhaus came with an order of Collins and Oldenburg to search for manuscripts of some French mathematicians like Desargues, Fermat, Roberval, and Pascal. In the case of Roberval there was barely anything in print, except some manuscript circulated among Parisian mathematicians [Hofmann 1974, p. 178f.]. Roberval wanted to get this manuscripts printed but he died on 27th October 1675, while work for a publication was still in progress. His possibly most important works fell to the Académie Royale des Sciences where Leibniz could inspect them and could also provide information to Tschirnhaus. Both friends agreed that these manuscripts were not worth publishing. Already in 1674 Leibniz had written a letter to Pascal's heir, Étienne Périer, and required a report on mathematical manuscripts which were still extant. One year later Leibniz received some of Pascal's manuscripts from Étienne's younger brothers who were then educated in Paris. Although it could only have been fragments, Leibniz made transcriptions to which he and Tschirnhaus added remarks. Leibniz sent the manuscripts back on 30th August 1676 with the clear recommendation to have them published. However, they were never published and today they are lost – Leibniz's notes are nearly all that is left of them.

Tschirnhaus had promised Oldenburg to publish a work concerning the resolution of equations in the Philosophical Transactions, but this did not happen. Now Oldenburg became impatient and wrote in December 1675 that he had not heard anything from Tschirnhaus. However, Leibniz had good reasons not to write an answer to Oldenburg during the next few weeks. He was not back in Paris before January 1676 since he was invited by Abbé Gravel to attend a conference concerning the neutrality of the city of Liège, taking place in Marchienne. Furthermore Leibniz had accepted with a heavy heart the offer of the Duke of Hanover to take up work there and could no longer remain in Paris [Hofmann 1974, p. 183].

What could Tschirnhaus have possibly told his friend Leibniz? We have already discussed that he was informed about the new mathematics of Newton by Collins, but not early enough to give Leibniz a hint. However, there is another reason to doubt the allegations of the *Commercium epistolicum* that Leibniz had heard important mathematical details from Tschirnhaus. We find this reason in the later correspondence of Leibniz. There it turns out that Tschirnhaus, who was always superior when algorithmic techniques in algebra were concerned, did not know the first thing about the meaning of the new infinitesimal calculus. Leibniz wrote in detail to Tschirnhaus concerning the meaning of the 'characteristica universalis' from which the symbols of the calculus eventually emanated, but Tschirnhaus would not be convinced: he considered the symbols of calculus to be pointless and obscure, and he apparently did not understand the analysis of Leibniz [Hofmann 1974, p. 185]. Tschirnhaus did not understand the quadrature based on indivisibles and he thought his own methods were more general than those of Pascal and Grégoire de Saint-Vincent, though they were not. In a letter of May 1678 Leibniz expressed his regret that Tschirnhaus showed so little enthusiasm concerning the new (Leibnizian) analysis due to his prejudices against the symbols. However, Tschirnhaus lacked the mathematical profundity to grasp the meaning of the new analysis of Newton or Leibniz. Leibniz could never have learned anything from Tschirnhaus at the beginning of 1676, but he had gained a loyal friend and an acolyte.

4.4 De quadratura arithmetica

When Leibniz eventually left Paris for Hanover on a proper detour via the Netherlands and England on the morning of the 4th October 1676 – it is a Sunday – to enter the services of Duke John Frederick, he left an important manuscript behind that is probably unique in the history of analysis. It is *De quadratura arithmetica circuli ellipseos et hyperbolae cujus corollarium est trigonometria sine tabulis* (On the arithmetical quadrature of the circle, the ellipse, and the hyperbola. A corollary of it is a trigonometry without tables) [Knobloch 1993]. We have already encountered the 'arithmetical quadrature of the circle'; it means the calculation of the area of the circle by means of infinite series with rational summands, cp. the $\pi/4$-series (4.5). A 'trigonometry without tables' means the series expansion of the trigonometric functions, so that (at least in principle) tables are no longer necessary [Knobloch 2002, p. 59]. This manuscript is the longest mathematical work which was ever written by Leibniz. He had made numerous preliminary studies and had condensed results which in part stemmed already from the beginning of his stay in Paris. Hence the harmonic triangle and the summation of the inverse figurate numbers are contained, as are the transmutation theorem and the $\pi/4$-series [Knobloch 1993, p. 9]. Tschirnhaus later read the final version.

Besides the contents, the history of this extensive manuscript being completely transcribed and commented on for the first time only in 1993 by Eberhard Knobloch is interesting. We summarise the genesis and the traditional history of this document following Knobloch [Knobloch 1993, p. 9-14]. There were four drafts of the manuscript: one for Huygens from October 1674 which was commented on by Huygens and given back to Leibniz; one for Jean Paul La Roque, the editor of the science periodical Journal des Sçavans, which was never sent off; one also never sent, for Jean Gallois from the end of 1675; and finally a hitherto unprinted version which was restricted to the arithmetical quadrature of the circle. We like to call the manuscript transcribed by Knobloch the 'final version'.

In April of the year 1691, hence nearly 15 years after he had left Paris and left the manuscript, Leibniz published *Quadratura arithmetica communis sectionum conicarum quae centrum habent, indeque ducta trigonometria canonica ad quantamcunque in numeris exactitudinem a tabularum necessitate liberata: cum usu speciali ad lineam rhomborum nauticam, aptatumque illi planisphaerium* (Common arithmetic quadrature of the conic sections which have a midpoint, and thereby can be deduced a canonical trigonometry up to an arbitrary numerical accuracy which no longer requires tables; with a particular application to the nautical rhumb line and to their adjusted planisphere) [Leibniz 2011, p. 103 ff.] under pseudonym O.V.E. (Godofredus Guillielmus Leibnitius)[9] in the Acta Eruditorum. However, in the table of

[9] The letters U and V were usually identified in Latin.

contents of the Acta Eruditorum there is a reference from O.V.E. to G.W.L. Here Leibniz wrote at the beginning [Leibniz 2011, p. 104]:

> *'I had compiled a little booklet on the arithmetical quadrature of the circle already in the year 1675 and since then have given it friends to read, but the material grew under my hands and I had no time to polish it for publication when still other occupations accrued. Most of all it seems to me not being worth the effort yet to lay down in all detail in the customary way what our new analysis can achieve with less effort.'*
>
> (Schon im Jahr 1675 hatte ich eine kleine Schrift über die arithmetische Quadratur zusammengestellt und seither habe ich sie Freunden zu lesen gegeben, doch wuchs mir das Material unter den Händen und mir fehlte die Zeit, es für die Veröffentlichung aufzupolieren, als [noch] andere Beschäftigungen hinzukamen. Vor allem aber erscheint es [mir] nicht recht der Mühe wert, jetzt [noch] in aller Ausführlichkeit auf die übliche Art darzulegen, was unsere neue Analysis mit wenig [Aufwand] leistet.)

The 'little booklet' can not have been the final version, as was proven by Knobloch [Knobloch 1993, p. 10f.], since the final version was not completed in 1675 - not even before autumn 1676. In his personal copy of the *Commercium epistolicum* Leibniz noted 1713 in the margin [Knobloch 1993, p. 10]:

> *'But this* [i.e. the arithmetical quadrature] *I had found already in the year 1674, but thereupon written a little work in 1675.'*
>
> (Aber diese hatte ich schon im Jahre 1674 gefunden, aber 1675 daraufhin ein kleines Werk verfasst.)

This 'little work' (opusculum) is quite obviously the 'little booklet' (opusculum) mentioned in the Acta Eruditorum, only that Knobloch has translated 'opusculum' as 'work' and Heß und Babin the same word as 'booklet'.

Knobloch has reconstructed the further history of the final version from the correspondence of Leibniz. Leibniz wrote to Gallois in September 1677, that he had left the manuscript in Paris in order to have it printed; he also wrote in the same tenor to the Helmstedt professor Hermann Conring on 29th March 1678, to Vincent Placcius at the beginning of July 1678, and on 18th September 1679 to Huygens. Leibniz's friend, the Abbé Soudry, was asked to oversee the printing, but he died in 1678. The manuscript then came into the hands of the Hofmeister (court-master) of Count Philipp Christoph von Königsmark, Friedrich Adolf Hansen, who passed it on to Christophe Brosseau, the Hanoverian Resident in Paris. Brosseau in turn gave it to Isaac Arontz, who travelled to Hanover, and on his way it got lost [Knobloch 1993, p. 11]. In a letter from the year 1683 to Brosseau Leibniz somehow resignedly wrote that he never had much luck with mailings from France. The loss was painful for Leibniz, since he wanted to achieve his admission to the Académie Royale des Sciences on the basis of this manuscript, as he wrote to Huygens in 1679 several times.

In February 1682 Leibniz's essay *De vera proportione circuli ad quadratum circumscriptum in numeris rationalibus expressa* (The true ratio of circle and circumscribed square expressed in rational numbers) [Leibniz 2011, p. 9-18] appeared in the Acta Eruditorum. In this work Leibniz presented his $\pi/4$-series and derived series without proof. With this series of Leibniz new trouble occurred. Jacques Ozanam had published this series in 1684 without mentioning Leibniz's name, against which Leibniz objected in a 1685 review and in a letter to Simon Foucher from the year 1686. When asked about this, Ozanam answered Foucher that he claimed the same rights as Leibniz concerning this series, since Leibniz had only incompletely introduced him into the quadrature of the circle [Knobloch 1993, p. 12f.].

In the year 1687 Leibniz had already found further allies concerning his new analysis, about whom we will have to talk later, namely the brothers Jacob and John Bernoulli from Basel.

In a letter of 25th December 1687 Jacob Bernoulli asked Leibniz to send him the geometry with which he had made such excellent discoveries as the quadrature of the circle and the measuring of other curves. The brother John wrote to Leibniz 1698 [Knobloch 1993, p. 13]:

> *'You would carry out a task useful and welcome to the public if you would issue the tract which you have composed on it.'*
>
> (Du würdest eine für die Öffentlichkeit nützliche und willkommene Aufgabe erledigen, wenn Du den Traktat herausgäbst, den du darüber verfaßt hast.),

and Leibniz answered

> *'My treatise on the arithmetical quadrature had met with approval at the time it was written. Now it would please the beginners in our methods more than it would please you.'*
>
> (Meine Abhandlung über die arithmetische Quadratur hätte damals Beifall finden können, als sie geschrieben wurde. Jetzt würde sie mehr Anfängern in unseren Methoden gefallen als dir.)

Hence this important work had to wait until our time to be completely printed for the first time.

The final version contains 51 theorems and 24 *scholia*[10], including Leibniz's criterion for the convergence of alternating series [Knobloch 2002, p. 60]. Completely surprising from the modern point of view is the nearly unbelievable

[10] A *scholion* (Greek) or *scholium* (Latinised) is an explanatory remark.

Fig. 4.4.1. Jacob Bernoulli [Niklaus Bernoulli, 1687] and John Bernoulli ([Painting: Joh. R. Huber about 1740] Alte Aula, University of Basel)

insight of Leibniz into the nature of infinitesimal quantities. Leibniz explains in theorem 6 the quadrature of curves entirely in the Riemannian sense, i.e. approximately 200 years before the theory of integration took its modern shape! Knobloch writes [Knobloch 1993, p. 15]:

> *'The 'very subtle' (spinosissima) theorem 6 lays the ground of the geometry of infinitesimals by means of analytical geometry. It shows that a curvilinear bounded area can be approximated with arbitrary accuracy by a rectilinear bounded staircase-shaped area. Arbitrary accuracy means:* **the error can be made smaller than any prescribed positive number**[11]*.'*
>
> (Der 'sehr spitzfindige' (spinosissima) Satz 6 gibt eine Grundlegung der Infinitesimalgeometrie mittels der analytischen Geometrie. Er zeigt, daß eine krummlinig begrenzte Fläche durch eine geradlinig begrenzte treppenförmige Fläche beliebig genau angenähert werden kann. Beliebig genau heißt: der Fehler kann kleiner als jede vorgegebene positive Zahl gemacht werden.)

This characterisation of infinitesimal quantities ('smaller than any prescribed positive number') is modern Weierstrassian thinking! Why did this characterisation not prevail in these days; why did Euler in the 18th century still compare the computation with infinitesimals with the computation with zeros, and why did even Leibniz in his correspondence always give different, heuristical explanations of the 'infinitely small' quantities to philosophers and

[11] My boldface.

mathematicians? One time an infinitely small quantity is simply 'negligible', for instance if Leibniz writes in 1701 against a critique of his infinitesimal calculus in the 'Journal de Trévoux' [Leibniz 2004, Vol. V, p. 350] (cited after [Volkert 1988, p. 98] with corrections):

> *'I add to this* [...] *that here one cannot grasp the infinite in a strong sense, but only, as one says in optics, that the rays of the sun come from a point infinitely far away and thus can be seen as parallel.'*
>
> (J'ajoûterai même á ce que cet illustre Mathématicien en a dit, qu'on n'a pas besoin de prendre l'infini ici á la rigeur, mais seulement comme lorsqu'on dit dans l'optique, que les rayons du Soleil viennent d'un point infiniment éloigné, et ainsi sont estimés parallèles.)

Even a less 'physical' explanation based on classical Greek mathematics is given, for example in the work *Responsio ad nonnullas difficultates a Dn. Bernardo Niewntiit circa methodum differentialem seu infinitesimalem motas* (Riposte to some of Mister Bernard Nieuwentijt's objections brought forward against the differential or infinitesimal method), which appeared in the Acta Eruditorum in July 1695 [Leibniz 2011, p. 273f.]:

> *'I namely agree with Euclid, Book 5, Definition 5*[12] *that homogeneous*[13] *quantities are comparable only if the one* [quantity], *if multiplied with a finite number, can surpass the other* [quantity]. *And what does not differ by such an amount I define as equal. This was done also by Archimedes and all others coming after him. And exactly that is meant if one says, that the difference* [of two quantities] *is smaller than an arbitrary given* [quantity]. *'*
>
> (Ich halte nämlich mit Euklid, [Elementa] Lib. 5, Defin. 5, homogene Größen nur dann für vergleichbar, wenn die eine [Größe], falls man sie mit einer [hier] aber endlichen Zahl multipliziert, die andere [Größe] übertreffen kann. Und was sich nicht um eine solche Größe unterscheidet, erkläre ich für gleich. Dies haben auch Archimedes und alle anderen nach ihm so gehalten. Und genau dies ist gemeint, wenn man sagt, dass die Differenz [zweier Größen] kleiner als eine beliebige gegebene [Größe] ist.)

---

[12]Meant is Definition 4 [Euclid 2013b, p. 114]: *'Magnitudes are said to* **have a ratio** *to one another which are capable, when multiplied, of exceeding one another.'* The background is the axiom of Archimedes: Given any two numbers $y > x > 0$ a natural number N can always be found, such that $N \cdot x$ is larger than the quantity y, i.e. that $Nx > y$ holds. The axiom of Archimedes excludes infinitesimals from the real numbers and this 'gap' was used by Leibniz in the same way as it is used today in nonstandard analysis.

[13]Quantities are homogeneous if they are 'similar in dimension', i.e. two lengths, two areas, two volumes, etc., but not an area and a volume.

Fig. 4.4.2. Bernard Nieuwentijt [Engraving: unknown artist, probably about 1700] and Pierre Varignon [Engraving: unknown artist, probably about 1700]

But why had Leibniz to defend his infinitesimals at all? After the first publication of Leibniz's new differential calculus under the title *Nova methodus pro maximis et minimis, itemque tangentibus, quae nec fractas, nec irrationales quantitates moratur, & singulare pro illis calculi genus* (New methods for the determination of maxima and minima as well as of tangents; a method which is neither hindered by fractions, nor by irrational quantities; and on the unique essence of the calculus necessary for it) in the Acta Eruditorum in October 1684 first critics entered the scenes for whom the foundations of the infinitesimal calculus seemed too weak. One of them was Bernard Nieuwentijt, born on 10. August 1654 in Westgraftdijk in North-Holland. He studied medicine and law in Leiden and Utrecht, became mayor of the city of Purmerend close to Amsterdam, and wrote various books on theological and philosophical topics [Nagel 2008, S. 200]. In the years 1694 and 1695 two books of Bernard Nieuwentijt were published which Leibniz got in June 1695. In these books the founders of the infinitesimal calculus were accused that a rigorous justification of the calculus was missing. Leibniz's above statement comes from his riposte to Nieuwentijt's accusations.

Sometimes 'infinitely small' meant a kind of convergence to zero for Leibniz. This point of view is expressed in a letter to Pierre Varignon (1654–1722) dated 2nd February 1702 [Leibniz 1985–1992, Vol. IV, p. 253]

> *'At the same time one has to consider that the incomparably small quantities, taken in the usual sense, are by no means unchanged and*

> *determined; rather, since they can be assumed arbitrarily small, they play the same role as the infinitely small in the strong sense in our geometrical considerations. For if an opponent would contradict our explanations our calculus shows that the error will be smaller than any determinable error, since it is in our power to keep the incomparably little, which can be an arbitrarily small quantity, small enough for this purpose.'*
>
> (Zugleich muß man jedoch bedenken, daß die unvergleichbar kleinen Größen, selbst im gebräuchlichen Sinne genommen, keineswegs unverändert und bestimmt sind, daß sie vielmehr, da man sie beliebig klein annehmen kann, in unseren geometrischen Überlegungen dieselbe Rolle spielen wie die unendlichkleinen im strengen Sinne. Denn wenn ein Gegner unserer Darlegungen widersprechen wollte, so zeigt sich durch unseren Kalkül, daß der Irrtum geringer sein wird als jeder bestimmbare Irrtum, da es in unserer Macht ist, das Unvergleichbarkleine, das man ja immer von beliebig kleiner Größe nehmen kann, für diesen Zweck klein genug zu halten.)

And Leibniz finally also identified 'infinitely small' with zero, as in the work *Theoria motus abstracti*, a work concerning the theory of motion (cited after [Lasswitz 1984, Vol. 2, p. 464]):

> *'And this is the foundation of Cavalieri's method, whereby its truth is evidently proved by thinking of certain so-called rudiments or beginnings of the lines and figures, smaller than any arbitrary assignable quantity.'*
>
> (Und dies ist das Fundament der cavalierischen Methode, wodurch ihre Wahrheit evident bewiesen wird, indem man gewisse sozusagen Rudimente oder Anfänge der Linien und Figuren denkt, kleiner als jede beliebige angebbare Größe.)

A number which is 'smaller than any arbitrary assignable quantity' can only be the number zero. This is something completely different from 'smaller than any prescribed positive number'

Why did Leibniz exclude his rigorous definition of an infinitesimal quantity from the final version of *De quadratura arithmetica* which he had already in 1676? He gave the answer in the final version himself! Leibniz saw the deterrent effect of the proof of theorem 6 on the reader. He had used the word 'scrupulositas' (over-accurate) and asked the reader to skip theorem 6 for the first reading [Knobloch 1993, p. 15]. Depending on the intellectual state of his correspondence partners Leibniz operated with adapted definitions which would not overstrain his corresponding partners.

In the final version of *De quadratura arithmetica* Leibniz also presented rules for the calculation with infinities (large and small). Eberhard Knobloch has analysed them in [Knobloch 2002]. In a bundle from the spring of 1673, called *Collectio mathematica* and reprinted as numbers 9, 10, 12, 14, 15, 16, 17 in Volume VII, 4 of the Academy Edition [Leibniz 2008a], Leibniz sought for an explanation of the notion 'indivisible'. He declared it as 'infinitely small quantity', but now had to define the term 'infinitely small'. What is 'infinitely small' supposed to mean? Leibniz gives two answers, based on two of his calculation rules [Knobloch 2008, p. 175]. The first answer stems from the equation

$$\text{finite} : \text{infinitely small} = \text{infinite}.$$

But what then is the meaning of 'infinite'? Leibniz gave an explanation in 1673: A quantity which is larger than any assignable number. Knobloch justly points out, that thereby a potential infinity is not enough; here in fact a cardinality beyond all natural numbers is requested, although Leibniz did not make this point a subject of discussion, of course. Note again the contrast to Leibniz's statements in *De quadratura arithmetica*, where he defined 'infinite' as a quantity larger than any prescribed number.

Concerning the second explanation Knobloch [Knobloch 2008, p. 176] remarks that it leads to nothing and was not further exploited by Leibniz, which is why we ignore it.

The outstanding definition of infinitely small quantities in the sense that they are smaller than any prescribed quantity, given by Leibniz in *De quadratura arithmetica*, comes with a new culture of proofs, whereby Archimedean precision served as a model to introduce the new infinitesimal quantities [Knobloch 2008, p. 183].

5 The Apparent Relaxation

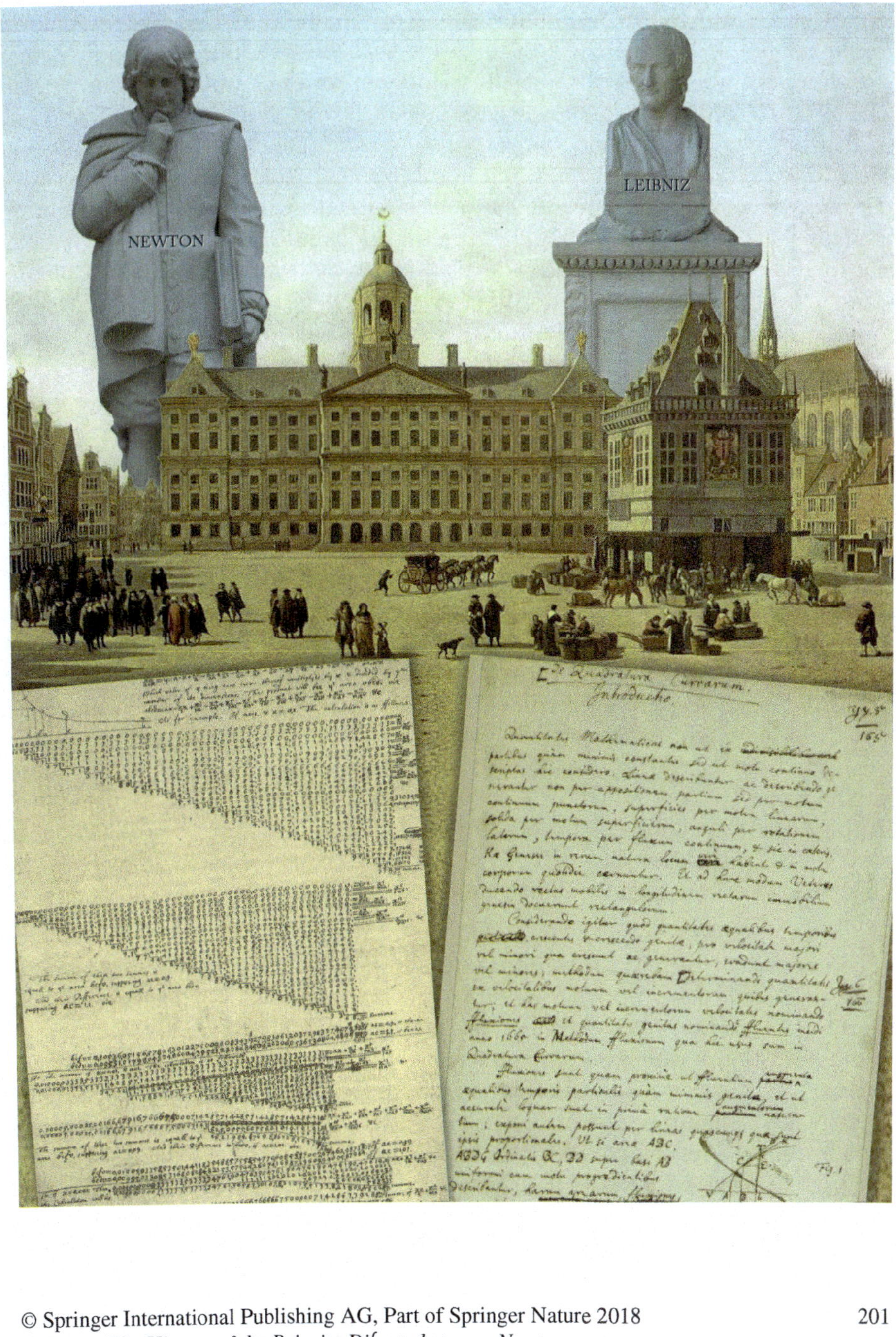

T. Sonar, *The History of the Priority Dispute between Newton and Leibniz*, https://doi.org/10.1007/978-3-319-72563-5_5

What did Leibniz and Newton know of each other at the beginning of the year 1676? Newton had heard not a thing from Leibniz since all of the correspondence took place with Oldenburg and Collins. Leibniz apparently knew the name of Newton, but perceived him only as one of several English mathematicians through the communications of Collins and Oldenburg. Leibniz could not know how far Newton had got with his calculus of fluxions, since Newton kept almost all of his manuscripts and did not publish them. At the end of 1675 Leibniz had knowledge about a few of Newton's results but no derivation whatsoever, and these results were all concerned with infinite series. Newton on the other hand did not know that excerpts of his manuscript *De analysi* and descriptions of his mathematical achievements were sent to Leibniz [Westfall 2006, p. 261f.]. This situation would change in the course of the year 1676 with two letters which in Newton's terms are called 'Epistola prior' and 'Epistola posterior' (first and second letter), respectively, marking the single contact between the two scientist before 1693, but would have far-reaching effects to turn the cold war into a hot one later.

5.1 The Beginning of the Correspondence: *Epistola prior*

On 12th May 1676 Leibniz wrote to Oldenburg [Turnbull 1959–77, Vol. II, p. 3ff.]. From Georg Mohr, wrote Leibniz, who got the information from Collins, he had received two series, namely the series (4.7) of the sine for a given arc z, and of the arc for given sine of x. Leibniz praised the rare elegance of these series; in fact he got them already a year earlier from Collins in the letter by Oldenburg of 22nd April 1675, but it seems that only now he realised their elegance. He may have forgotten these series, as A. Rupert Hall has suspected [Hall 1980, p. 63], but perhaps he considered them not important at the time. He now asked for a proof of this series representation; in return he promised to send his own work in this direction. We have to note here that Leibniz did not know how to derive the series of the sine in May 1676. In any case, Oldenburg and Collins wanted to supply Leibniz with the necessary information and they hastened to get this informations directly from Newton.

Those were hard times for the English. James Gregory, the only mathematician who could keep up with Newton's genius, had died of a stroke in Edinburgh in October 1675. Newton was again struggling with sceptics of his theory of colours which became increasingly burdensome, and Oldenburg was accused of treachery and alien espionage by Hooke. Furthermore it seemed that Newton had got out of mathematical studies, as Collins yet remarked in a letter to Gregory dated 29th October 1675 [Turnbull 1959–77, Vol. I, p. 356]. Collins wrote that he had not heard anything of Newton for 11 or 12 months; he was probably involved in chemical studies.

The possibility of publication, at that time unthinkable for Newton anyway, was strongly restricted for English mathematicians: The London publishers had wrongly estimated the costs and profits of printing and selling mathematical books, in particular in the case of Barrow's books [Hall 1980, p. 63], and no publisher accepted a printing contract without advance payment; even Collins with his good contacts could not do anything to change the situation. The reservations about Leibniz had grown, too, but Leibniz had not yet felt that.

It seems almost curious against this background that Oldenburg and Collins immediately administered to Leibniz's needs. Collins began summarising Gregory's mathematical achievements as far as manuscripts and transcripts were available to him. This compilation is known under the title *Historiola*, which is comprised of 50 pages and will be looked at by Leibniz during his second trip to London in October. Excerpts of the *Historiola* can be found in [Turnbull 1959–77, Vol. II, p. 18ff.]. The document was probably too long in the eyes of Oldenburg; in any case Collins produced a shorter version, the *Abridgement* [Turnbull 1959–77, Vol. II, p. 47ff.] of 24th June 1676. Leibniz's request was communicated to Newton who now turned again to mathematics and started re-working his old manuscripts (the 'magic year' 1666 is a decade in the past!), thereby coming to new and interesting insights. Newton also composed a letter for Leibniz, to be sent via Oldenburg, with the *Epistola prior* beginning as follows [Turnbull 1959–77, Vol. II, p. 20ff.]:

> '*Most worthy Sir,*[1]
> *Though the modesty of Mr Leibniz, in the extracts from his letter which you have lately sent me, pays great tribute to our countrymen for a certain theory of infinite series, about which there now begins to be some talk, yet I have no doubt that he has discovered not only a method for reducing any quantities whatever to such series, as he asserts, but also various shortened forms, perhaps like our own, if not better. Since, however, he very much wants to know what has been discovered in this subject by the English, and since I myself fell upon this theory some years ago, I have sent you some of those things which occured to me in order to satisfy his wishes, at any rate in part.*'

In the *Epistola prior* Newton presents his binomial theorem in the form

$$\begin{aligned}&(P+PQ)^{m/n}\\&=P^{m/n}+\frac{m}{n}AQ+\frac{m-n}{2n}BQ+\frac{m-2n}{3n}CQ+\frac{m-3n}{4n}DQ+\text{etc.}\end{aligned}$$

without proof, and nine examples follow showing how one can compute the coefficients $A, B, C, \dots$ in particular cases. The letter contains yet other examples, but Newton writes [Turnbull 1959–77, Vol. II, p. 35]:

[1] The English translation starts at [Turnbull 1959–77, Vol. II, p. 32].

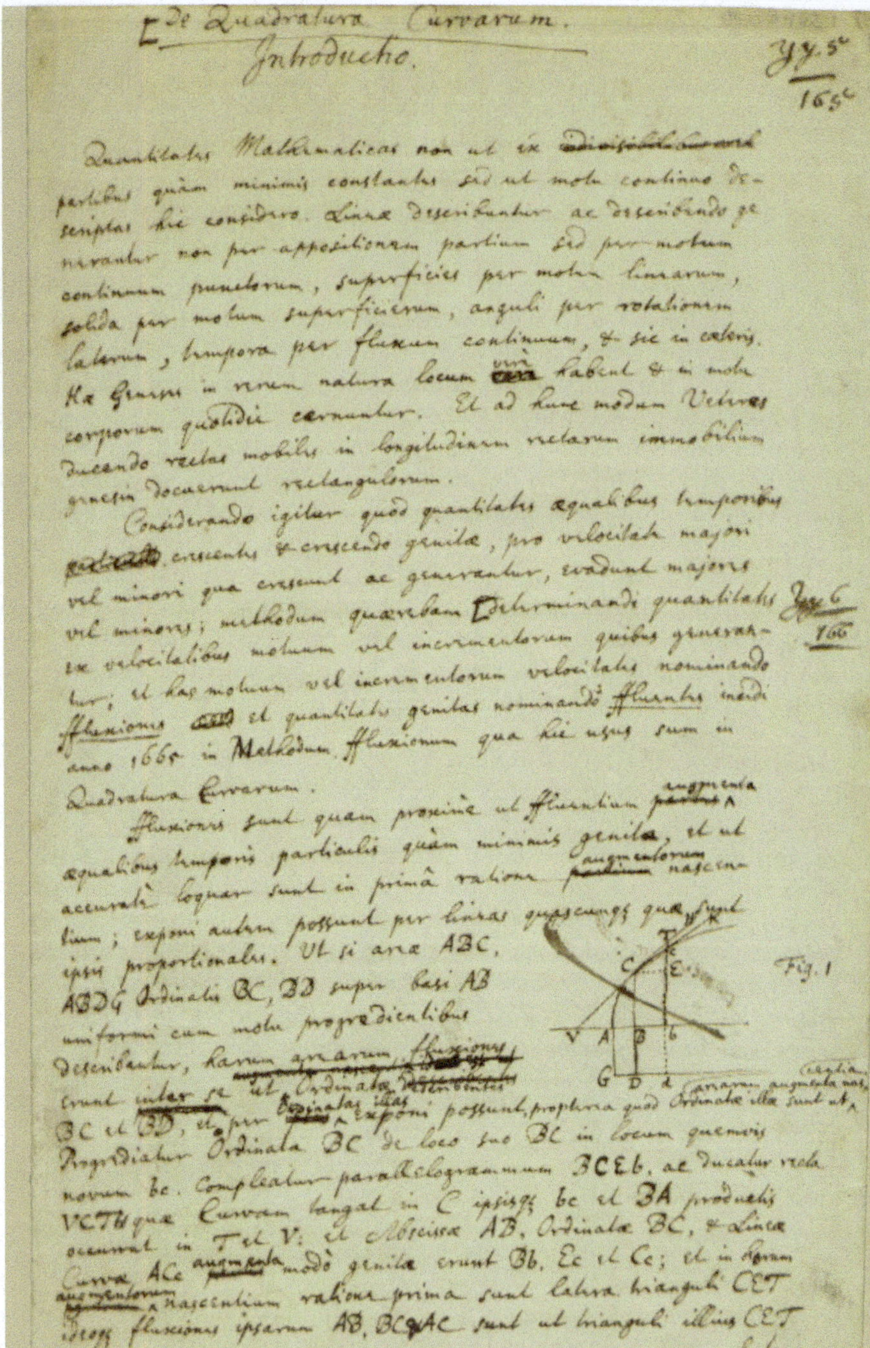

De Quadratura Curvarum.

Introductio.

Yy.5c / 165c

Quantitates Mathematicas non ut ex partibus quàm minimis constantes sed ut motu continuo descriptas hic considero. Lineæ describuntur ac describendo generantur non per appositionem partium sed per motum continuum punctorum, superficies per motum linearum, solida per motum superficierum, anguli per rotationem laterum, tempora per fluxum continuum, & sic in cæteris. Hæ Geneses in rerum natura locum vere habent & in motu corporum quotidie cernuntur. Et ad hunc modum Veteres ducendo rectas mobiles in longitudinem rectarum immobilium genesin docuerunt rectangulorum.

Considerando igitur quod quantitates æqualibus temporibus crescentes & crescendo genitæ, pro velocitate majori vel minori qua crescunt ac generantur, evadunt majores vel minores; methodum quærebam determinandi quantitates ex velocitatibus motuum vel incrementorum quibus generantur; et has motuum vel incrementorum velocitates nominando fluxiones et quantitates genitas nominando fluentes incidi anno 1665 in Methodum fluxionum qua hic usus sum in Quadratura Curvarum. (Yy.6 / 166)

Fluxiones sunt quam proxime ut fluentium augmenta æqualibus temporis particulis quàm minimis genita, et ut accurate loquar sunt in primâ ratione augmentorum nascentium; exponi autem possunt per lineas quascunque quæ sunt ipsis proportionales. Ut si areæ ABC, ABDG Ordinatis BC, BD super basi AB uniformi cum motu progredientibus describantur, harum arearum fluxiones erunt inter se ut Ordinatæ describentes BC et BD, et per Ordinatas illas exponi possunt, propterea quod Ordinatæ illæ sunt ut arearum augmenta nascentia. Progrediatur Ordinata BC de loco suo BC in locum quemvis novum bc. Compleatur parallelogrammum BCEb, ac ducatur recta VCTH quæ Curvam tangat in C ipsisque bc et BA productis occurrat in T et V: et Abscissæ AB, Ordinatæ BC, & Lineæ Curvæ ACc augmenta modò genita erunt Bb, Ec et Cc; et in horum augmentorum nascentium ratione prima sunt latera trianguli CET, ideoque fluxiones ipsarum AB, BC et AC sunt ut trianguli illius CET latera

Fig. 1

Fig. 5.1.1. Introductio ad quadraturam curvarum, Original page of the Newton manuscript (University of Cambridge, Digital Library)

> *'How the areas and lengths of curves, the volumes and surfaces of solids or of any segments of such figures, and their centres of gravity are determined from equations thus reduced to infinite series, and how all mechanical curves may also be reduced to similar equations of infinite series, and hence problems about them solved just as if they were geometrical, all this would take too long to describe. Let it suffice to have reviewed some examples of such problems; ...'*

Hence, in the *Epistola prior* again no proofs were given but only results which were known in one or another form. Leibniz researcher Hofmann has deduced from this that '*... and yet everything was done to prevent Leibniz from, as it were, improperly penetrating the world of Newton's thought*', but that seems to be a verdict much too harsh in my eyes. A. Rupert Hall has presented three objections against this verdict which are worth being noted [Hall 1980, p. 65f.]. Firstly: Why should Newton have written a complete treatise to a person who was a stranger to him? Secondly: Leibniz wanted to get information about the theory of infinite series but not about the calculus of fluxions, of which even Oldenburg knew nothing. Thirdly: Newton's *Introductio ad quadraturam curvarum* (Introduction to the quadrature of curves) was printed only in 1710 and was definitely written later than 1676. Thus, Newton could not have written anything on the term-wise integration of infinite series.

The *Epistola prior* reached Oldenburg on 23rd June 1676, was read at a meeting of the Royal Society on 25th June and approved, and was sent to Leibniz on 5th August 1676, who received the document on 24th August[2]. Oldenburg did not send the letter by ordinary mail but entrusted it to the young German mathematician Samuel König who was travelling from London to Paris. Leibniz was not at home and hence König deposited the letter at a German apothecary in Paris where Leibniz found it on 24th August. Leibniz replied with a hastily jotted letter on 27th August and remarked that the letter had to wait for him for a few days. Unfortunately, Oldenburg omitted this remark when he copied the letter into the 'letter book' of the Royal Society, probably because it seemed not important to him [Hofmann 1974, p. 232].

The *Epistola prior* was printed for the first time in the third volume of John Wallis's *Opera mathematica* in the year 1699. There we find the date of dispatch falsely given as 6th July (26th June in the Julian calendar)! In the later indictment *Commercium epistolicum D. Johannis Collins et aliorum de Analysi promota* of the Royal Society from 1712 (date of print) this wrong date was adopted; it still was adopted in the second edition of the *Commercium epistolicum* in 1722 and also in Newton's anonymous review *An Account of the Book entituled Commercium Epistolicum Collinii & aliorum* [Hall 1980, p. 276] published in the year 1715.

[2]Hall notes 16th August 1676 as the date of receipt [Hall 1980, p. 64], but here we rather trust Hofmann [Hofmann 1974, p. 232] who was a true Leibniz scholar.

It seems not very likely that Newton in 1715 should not have looked up the date when the letter was actually sent in the papers of the Royal Society (he was their president since 1703). Hence, in the *Commercium epistolicum* it was suggested that it took Leibniz six weeks to write a reply to the *Epistola prior*! Newton, who was the actual editor of the *Commercium epistolicum*, apparently not only wanted to leave the reader in the belief that the letter was sent to Leibniz on 6th July, but also that Collins's *Historiola* was sent to Leibniz on 5th August. However, the *Historiola* was never sent to Leibniz; only during his second journey to London in October 1676 could he have glimpsed it.

Two drafts of the hastily jotted answer of Leibniz are known. Since the first apparently did not appeal to Leibniz he began writing the second. It seems that he had recognised the meaning of a correspondence with Newton and had feared that a longer delay in the reply to the *Epistola prior* would place him in a bad light in England. The letter Leibniz eventually sent to Oldenburg on 27th August [Turnbull 1959–77, Vol. II, p. 57ff.] exhibited bad handwriting, contained errors in formulae, and Collins could not decipher some things in his transcription intended for Newton. As a result a document flawed by serious errors went to Newton [Hofmann 1974, p. 233]. To make things worse, Leibniz had written [Turnbull 1959–77, Vol. II, p. 65]:

> *'I received your letter of a month ago (dated 26 July) only yesterday 26 August, ...'*

to create the impression that he had practically answered *immediately*. That was not true; however, only three days lay between the receipt of the letter and the reply. He then went on [Turnbull 1959–77, Vol. II, p. 65]:

> *'Your letter contains more numerous and more remarkable ideas about analysis than many thick volumes published on these matters. For this reason I thank you as well as the very distinguished men, Newton and Collins, for wanting me to participate in so many excellent speculations. Newton's discoveries are worthy of his genius, which is so abundantly made manifest by his optical experiments and by his catadioptrical tube. His method of obtaining the roots of equations and the areas of figures by means of infinite series is quite different from mine, so that one may wonder at the diversity of paths by which one can reach the same conclusion.'*

Here Leibniz clearly stated that Newton's methods of quadrature by means of infinite series were completely different from his own and he did not withhold praise for Newton. He wrote furthermore that his methods were based on a general theory which stemmed from certain transformations. A general theory of such transformations, leading to rational terms and hence opening the way to a series expansion by division similar to the work by Mercator, should be of the highest importance for analysis. Leibniz also gave an example and wrote quite openly; perhaps he had the hope that the English would then communicate more details to him [Hofmann 1974, p. 235]. The example in question is the quadrature of the circle $y^2 = 2ax - x^2$. By means of the transformation $ay = xz$, stemming from the transmutation theorem[3], Leibniz succeeded in arriving at a series representation with rational summands in which he was able to integrate term-wise. However, Leibniz was not very open here since he kept the origin of the transformation $xz = ay$ secret, because he did not want to reveal the transmutation theorem [Hofmann 1974, p. 236]. Leibniz is not very explicit even in the description of other results as Hofmann noted [Hofmann 1974, p. 237]:

> '*Unfortunately Leibniz has expressed himself in an over-obscure manner in this passage* [where he mentioned the series expansion of the hyperbola] *since he wished to preserve the secret of his way of deduction.*'

Leibniz wrote he had derived the series

$$y = x - \frac{x^2}{2!} + \frac{x^3}{3!} \mp \ldots$$

from $x = \ln \frac{1}{1-y}$, and if $x = \ln(1+y)$ this would yield

$$y = x + \frac{x^2}{2!} + \frac{x^3}{3!} + \ldots.$$

We recognise here the series of $e^x - 1$ in the second case, and of $1 - e^{-x}$ in the first case, and if we replace in the second case $1 + y$ by $1/(1 - y)$ the second series becomes the first.

In the later indictment *Commercium epistolicum D. Johannis Collins et aliorum de Analysi promota* of the Royal Society from 1712 (date of print) the methods communicated by Leibniz were simply depicted as being not new at all. These methods were said to be the same which could be found in the writings of Gregory and already in the work of Barrow. Furthermore it was claimed that Leibniz had taken these methods directly from Newton and had only interchanged the signs of x and y. That Leibniz had derived the series expansion of $e^x - 1$ in a manner completely different from Newton could not be seen from Leibniz's letter [Hofmann 1974, p. 237f.].

[3]Refer to (4.2) and note $dy/dx = (a-x)/(\sqrt{2ax - x^2})$, hence $z = y - dy/dx \cdot x = \sqrt{2ax - x^2} - (ax - x^2)/\sqrt{2ax - x^2} = ax/y$, yielding $yz = ax$. Note that in the writings of Leibniz the axes x and y are always interchanged, because he drew x to the right but y downwards. Interchanging in our result the roles of x and y then $xz = ay$ follows.

Newton had written in the *Epistola prior* [Turnbull 1959–77, Vol. II, p. 39]:

> '*From all this* [previously described] *it is to be seen how much the limits of analysis are enlarged by such infinite equations: in fact by their help analysis reaches, I might almost say, to all problems, the numerical problems of Diophantus*[4] *and the like excepted.*'

In his reply Leibniz can not follow Newton's opinion that the theory of series would solve all problems, except Diophantine and related ones [Turnbull 1959–77, Vol. II, p. 71]:

> '*What you and your friends seem to say, that most difficulties (Diophantine problems apart) are reduced to infinite series, does not seem so to me. For there are many problems, in such a degree wonderful and complicated, such as neither depend upon equations nor result from squarings, as for instance (among many others) problems of the inverse method of tangents which even Descartes admitted to be beyond his power.*'

The 'inverse method of tangents' denotes the solution of differential equations in geometric language: Find a function from given properties of its tangent (i.e. the derivative of a function).

In the later indictment *Commercium epistolicum D. Johannis Collins et aliorum de Analysi promota* of the Royal Society from 1712 (date of print) it is stated [Turnbull 1959–77, Vol. II, p. 75, Remark (33)]:

> '*If differential equations had already become known to Mr Leibniz, he would not have said that problems in the inverse method of tangents do not depend on equations.*'

As Leibniz noted in the margins of his copy of the *Commercium epistolicum* he had used the word 'equation' in the commonplace sense and not in the sense of 'differential equation'. In the same sense Newton had used the word 'equation' [Hofmann 1974, p. 241].

However, if one reads further in Leibniz's letter it becomes immediately clear that he surely had advanced to the solution of differential equations. In particular he wrote on the problem of de Beaune[5], asking for a function with a constant subtangent everywhere. Leibniz had found this problem in a letter of de Beaune to Descartes [Turnbull 1959–77, Vol. II, S. 71]:

---

[4] Diophantine problems are problems in number theory for which Newton developed only little interest – in contrast to Leibniz [Turnbull 1959–77, Vol. II, p. 46, Remark (18)].

[5] Florimond de Beaune (1601–1652) was a French amateur mathematician who was appreciated highly by Descartes.

> *'I myself, however, on the day, indeed in the hour, when I first began to seek it, solved it at once by a sure analysis.'*

In fact, Leibniz had found the differential equation

$$\frac{y}{C} = \frac{dy}{dx}$$

and its solution $x = C \ln y$, where C denotes the constant length of the subtangent. The function looked for is hence an exponential function.

Leibniz wrote [Turnbull 1959–77, Vol. II, p. 64] '... quarum una est huius naturae ...' (... [curves,] of which one has this nature ...), but unfortunately Collins falsely transcribed 'huius' as 'ludus' (game) and this faulty transcription found its way into Wallis's third volume of his *Opera mathematica*, and from there eventually into the *Commercium epistolicum*. Newton used this false transcription for an attack on Leibniz [Hofmann 1974, p. 241].

Obviously Leibniz did show the *Epistola prior* as well as his reply to his friend Tschirnhaus. Thereupon the latter composed his own letter to Oldenburg dated 1st September 1676 which he apparently did not show to Leibniz, since it included some passages which Leibniz would not have let pass [Hofmann 1974, p. 250]. Promptly Tschirnhaus's letter was later cited in the *Commercium epistolicum* in order to use it against Leibniz. However, the letter was cited after a grossly faulty transcript made by Collins [Turnbull 1959–77, Vol II, p. 84f.]. Collins could not read some words and even formulae and sent the copy to Wallis who was able to correct at least some of Collins's mistakes. At the beginning of his letter Tschirnhaus remarked that he had waited for quite a long time for an answer to his letter written months ago. He was grateful for being included in the correspondence between Leibniz and the English. He had, wrote Tschirnhaus, read the letter from England (the *Epistola prior*) only superficially but he had missed the infinite series which Leibniz had found for the quadrature of the circle segment. Then he praised Leibniz and his mathematical achievements in glowing terms.

In the later indictment *Commercium epistolicum D. Johannis Collins et aliorum de Analysi promota* of the Royal Society from 1712 (date of print) it was stated that it was highly unbelievable that Tschirnhaus had not seen the related series of the inverse tangent of Gregory. This statement in the *Commercium epistolicum* is provably wrong since we only find the series for the tangent in the *Abridgement* [Turnbull 1959–77, Vol. II, p. 47ff.] of 24th June 1676. In the *Historiola*, of which the English wrongly stated that it was sent to Leibniz, this whole section of the *Abridgement* is missing [Hofmann 1974, p. 250f.].

Tschirnhaus also praised Newton's theory of series, but, so he wrote, he believed in the existence of a universal method so that one could get away completely without the expansion in series. Similar remarks would apply to Gregory's results. Statements of this kind would have never been made by Leibniz, in particular not in that manner [Hofmann 1974, p. 251]! It is this passage in Tschirnhaus's letter that gave rise to a cutting response of Newton to Leibniz on the 3rd November 1676; in Newton's eyes Tschirnhaus was presumably a mere a straw man of Leibniz who did not deserve to get an answer. Tschirnhaus further reported on problems in the solution of equations – very interesting for Collins as well as Newton and Wallis who received a transcript – and finished his letter with the announcement of a journey to Italy and the offer to Oldenburg to gladly carry out any orders there for him.

When Wallis received the faulty transcript of the letter of Tschirnhaus he was just working on his book on algebra and so he was grateful to receive a copy of this letter. For a thorough history of Wallis' book on algebra see [Stedall 2002]. Wallis answered Collins from Oxford and now Collins started to compile material for an answer to Tschirnhaus. This answer of 10th October 1675 contained a great praise for Leibniz's excellent methods of transformation and his series of the circle sector which was not among the communicated series of Newton. Collins stated the bad convergence of Leibniz's series which he had learned from a letter of Wallis of 26th September 1676. Admittedly Newton's methods were also useable in a fairly general sense and Collins now wrote in detail on the problem of Davenant to find the first four successive terms of a geometrical progression if the sum of their squares and their cubes are given. Finally Oldenburg thanked Tschirnhaus for the offer to carry out orders in Italy, and he ordered some books of mathematical contents. This letter to Tschirnhaus marks the end of the correspondence between Oldenburg and Tschirnhaus [Hofmann 1974, p. 257]. Although the letter arrived in Paris on schedule it was never answered, since Tschirnhaus probably found no time in Rome and Oldenburg died in September 1677. Whether Tschirnhaus ever came in closer contact with mathematicians of the Royal Society again is not known.

For us the letter is three-foldly important. Firstly it shows clearly that Leibniz could really learn nothing from Tschirnhaus concerning the achievements of the English in the new mathematics; secondly it shows the extremely (and unprejudiced) attitude of Collins towards Leibniz's results, and thirdly it contains the clear statement that the infinite series of the circle sector was not submitted to Leibniz.

5.2 The Second Journey to London

Leibniz left Paris on Sunday, 4th October 1676, in a stagecoach; the Hanoverian, in whose service he was officially since the beginning of the year, wanted to see his new councillor at last *in persona* [Antognazza 2009, p. 176].

On the evening of 10th October Leibniz arrived in the French port of Calais. The weather is so tempestuous that the post ship could not dare to depart, as Leibniz wrote [Müller/Krönert 1969, p. 45]:

> '... *where I had to lay still for 5 days because the mail ship was hindered to depart by storm and contrary wind.*'
>
> (... alda ich 5 tage stillliegen müßen bis das paquetbot so durch sturm und contrari wind verhindert worden fortgehen können.)

Hence he boarded ship for the passage to Dover only on 15th October where Leibniz stayed one night and on the next morning he travelled on to London, where he eventually arrived on 18th October. He stayed for only a little more than one week [Müller/Krönert 1969, p. 45]:

> *'I stayed in London for little more than a week, because I wanted to enquire about some very important things as I then in fact did.'*

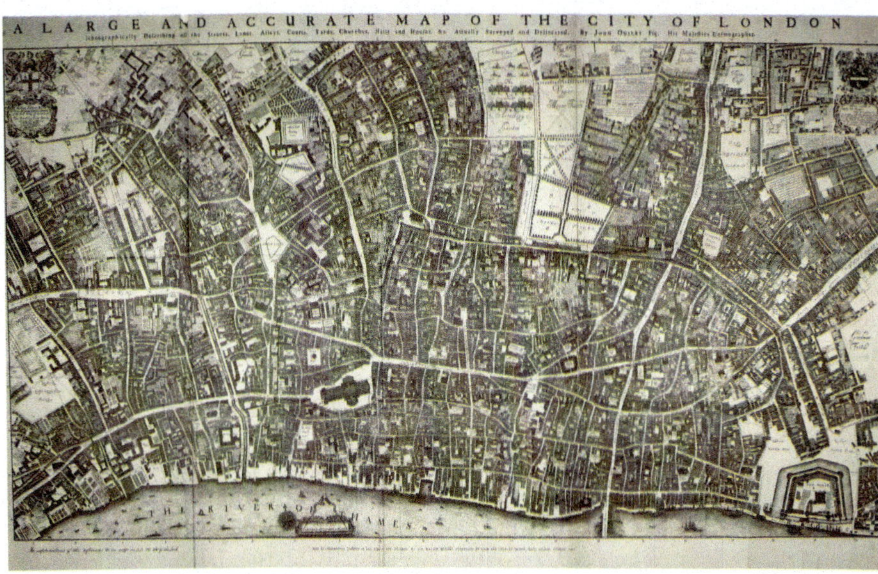

Fig. 5.2.1. Map of London by John Ogilby of the year 1676, edited by Geraldine Edith Mitton 1908

(In Londen bin ich etwas über eine woch geblieben; dieweil ich einiger sehr wichtigen dinge wegen mich erkundigen wollen, wie ich dann auch gethan.)

First of all he visited his fellow countryman Henry Oldenburg, of course; we do not know whether he met other fellows of the Royal Society except Oldenburg and Collins [Hofmann 1974, p. 277]. It is certain, however, that he did neither meet Newton nor Wallis. The regular meetings of the Royal Society had not yet been resumed, so that Leibniz can present his calculating machine only to Oldenburg but not during a meeting of the Royal Society. Leibniz did also present some mathematical works on algebraic themes which he had already promised to show a while ago. Collins granted him access to works of Newton and Gregory; amongst others Newton's *De analysi*. Hofmann has analysed in detail which excerpts Leibniz did make from this and other works [Hofmann 1974, p. 278f.]. It is very interesting that Leibniz's excerpts only concern Newton's series expansions; he did not make any notes from the sections containing infinitesimal methods. We hence may safely deduce that these sections did not contain anything new to Leibniz.

Directly after having made excerpts from *De analysi* Leibniz started excerpting from a letter of Newton to Collins of the year 1672 [Turnbull 1959–77, Vol. I, p. 229f.]. It involved the application of logarithmic scales in the solution of equations. On this occasion Leibniz excerpted also Newton's remark concerning one of Gregory's series for the volume of an ellipsoid which is intersected by two pairs of parallel planes which are orthogonal to each other. The volume is given by the double integral

$$2\int_0^a \int_0^b r\sqrt{1-\frac{y^2}{r^2}-\frac{x^2}{c^2}}\,dx\,dy,$$

if the ellipsoid is described by $x^2/r^2 + y^2/c^2 + z^2/r^2 = 1$. Gregory had succeeded in expanding the integrals into a double series in a and b and Collins had sent this series to Newton in a now lost letter of June or July 1672 [Turnbull 1959–77, Vol. I, p. 214f.]. Gregory wanted to know whether Newton would be aware of another, simpler series, but that was not the case. Newton had found exactly the series of Gregory and wrote [Turnbull 1959–77, Vol. I, p. 229]:

'I tried two or three others, but could find none more simple.'

More important in our context are Leibniz's excerpts from the *Historiola* because they show that Leibniz did not know it before [Hofmann 1974, p. 279]. In the context of the *Historiola* Leibniz also took excerpts from the corresponding letters of Gregory. On the cover sheet of the *Historiola* we find a request to Leibniz to hand it back to Collins after inspection.

In the later indictment *Commercium epistolicum D. Johannis Collins et aliorum de Analysi promota* of the Royal Society from 1712 (date of print) one could read that Leibniz would have learned his method for the computation of tangents from a letter of Newton to Collins of December 1672 which was included in the *Historiola*. In Leibniz's copy of the *Commercium epistolicum* he remarked that this statement was nonsense ('ineptum') [Hofmann 1974, p. 280]. In the second edition of the *Commercium epistolicum* published in 1722 it is stated that the *Historiola* was sent to Leibniz to Paris in 1676; we have already falsified this statement.

Why did Collins, being suspicious to a certain degree against Leibniz, grant his German visitor so freely access to the *Historiola* and Newton's manuscript *De analysi*? The reason is that Collins and Leibniz met for the first time during Leibniz's second journey to London and the visitor made quite an impression on Collins. Collins, however, was seriously ill. In September 1676 he had contracted a dangerous and painful blood poisoning which rendered his right arm unusable for some time so that he could not write. Leibniz could read English, but he was far from speaking it fluently; on the other hand Collins's Latin was insufficient for a conversation with Leibniz [Aiton 1991, p. 105f.]. In a letter dated 3rd November 1676 Collins wrote to Thomas Strode, a mathematician who corresponded with Collins and Gregory [Turnbull 1959–77, Vol. II, p. 109]:

> '... *The admirable Monsieur Leibnitz a German but a Member of the R S scarce yet middle aged was here last Weeke, being on his returne from Paris to the Court of the Duke of Hannover by whome he was importuned to come away and refuse such Emoluments as were offered him at Paris but during his stay here which was but one Weeke I was in such a Condition I could have but little conference with him; for being troubled with a Scorbutick humour or saltnesse of blood, and taking remedies for it they made me ulcerous and in an uneasy condition: however by his Letters and other Communications I presume I perceive him to have outtopt our Mathematicks quantum inter Lenta*[6] *& c his Combinatory tables are specious and not numerical* ...'

The excellent impression of Leibniz on Collins is also documented by a jointly written manuscript which Leibniz took with him on his return journey. It is a

[6]Following [Turnbull 1959–77, Vol. II, p. 109, Remark (4)] this is a somehow faulty citation from Vergil's Bucolica, Eclogue 1, verse 24f [Vergil 2001, p. 8] (my translation from German):

> *'But this* [city of Rome] *has her head so high above other cities as cypresses have theirs above shrubs of wild sage'* (verum haec tantum alias inter caput extulit urbes, quantum lenta solent inter viburna cupressi.)

In this context it is used to express that Leibniz is as high above mathematics as Rome is above other cities.

Fig. 5.2.2. Cityscape of Amsterdam (De Dam) about the end of the 17th c. Such a view was presumably offered to Leibniz when he visited Jan Hudde ([Painting: Gerrit Adrianensz, Berckheyde] Gemäldegalerie Alter Meister, Dresden)

simply folded page in folio and Leibniz has written *Colliniana* as a title. On the four pages of this folded page we find computations, diagrams, and also text, and Hofmann has conjectured that this page was used to clarify things in written form due to the mutual language problems [Hofmann 1974, p. 287f.].

On 29th October Leibniz boarded a ship in London which left the harbour only on 31st October and berthed in Gravesend on the same day, where the skipper waited for cargo for another four days. Storms prevented the departure into the Channel from 5th to 10th November so that the ship was tied up close to Fort Sheerness. Finally, on 11th November, the ship set sail on route to Rotterdam [Müller/Krönert 1969, p. 45f.]. Leibniz arrived in Amsterdam on 13th November and made the acquaintance of some men, including the mathematician and mayor of Amsterdam, Jan Hudde, with whom he discussed mathematical problems. Starting from Amsterdam Leibniz went on tour until 24th November. This tour led him to Haarlem, Leiden, Delft, and finally to The Hague. In Delft he visited the Dutch natural scientist and microscope maker Antoni van Leeuwenhoek (1632–1723). Simple microscopes existed even before van Leeuwenhoek, but van Leeuwenhoek was a ingenious lens grinder and he improved the existing instrument so much that up to a 270-fold magnification could be achieved. With his instrument he observed bacteria in pond water and in human saliva, and so he became a member of the Royal Society in 1680. He is also known as the discoverer of spermatozoa of humans and of insects.

Fig. 5.2.3. Antoni van Leeuwenhoek ([Painter: Jan Vercolje, about 1680, section] Museum Borhave Leiden)

On 18th November, Leibniz was the guest of Baruch (Benedictus) de Spinoza for some days, with whom he discussed the latter's *Ethica*, the theory of motion of Descartes, the characteristica universalis, and theological problems, mainly the question of the existence of God [Müller/Krönert 1969, p. 46], [Antognazza 2009, p. 178].

On 24th November Leibniz was back in Amsterdam but his health was weakened [Müller/Krönert 1969, p. 46]:

> *'The ship on which I slept for 10 nights ... has given an umbrage to my health as have the night boats on which I travelled from time to time in Holland to use the night and gain time. In addition there was the wet and cold air in Holland and change in diet, so that, when back in Amsterdam, I found myself not well, completely without appetite and not without some temperature and faintness. Thereupon I kept myself indoors for several days, kept myself warm, and brought myself nearly up again with a diet, ...'*
>
> (Alleine es hat mir das schiff darauff ich etwa 10 nacht geschlaffen ... einen anstoß an meiner gesundheit verursacht sowohl als die Nachtscheüten darauff ich in Holland hin und wieder gefahren, umb der Nacht mich zu bedienen und Zeit zu gewinnen, dazu dann die naße und kalte lufft in Holland, und verenderung der kost geholffen, also daß, alß ich wieder nach Amsterdam, als in mein centrum kommen, ich mich sehr unbaß, ganz ohne appetit und nicht ohne einiger hize und mattigkeit befunden, habe mich darauff vor etliche tage eingesperret, warm gehalten, und durch diaet wieder fast zu recht gebracht, ...)

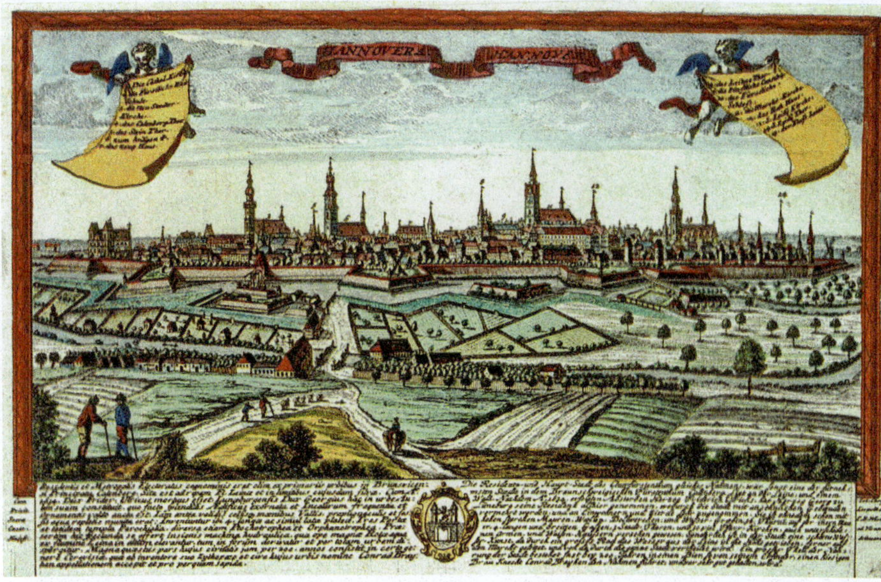

Fig. 5.2.4. Sight of Hanover from the north-west about 1730 ([Copperplate by F. B. Werner] Historisches Museum Hannover (Hanover))

Leibniz left Amsterdam at the end of November and arrived in Hanover sometime between December 10th and 15th, 1676. More than two months had passed since he left Paris. He immediately started to work and, as the librarian, took over the ducal library. He is the successor of the previous librarian Tobias Fleischer in whose flat in the library of the Leine palace he moved. He stayed in this flat until 1688. The library was founded only in 1665 by Duke John Frederick, and when Leibniz took over it contained 3310 Volumes and 158 manuscripts [Müller/Krönert 1969, p. 47]. Presumably still in December Leibniz made the acquaintance of some of his Hanoverian colleagues: the privy counsellor and chamber counsellor Otto Grote, the privy counsellor and vice chancellor (from 1677) Ludolf Huge and the privy and chamber counsellor Hieronymus von Witzendorf, who became president of the chamber and later inspector of mines in Clausthal.

At the beginning of 1677 Leibniz presented a work plan for the enlargement of the library to the Duke and declared his intention to get news on all important scientific developments by means of correspondence with learned men in Germany, Italy, France, England, and Holland. It is here where the unbelievably extensive exchange of letters starts which Leibniz continued until the end of his life. His correspondence was global indeed; he even corresponded with Jesuits in China. On 1st July Leibniz received the second letter of Newton via Oldenburg, the *Epistola posterior*.

5.3 The Correspondence Ends: *Epistola posterior*

Collins was convinced that Leibniz's series of the circle sector was a discovery independent of the English, but he was not quite sure whether the series was correct. He did not see that this series was but a special case of Gregory's series of the inverse tangent. Wallis, to whom Collins had sent a copy of Leibniz's letter, noted the slow convergence of the $\pi/4$-series but remarked that the series would appear to be correct. For the time being Newton did not answer.

Collins was afflicted with his blood poisoning, Newton had not reacted, but while reading the copies of the letters he had received he had gained the impression that he could not seriously take Tschirnhaus into account as a discussion partner. According to Hofmann [Hofmann 1974, p. 259f.] suspicion grew in Newton while reading Leibniz's reply to the *Epistola prior*, that Leibniz was a plagiarist who succeeded in reconstructing some concepts of the new infinitesimal calculus from Newton's letter. Tschirnhaus's cheeky letter should additionally have strengthened Newton's suspicion that Leibniz had not really penetrated the central ideas of the new analysis. On top came Collins's translation error *ludus naturae* which might have sounded like metaphysics in Newton's ears. Aiton has also conjectured [Aiton 1991, p. 123] that at the time of the writing of his second letter Newton already saw Leibniz as a plagiarist. I suspect these presumptions are parts of a conspiracy theory which can be falsified by the content of the *Epistola posterior*. Since we know the 'hot' war in the priority dispute starting in 1690 and since there we see a vindictive, hateful, and devious Newton it seems to lie close to insinuate his worst intentions also in 1676. But this is unfair and simply wrong, and the *Epistola posterior* gives no indications for an enhanced distrust concerning Leibniz.

The *Epistola posterior* is a small treatise of 19 pages starting with very friendly words of Newton [Turnbull 1959–77, Vol. II, p. 130]:

> *'Leibniz's method for obtaining convergent series is certainly very elegant, and it would have sufficiently revealed the genius of its author, even if he had written nothing else. But what he has scattered elsewhere throughout his letter is most worthy of his reputation – it leads us also to hope for very great things from him. The variety of ways by which the same goal is approached has given me the greater pleasure, because three methods of arriving at series of that kind had already become known to me, so that I could scarcely expect a new one to be communicated to us. One of mine I have described before; I now add another, ... And an explanation of this will serve to lay bare, what Leibniz desires from me, the basis of the theorem set forth near the beginning of the former letter.'*

The first of the three methods consisted in a stepwise substitution and was communicated to Leibniz already in the *Epistola prior*, the second consisted in a numerical induction and interpolation – a procedure that Newton soon abandoned – and the third, which was preferred by Newton, consisted of the inversion of series and comparison of coefficients [Hofmann 1974, p. 261]. Newton wrote further how he came to invent the second method after having read Wallis's *Arithmetica infinitorum* in which interpolation and induction were described. From the start this second method certainly seemed not trustworthy to him. Wallis's interpolation is not interpolation as we understand it now, but instead a somewhat courageous technique which operated vaguely as follows. If the integrals of x^p and x^{p+1} are known then (hopefully) the integral of $x^{p+1/2}$ should be computable from the two known ones by 'interpolating' the exponents. Wallis's induction also is not the mathematical induction we use today but an 'incomplete' one. If a mathematical statement holds for $n = 1, 2, 3, 4, 5$ and if one 'sees' a pattern, then Wallis accepted the statement as holding true for all n. It is not surprising that these methods of proof did not appear trustworthy to the astute Newton.

Then, so wrote Newton, he noted that the square of the series $1 - \frac{1}{2}x^2 - \frac{1}{8}x^4 - \frac{1}{16}x^6 - \dots$ was just $1 - x^2$ and that the third power of the series $1 - \frac{1}{3}x^2 - \frac{1}{9}x^4 - \frac{5}{81}x^6 - \dots$ also yielded $1 - x^2$. Therefore he had proved that these series are radical terms of $1 - x^2$ and hence he had worked out another method to arrive at series based on extracting roots.

Newton wrote that he could barely remember the discovery of the binomial theorem but when he worked through his manuscripts when writing the *Epistola prior* he started to remember it vividly. Newton did attach great importance to having made this discovery *before* Mercator's *Logarithmotechnia* appeared in print [Turnbull 1959–77, Vol. II, p. 133]:

> *'Yet at the very time when this book* [the *Logarithmotechnia*] *appeared, a certain compendium of the method of these series was communicated by Mr Barrow (then professor of mathematics) to Mr Collins; ...'*

The 'certain compendium' is *De analysi*, of course. Since in 1665 he calculated the logarithm with extreme accuracy as the area under the hyperbola from a series, cp. figure 5.3.1, he now wrote [Turnbull 1959–77, Vol. II, p. 133]:

> *'I am ashamed to tell to how many places I carried these computations, having no other business at that time: for then I took really too much delight in these inventions.'*

Newton described Collins's wish that Newton may publish his discoveries. Five years before, Newton wrote, when he composed a treatise on the refraction of light and on colours, thereby responding to pressure of his friends, he also thought about his series again and wrote a manuscript in order to publish it. No doubt, this manuscript was *De methodis*. But then, Newton went on,

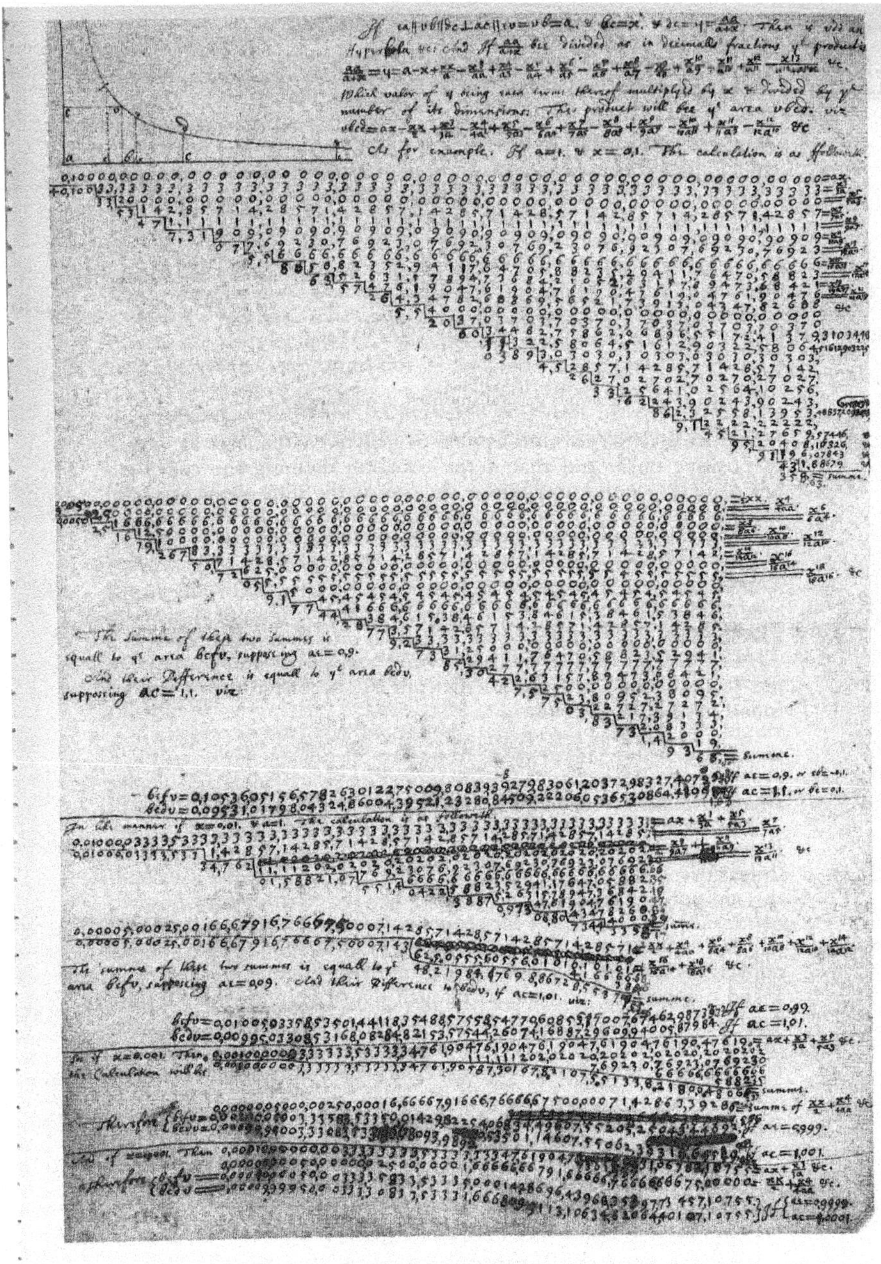

Fig. 5.3.1. Newton's calculation of the area under the hyperbola (= logarithm) of the year 1665
(Cambridge University Library)

after sending his manuscript on the theory of light to the Royal Society, *'something unexpected'* happened which kept him from a publication of *De methodis* [Turnbull 1959–77, Vol. II, p. 133]:

> *'... quite deterred me from the design and caused me to accuse myself of imprudence, because, in hunting for a shadow hitherto, I had sacrificed my peace, a matter of real substance.'*

We recognise clearly here the effects of the struggle concerning Newton's manuscript on the theory of light which must have been quite insulting for him.

James Gregory was also mentioned by Newton but Newton made it crystal clear that Gregory used one of Newton's series as a starting point for his discoveries [Turnbull 1959–77, Vol. II, p. 133]:

> *'About that time, from just a single of my series which Collins had sent him, Gregory after much reflection, as he wrote back to Collins, arrived at the same method, and he left a treatise on it which we hope is going to be published by his friends. Indeed, with his strong understanding he could not fail to add many discoveries of his own, and it is in the interest of mathematics that these should not be lost.'*

The treatise of Gregory described here unfortunately is lost today but we may assume with some justification that it was a description of the differential and integral calculus, a discovery with which Gregory forestalled even Newton. Newton's presentation is not exactly correct at this point since he knew already in 1670 that Gregory had discovered the binomial theorem as also the Taylor polynomial before 1668 and independently of him [Hofmann 1974, p. 262].

We can not discuss all of the *Epistola posterior* in detail but I wanted to show the *tone* of this letter in a few extracts. No accusations of plagiarism can be found in the entire letter: on the contrary, the works of Leibniz and Gregory are treated with high praise.

Of some interest are certainly the two anagrams which may still support the suspicion that Newton was so scared of Leibniz as a competitor that he encrypted some important references. Newton wrote [Turnbull 1959–77, Vol. II, p. 134]:

> *'And the same is true in questions of maxima and minima, and in some others too, of which I am now speaking. The foundation of these operations is evident enough, in fact; but because I cannot proceed with the explanation of it now, I have preferred to conceal it thus: 6accdae13eff7i3l9n4o4qrr4s8t12vx.'*

Newton has revealed the meaning of this anagram in a letter of 1693 to Leibniz, after Wallis had requested to reprint it together with the resolution in his treatise on algebra. Newton wrote to Leibniz in 1693 [Turnbull 1959–77, Vol. III, p. 285]:

> *'given an equation involving any number of fluent quantities to find the fluxions, and conversely.'*
>
> (Data aequatione quantitates quotcunque fluentes involvente invenire fluxiones, & vice versa.)

Why Newton included this anagram at this point will stay his secret. What could Leibniz have learned from the sentence 'given an equation involving any number of fluent quantities to find the fluxions, and conversely'?

Newton went on with the description of some problems he had solved. Close to the end of the letter the second anagram followed [Turnbull 1959–77, Vol. II, p. 129 and p. 148]:

> *'Nevertheless – lest I seem to have said too much – inverse problems of tangents are within our power, and other more difficult than those, and to solve them I have used a two-fold method of which one part is neater, the other more general. At present I have thought fit to register them both by transposed letters, lest, through others obtaining the same result, I should be compelled to change the plan in some respects. 5ac cdae10effh11i4l3m9n6oqqr8s11t9v3x: 11ab3cdd10eaeg10ill4m7n6o3p3q 6r5s 11t8vx, 3acae4egh5i4l4m5n8oq4r3s6t4vaaddaeeeeeiijmmnnooprrr sssssttuu'*

The resolution, printed in the third volume of Wallis's collected works and in the *Commercium epistolicum*, reads

> *'One method consists in extracting a fluent quantity from an equation at the same time involving its fluxion; but another by assuming a series for any unknown quantity whatever, from which the rest could conveniently be derived, and in collecting homologous terms of the resulting equation in order to elicit the terms of the assumed series.'*

We refrain from giving the Latin sentence which actually was encoded in the anagram; it can be found in [Turnbull 1959–77, Vol. II, p. 159, Remark (72)]. Could Leibniz have been able to reconstruct this paragraph from the anagram? Certainly not! If one counts the characters in the original Latin sentence one finds that two i (or j) are missing, but one s is superfluous! The transcript which was sent to Leibniz hence contained some lapses of the pen. The paragraph relates to two methods for the solution of ordinary differential equations, hence equations containing x, y, and dy/dx. In Newton's later notation the quantities involved are x, y, and $\dot{y}/\dot{x}$; in the year 1676 Newton still wrote x, y, q/p. The first method is applicable if one of the quantities x or y does not appear explicitly in the equation; method 2 is concerned with the general case.

Looking at this (second) anagram one can imagine that Newton wanted to secure his priority since he did not know that Leibniz also was able to solve differential equations. To derive evidence against Newton from this is unjustified in my eyes. Had not Leibniz deliberately kept the details of his transmutation theorem secret?

What we have discussed here is the first draft of the letter written in Newton's hand (which was translated by Turnbull in [Turnbull 1959–77, Vol. II, p. 130-149]) and which deviates slightly from the final version which was sent to Leibniz. Hofmann has found that some very friendly passages on Leibniz and Tschirnhaus were removed by Newton in the final version. The first draft ends with greetings to Leibniz and Tschirnhaus – they were removed [Hofmann 1974, p. 272]. Indeed, we know of a letter of Newton to Oldenburg written 26th October (jul.) 1676 in which Newton asked for some changes in the *Epistola posterior*. The letter ends with [Turnbull 1959–77, Vol. II, p. 162f.][7]:

> *'I fear I have been somewhat too severe in taking notice of some oversights in M. Leibnitz letter considering ye goodness & ingenuity of ye Author & yet it might have been my own fate in writing hastily to have committed ye like oversights. But yet they being I think real oversights I suppose he cannot be offended at it. If you think anything be expressed too severely pray give me notice & I'll endeavour to mollify it, unless you will do it with a word or two of your own. I believe M. Leibnitz will not dislike ye Theorem towards ye beginning of my letter page. 4 for Squaring Curve lines Geometrically. Sometime when I have more leisure it's possible I may send him a fuller account of it: explaining how it is to be ordered for comparing curvilinear figures with one another, & how ye simplest figure is to be found with which a propounded curve may be compared. Sr I am*
> *Your humble Servant*
> *Is. Newton'*

And this is followed by [Turnbull 1959–77, Vol. II, p. 163]:

> *'Pray let none of my mathematical papers be printed without my special license.'*

Hofmann deduced a changing tone in the course of the *Epistola posterior*. Starting with a friendly opening he states that Newton's attitude towards Leibniz changes in the course of the letter. Hofmann concludes [Hofmann 1974, p. 273]:

[7]Newton started the letter in Latin, then switched to English.

> *'This letter is thus a curious blend of a singular conflict of emotions and compactly described scientific results – ...'*

Not by any stretch of my imagination can I share this assessment and I am not alone [Hall 1980, S. 67]. The *Epistola posterior* is certainly not a complete description of Newton's achievements, but it is 19 pages long after all! Why should Newton have written such a long letter at all to Leibniz if he actually wanted to hide everything from him and if he suspected him being a plagiarist? Yes, Newton was a most irritable, malicious, and unpredictable person at the end of his life; but at *this* time, 1676, he can not be reproached in his dealings with Leibniz!

The *Epistola posterior*, the first draft of which stems from the 3rd November 1676 (24th October jul.), left London fairly late on 22nd May 1677 and arrived in Hanover on 1st July 1677 [Müller/Krönert 1969, p. 50][8]. Leibniz answered it practically immediately still on 1st July. The statement of the date 21st June in [Turnbull 1959–77, Vol. II, S. 211] is wrong and relies on a classical mistake: the Julian calendar was used in Hanover until 1700 and Leibniz noted the 21st June (jul.) on his letter. Following the Gregorian calendar this is 1st July, but in London the 21st June was assumed a Gregorian date and therefore it was dated 10 days back. Leibniz wrote more than friendly about Newton and Newton's results. He explained his tangent method which was derived independently of the one by Sluse. It holds $d(y^2) = 2y\,dy$ and so on for all other powers; the product rule is also contained in the letter, even the case of irrational quantities like the function of a function would not cause any problems, and Leibniz gave examples. Leibniz also demonstrated the use of series in the calculation of tangents in case of a function $f(x, y) = 0$. It is likely that the last part of Leibniz's letter is missing, as Turnbull [Turnbull 1959–77, Vol. II, p. 225, Remark (1)] has noted. Ten pages in the handwriting of a copyist have survived but Leibniz's signature is not present, so that at least one page is probably missing. Oldenburg wrote to Leibniz on 22nd July 1677 (12th July jul.) that he hoped that Leibniz would have received the *'veritable Newtonian treasure'*.

In the letter in response to the *Epistola posterior* Leibniz asked Newton for the explanation of some unclear point, but already on 22nd July Leibniz wrote again to Oldenburg [Turnbull 1959–77, Vol. II, p. 231f.] that one of the ambiguities had disappeared by reading Newton's letter for the second time. As a reply to his two letters Leibniz received a letter from Oldenburg dated 19th August 1677 which contained the following paragraph [Turnbull 1959–77, Vol. II, p. 235]:

> *'Since I wrote to you by Mr Sembin of Heidelberg, I have received two letters from you, both of which set out your thoughts on that long letter of Mr Newton's, which was previously sent to you. There is no reason*

[8]Hofmann writes in [Hofmann 1974, S. 274] of end of July.

> *why you should expect a reply to them so soon from the aforesaid Newton or even from our friend Collins, since they are out of town and are preoccupied with various other affairs.'*

This letter is not be understood as unfriendly. The diplomatic Oldenburg just wanted to make sure that Leibniz would not interpret the silence of London wrongly. Shortly after Oldenburg was dead, but even if Oldenburg had lived longer Newton certainly would not have answered again. First of all he was 'pissed off' by the discussions and attacks concerning his theory of light, additionally he had turned to very different topics and he did not desire talking of mathematical things which for him lay more than 10 years in the past. We can also argue that Newton learned from a letter of Leibniz to Oldenburg of 28th November 1676 that Leibniz was in London before and that Collins and Oldenburg had met him. As Hofmann suspects [Hofmann 1974, p. 273], this may (perhaps) fired his suspicions.

5.4 The Front Line in the Year 1677

We have to note here: The cold war prevailed between Leibniz (and Huygens) on one side, and some members of the Royal Society like Hooke, Collins, Pell, and even Oldenburg, but not between Newton and Leibniz. The second visit to London had taken Collins completely *for* Leibniz; Oldenburg was also very satisfied with Leibniz since he had now submitted his improved calculating machine. The cold war seemed finished.

Leibniz's reply letter to the *Epistola posterior* must have clearly shown Newton that Leibniz had arrived at a new kind of mathematics which was just as good as his own. Newton, the intellectually consistent thinker, may probably have felt repelled by the somewhat pragmatic and rapid kind with which Leibniz did mathematics; Huygens did also not like this manner [Hall 1980, p. 71]. The d-notation, so very important a symbol, which was made public by Leibniz for the first time in his reply letter, probably caused problems with Newton who never understood the meaning of a 'characteristica universalis'. We do not have any written evidence on Newton's thoughts about Leibniz's letter. In Newton's opus magnus, the *Philosophiae Naturalis Principia Mathematica* of 1687, we however find a great praise of Leibniz's mathematics. Until then the front was at rest. There is no war yet and the two warriors have other things to do.

6 The Aftermath of the *Principia Mathematica*

T. Sonar, *The History of the Priority Diſpute between Newton and Leibniz*, https://doi.org/10.1007/978-3-319-72563-5_6

6.1 The Warrior's Careers up to 1687

From 1677 Leibniz is serving the Hanoverians and is occupied with judicial, political, historical and administrative work. He corresponds on a worldwide level, returns from time to time to mathematical and physical questions, but has no longer partners for correspondence within the Royal Society. The year 1684 sees the first publication of Leibniz's differential calculus in the Acta Eruditorum.

Meanwhile, Newton lives his reclusive and solitary life as Lucasian professor in Cambridge. His fame is growing in England but for the time being there are no further contacts with Leibniz. Then a real thunderbolt: in 1687 his epoch-making work *Philosophiae Naturalis Principia Mathematica* (Mathematical principles of natural philosophy) appears for the first time; it is no less than the natal hour of modern physics.

6.1.1 The Privy Councillor Leibniz – Stranded in Hanover

Already in February 1677 Leibniz asked his Duke John Frederick to introduce him officially into his rights and obligations as a privy councillor and to increase his salary [Müller/Krönert 1969, p. 48]:

> *'Since I can honestly desire no less, after having had decent appointments already 10 years ago and being only a librarian would have better suited me in the 20th year of my life. Hence I have to be careful what matches my age and my work in the eyes of those who had an eye on me in the past and in particular that I do not have to feel ashamed if I have to compare myself with those who were not my equals but have come to better positions now: Your Serene Highness knows that the people working with books alone are sparseley respected and are considered being inept to do other things: all the more it would damage my reputation since in the past I was involved in judicial business. Moreover, the profession of a librarian is only respected by dukes of tremendous minds like Your Serene Highness, but at this court it is not, as the library is not public.'*
>
> (Denn ich zuvörderst ehrenthalben nicht weniger begehren kan, nachdem ich bereits vor 10 jahren gradum angenommen, anständige vocationen gehabt, und die bloße verrichtung eines Bibliothecarii mir beßer im 20ten jahr meines alters als iezo angestanden hätte, also bis dahin bedacht seyn muß, wie ich meinem alter und profectibus nach also stehen möge, daß ich deßen bey denen so vor diesen einiges absehen auff mich gehabt, und sonderlich bey denen die vor jahren mir nicht gleich gewesen, aniezo aber zu dergleichen stellen gelanget seyn, nicht

Fig. 6.1.1. Leibniz's employers until 1698: Duke John Fredrick ([unknown artist after Jean Michelin, about 1670–1680] Historisches Museum Hannover (Hanover)); Duke Ernest Augustus (Residenzmuseum Celler Schloss (Celle Castle))

> zu schämen habe: so wißen auch ferner E. Hochfürstl. Durchlt daß die jenigen so bloß und allein mit Büchern umbgehen, wenig geachtet und gemeiniglich als zu andern dingen untüchtig gehalten werden; so dann mir umb soviel desto mehr schaden solte, weil man vor diesen eine andere meinung vor mir gehabt, ich auch würcklich in judiciis und geschäfften gewesen; überdieß die charge eines Bibliothecarii nur bey denen Fürsten geachtet wird so wie E. Hf. D. eines ungemeinen verstandes, an diesem Hoff aber gar nicht beständig noch stetswehrend, so wenig als die Bliothec publick ist.)

In a memorandum to the Duke dated 2nd October Leibniz had to repeat his request concerning his public introduction as privy councillor. Nevertheless, in John Frederick Leibniz found a ruler who always kept an open ear for the numerous plans of his counsellor and librarian. Leibniz served as an advisor in political and juridical questions so that a closer personal relationship developed between him and the Duke. John Frederick had converted to catholicism while traveling in Italy and had married a catholic, but he granted free practice of religion. He also permitted the building of a new, grand, protestant church in Hanover; the Neustädter church St John, in which Leibniz eventually found his final resting place. Beside the numerous private correspondence many works for the Duke flowed from Leibniz's quill. Between June and October 1677 a manuscript took shape to strengthen the position of those German earls who were not electorial princes [Antognazza 2009, p. 204f.]. Leibniz did not appear as the author but used the synonym 'Caesarinus Fürstenerius'. The manuscript finally grew into a book which was eventually published in the form of a French dialogue *Entretien de Philarete et d'Eugene* (Conversation between Philarete and Eugene) and enjoyed 16 editions within 14 years — being adapted to the then current political situation by Leibniz. A whole flood of proposals from Leibniz poured towards the duke: a cipher machine, a new type of carriage, the mechanisation of silk production, better clocks, pharmacological cures, and his own calculating machine, of course [Walsdorf 2014], for which Leibniz unsuccessfully had tried to get the Parisian mechanic Olivier to Hanover [Antognazza 2009, p. 206]. In 1669 Henning Brand (about 1630–after 1692), a German pharmacist and alchemist, had found phosphorus in the attempt to turn a distillation of urine into gold. Leibniz discussed the production of phosphorus in a correspondence with Johann Daniel Crafft (1624–1697), a physicist and chemist. Significant for the ducal family as well as for Leibniz was the latter's proposal of August 1677 to trace the possible links of the House of Welf (Guelf, Guelph) with the ancient Italian family of the House of Este.

At the end of 1677 Leibniz was finally publicly advanced to become a privy councillor (Hofrat) which was connected with a salary increase. In January 1678 he tried to convince Tschirnhaus to come to Hanover but he was not successful. Already at this point Hanover is much too small for privy councillor Leibniz and he tried at least temporarily to flee the intellectually dull milieu by means of travels. When Spinoza died on 21st February 1677 Leibniz proposed a journey to Holland to examine the philosopher's unpublished papers but the idea remained a mere proposal. Only in summer 1678 can Leibniz leave the boundaries of the dukedom; as ducal librarian he travelled to Hamburg in order to buy up the extensive library of the physician and linguist Martin Fogel (1634–1675) who had died in 1675.

A large source of income for the duke was the silver mines in the Harz Mountains. Time and again these silver mines had problems with the penetration of water and Leibniz believed that he could lead away the water by means of pumps driven by windmills. Traditionally the pumps were driven by water power, but in dry years the energy of the water was either not available or not sufficient; that happened in the years 1678-79 when the income of the duke dropped about 50%. Leibniz visited the Harz Mountains between September and October 1679 for the first time [Antognazza 2009, p. 212]. Many further visits will follow, but Leibniz will fail – he worked with horizontal wind mills, not knowing that their efficiency was much too small to deliver the power needed. In addition the mining authorities, not willing to tolerate a lateral entrant, created many obstacles to prevent Leibniz's plans from becoming reality.

Fig. 6.1.2. Sophia of Hanover [unknown artist] and Sophia Dorothea of Ahlden with her children ([attributed to Jaques Vaillant, about 1690] Bomann Museum Celle, BM 120)

In 1678 Leibniz made great progress in physics, namely in what he called dynamics, and he wrote the manuscript *De corporum concursu* (On the collision of bodies). The fact that Leibniz grappled intensely with dynamics (that is the investigation of the forces when bodies move) will be of particular importance in the history of the priority dispute as far as consequences of Newton's *Principia mathematica* are concerned. In 1686 *De corporum concursu* gave rise to *Brevis demonstratio erroris memorabilis Cartesii et aliorum circa legem naturae* (Short evidence of a remarkable error of Descartes and others with regard to a law of nature) in which Leibniz refuted Descartes's claim that the quantity mv, being the product of mass m and velocity v, was the 'force' inherent in a body. Instead, Leibniz proposed the product mv^2 which is indeed proportional to what we now call kinetic energy [Antognazza 2009, p. 239]. It seems tragic that the notion of 'force' at that time was not clearly carved out. Scientists talked about 'force', 'strength', 'action', and 'effect' [Szabó 1996, p. 65] meaning always the same but leading to misinterpretations. Later on Leibniz wrote about the 'dead' and the 'living' force and meant mv and mv^2, respectively. The discussions focusing on the notion of force is described in detail by István Szabó in [Szabó 1996, Kapitel II.A, p. 47ff.]. Important to us is the dispute with the Cartesians which broke out when Newton's *Principia* was published in 1687. We will have to illuminate this dispute.

When John Frederick died on 28th December 1670 aged 54 while travelling in Italy Leibniz's time serving an understanding and insightful Duke ended. John Frederick's younger brother Ernest Augustus, having no understanding

Fig. 6.1.3. The castle of Ahlden about 1654 [Merian 1654]

for the high-flying plans of the librarian of his predecessor, became the new duke. Although a Lutheran this duke viewed religion as being purely a question of politics. When his brother George William, Duke of Celle, terminated his engagement with Sophia of the Palatinate Ernest Augustus willingly stepped in and married her – at least she was the daughter of the 'Winter King' Frederick V., Elector Palatine, and his wife Elisabeth Stuart, daughter of the King of England, James I, and Scotland (as James VI). Hence she had good and proper blue blood in her veins and Ernest Augustus hoped to move closer to the ninth elector's dignity due to this tie.

Sophia not only looked dazzling; in contrast to her husband she was intelligent, witty, and funny. She gave birth to six sons and a very remarkable daughter, Sophia Charlotte. For Ernest Augustus anything would be justified to enlarge his duchy and eventually to become the ninth elector's dignity. His brother George William had left Sophia for a mistress who he married and with whom he had a daughter, Sophia Dorothea of Celle (1666–1726). Ernest Augustus forced his oldest son George Louis (1660–1727) to marry his 16 years old cousin Sophia Dorothea. The marriage turned out to end in a personal catastrophe for the wife: deeply in love with the Swedish count Philip Christoph von Königsmark Sophia Dorothea wanted to escape with him, but their flight was betrayed. Von Königsmark disappeared in the night of the 1st July 1694 and was never seen again, Sophia Dorothea was arrested after a divorce proceeding and locked in a castle in Ahlden (Ahlden House) on the river Aller until the end of her life. She was not allowed to see her two children again.

Called 'Princess of Ahlden' Sophia Dorothea found her last resting place after 30 years of house arrest eventually in the Fürstengruft (prince's crypt) in Celle. When George William died in 1705 Celle fell to Hanover.

As a further move Ernest Augustus married his daughter Sophia Charlotte to Frederick III of the Hohenzollern dynasty, elector of Brandenburg, on the 8th October 1684. When Frederick III became the first King in Prussia as Frederick I in 1701, Sophia Charlotte became the first crowned Queen of Prussia. Finally Ernest Augustus arrived at the goal of his dreams: in 1692 he was conferred the ninth elector's dignity.

About June 1678, still under duke John Frederick, Leibniz described his duties in a letter to Hermann Conring as follows [Müller/Krönert 1969, p. 52]:

> *'... I admit that since I am accepted under the privy councillors I have other duties to care for: I have to study court files, pass sentences, and from time to time give political opinion on behalf of the sovereign. Nevertheless does the generous sovereign in his goodwill shown to me not demand that I devote all of my time to affairs of state, but left it to me to stay away from the meetings as often as it may occur necessary because of my other works. Moreover, since the sovereign gives me private orders on occasion, and since the management of the library behoves me and I ought to keep up a continuous correspondence with scholars, it doubtlessly seems justified when I claim a liberal treatment. In fact I do not want to be condemned to the Sisyphean task of court affairs as poring over a rock, even if greater wealth and the highest honours would be promised.'*
>
> (... ich gebe zu, daß ich, seitdem ich unter die Hofräte aufgenommen bin, andere Aufgaben zu erfüllen habe: ich muß Gerichtsakten studieren, Urteile fällen und gelegentlich auch auf Anordnung des Fürsten politische Gutachten abgeben. Dennoch verlangt der hochherzige Fürst in seinem mir erwiesenen Wohlwollen nicht, daß ich meine Zeit vollständig den Staatsgeschäften widme, sondern hat es mir freigestellt, den Sitzungen fernzubleiben, so oft es mir wegen meiner anderen Arbeiten nötig erscheint. Da der Fürst überdies gelegentlich private Aufträge für mich hat, mir die Leitung der Bibliothek obliegt und ich ständigen Briefwechsel mit Gelehrten pflegen soll, ist es zweifellos berechtigt, wenn ich Anspruch auf eine liberalere Behandlung erhebe. Tatsächlich möchte ich nicht verurteilt sein, einzig und allein die Sysiphusarbeit der Gerichtsgeschäfte wie einen Felsblock wälzen zu müssen, und wenn mir dafür auch der größte Reichtum und die höchsten Ehren versprochen würden.)

In autumn 1679 Leibniz is in the Harz Mountains for the first time to conclude a contract with the mining authority on his 'Windkunst' (wind art) to drain the pit 'Dorothea Landeskron'. The expenses of this experiment will be borne by Leibniz himself [Müller/Krönert 1969, p. 57]. Already in April 1680 the new duke decided not to carry out the experiment at 'Dorothea Landeskron' but at the pit 'Catharina'. At the beginning of 1680 Leibniz started his work on the history of the Welfs (House Brunswick-Lüneburg). During his work he founded what we now call the science of history; he collected an unbelievable mass of material and in the year of his death 1716 he will only have arrived at the year 1005.

In March and April 1681 Otto Mencke (1644–1707), professor of moral and politics at Leipzig University, conducted a conversation with Leibniz concerning the publishing of a scientific journal with the title 'Acta Eruditorum' based on the models of the French 'Journal des Sçavans' and the English 'Philosophical Transactions'.

The Acta Eruditorum appeared in Leipzig in 1682 for the first time and quickly developed into Leibniz's company magazine. During his lifetime he published more than 100 of his mathematical and scientific works in the Acta.

The activities in the Harz Mountains were continued 1682 and 1683. In 1683 Leibniz got to know Duke Anthony Ulrich of Brunswick-Wolfenbüttel by whom he received much understanding and recognition. On 6th December Ernest Augustus terminated the proportionate payments of the court and the trades to Leibniz's wind art in the Harz Mountains [Müller/Krönert 1969, p. 71]. Leibniz thereupon decided to continue at his own expense. From 1684 an intimate and nearly amicable relationship developed between Leibniz and duchess Sophia. Leibniz spent several weeks in the Harz Mountains where he still wrestled with his wind art and where initial experiments took place.

In October 1684 Leibniz's work *Nova methodus pro maximis et minimis, itemque tangentibus, quae nec fractas, nec irrationales quantitates moratur & singulare pro illis calculi genus* (New method for the determination of maxima and minima as for tangents; a method neither hindered by fractional nor irrational quantities; and a unique kind of the required calculus) [Leibniz 2011, p. 51ff.]. This publication marks the first appearance of Leibniz's d and some of the rules of the Leibnizian calculus. We recall that Leibniz had sent his d and some of the rules of calculus to Newton in the answer to the latter's *Epistola posterior*, but now his calculus was laid open for all readers of the Acta Eruditorum. Besides some basic rules the printed work contained four examples of how to apply the new calculus, including a problem concerning the computation of extreme values, an implicit computation of tangents, and the solution of a differential equation, the so-called Problem of de Beaune. Unfortunately, some printing errors occurred – for example a sign error in the quotient rule, the mixing up of 'convexity' and 'concavity', and errors in some derivatives. The printer certainly had some trouble with the manuscript, since Otto Mencke wrote to Leibniz [Hess 1986, p. 64]:

> *'The Schediasma de methodo Tangentium etc. will be found by my high master in October and the book printer had cut some characters in wood. Whether it was well done I might like to hear.'*

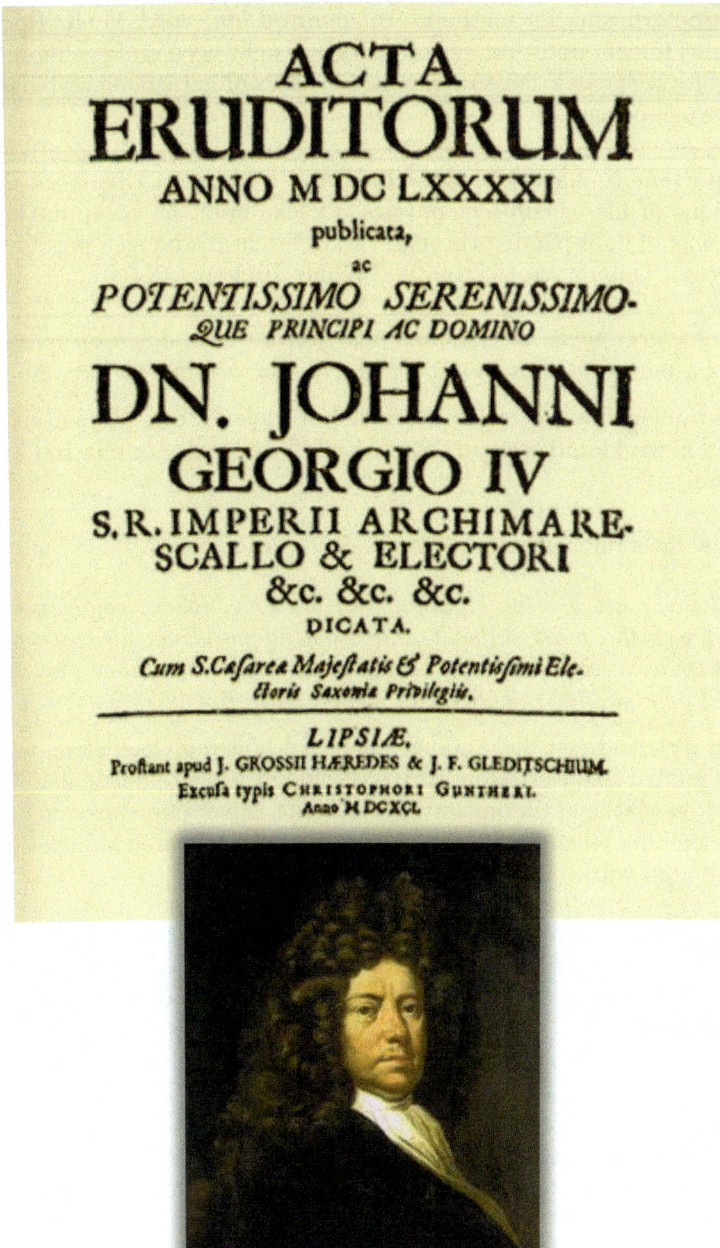

ACTA
ERUDITORUM
ANNO M DC LXXXXI
publicata,
ac
POTENTISSIMO SERENISSIMO-
QUE PRINCIPI AC DOMINO
DN. JOHANNI
GEORGIO IV
S.R. IMPERII ARCHIMARE-
SCALLO & ELECTORI
&c. &c. &c.
DICATA.
Cum S. Cæsareæ Majestatis & Potentissimi Ele-
ctoris Saxoniæ Privilegiis.

LIPSIÆ,
Prostant apud J. GROSSII HÆREDES & J. F. GLEDITSCHIUM.
Excusa typis CHRISTOPHORI GUNTHERI.
Anno M DCXCI.

Fig. 6.1.4. Otto Mencke (1644–1707) was professor of moral and politics in Leipzig. His greatest achievement surely is the publishing of the first German scholarly journal Acta Eruditorum which his son Johann Burckhard Mencke (1674–1732) continued after his father's death. The journal ceased in 1782.

(Das Schediasma de methodo Tangentium etc. wird M. h. Herr im Octobri finden, undt hat der buchdrücker deßwegen einige characteres in Holtz schneiden lassen müssen. Obs so wol gerathen, verlange ich zu vernehmen.)

It is important to remark that Leibniz had already in this work described applications of his calculus in physics, for example the computation of the shortest way of light passing through two different media as a proof of the law of refraction. Leibniz wrote proudly [Leibniz 2011, p. 60]:

> *'What other highly learned men had tried to hunt down on many detours a man versed in this calculus will now achieve in three lines.'*
>
> (Was andere, hochgelehrte Männer auf vielen Umwegen zu erjagen suchten, wird künftig ein in diesem Kalkül Bewanderter mit drei Zeilen leisten.)

And a few lines further on he wrote [Leibniz 2011, p. 61]:

> *'And these are but the beginnings of a much higher geometry which comprises the most difficult and beautiful problems also of applied mathematics which can not easily be worked out without our differential or a similar calculus with comparable effortlessness.*
>
> (Und dies sind erst die Anfänge einer viel höheren Geometrie, welche die denkbar schwierigsten und schönsten Probleme auch der angewandten Mathematik umfasst, die so leicht keiner ohne unseren differentialen oder einen ähnlichen Kalkül mit vergleichbarer Mühelosigkeit bearbeiten wird.)

Doubtless, *Nova methodus* contained very abstract mathematics in regard to the time it was written. Additionally, it was written in a tight manner and the (few) readers certainly had some problems with it. It is an interesting question why Leibniz, being in the possession of not only his differential, but also his integral calculus well before 1684, decided to publish only the basic rules of his differential calculus in so succinct a form just in 1684. This question was thoroughly examined by Hess in [Hess 1986] and two essential motives were found. First, Leibniz had committed himself to Mencke to participate constructively in the Acta Eruditorum. He had already published six manuscripts there and a sense of pressure was felt to continue in order to fill the journal. On the other hand there were problems with Tschirnhaus and these problems seem to be the true reason to publish *Nova methodus.*

Ehrenfried Walter von Tschirnhaus had returned to Paris at the beginning of 1682 to effect his inclusion into the Académie des Sciences and thereby achieve a pension from the king [Hess 1986, p. 73]. To underpin his plan he published a series of journal articles with mathematical contents in the Acta Eruditorum;

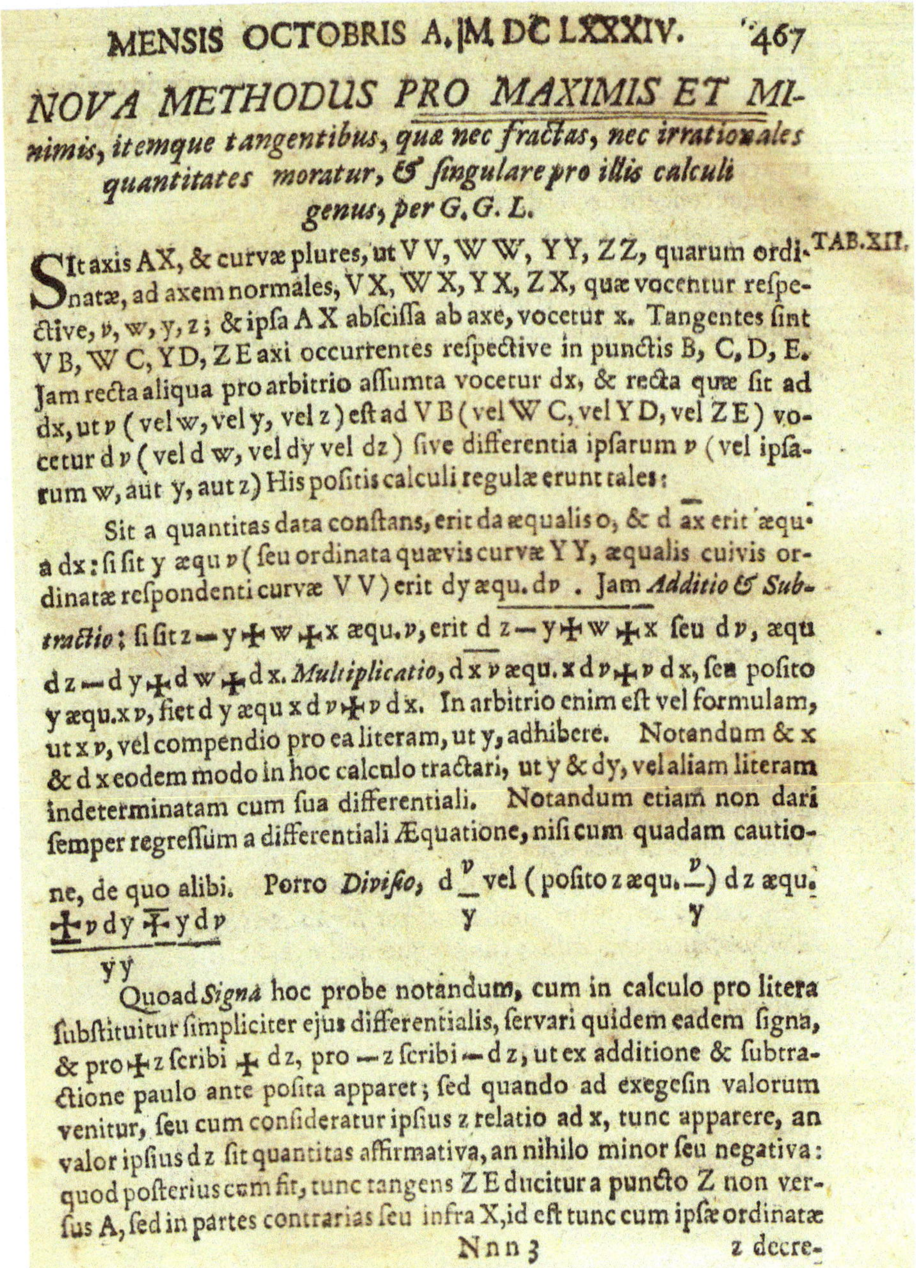

MENSIS OCTOBRIS A. M DC LXXXIV. 467

NOVA METHODUS PRO MAXIMIS ET MINIMIS, itemque tangentibus, quæ nec fractas, nec irrationales quantitates moratur, & singulare pro illis calculi genus, per G. G. L.

TAB. XII.

Sit axis AX, & curvæ plures, ut VV, WW, YY, ZZ, quarum ordinatæ, ad axem normales, VX, WX, YX, ZX, quæ vocentur respective, v, w, y, z; & ipsa AX abscissa ab axe, vocetur x. Tangentes sint VB, WC, YD, ZE axi occurrentes respective in punctis B, C, D, E. Jam recta aliqua pro arbitrio assumta vocetur dx, & recta quæ sit ad dx, ut v (vel w, vel y, vel z) est ad VB (vel WC, vel YD, vel ZE) vocetur dv (vel dw, vel dy vel dz) sive differentia ipsarum v (vel ipsarum w, aut y, aut z) His positis calculi regulæ erunt tales:

Sit a quantitas data constans, erit da æqualis 0, & d $\overline{ax}$ erit æqu. a dx: si sit y æqu v (seu ordinata quævis curvæ YY, æqualis cuivis ordinatæ respondenti curvæ VV) erit dy æqu. dv. Jam *Additio & Subtractio*; si sit z − y + w + x æqu. v, erit d $\overline{z - y + w + x}$ seu dv, æqu dz − dy + dw + dx. *Multiplicatio*, d $\overline{xv}$ æqu. x dv + v dx, seu posito y æqu. xv, fiet dy æqu x dv + v dx. In arbitrio enim est vel formulam, ut xv, vel compendio pro ea literam, ut y, adhibere. Notandum & x & dx eodem modo in hoc calculo tractari, ut y & dy, vel aliam literam indeterminatam cum sua differentiali. Notandum etiam non dari semper regressum a differentiali Æquatione, nisi cum quadam cautione, de quo alibi. Porro *Divisio*, d $\frac{v}{y}$ vel (posito z æqu. $\frac{v}{y}$) dz æqu. $\frac{\pm v\,dy \mp y\,dv}{yy}$

Quoad *Signa* hoc probe notandum, cum in calculo pro litera substituitur simpliciter ejus differentialis, servari quidem eadem signa, & pro + z scribi + dz, pro − z scribi − dz, ut ex additione & subtractione paulo ante posita apparet; sed quando ad exegesin valorum venitur, seu cum consideratur ipsius z relatio ad x, tunc apparere, an valor ipsius dz sit quantitas affirmativa, an nihilo minor seu negativa: quod posterius cum fit, tunc tangens ZE ducitur a puncto Z non versus A, sed in partes contrarias seu infra X, id est tunc cum ipsæ ordinatæ

Nnn 3 z decre-

Fig. 6.1.5. From the year 1684: Titlepage of *Nova methodus* [Acta Eruditorum]

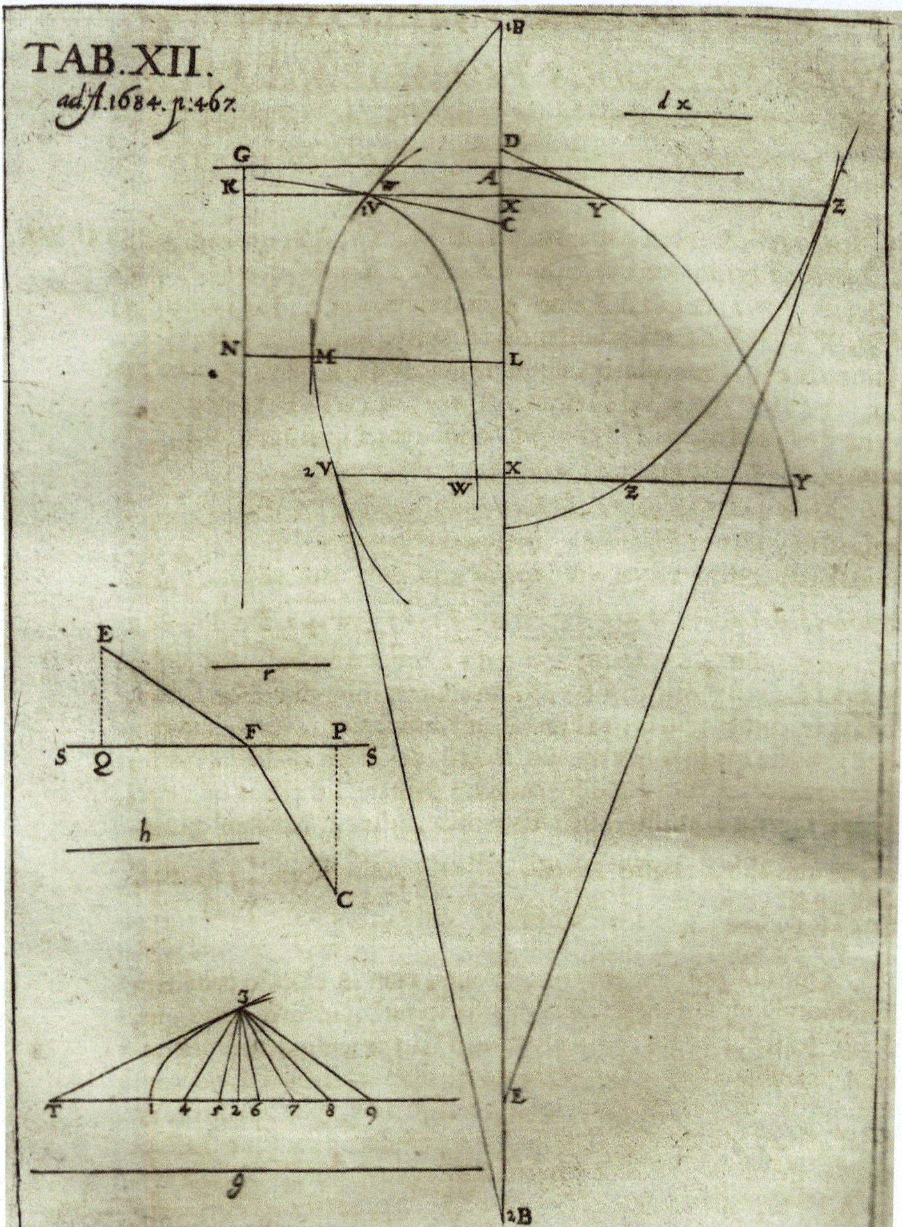

Fig. 6.1.6. Diagram in *Nova methodus*
[Acta Eruditorum]

as in December 1682 *Nova methodus tangentes curvarum expedite determinandi*, in March 1683 *Nova methodus determinandi maxima et minima* and in October 1683 *Methodus datae figurae ..., aut quadraturam, aut impossibilitatem ejusdem quadraturae determinandi* (Method to either determine the quadrature of a given figure or to determine the impossibility of that same quadrature). Although Tschirnhaus was frequently warned against errors and immaturity by Leibniz there were genuine Leibnizian results regarding the infinitesimal calculus especially in the work of October 1683 and they were reproduced with mistakes. Leibniz had written a letter to Tschirnhaus but the latter did not receive it and could not answer it. Hence, Leibniz was forced to claim his priority publicly in the spring of 1684 and to correct Tschirnhaus's mistakes. Thus was published Leibniz's manuscript *De dimensionibus figurarum inveniendis* (How one achieves measuring figures) [Leibniz 2011, p. 39ff.] in May 1684 in the Acta Eruditorum. Incidentally, in this manuscript Leibniz was very cautious; after all Tschirnhaus was a good friend. Nevertheless the publication of this manuscript led to a letter of justification by Tschirnhaus, which was rejected for publication in the Acta Eruditorum.

In *De dimensionibus* Leibniz had described the two basic tasks of his calculus, namely differentiation and integration, in a very geometrically hedged manner. Hence, a truly content-related refinement of his priority claim was missing and exactly this *Nova methodus* of October 1684 was meant to achieve in the case of the differential calculus. It is conceivable and probable that Leibniz also planned to publish his integral calculus but he was so occupied with his activities in the Harz Mountains that for the time being the short description of the differential calculus remained the only publication. Another theory regarding the scarceness of *Nova methodus* stems from Hofmann [Hofmann 1966, p. 219]:

> *'It was intentionally written sketchily so that a clueless reader could only laboriously, if at all, advance to the basic thoughts of the symbolism.'*
>
> (Sie war absichtlich so knapp gefaßt, daß ein ahnungsloser Leser nur mühsam, wenn überhaupt, zu den Grundgedanken des Symbolismus vordringen konnte.)

I can not agree with this thesis. Despite the disturbing printing errors and the not very didactical kind of presentation in *Nova methodus*, a reference to this work can be found already one year later in a book of the Scottish mathematician John Craig (1666–1731) entitled *Methodus figurarum lineis rectis et curvis comprehensarum quadraturas determinandi* (Method to determine the quadrature of figures which are fenced in between straight and curved lines). To determine the tangent at a curve Craig appealed to Leibniz's *Nova methodus* although the application of Leibniz's method clearly showed that Craig had not much penetrated the foundations of the differential calculus [Hess 1986, p. 74]. To prepare an opinion regarding Craig's book, Leibniz composed

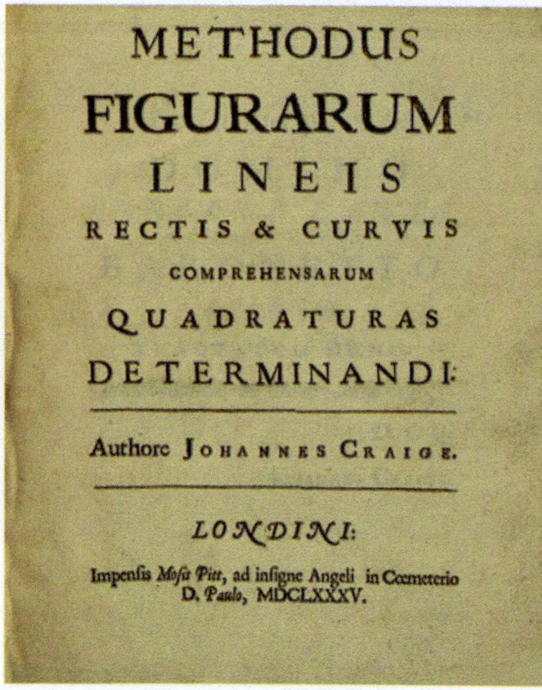

METHODUS
FIGURARUM
LINEIS
RECTIS & CURVIS
COMPREHENSARUM
QUADRATURAS
DETERMINANDI.

Authore JOHANNES CRAIGE.

LONDINI:

Impensis *Mosis Pitt*, ad insigne Angeli in Cœmeterio
D. *Paulo*, MDCLXXXV.

Fig. 6.1.7. Titel of John Craig's *Methodus figurarum* of 1685

the manuscript *De geometria recondita et analysi indivisibilium atque infinitorum, Addenda his quae dicta sunt in Actis a. 1684, Maji p. 233; October. p. 264; Decemb. P. 586* (The profound geometry and analysis of the indivisible and the infinite. Supplements to what is carried out in the Acta of the year 1684, May p. 233; October p. 264; December p. 586) [Leibniz 2011, p. 69ff.], in which he contributed decisively to the explanation of his infinitesimal methods. At the beginning he wrote [Leibniz 2011, p. 69]:

> *'I see that a number of things that I have published in these Acta regarding the advancement of geometry is appreciated to no less extent by some scholars; some however, may it be due to mistakes of the author, may it be due to other reasons, is not sufficiently understood. Hence I have regarded it worthwhile to add here something which can illuminate the hitherto published.'*
>
> (Ich sehe, dass etliches von dem, was ich in diesen Acta zum Fortschritt der Geometrie veröffentlicht habe, von einigen Gelehrten in nicht geringem Maße geschätzt und sogar allmählich in Anwendung gebracht wird, einiges jedoch, sei es aufgrund von Fehlern des Verfassers, sei es aus anderen Gründen von manchen nicht ausreichend verstanden worden ist. Daher habe ich es für der Mühe wert erachtet, hier etwas hinzuzufügen, was das bisher Veröffentlichte erhellen kann.)

Thus Leibniz himself contributed decisively to a publicisation of his infinitesimal calculus [Hess 1986, p. 74f.]. Whichever way me may look at it: *Nova methodus* marks the natal hour of the *modern* differential calculus. From October 1684 Leibniz's symbolism, basic to the differential calculus, is 'out in the world' and it will soon start a triumphal march which in this comprehensive form Leibniz arguably could have guessed. The complete penetration of mechanics by infinitesimal mathematics, already begun by Leibniz, will be completely accomplished in the 18th century by Leonhard Euler. The mechanics of fluids, wave optics, elasticity theory, and every other physical theory within continuum mechanics is not conceivable today without Leibniz's infinitesimal mathematics. Modern MP3-players, the design of our cars and aircrafts, the construction of our engines and turbines, and the modern communication techniques are based on Leibniz's analysis. This all started with *Nova methodus*.

At the beginning of the year 1685 only moderately successful series of experiments continued in the Harz Mountains. The duke who had started to invest again after 1683 decreed on 23rd March from Venice to discontinue the work in the Harz Mountains [Müller/Krönert 1969, p. 74]. In the years from 1680 to 1686 Leibniz undertook 31 journeys to the Harz Mountains and spent there 165 weeks altogether [Antognazza 2009, p. 227]. On 10th August Leibniz received the ducal order to write the history of the House of Welf up to the present times. In return the duke confirred the title of a privy councillor for life and Leibniz's salary was transformed into a lifelong pension. Leibniz was also freed from the usual duties of office work. The ambitious duke was thinking of a short, sparse history of his house which was supposed to prove conclusively that his dynastic roots reached far back in history. He wanted to cut a figure and to smooth the way towards an elector's dignity. It is certain that Leibniz also had a short work in mind but this hope was not fulfilled. During the next two years Leibniz dedicated his whole energy to the historical work; he searched for materials in archives and libraries in Brunswick, Wolfenbüttel, Lüneburg, and Celle and examined the literary remains of the master of the chamber (Kammermeister) Hoffmann, deceased in March 1680, in which the fruits of a 20-year long research on the Brunswick-Lüneburg Welfs could be found [Antognazza 2009, p. 231]. Even in his correspondence the great general historical interests of Leibniz was reflected in these years.

Leibniz also worked on a reunion of the confessions, on his characteristica universalis, on a 'scientia generalis', on his dynamics, his metaphysics, and – still – on his ideas to drain the mines in the Harz Mountains. Antognazza writes about an 'ocean of Leibniz's private papers' [Antognazza 2009, p. 238] in the years from 1682 to 1686 and in particular about an 'annus mirabilis par excellence' [Antognazza 2009, p. 239]. The work *Brevis demonstratio erroris memorabilis Cartesii et aliorum circa legem naturae* in which he identified the product of mass and the square of the velocity (hence what we nowadays call kinetic energy up to a factor of 1/2) as a conserved quantity of motion was sent to Otto Mencke for publication in the Acta Eruditorum on 16th January 1686. On behalf of Antoine Arnauld he wrote *un petit discours de Metaphysique* (a small discourse on metaphysics) which marks the beginning of the philosophical key correspondence with Arnauld [Antognazza 2009, p. 240]. Manuscripts concerning logic and metaphysics were written and questions of religion and law were treated. In mathematics he achieved genuine break-throughs as well. In June 1686 the works *Meditatio nova de natura anguli contactus et osculi, horumque usu in practica mathesi, ad figuras faciliores succedaneas difficilioribus substituendas* (New considerations concerning the nature of the contact and the osculation angle for the use in applied mathematics with the aim to replace complicated figures by representative easier ones) [Leibniz 2011, p. 63ff.] and *De geometria recondita*, of which we have already spoken on page 238. In this work the integral sign $\int$ appeared for the first time when Leibniz wrote [Leibniz 2011, p. 76]:

> '.... *then it is immediately clear that following my method* $p\,dy = x\,dx$ *is valid, ... If one transforms this equation of differentials into the succinct one, it holds* $\int p\,dy = \int x\,dx$.'
>
> (..., dann ist sofort klar, dass nach meiner Methode $p\,dy = x\,dx$ ist, ... Wenn man diese differentiale Gleichung in die summatorische verwandelt, gilt $\int p\,dy = \int x\,dx$.)

In November 1687 Leibniz started out on a long journey to South Germany, Austria, and Italy, which will bring him back to Hanover not until June 1690. Officially the issue is the search for documents concerning the history of the House of Welf in archives; in particular the matter of the origin of the North Italian margrave Albert Azzo II (996–1097) who was seen as the ancestor of not only the House of Welf, but also of the much more important House of Este [Antognazza 2009, p. 281]. Since doubts were declared by French and German historians concerning Azzo II being really the ancestor of the House of Este Leibniz wanted to form his own opinion. In fact, he was able to confirm the connection between the Houses of Welf and Este by studying old documents. On 8th May 1688 Leibniz arrived at Vienna where Leopold I, emperor of the Holy Roman Empire, resided. Leibniz tried without success to realise his old dream of an appointment at the imperial court.

In Vienna an event occurred which laid a fuse to the powder peg of the priority dispute. In the Acta Eruditorum of June 1688 Leibniz saw a book review by the Leipzig professor of mathematics Christoph Pfautz (1645–1711). The book discussed is the *Philosophiae naturalis principia mathematica*, *Principia* in short, the epoch-making oeuvre of Isaac Newton! By this book review Leibniz felt encouraged to publish his own ideas concerning physics and published hastily three manuscripts in the Acta Eruditorum, *De lineis opticis et alia* (On optical lines and other things) in January 1689, *Schediasma de resistentia medii et motu projectorum gravium in medio resistente* (Short writing on the resistance of the medium and the motion of thrown heavy bodies in a resistant

Fig. 6.1.8. Duke Welf IV, ancestor of the House of Welf [Ideal portrait in the Weingartener Stifterbüchlein about 1510] and emperor Leopold I [Painting by Benjamin v. Block, 1672]

medium), also in January 1689, and *Tentamen de motuum caelestium causis* (Attempt on the cause of the heavenly motions) in February 1689. The latter work is short and serves only as an introduction to the following two. There he wrote [Meli 1993, p. 7]:

> *'So, when I was examining the Proceedings for June of this year* [in the Acta Eruditorum] *I came across an account of the celebrated Isaac Newton's Mathematical Principles of Nature. This account I have read eagerly and with much enjoyment ...'*

Leibniz claimed – and will claim further on – that he had not read the *Principia* before he wrote his three manuscripts. That was a downright lie!

6.1.2 Isaac Newton – The Hermit in Cambridge

At the beginning of the year 1675 Isaac Newton visited London. He desired the royal dispensation to keep his fellowship in Cambridge without the necessity to become a priest of the Church of England and his wish was granted. Newton's refusal to be ordained can be understood if one studies his theological works. Being a deeply religious broody person Newton can not share the belief in

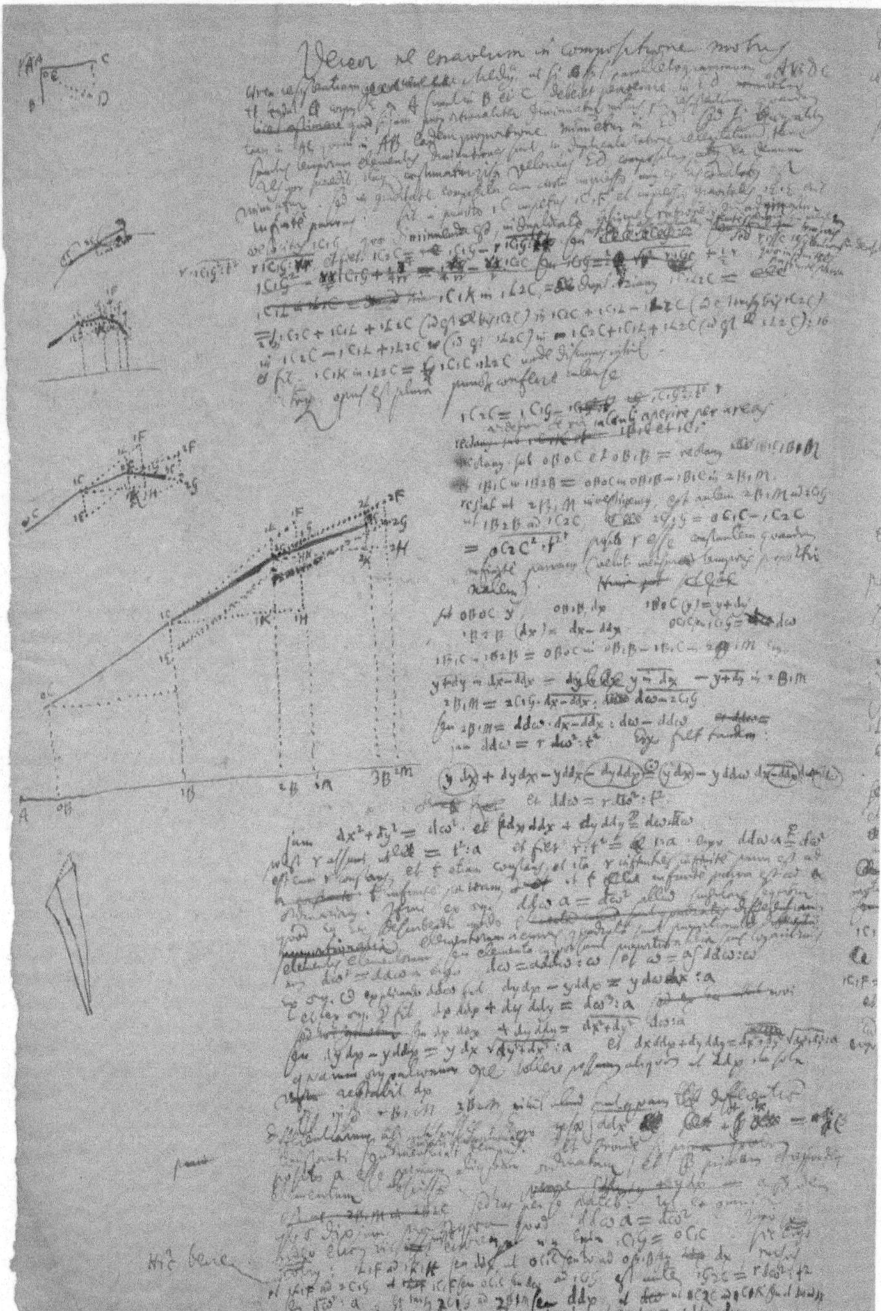

Fig. 6.1.9. Manuscript page of *Schediasma* (Gottfried Wilhelm Leibniz Library - Niedersächsische Landesbibliothek Hannover (Hanover), LH XXXV IX 6 Bl 3v)

Fig. 6.1.10. Council of Nicaea. Arius is seen under the feet of the emperor Konstantin and the bishops. Icon in the cloister Mégalo Metéoron in Greece [Photo: Jjensen 2008, detail]

the Holy Trinity – God, Jesus, and Holy Ghost – after having studied many works of the church fathers; and he takes it seriously. Presumably already since 1670 Newton had become an anti-trinitarian who thought of the Holy Trinity as being a diabolic seduction introduced by perverters of the Holy Scripture in the fourth century AD [Iliffe 2007, p. 72]. He accused Athanasius of Alexandria (about 298–373), the father of orthodox christendom, of having introduced wrong texts into the bible. Newton's belief was Arianism, seen as heretic, which took its name from the christian presbyter Arius of Alexandria (about 260–336) and Newton even identified himself intellectually with Arius [Westfall 2006, p. 318]. He also believed that God had chosen him, Isaac Newton, to figure out the truth concerning the decline of Christendom and he considered his works in this regard as being the far most important ones – much more important than his mathematical and scientific works. In the whole period between 1675 and 1685 Newton was busy with the interpretation of the Book of Revelation of Saint John, and he 'proved' his conclusions in the same astute manner which he applied a little later in his *Principia* [Iliffe 2007, p. 74]. If one once recognises this activity of Newton (one may not and can not understand it readily) then it is clear that Newton never could have become a priest of the Church of England. He could not even talk about his convictions since they comprised a serious heresy.

Fig. 6.1.11. Newton on a stamp of the German Federal Mail (Deutsche Bundespost) from 1993. In the background the experimentum crucis can be seen

On 18th February 1675 Newton attended a meeting of the Royal Society for the first time where he registered in the membership directory and hence confirmed his full membership. The meeting turned out very favourably for Newton since he was received with great admiration so that he even attended the following two meetings [Westfall 2006, p. 267f.]. The disputes concerning his theory of light continued lively in these days. This time it is Francis Hall, latinised Franciscus Linus (1595–1675), who wrote agressively against Newton's theory in letters to Oldenburg. Linus was an English Jesuit who worked as a professor for Hebrew and mathematics at the Jesuit College in Liège. He had already gained sad notoriety in England when he attacked Robert Boyle's gas law. The conflict between Linus and his disciples with Newton lasted until 1678 [Westfall 2006, p. 267], but even Robert Hooke was on Newton's side this time. Newton again explained his 'experimentum crucis' and sent two manuscripts to the Royal Society on 7th December 1675: *Discourse of Observations* and *An Hypothesis explaining the Properties of Light discoursed of in my severall Papers* [Turnbull 1959–77, Vol. I, p. 362ff.]. The first treatise is almost identical with the first three parts of the second book of Newton's *Opticks* which will be published only in 1704.

Both manuscripts made a great impression in the Royal Society and one wanted to publish the *Discourse* immediately, but Newton refused the offer. As was to be expected Hooke went through the roof on seeing Newton's remark in *An Hypothesis* that the diffraction of light as described by Hooke was nothing else than a kind of refraction, was known for a long time, and was already described by Francesco Maria Grimaldi (1618–1663). Hooke claimed that all that was already contained in his *Micrographia* and that Newton had only conducted some of the details [Westfall 2006, p. 272]. Thereupon Newton went through the roof. With date of 31st December 1675 a letter was dispatched to Oldenburg in which one can feel the anger against Hooke growing beyond all bounds [Turnbull 1959–77, Vol. I, p. 404ff.]. A second letter of 30th January 1676 was only barely moderate in tone. Incomprehensibly, Oldenburg read only a passage of Newton's first letter to the Royal Society, but not the parts concerning Hooke. Hooke was not even personally informed by Oldenburg about these parts so that he was greatly amazed when he became aware of them at a meeting on 20th January. Now Hooke was convinced that Oldenburg wanted to fuel a conflict between him and Newton and that he deliberately had provided Newton with wrong informations [Westfall 2006, p. 273]. He took the lead and wrote directly to Newton [Turnbull 1959–77, Vol. I, p. 412f.]:

> *'Robert Hooke – These to my much esteemed friend, Mr Isaack Newton, at his chambers in Trinity College in Cambridge.'*

Hooke suspected, he wrote, that Newton was deliberatley misinformed by 'sinister practice'. He praised Newton's excellent treatises and proposed a correspondence concerning philosophical (that is, natural philosophy) issues. Newton replied utterly friendly on 15th February, accepted the proposal of a correspondence, and praised Hooke's contributions to optics [Turnbull 1959–77, Vol. I., p. 416f.]:

> *'What Des-Cartes did was a good step. You have added much several ways, & especially in taking ye colours of thin plates into philosophical consideration. If I have seen further it is by standing on ye shoulders of Giants.'*

However, behind the gentleness of the letters one can feel no wish to actually take up a correspondence on either side if one reads between the lines. Whoever the 'Giants' were that Newton said he would stay on their shoulders – surely Hooke was not meant.

Franciscus Linus died in November 1675 but the conflict concerning the theory of light did not stop with his death. A pupil of Linus's, John Gascoines, filled the gap his teacher had left and pestered Newton further, although the Royal Society had successfully repeated the experimentum crucis in the spring of 1676. Around the time when Newton wrote the *Epistola prior* another English Jesuit, Anthony Lucas, entered the stage who was instrumentalised by Gascoines [Westfall 2006, p. 274]. Lucas conducted (lousy) experiments with prisms, too, and jarred on Newton's nerves with his letters. When Lucas did not stop his efforts Newton was convinced that the 'Liègois' (=papists) had forged a plot against him. He became more and more angry but planned to publish a larger work on optics after the third letter of Lucas had arrived in February 1677. We have some evidence that Newton worked on this project during the whole year 1677 and in December he even informed Robert Hooke, who had taken over the duties of the secretary of the Royal Society after Oldenburg had died. But then a fire broke out in Newton's rooms and destroyed the manuscripts [Westfall 2006, p. 276f.].

There is no question that the fire really took place. Fourteen years after this event a student of St John's College in Cambridge, Abraham de la Pryme, reported on the fire. In days of yore one was prone to interpret the student's report as being the description of the fire in 1693 (cp. [Rosenberger 1987, p. 278]) but de la Pryme's story was written 18 months before the fire of 1693 and Newton's correspondence rested completely from December 1677 until February 1678. Newton must have been in a devastated state of mind. Not only had the fire destroyed his important manuscript on optics, not only was he pestered by Gascoines and Lucas with their letters describing utterly insufficient experiments which were positioned against his theory, no, to make matters worse he started losing his long-standing roommate John Wickins! In 1677 Wickins stayed only $13\frac{1}{2}$ weeks in College, in 1678 only just 6 weeks [Westfall 2006, p. 278, footnote 122]. Wickins finally withdrew from College only in 1684 but to this point he hardly stayed more than a few days in Cambridge. However, even in 1693, the year of the alleged fire, we have to report on a dramatical personal break: the separation from Fatio de Duillier. In the state of mind Newton was in, he drafted two arrogant, brutal letters [Turnbull 1959–77, Vol. II, p. 254f.], [Turnbull 1959–77, Vol. II, p. 262f.] bursting with paranoia. Once again Lucas sent a letter which Newton did not answer. A last letter awaited him in London but Newton was weary of the confrontation [Westfall 2006, p. 279]:

> *'Mr Aubrey,*
> *I understand you have a letter from Mr Lucas for me. Pray forbear to send me anything more of that nature.'*

This marks the end of the unfortunate correspondence concerning the theory of colours. Newton retreated to 'inner exile'. He terminated the correspondence with Collins and hid himself away like a hermit in his rooms in Trinity College to pursue his alchemistic experiments and his theological works.

One can write big books on Newton's alchemistic works, see [Dobbs 1991]. He started with experimental and theoretical works on chemistry probably already in 1669 but then abandoned chemistry in favour of alchemy. As John Harrison's list of books in Newton's library reveals there were 138 books on alchemy and 31 on chemistry, alltogether they comprise a tenth of the whole library, not counting the numerous manuscripts with alchemical contents.

At the end of the year 1676 Newton was so deeply immersed in his theological and alchemical works that it almost seems surprising that he took his time to write the *Epistola prior* and the *Epistola posterior*! When he brutally terminated the disputes concerning optics in mid-1678 almost a decade of silence followed. In the 1680s Newton employed a secretary or rather an assistant, who was distantly related to him: Humphrey Newton. Humphrey later wrote that Newton was careless in the handling of money since he left a box full of Guineas[1], perhaps 1000 of them, unlocked near the window [Westfall 2006, p. 335]. Newton was again so deeply immersed in his work that he frequently forgot to eat and the Master of Trinity feared between 1677 and 1683 that he would kill himself by means of his studies. Humphrey saw Newton laugh exactly once in five years.

Isaac Newton's mother Hannah had hurried to the sickbed of her son Benjamin Smith in late spring of 1679 to care for him. Apparently she got infected and developed an evil fever. Now Newton travelled to Woolsthorpe to nurse his mother. He noted [Westfall 2006, p. 339]:

> *'sate up whole nights with her, gave her all her Physick himself, dressed all her blisters with his own hands & made use of that manual dexterity for wch he was so remarkable to lessen the pain wch always attends the dressing the torturing remedy usually applied in that distemper with as much readiness as he ever had employed it in the most delightfull experiments.'*

However, it was pointless; eventually his mother died. Only a short time later another loss which we have already mentioned came along. The long-standing roommate John Wickins, having been in Cambridge only for short amounts of time during the last years, decided in 1683 to abandon his fellowship and to get married. Officially the fellowship ended in 1684 but Wickins is in College for the last time in spring 1683. We know nothing about the nature of the 20 year long friendship between Newton and Wickins. Did it terminate with a breach? Were there homoerotic feelings involved? Directly after Wickins left College 1683 Newton disappeared with unknown destination for about one month. In May 1683 he was back in College for a few days, then he disappeared again for a week [Westfall 2006, p. 343].

Deeply immersed in alchemistic and theological studies Newton still found great pleasure in mathematical occupations. As the Lucasian professor it is his duty to represent mathematics after all. About the year 1680 he became seriously interested in classical geometry and started to learn what the ancients knew. He now began to be dissatisfied with the mathematics of René Descartes, wrote a manuscript titled *Errors in Descartes' Geometry* [Westfall 2006, p. 378ff.], and called Descartes's geometry *'the Analysis of the Bunglers in Mathematicks'*. The great Descartes, godfather not only to the creation of the calculus of fluxions but also to considerations concerning physics, was now rigorously rejected by Newton. This rejection was particularly important to decisively determine the content and the form of his *Principia*. Huygens's *Horologium oscillatorium* surely was important for Newton's clear orientation towards classical geometry as well. The *Horologium oscillatorium* proofs and

[1]The Guinea was an English gold coin in circulation from 1663. Its initial value was 1 Pound Sterling = 20 shillings.

reasoning were carried out completely in the geometric tradition of the ancients. Whether there was also a correlation of the worship of the ancients in Newton with his theological and alchemistic studies can only be suspected.

Already Pappos of Alexandria (about 290–about 350) whose *Collectio Mathematica* had been edited and published by Frederico Commandino in Pesaro in 1588 distinguished between 'analysis' and 'syntheses'. Analysis (resolutio) designates the process of starting from a supposed mathematical theorem and working stepwise backwards until a true result occoured. Synthesis (compositio) in contrast describes the process starting at known truths and working stepwise upward until the supposed theorem emerges. In this sense synthesis is the reversal of analysis. Synthesis was thought of being the more rigorous of the two procedures [Guicciardini 2009, p. 34], [Guicciardini 2002, p. 308ff.]. Between autumn 1683 and the early winter of 1684 Newton's *Lucasian Lectures on Algebra* were finished; printed in 1707 as *Arithmetica Universalis*. Here Newton is outrightly 'anti-Descartes' since he rejected Descartes's method of establishing equations from the intersection of curves. Since Newton was delighted by the mathematics of the ancients he now wanted to put his fluxional calculus on the sound foundation of synthesis, hence conducting proofs in geometric manner. The plan is to create a 'synthetic method of fluxions' [Guicciardini 2009, p. 217].

About 1680 Newton wrote *Geometria Curvilinea* [Whiteside 1967–81, Vol. IV, p. 420ff.]. The ancients had examined straight lines, planes, circles, and conic sections. Other crooked curves could not be studied by the methods of Euclid's *The Elements* without introducing infinitesimals. Newton wrote [Whiteside 1967–81, Vol. IV, p. 423]:

> *'Observing therefore that numerous kinds of problems which are usually resolved by* [an algebraic] *analysis may (at least for the most part) be more simply effected by synthesis, I have written the following treatise on the topic. At the same time, since Euclid's elements are scarcely adequate for a work dealing, as this, with curves, I have been forced to frame others. He* [Euclid] *has delivered the foundations of the geometry of straight lines. Those who have taken the measure of curvilinear figures have usually viewed them as made up of infinitely many infinitely-small parts. I, in fact, shall consider them as generated by growing, arguing that they are greater, equal or less according as they grow more swiftly, equally swiftly or more slowly from their beginning. And that swiftness of growth I shall call the fluxion of a quantitiy.'*

Newton made a really great step here; away from the infinitesimals as infinitely small quantities and towards quotients of vanishing quantities. With the help of this quotients Newton could exactly describe how fast a vanishing quantity grew or got smaller in relation to another one. These quotients are already fairly close to genuine limits even if Newton's definition left something to desire in terms of modern rigour. From 1680 on Newton developed this idea further and used it also in his *Principia* [Guicciardini 2009, p. 218].

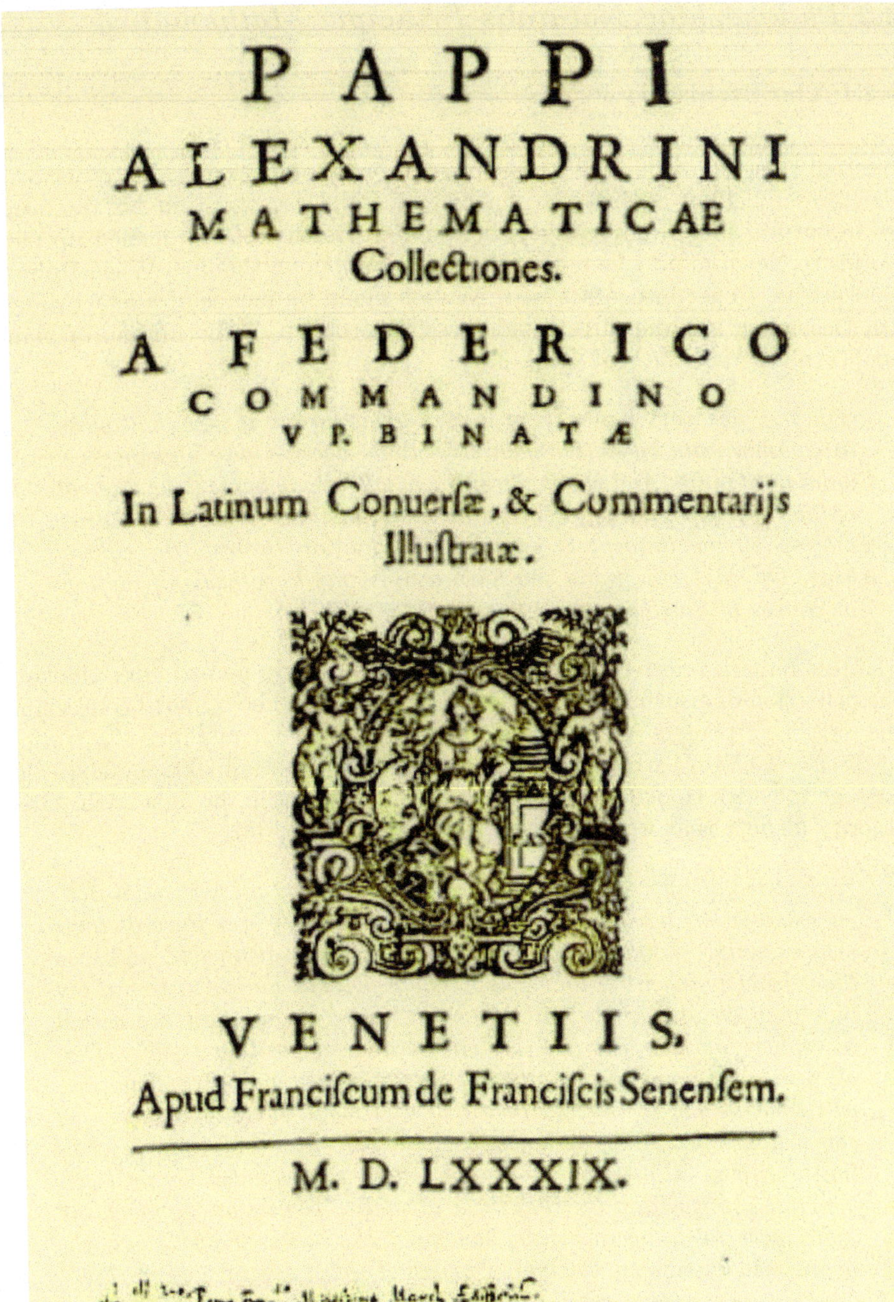

PAPPI
ALEXANDRINI
MATHEMATICAE
Collectiones.
A FEDERICO
COMMANDINO
VRBINATÆ
In Latinum Conuersæ, & Commentarijs
Illustratæ.

VENETIIS.
Apud Franciscum de Franciscis Senensem.

M. D. LXXXIX.

Fig. 6.1.12. Title page of *Collectio Mathematica*, 1589

6.2 *Philosophiae Naturalis Principia Mathematica*

6.2.1 The Prehistory

Westfall [Westfall 2006, p. 381] has pointed out that the coincidence between Newton's rebellion against the mathematics of Descartes and his discoveries in natural philosophy can not be ignored. Descartes had opened up new worlds to Newton; now he rejected Descartes all along the line. When Robert Hooke tried to get in contact with Newton again by means of a letter dated 4th December 1679 he directly addressed a problem of the motion of planets [Turnbull 1959–77, Vol. II, p. 297]:

> *'For my own part I shall take it as a great favour if you shall please to communicate by Letter your objections against any hypothesis or opinion of mine, And particularly if you will let me know your thoughts of that of compounding the celestiall motions of the planetts of a direct motion by the tangent & an attractive motion towards the centrall body, Or what objections you have against my hypothesis of the laws or causes of Springinesse.'*

The last remark referred to the law which is nowadays named after Hooke – the deflection of a spring is proportional to the force. The remarks concerning the motion of planets referred to Hook's paper *Attempt to Prove the Motion of the Earth* of 1674 which was reprinted in 1679 [Westfall 2006, p. 382]. The passage to which Hooke was refering to is so important in the light of the later priority dispute that we cite it here [Westfall 2006, p. 382]:

> *'This depends upon three Suppositions. First, That all Coelestial Bodies whatsoever, have an attraction or gravitating power towards their own Centers, wherebye they attract not only their own parts, and keep them from flying from them, as we may observe the earth to do, but that they do also attract all other Coelestial Bodies that are within the sphere of their activity* [...] *The second supposition is this, That all bodies whatsoever that are put into a direct and simple motion, will so continue to move forward in a streight line, till they are by some other effectual powers deflected and bent into a Motion, describing a Circle, Ellipsis, or some other more compounded Curve Line. The third supposition is, That these attractive powers are so much the more powerful in operating, by how much the nearer the body wrought upon is to their own Centers. Now what these several degrees are I have not yet experimentally verified* [...] .

Was Hooke already aware of the nature of gravitation? Was he already in the possession of the law of gravitation that the attracting force between two bodies change with the square of their distance? Later on Hooke will claim this as a fact! It is clear that for the first time a researcher of distinction broke here with the traditional view of the attraction of planets. It is also clear that for the first time we find here a description of the true dynamics of celestial motions. Centrifugal forces are not mentioned, however; a planet is forced by gravitation of the sun from its tangential path to a circular or elliptical orbit. At the time of Hooke sending this letter Newton was not yet thus far in his understanding of the motion of planets. So far he assumed a state of equilibrium between the (Huygensian) centrifugal force and a force of attraction acting inwardly, but now Hooke wrote about a permanent motion of imbalance in which the force of attraction bends the straight line on which the planet would move [Westfall 2006, p. 383].

Newton replied to Hooke with a letter dated 8th December 1679 [Turnbull 1959–77, Vol. II, p. 301]. He wrote that he had not seen Hooke's paper (what seems unbelievable) but described an experiment to prove the rotation of the earth. Since the tangential velocity at the tip of a high tower is larger than at the surface of the earth a stone falling from the tip should land east of the tower[2]. Then Newton added a drawing of the trajectory of a falling stone if the earth was absent as a spiral winding around the centre of the earth.

Newton's reflections were simply wrong and Hooke saw that immediately. In a letter of 19th December 1679 [Turnbull 1959–77, Vol. II, p. 304ff.] he wrote very friendly and cautiously that, following his theory, the stone would never fall into the centre and that the trajectory would not be a spiral but that the stone would move in an elliptical orbit. Newton was caught having committed an error in his hasty reply to Hooke's letter, but he accepted Hooke's correction in his reply letter of 23rd December 1679 [Turnbull 1959–77, Vol. II, p. 307f.], although in a somewhat sour tone. In Westfall's words [Westfall 2006, p. 385]:

> *'His reply, as he accepted Hooke's correction, was as dry as a piece of burned bacon.'*

But Newton remarked [Turnbull 1959–77, Vol. II, p. 307], cp. Fig. 6.2.2:

> *'Yet I imagin ye body will not describe an Ellipsoeid but rather such a figure as is represented by AFOGHIKL & c.'*

Then Newton added a few remarks on the appearance of the trajectory in case of variable gravitation. He had not to wait long to once again receive a correction from Hooke; the letter arrived on 16th January 1680 [Turnbull 1959–77, Vol. II, p. 309]:

> *'Sir*
> *Your Calculation of the Curve by a body attracted by an aequall power*

[2]Classical arguments against the rotation of the earth stated that a falling stone would land west of the tower because the stone would remain behind the rotation of the earth during the fall. But this could not be observed.

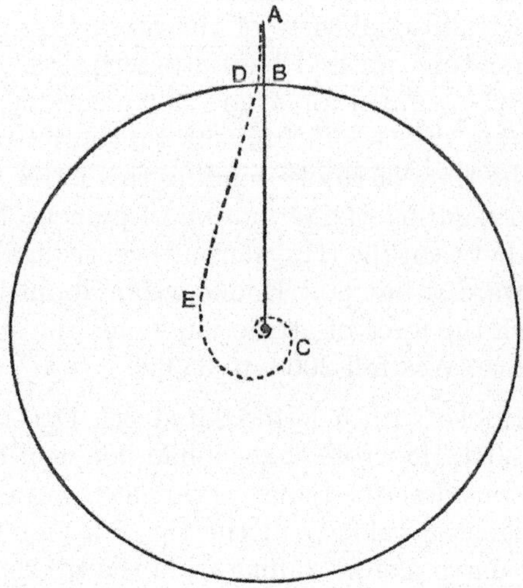

Fig. 6.2.1. Newton's sketch of the trajectory of a falling stone into the earth's centre ([Westfall 2006, p. 384] © Cambridge University Press)

at all Distances from the center Such as that of a ball Rouling in an inverted Concave Cone is right and the two auges[3] *will not unite by about a third of a Revolution. But my supposition is that the Attraction always is in a duplicate proportion to the Distance from the Center Reciprocall, and Consequently that the Velocity will be in a subduplicate proportion to the Attraction and Consequently as Kepler Supposes Reciprocall to the Distance.'*

Doesn't that mean that Hooke *knew* that the force of attraction changes with the square of the distance and in fact inversely proportional: $F = \text{const.}/r^2$? So it seems but it isn't so. Hooke rather started from Galileo's relation

$$v^2 = 2as,$$

v=velocity, a=acceleration, s=path, which is valid only in case of uniform acceleration from rest. This relation was blindly applied to all possible problems by Hooke [Westfall 2006, p. 387], even to those in which the acceleration is not uniform. Moreover, Hooke had employed the erroneous, original velocity assumption of Kepler in which the velocity everywhere on the orbit of a planet is

[3] The auges (apses) are the two main apexes of the elliptical orbit of a planet around a central star. The point with the largest distance to the star is called *apoapsis*, the point with the least distance *periapsis*.

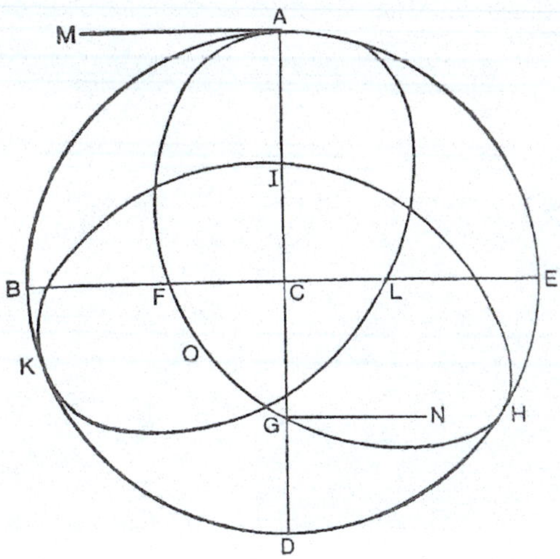

Fig. 6.2.2. Newton's new sketch of the trajectory of a falling stone ([Westfall 2006, p. 386] © Cambridge University Press)

inversely proportional to its distance[4] [Sonar 2011, p. 181]. Nonetheless Hooke will later accuse Newton of plagiarism and will invoke the correspondence of 1679-80 as proof of his priority.

On 27th January 1680 Hooke again wrote to Newton [Turnbull 1959–77, Vol. II, p. 312f.]. He had conducted an experiment which clearly proved the rotation of the earth and raised an open problem: what is the shape of a planet's orbit if it is attracted by a gravitational force acting in inverse proportion to the square of distance.

We do not have to care about Hooke's experiment; it is now acknowledged that he had misunderstood it [Westfall 2006, p. 387, footnote 143]. However, the question concerning the shape of the orbit of a planet if a gravitational law of the form $F = \text{const.}/r^2$ is assumed stimulated Newton tremendously. Following an ingenious idea he reversed the question: assuming an elliptical orbit he wanted to show that the gravitation can only have the form of a force being inversely proportional to the square of the distance. Newton found the proof in the first half of the year 1680 [Westfall 2006, p. 388] but did not send it to Hooke. The impetus coming from Hooke and Newton's experiences in alchemy provided in 1679-80 the reason for a complete revision of Newton's natural philosophy. In fact, the problem of the motion of planets as well as alchemy had much to do with long-distance effects [Westfall 2006, p. 390].

[4] Kepler had used this assumption to derive his second law but had not maintained it permanently.

Fig. 6.2.3. John Flamsteed [Painting: unkown painter, about 1719]; Portrait about 1720 after a painting of Godfrey Kneller, 1702 (Wellcome Library, London)

About the turn of the year 1680/81 a comet could be seen by daylight with the bare eye. Since the comet was sighted early in November 1680, then vanished at the end of November and appeared again much brighter in mid-December, almost all astronomers in Europe spoke about *two* comets – all but one. John Flamsteed (1646–1719) became the first Astronomer Royal, hence was the royal court astronomer of the newly founded Royal Greenwich Observatory.

Flamsteed expressed the opinion that the two big comets were but one. He supposed that the sun was responsible for the change of direction of the comet and that was a true breakthrough. Flamsteed immediately wrote three letters to his friend James Crompton, fellow of Jesus College Cambridge, and asked him to get a comment from Isaac Newton. The letters were dated 25th December 1680 [Turnbull 1959–77, Vol. II, p. 315], 13th January 1681 [Turnbull 1959–77, Vol. II, p. 319f.] and 22nd February 1681 [Turnbull 1959–77, Vol. II, p. 336] of which Newton probably received only excerpts. Newton himself could observe the comet on 22nd December 1680 in Cambridge over the chapel of King's College and he drew a sketch [Westfall 2006, p. 392]. Almost daily he followed the comet, initially with the bare eye, than with the help of a telescope. Newton started again to work on his reflecting telescope and from this work part IV of the second book of his *Opticks*, published in 1704, will emerge.

Newton answered Flamsteed in two long letters [Turnbull 1959–77, Vol. II, p. 340ff, p. 363ff.], the second was directly addressed to Flamsteed. As can be seen from the letters Newton had not yet the idea that comets move like planets in the solar system. He therefore did not possess his theory of universal gravitation at this time [Westfall 2006, p. 395]. From 1680 on Newton recorded systematically the observation of comets; he tried to classify them and started

Fig. 6.2.4. The Royal Obseratory Greenwich
[The Penny Magazine, Volume II, Number 87, August 10, 1833]

to accept the laws of motion of planets even for comets. We will give a detailed report on the serious conflict between Newton and Flamsteed which did not yet emerge in the 1680s in section 7.4.2.

In the year 1682 Edmond Halley (1656–1742) observed a comet which will later be known under the name Halley's comet. Newton also observed this comet and recorded its path. Halley will everlastingly alter Newton's life – and thereby the development of physics.

In January 1684 three men met in the rooms of the Royal Society to talk: Edmond Halley, Christopher Wren, and Robert Hooke. Hooke claimed he would be able to deduce all laws of the planetary motion from the inverse square distance rule. Halley admitted that his own attempts had failed, Wren was sceptical about Hooke's claim. Since the question of the deduction of Kepler's laws of planetary motion from the principles of dynamics was a burning and long discussed problem Wren offered the presentation of a book award. Whoever presented the proof of a connection between the laws of the planets and the inverse quadratic distance formula should win the prize [Westfall 2006, p. 403].

Fig. 6.2.5. Edmond Halley, Bust in the museum of the Royal Greenwich Observatory, London [Photo: Klaus-Dieter Keller; Bust: Henry Alfred Pegram, 1904] and as painting by Thomas Murray, about 1687

Halley had met Newton already in the year 1682 and discussed comets with him. Now seven months had to pass after the meeting of Halley, Hooke, and Wren until Halley visited Newton in Cambridge. Although Hooke had claimed the required proof but would only show it when others had learned to appreciate him the three men had not come closer to a solution. Halley wanted nothing else than to consult the man he thought to be an expert – Newton. Concerning the visit to Newton there is an account by Abraham de Moivre [Westfall 2006, p. 403]:

> *'In 1684 Dr Halley came to visit him at Cambridge; after they had been some time together, the D^{r} asked him what he thought the Curve would be that would be described by the Planets, supposing the force of attraction towards the Sun to be reciprocal to the square of their distance from it. Sr Isaac replied immediately that it would be an Ellipsis. The Doctor struck with joy & amazement, asked him how he knew it. Why, saith he, I have calculated it; whereupon Dr Halley asked him for his calculation; without any further delay, Sr Isaac looked among his papers but could not find it, but he promised him to renew it, & then to send it him ...'*

Since the sought manuscript was found among Newton's posthumous papers we may assume with Westfall that the account given by de Moivre and told to him by Newton is a myth. After his earlier experiences, Newton would not give freely away such a manuscript. He probably wanted time to at least be able to check the computations thoroughly. In any case, the account by de Moivre contained the myth of the origin of the *Principia*.

6.2.2 The Formation Phase

In November 1684 Halley got a little more than the required derivation, namely a small treatise titled *De motu corporum in gyrum* (On the motion of bodies in orbit). Not only did Newton show that an elliptic orbit necessitates a force inversely proportional to the square of the distance, but he also sketched the proof of the original problem: a force being inversely proportional to the square of the distance necessitates an orbit in form of a conic section which turns out to be an ellipse if the orbital velocity remains below a certain threshold. Starting from postulated principles of dynamics Newton additionally deduced Kepler's second and third law.

1.	The orbit of a planet is an ellipse with the sun in one of the two foci.
2.	A straight line joining a planet and the sun sweeps out equal areas in equal intervals of time.
3.	The square of the orbital period of a planet is proportional to the cube of the semi-major axis of its orbit.

Table 6.1. The three Kepler's laws

Halley saw immediately that the manuscript he held in his hands contained a revolutionary advance in the theory of celestial mechanics. Swiftly he travelled to Cambridge to talk to Newton and on 20th December he reported to the Royal Society [Westfall 2006, p. 404]:

> *'Mr. Halley gave an account, that he had lately seen Mr. Newton at Cambridge, who had shown him a curious treatise, De motu; which, upon Mr. Halley's request, would he said be sent to the Society to be entered upon their register.'*

As usual, a lot of time elapsed until Newton sent the promised manuscript to the Royal Society. Newton was simply gripped by problems of celestial mechanics and he worked feverishly to solve them. In the manuscript Halley received in November 1684, Newton treated four theorems and five problems concerning the motion of bodies in inviscid media; now Newton worked on a theory of motion under friction. He wrote to Flamsteed to receive data concerning the motion of planets and then withdrew from his surroundings [Westfall 2006, p. 405]. Halley later would say that he was the Ulysses who produced this Achilles; Halley did not drag the *Principia* from a secluded Newton, but unintentionally gave an impulse which drew Newton to scientific peak performance. In the time between August 1684 and spring 1686, only isolated activities of Newton are known – letters to Flamsteed, a visit at Woolsthorpe on family grounds, and the rejection of a request to organise a philosophical society in Cambridge. Otherwise he was completely obsessed by the work on what would later lead to *Principia*; he even let rest his alchemy works [Westfall 2006, p. 406]. Newton's assistant Humphrey Newton gave a lively description of Newton's behaviour in these days [Westfall 2006, p. 406]:

> *'So intent, so serious upon his Studies, y^t he ate very sparingly; nay, ofttimes he has forget to eat at all, so y^t going into his Chamber, I have found his Mess untouch'd of wch when I have reminded him,* [he] *would reply, Have I; & then making to ye Table, would eat a bit or two standing ... At some seldom Times when he design'd to dine in ye Hall, would turn to ye left hand, & go out into ye street, where making a stop, when he found his Mistake, would hastily turn back, & then sometimes instead of going into ye Hall, would return to his Chamber again ... When he has sometimes taken a Turn or two* [in the garden], *has made a sudden stand, turn'd himself about, run up ye Stairs, like another Alchimedes* [sic], *with an εὔρεκα, fall to write on his Desk standing, without giving himself the Leasure to draw a Chair to sit down in.'*

Up to 1684 we know Newton only as the ingenious scientist who could not finish any one of his numerous projects. Mathematics, mechanics, alchemy, theology – in all areas Newton's genius shone, but sooner or later he lost his interests in all these matters. Now he had found a task thrilling him so much that he got through from the start to the end, and in the end there arose a masterpiece which should change the world, the *Principia*. In the course of his work he developed his mathematics further and worked more intensely on dynamics as never before. Thereby he succeeded in concentrating his mathematical and scientific interests in a way not experienced before. Newton's *Principia* mark no less than the birth of modern physics.

Physics is not so much in the centre of our considerations as are the mathematical foundations. Newton took up again his theory of first and last ratios which he had created to get rid of the notion of fluxions. The reader naively reaching for the *Principia* expecting a fireworks of mathematical proofs resting on Newton's infinitesimal calculus will, however, be disappointed. The proofs in the *Principia* are classical geometrical proofs! Here Newton very much reminds one of Archimedes who derived his results by means of his law of the lever weighing indivisibles to afterwards look for geometric arguments making his proofs acceptable for the ancients [Sonar 2011, p. 71ff.]. Newton even acknowledged his reference to the ancients. In the aftermath of the priority dispute Newton wrote in 1715 the anonymous review *An Account of the Book entituled Commercium Epistolicum Collinii & aliorum* [Hall 1980, p. 263ff.] of the *Commercium epistolicum* of the Royal Society and he remarked [Hall 1980, p. 296]:

'By the help of the new Analysis Mr. Newton found out most of the Propositions in his Principia Philosophiae: but because the Ancients for making things certain admitted nothing into Geometry before it was demonstrated synthetically, he demonstrated the Propositions synthetically, that the Systeme of the Heavens might be found upon good Geometry. And this makes it now difficult for unskilful Men to see the Analysis by which those Propositions were found out.'

With the exception of two, no manuscripts were found in Newton's estate which would underpin this assertion [Westfall 2006, p. 424]. However, under the impression of the geometry of the ancients Newton had given up fluxions in favour of first and last ratios and was of the firm conviction that the first and last ratios would represent the natural extension of classical geometry. However, in a kind of 'mathematical prefix' Newton casted the mathematical foundations of the *Principia* into eleven lemmas on the first and last ratios which can be found in Section I of Book I. Under the heading 'The method of first and ultimate ratios, for use in demonstrating what follows' we find [Newton 1999, p. 433]:

'Quantities, and also ratios of quantities, which in any finite time constantly tend to equality, and which before the end of that time approach so close to one another that their difference is less than any given quantity, become ultimately equal.'

The proof by Newton is the following:

'If you deny this, let them become ultimately unequal, and let their ultimate difference be D. Then they cannot approach so close to equality that their difference is less than the given difference D, contrary to the hypothesis.'

By means of illustration let us consult Lemma II which refers to figure 6.2.6:

'If in any figure AacE, comprehended by the straight lines Aa and AE and the curve acE, any number of parallelograms Ab, Bc, Cd, ... are inscribed upon equal bases AB, BC, CD, ... and have sides Bb, Cc, Dd, ... parallel to the side Aa of the figure; and if the parallelograms aKbl, bLcm, cMdn, ... are completed; if then the width of these parallelograms is diminished and their number increased indefinitely, I say that the ultimate ratios which the inscribed figure AKbLcMdD, the circumscribed figure AalbmcndoE, and the curvilinear figure AabcdE have to one another are ratios of equality.'

The lemma is demonstrated as follows:

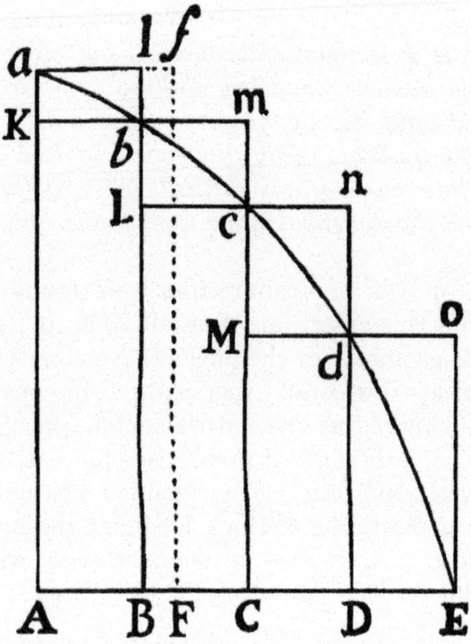

Fig. 6.2.6. Figure refering to Lemma II in the *Principia*

> *'For the difference of the inscribed and circumscribed figures is the sum of the parallelograms Kl, Lm, Mn, and Do, that is (because they all have equal bases), the rectangle having as base Kb (the base of one of them) and as altitude Aa (the sum of the altitudes), that is, the rectangle ABla. But this rectangle, because its width AB is diminished indefinitely, becomes less than any given rectangle. Therefore (by lem. 1) the inscribed figure and the circumscribed figure and, all the more, the intermediate curvilinear figure become ultimately equal. Q.E.D.'*

As a first impression these two lemmas of Section I of Book I may be sufficient to get a first glimpse on the method of first and last ratios which Newton wanted to be understood as mathematical foundation of his *Principia*.

6.2.3 Leibniz in Newton's Thought

In June 1684 Newton received a letter from James Gregory's nephew, David Gregory (1659–1708).

David Gregory became professor of mathematics at the University of Edinburgh in 1683. It was the chair which his uncle had occupied. The reason for Gregory's letter to Newton was the book *Excercitatio geometrica de dimensione figurarum* (Geometrical exercises on the measurement of figures) in

Fig. 6.2.7. Sculpture of David Gregory and Marischal College of the University of Aberdeen, at which he studied until 1675
[Photo: colin f m smith 2003]

which David had published the works of his uncle concerning infinite series and quadrature. David acknowledged Newton's contribution to these areas without limitations and hence did not pose a risk to Newton, but Newton was nevertheless alarmed. Again a book concerning his infinitesimal mathematics was published as Mercator's *Logarithmotechnia* before, but he himself had published nothing to it [Westfall 2006, p. 400]. Newton thereupon began to draft the plan for a presentation in six chapters titled *Matheseos universalis specimina* (Examples of a universal mathematics). He wanted to publish letters exchanged between him and James Gregory to ensure his priority but from the start he 'forgot' about James Gregory and dwelled on Leibniz instead! Although he had ceased the correspondence with Leibniz in 1677 and allegedly had no further interest in mathematical problems, Leibniz had arguably never left his mind. Newton replied to David Gregory with a polemically coloured writing against Leibniz. Newton never completed *Matheseos universalis specimina* but chapter 4, in which he treated the calculus of fluxions, is particularly interesting [Westfall 2006, p. 401]. There Newton explained both anagrams of the *Epistola posterior* and compared his fluxional calculus with Leibniz's differential calculus. Leibniz had already published the arithmetical quadrature of the circle in *De vera proportione circuli ad quadratum circumscriptum in numeris rationalibus expressa* (The true ratio of circle to circumscribed square in rational numbers) in the Acta Eruditorum of 1682, but that Newton was aware of this work when he started writing *Matheseos universalis specimina* is not likely. Of course, Newton described his refusal of modern analysis and analysts where implicitly also Leibniz was meant. How strong his aversion against Descartes had grown in this time can be seen from the sentence [Westfall 2006, p. 401], [Whiteside 1967–81, Vol. IV, p. 571]:

> *'On these matters I pondered nineteen years ago, comparing the findings of and Hudde with each other.'*

In the blank passage *only* the name of Descartes can fit in, but Newton was arguably incapable to even write down this name!

Newton lost interest in *Matheseos universalis specimina* and undertook a new attempt under the title *De computo serierum* (On the computation of series). However, he also cancelled the work on this manuscript [Westfall 2006, p. 401].

In working on the *Principia*, probably at the beginning of 1686, Newton must have become aware that Leibniz had already begun to publish on the differential calculus. Newton now anxiously tried to secure his priority and inserted a scholium after Lemma II in Book II where he refers to his correspondence with Leibniz in the year 1676. Newton wrote [Westfall 2006, p. 426]:

> *'The most eminent man* [Leibniz] *wrote back that he also had fallen into a similar method, and he communicated his method, which scarcely differed from mine except in the words and notations it used. The foundation of both methods is contained in this lemma.'*

We will further discuss lemma II at another place on page 460; here we remark only that this lemma is concerned with the differentiation of a product of two functions.

Apparently, Newton could not let go the discussion with Leibniz of the years 1676–77 in the following years. The priority dispute – not yet fought openly – probably smouldered in Newton's thoughts.

6.2.4 The *Principia* Published

On 1st May 1686 Halley read a *Discourse Concerning Gravity* (Speech concerning gravity) to the Royal Society and informed the Royal Society about Newton's new work being almost ready to be printed. One week later Newton delivered [Westfall 2006, p. 444f.]:

> *'Dr. Vincent presented to the Society a manuscript treatise entitled Philosophiae Naturalis principia mathematica, and dedicated to the Society by Mr. Isaac Newton, wherein he gives a mathematical demonstration of the Copernican hypothesis as proposed by Kepler, and makes out all the phaenomena of the celestial motions by the only supposition of a gravitation towards the center of the sun decreasing as the squares of the distances therefrom reciprocally.*
>
> *It was ordered, that a letter of thanks be written to Mr. Newton; and that the printing of his book be referred to the consideration of the council; and that in the meantime the book be put into the hands of Mr. Halley, to make a report thereof to the council.'*

Three weeks passed without anything happening and Halley, just elected new secretary of the Royal Society, was alarmed. Although the governing council alone was in charge he let the members decide on the printing of the *Principia* during a meeting of 29th May. That bore a high personal financial risk. Although he grew up in a wealthy family the death of his father in 1684 left him impoverished. As secretary he earned a salary of 50 pounds a year but he had to care for his wife and family [Westfall 2006, p. 445].

With date of 1st June 1686 Halley informed Newton about the upcoming print [Turnbull 1959–77, Vol. II, p. 431]. But he added:

> *'There is one thing more that I ought to informe you of, viz, that Mr Hook has some pretensions upon the invention of ye rule of the decrease of Gravity, being reciprocally as the squares of the distances from the Center.'*

Newton's reply of 6th June (naturally!) was concerned with Hooke right from the start [Turnbull 1959–77, Vol. II, p. 433]:

> *'I thank you for wt you write concerning Mr Hook, for I desire that a good understanding may be kept between us. In the papers in your hands there is noe one proposition to which he can pretend, & soe I had noe proper occasion of mentioning him and others.'*

Newton again recapitulated the events of the year 1676 when he found the key to the theory of gravitation through the correspondence with Hooke. However, Halley's hope that Newton would not again step in a quarrel with Hooke was not fulfilled. It fumed slowly but continuously in Newton for three weeks before he boiled over.

On 30th June Newton wrote to Halley [Turnbull 1959–77, Vol. II, p. 435f.] a sole bitter complaint concerning Hooke's behaviour. He should acknowledge Hooke's achievements? What achievements? There was nothing to be acknowledged! In the manuscript of the *Principia* he deleted an appreciative remark concerning Hooke's concept of the attractive force. In a subsequent part, Book III, in which celestial mechanics is concerned, Newton deleted 'Cl[arissimus] Hookius' (the very famous Hooke) and replaced it by a simple 'Hooke'. In a following revision of the manuscript the reference to Hooke was dropped completely. Newton threatened Halley to even suppress Book III completely [Westfall 2006, p. 449]. However, Halley turned out being a brilliant diplomat. In a long letter of 9th July 1686 [Turnbull 1959–77, Vol. II, p. 441-443] he positioned himself and the Royal Society at Newton's side, tried to remove Hooke from the line of fire, and begged Newton fervently not to do without Book III. In effect, Newton calmed down a bit but inside him something was gnawing. How very becoming would it have been for Newton if he would have been generous at this point, but he was not. The glance inside his tortured soul which we can take from his letters to Halley reveals not much that is pleasant.

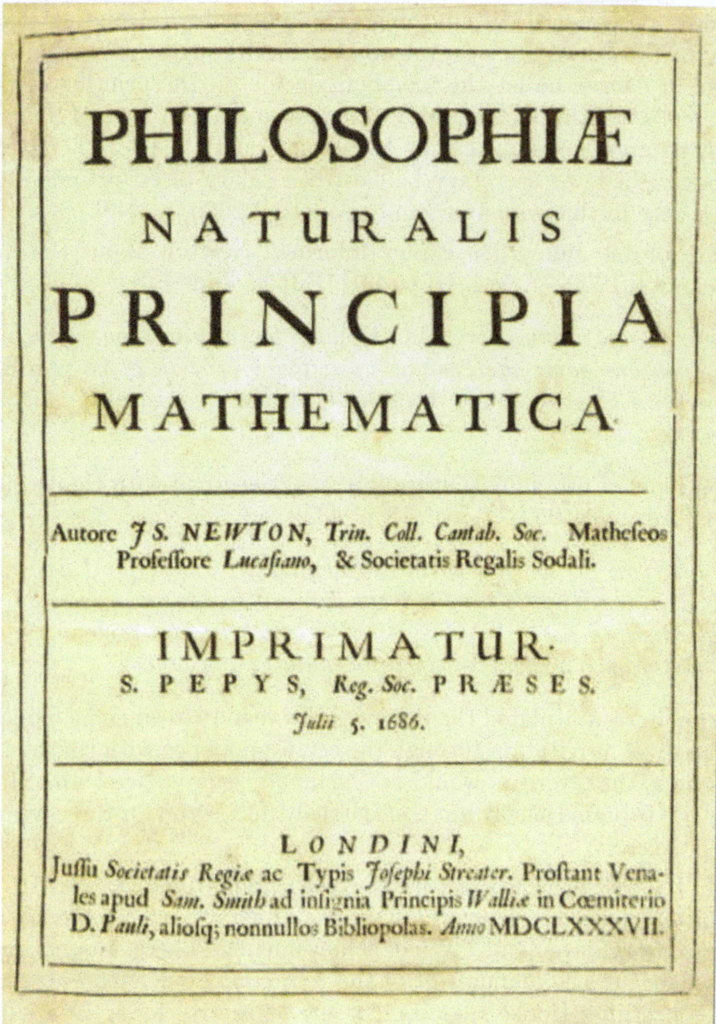

PHILOSOPHIÆ
NATURALIS
PRINCIPIA
MATHEMATICA.

Autore *J S.* NE*WTON*, *Trin. Coll. Cantab. Soc.* Matheſeos Profeſſore *Lucaſiano*, & Societatis Regalis Sodali.

IMPRIMATUR.
S. PEPYS, *Reg. Soc.* PRÆSES.
Julii 5. 1686.

LONDINI,
Juſſu *Societatis Regiæ* ac Typis *Joſephi Streater*. Proſtant Venales apud *Sam. Smith* ad inſignia Principis *Walliæ* in Cœmiterio D. *Pauli*, alioſq; nonnullos Bibliopolas. *Anno* MDCLXXXVII.

Fig. 6.2.8. Title page of the *Principia* by [Isaac Newton, 1687]

Meanwhile the financial situation of the Royal Society had dramatically deteriorated. A pronouncedly expensive book on the natural history of fishes had been published recently and in this situation Newton's *Principia* had to be printed. There was no money any longer but had not Halley taken the responsibility for the project without considering the council? He himself should pay the printing then [Westfall 2006, p. 453]! On 15th July 1686 the imprimatur was given by the president of the Royal Society, Samuel Pepys. Hence the *Principia* was published in mid-July 1687 as if the Royal Society were the editor. The true editor, however, was Halley.

6.2.5 Mr Leibniz Lays a Fuse

Already in section 6.1.1 it was reported that Leibniz was in Vienna in June 1688 and read the book review of the *Principia* by Christoph Pfautz in the Acta Eruditorum. Leibniz hastily begun writing three manuscripts to explain his own ideas concerning physics. Besides a publication in January 1689 which served as an introduction the two further publications were *Schediasma de resistentia medii et motu projectorum gravium in medio resistente* in January 1689, taking up themes from the second book of Newton's *Principia*, and *Tentamen de motuum caelestium causis* in February 1689. This is the time when Leibniz set out from Vienna in direction of Venice, cp. page 241. The latter work treated topics from the third book of Newton's *Principia*.

Concerning the *Schediasma* one can find notes reaching back to Leibniz's Parisian years but this is not true in case of the *Tentamen* [Antognazza 2009, p. 296]. All evidences from manuscripts and notes show that this work on cosmological theories resulted from the thoughts Leibniz had when he read Newton's *Principia* and which were put on paper 1688 in Vienna. Contrary to his affirmations that he had read the *Principia* for the first time only after 14th April 1689 when he came to Rome Leibniz had in fact already read it and had worked through it.

In the first of the three papers, *De lineis opticis et alia* (On optical lines and others), Leibniz stated that he had not known the *Principia* beforehand. This assertion was repeated in the *Tentamen*. Why would Leibniz have published the dozen pages *Tentamen* on the motion of planets long after Newton had published some hundred pages concerning that theme two years earlier without Leibniz having read the *Principia* previously [Meli 1993, p. 7f.]? With date of 8th February 1690 Huygens requested to know from Leibniz whether the latter's opinions had changed after subsequently having read the *Principia*. After all, Leibniz himself had written that he had only seen the book review by Pfautz when he wrote *Tentamen*. Leibniz received the letter only late in September, but Huygens wrote again to Leibniz on 24th August concerning the same point: did Leibniz change his opinions as expressed in the *Tentamen* after reading the *Principia*? Leibniz prepared an answer – but did not send it [Meli 1993, p. 8]! In this draft he wrote that he had seen the *Principia* only in Rome, hence between April and December 1689. In fact, Leibniz's opinions on the theory of planetary motion did not show any influence of Newton's theories even later on. No wonder, since he knew the *Principia* beforehand and therefore had no reason to change his views.

At the end of *De lineis opticis et alia* Leibniz wanted to express clearly that he got his ideas independently of Newton [Meli 1993, p. 8]:

> *'Conclusions about the resistance of the medium, which I have put on a special sheet, I had reached to a considerable extent twelve years ago in Paris, and I communicated some of them to the famous Royal*

Academy. Then, when I too had chanced to reflect on the physical cause of celestial motions, I thought it worth while to bring before the public some of these ideas in a hasty exploration of my own, although I had decided to suppress them until I had the chance to make a more careful comparison of the geometrical laws with the most recent observations of astronomers. But (apart from the fact that I am tied by occupations of quite another sort) Newton's work stimulated me to allow these notes, for what they are worth, to appear, so that sparks of truth should be struck out by the clash and sifting of arguments, and that we should have the penetration of a very talented man to assist us.'

In the case of the work on bodies moving in resisting media Leibniz could fall back on a manuscript he had sent to the Academy in Paris but which was not publicly accessible. He had hoped thereby to show his integrity. Manuscripts on which Leibniz could fall back did however not exist in the case of the *Tentamen*. His remark, that he actually wanted to wait to compare his theory with observations would suggest that he had found the theory earlier.

The whole book [Meli 1993], containing also a reconstruction of the *Tentamen* and Leibniz's notes concerning the *Principia*, is devoted to the search for the true circumstances of the publication of the three papers by Leibniz. Its author comes to a clear conclusion [Meli 1993, p. 9]:

'Leibniz formulated his theory in autumn 1688, and the Tentamen was based on direct knowledge of Newton's Principia, not only of Pfautz's review.'

6.3 The Reception of the *Principia*

6.3.1 The Situation in England

Three editions of the *Principia* were published in Newton's lifetime. Today nobody can say with certainty how large the first edition of the *Principia* was [Cohen 1971, p. 138]. Estimates vary between 250 and 400 copies[5]. Some were designated for export to the continent. Halley communicated on 15th July 1687 that the printing process of the *Principia* was finished [Turnbull 1959–77, Vol. II, p. 481]:

[5]Following [Pask 2013, p. 476] the number of printed copies of the second edition amounted to 750 and that of the third edition 1200. An edition of 250 copies was regarded as being large in these days which can serve to document the interest in the *Pricipia*; but that does not say anything about the number of readers who could truly understand the work.

'Honoured Sir,

I have at length brought your Book to an end, and hope it will please you.'

and he also informed the author that he would give certain copies to recipients chosen by Newton [Turnbull 1959–77, Vol. II, p. 481]:

'I will present from you the books you desire to the R. Society, Mr Boyle, Mr Pagit, Mr Flamsteed and if there be any elce in town that you design to gratifie that way; and I have sent you to bestow on your friends in the University 20 Copies, which I entreat you to accept. In the same parcell you will receive 40 more, wch, having no acquaintance in Cambridg, I must entreat you to put into the hands of one or more of your ablest Booksellers to dispose of them: I intend the price of them bound in Calves leather and lettered to be 9 shillings here, those I send you I value in Quires at 6: Shill to take my money as they are sold, or at 5.sh a price certain for ready or elce at some short time; ...'

What did the recipients do with the book and who read the *Principia* at all? There is a nice anecdote about a Cambridge student observing Newton passing by and saying: *'There goes the man who has writt a book that neither he nor any one else understands.'* [Guicciardini 1999, p. 170]. The number of people who could understand a book like the *Principia* was exceptionally small and is not very large even today. Unfortunately, we do not know how Christopher Wren and John Wallis – both eminent mathematicians – reacted to the *Principia.* Wallis was very much interested in problems of natural philosophy and himself worked on the resistance of bodies moving in air. After the *Principia* was published Wallis never came back to problems of applied mathematics; perhaps a sign how large the impression was that Newton's investigations made on him [Guicciardini 1999, p. 172]. Robert Hooke, having motivated Newton to think about the inverse distance square law, was not able to follow the explanations in the *Principia.* The situation was better in case of Edmond Halley who stayed in direct contact with Newton during the period of origin of the *Principia.* In the year 1686 Halley had read a kind of introduction to the forthcoming *Principia* to the Royal Society. This introduction was titled *A Discourse concerning Gravity, and its Properties, wherein the Descent of Heavy Bodies, and the Motion of Projects is briefly, but fully handled: Together with the Solution of a Problem of great Use in Gunnery* and it was published in the Philosophical Transactions. The matter Halley was talking about was obviously at his command. In a letter to king James II which reads like a book review and which the latter received together with a copy of the *Principia* Halley presented the foundations of Newton's cosmology from Book III and also spoke about applications of the theory of the moon to navigation. Halley was a master also in this area since he not only was an able astronomer but also a mariner who had participated in scientific expeditions at sea. The theory of tides was also well understood by Halley since he had studied the tides in Tonkin himself. After the *Principia* was published Halley several times read papers to the Royal Society to popularise Newton's physics. Moreover, Halley had published a book review on the *Principia* in the Philosophical Transactions of the year 1687 in which he described the development and the content of Newton's work clearly and with great insight [Newton 1999a, p. 581ff.], [Cohen 1971, p. 148ff.].

Although many in the sphere of the Royal Society were prepared to view Newton's magnus opus as *the* foundation of a new natural philosophy most of them sadly were *'Philosophers without Mathematicks'*. Most of them had great problems with the *Principia*.

In this situation Newton recommended the study of Huygens's *Horologium oscillatorium*. To Richard Bentley, Master of Trinity College who was to play an important role in the new edition of the *Principia* (cp. page 396ff.), he gave concrete reading instructions to get prepared for a study of the *Principia*. He recommended Euclid's *The Elements*, books by Jan de Witt (1623–1672) and Philippe de la Hire (1640–1718) on conic sections. For algebra he recommended Erasmus Bartholinus's *Selecta Geometrica* and then Descartes's *Géométrie* in the Latin edition of van Schooten which was augmented by comments and some other works, and for basic knowledge in astronomy relevant works of Pierre Gassendi (1592–1655) and of Nicolaus Mercator. Then Newton wrote [Turnbull 1959–77, Vol. III, p. 156]:

> *'These* [works] *are sufficient for understanding my book: but if you can procure Hugenius's Horologium oscillatorium, the perusal of that will make you much more ready.*
>
> *When you have read the first 60 pages, pass on to ye 3d Book & when you see the design of that you may turn back to such Propositions as you shall have a desire to know, or peruse the whole in order if you think fit.'*

The first 60 pages of the *Principia*, i.e. the first three sections in Book I, were thus thought intelligible to readers with only a poor mathematical background by Newton. They should be read before studying the cosmological discourse in Book III. Hence in the 18th century the first three section of Book I became the standard reading matter at British universities [Guicciardini 1999, p. 176]. Be it as it may, it seems as if Richard Bentley never seriously studied the first 60 pages but acquired a good qualitative understanding of Newton's *'System of the World'*, i.e. of Book III.

What happened to Bentley also happened to the philosopher John Locke (1632–1704) who became one of the few friends of Newton, as we will discuss in section 7.2. Two months after it was published Locke read the *Principia*, made excerpts and notes [Guicciardini 1999, p. 176], but had great problems

Fig. 6.3.1. Richard Bentley (1662–1742) was a classical philologist, textual critic, theologian, and from 1700 on Master of Trinity College. He became a member of the Royal Society in 1695 and a year later received the degree of a doctor of divinity ([Engraving by George Vertue after J. Thornhill, about 1710] Wellcome Images/Wellcome Trust, photo L0021234) and main entrance of St John's College (Cambridge), the place where Richard Bentley studied [Photo: Alexander Czuperski 2007]

to understand what he had read. He received his own copy by Newton in 1691 in which Newton had added corrections in the margins. In 1688 an anonymous book review of the *Principia* appeared in the journal 'Bibliothèque Universelle' which is nowadays attributed to John Locke [Cohen 1971, p. 147f.]. Locke had no mathematical education, therefore he could barely say anything concerning the contents of Books I and II. Therefore the book review [Newton 1999a, p. 584ff.] (which barely deserves this name) was restricted to Book III and the translation of the Latin headings of the sections in Books I and II. After all, Locke had insured himself that the mathematics in the *Principia* was correct by having asked Christiaan Huygens.

6.3.2 Huygens as Recipient

As was the case in England the number of knowledgable readers of the *Principia* on the continent was small. The old master of natural philosophy, Christiaan Huygens, received the *Principia* at the end of 1687 after having heard already something about Newton's work from Fatio de Duillier. Besides the theory of the tides with which Huygens had occupied himself, his attention was mainly drawn to the first three sections of Book II in which Newton had treated the motion of bodies in resistant media [Guicciardini 1999, p. 121].

Huygens recognised Newton's achievements clearly but had problems with the mathematics and the theory of gravitation. In a letter to Hudde of 1688 he wrote [Guicciardini 1999, p. 122]:

> *'Professor Newton in his Book entitled Philosophiae Naturalis Principia Mathematica has stated several hypotheses of which I cannot approve.'*

And to Leibniz in 1690 [Guicciardini 1999, p. 122]:

> *'As for the cause of the ebb and flow given by Newton, I am not at all satisfied. I am dissatisfied also with all the other theories which he bases on the principle of attraction, which seems to me absurd* [...] *And I was often surprised to see how he could make such an effort to carry on so many researches and difficult calculations which have as foundation this very principle.'*

Huygens was a Cartesian and had published *Discours de la cause de la pesanteur* (Treatise on the reason of gravity) as an appendix to his *Traité de la lumière* in which he had developed his own theory of gravitation based on cartesian vortices of subtle matter. In this *Traité* he opposed Newton's particle theory of light, in the *Discours* he opposed Newton's gravitation. Mind you, Huygens accepted the validity of Newton's derivation of a force being proportional to the inverse squared distance which swept away Descartes's vortices, but the 'occult' long-distance force of Newton's gravitation could not be accepted. Already in 1669 had Huygens presented an elaborate theory of terestial attraction to the Académie Royale des Sciences in which the cartesian vortex was replaced by small particles moving around the earth on spherical surfaces in all directions [Koyré 1965, p. 116].

In his *Pensées privées* (Private thoughts) Huygens had written in 1686 [Koyré 1965, p. 117]:

> *'Planets swim in matter. For, if they did not, what would prevent the planets from fleeing, what would move them? Kepler wants, but wrongly, that it should be the sun.'*

In 1688 he still had in mind to rescue his vortex theory. He stated [Koyré 1965, p. 117]:

> *'Vortices destroyed by Newton. Vortices of spherical motion in their place.*
>
> *To rectify the idea of vortices.*
>
> *Vortices necessary;* [without them] *the earth would run away from the sun; but very distant the one from the other, and not, like those of M. Des Cartes, touching each other.'*

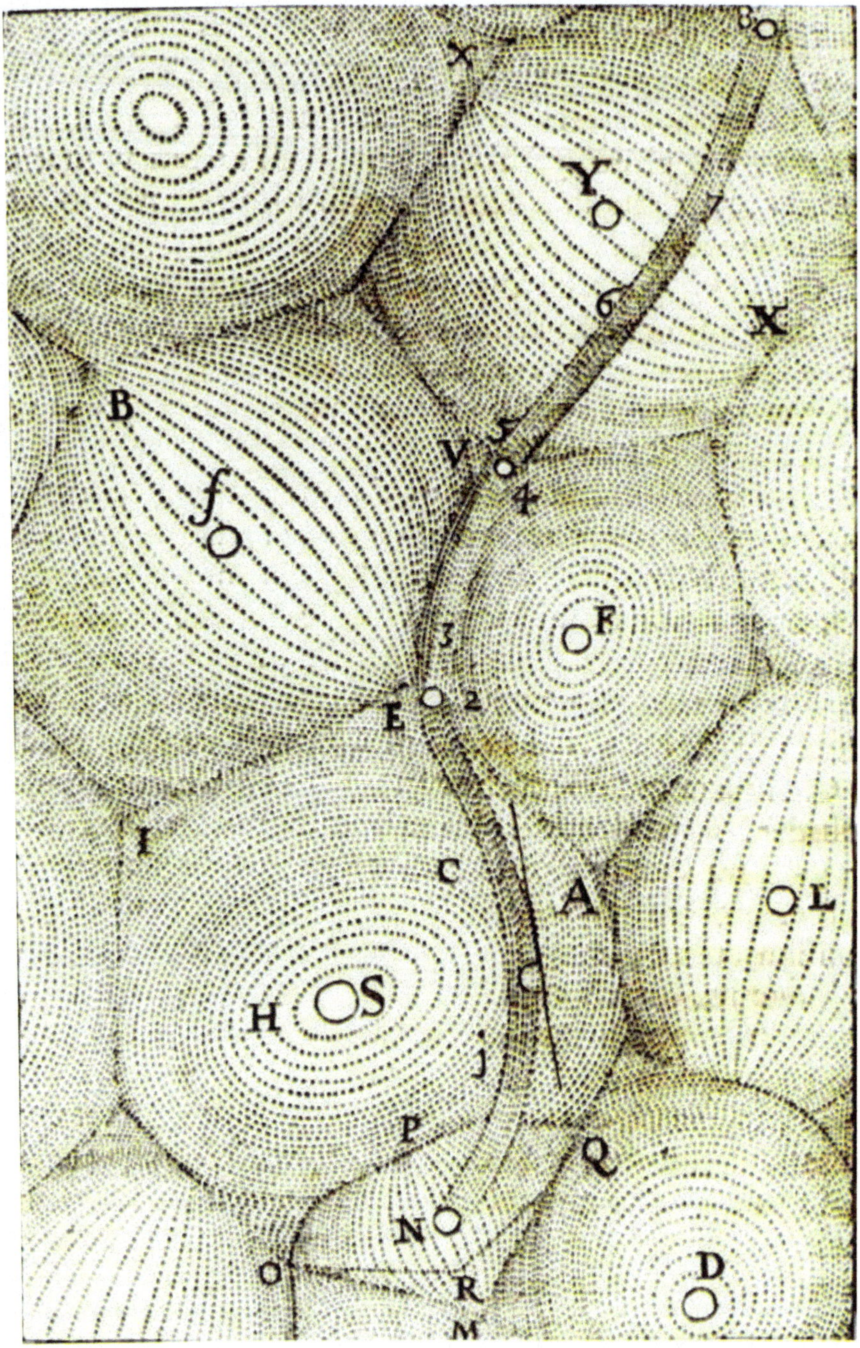

Fig. 6.3.2. The system of vortices in the cosmos in the *Principles of Philosophy* by René Descartes ([Descartes, Principia Philosophiae, 1656, S. 72] Bayerische Staatsbibliothek, Munich, Sign.: 858338 4 Ph.u. 44 858338 4 Ph.u. 44). The 'Wirbel des ersten Himmels' (vortex of the first heaven) [Descartes 2005, p. 253] is AYBM, in the centre of which is the sun S. This central vortex (and all the others too, of course) was 'destroyed' by Newton in Huygens's view. The history of vortex theory can be found in [Aiton 1972a] and [Descartes 2005]

For Huygens the idea of Newton's long-distance force was so little acceptable that he replaced the vortex AYBM in figure 6.3.2, which was 'destroyed' by Newton, by a system of smaller vortices and he even presented this system to the Royal Society during his stay in London in 1689 [Koyré 1965, p. 117]. It goes without saying that a well-educated, distinguished, and courteous man like Christiaan Huygens would not attack Newton directly or even publicly. It was simply the question for the reasons of gravitation in which Huygens could not be in accord with Newton. For Aristotle 'gravity' was a basic quality of some bodies, as 'levity' was the quality of others. For Archimedes 'gravity' was the quality of all bodies, for Copernicus the expression of an inner tendency of homogenous parts to unite and to form a whole, and for Kepler it was the effect of mutual attraction between the parts and the whole. Not so for Descartes and Huygens. In Descartes's physics [Descartes 1983] gravitation is the impact of an external influence. Bodies are heavy because they are pressed towards the earth by other 'bodies'; more precisely: by the pressure of a vortex of subtle matter moving with enormous speed around the earth [Koyré 1965, p. 118]. Huygens could affiliate with such a mechanical concept. He even conceived an experiment with which he wanted to prove his vortex theory [Koyré 1965, p. 119ff.].

Huygens never became a Newtonian; he could neither accept Newton's gravitation nor the particle theory of light until his death. In his last work *Cosmotheoros* (viewer of the world) published posthumously in 1698 he stated [Koyré 1965, p. 123]:

> '*I think that every Sun* [that is, every star] *is surrounded by a certain vortex of matter in quick movement, but that these vortices are very different from the Cartesian vortices as well in respect to the space that they occupy as in respect to the manner in which their matter is moving.*'

6.3.3 Leibniz as Recipient

We have already discussed in section 6.2.5 how Leibniz reacted after having read Newton's *Principia*: he wrote three works to explain his own natural philosophy. Thereby Leibniz claimed having not yet read the *Principia* at that time but only Pfautz's book review[6] in the Acta Eruditorum [Newton 1999a, p. 588ff.]. This book review deserves its name since it gives clear and objective information concerning the content of the work.

What was contained in the *Schediasma de resistentia medii et motu projectorum gravium in medio resistente* and the *Tentamen de motuum caelestium*

[6]This book review was published anonymously; today there is no doubt that Pfautz was the author.

causis which were published in January and February 1689, respectively? Did Leibniz take results or derivations from Newton's *Principia*? Far from it! Leibniz formulated profoundly original thoughts and he used his calculus throughout, so that he started something which was successfully finished by Leonhard Euler only in the 18th century: the penetration and foundation of mechanics and celestial mechanics with, and on the basis of the new differential and integral calculus. This step which, as we saw, Newton tried to avoid under all circumstances in his *Principia* is one of the essential achievements of Leibniz in the *Schediasma* and the *Tentamen*. Calculus does not show explicitly in the *Schediasma* but Leibniz left no doubt that the results were in fact derived by means of his calculus [Guicciardini 1999, p. 147]. The derivations apparently were lacking elegance or conclusiveness in Leibniz's view.

In the *Schediasma* Leibniz treated problems which were also solved by Huygens and Newton. We follow Aiton in the reconstruction of the mathematics underlying the *Schediasma* [Aiton 1972]. For Leibniz there existed two kinds of resistance if a body moved in a resistant medium: the absolute and the pertinent resistance. The absolute resistance results from the friction of fluid particles with the surface of the body and is independent of the velocity of the particles (i.e. the absolute resistance is independent of the velocity and depends only on the viscosity of the fluid). The dynamical effect of the absolute resistance, hence the decrease of the velocity due to friction of the fluid particles, is proportional to the number of particles and hence to the distance travelled per time unit, and therefore proportional to the velocity. The pertinent resistance arises from the influence of the fluid on the body in the sense of the impact of fluid particles on the body. The pertinent resistance is proportional to the velocity but the dynamical effect is also proportional to the number of elements or the distance travelled, so that the dynamical effect of the pertinent resistance is overall proportional to the square of the velocity.

In article I Leibniz treated the uniform motion of a body lagging through friction proportional to the distance travelled. In proposition 1 Leibniz recorded that the decrease of forces is proportional to the increase of distances, following proposition 2 the velocities are proportional to the distances. Assuming the increment of the distances being the same the decrease of the forces will be the same by proposition 1. If the decrement of the forces is the same the decrease of velocity will be the same also. 'Force' is meant here in the sense of the 'living force' (vis viva), i.e. the force is proportional to the square of the velocity. However, *another* concept of force is meant in proposition 1 that Leibniz admitted in a letter to John Bernoulli of March 1696.

If in figure 6.3.3 AE is the initial velocity and the whole distance travelled in the medium is AB then the already travelled distance is AM and the distance still to be travelled MB. The remaining velocity then is MC (or AF), the 'lost' velocity FE, and then ECB will be a straight line.

Now follows proposition 3: if the remaining distances (MB or LT) are like numbers then the time periods (ML or BT) will be like logarithms. For if the

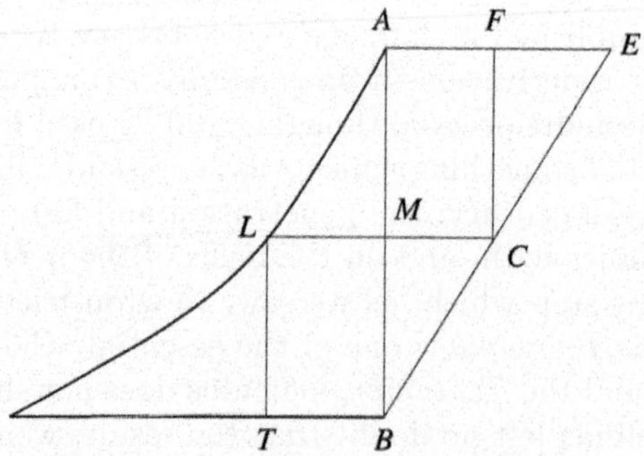

Fig. 6.3.3. Leibniz's velocity diagram

elements (=differentials) of the distances are in geometric progression then the remaining distances will also be in that progression. Then it follows from proposition 2 that even the remaining velocities will behave in that way. Hence the increments of the time periods are the same and therefore the time periods themselves are in arithmetic progression.

We should not become desperate if the conclusions above remain dark and obscur. Leibniz has given a symbolic derivation to correct mistakes which had crept into *Schediasma* in his work *Additio ad schediasma de medii resistentia* (Amendment to the short paper on the resistance of the medium) in the Acta Eruditorum of 1691. If the velocity is v, the maximum velocity a, the time t, the distance travelled s, and the maximum distance travelled b, then it follows from proposition 1

$$-\frac{dv}{a} = \frac{ds}{b}, \tag{6.1}$$

where the negative sign is missing in Leibniz's work [Aiton 1972, p. 262, footnote 25]. If v_1 and s_1 are two corresponding values of velocity and distance travelled then it follows

$$\frac{v - v_1}{a} = \frac{s_1 - s}{b}. \tag{6.2}$$

Now the initial velocity is $v_1 = a$ and the initially distance travelled $s_1 = 0$ so that

$$\frac{a - v}{a} = \frac{s}{b} \tag{6.3}$$

is valid, and that is just the statement of proposition 2. The derivative of the distance with respect to time is the velocity, hence in this case

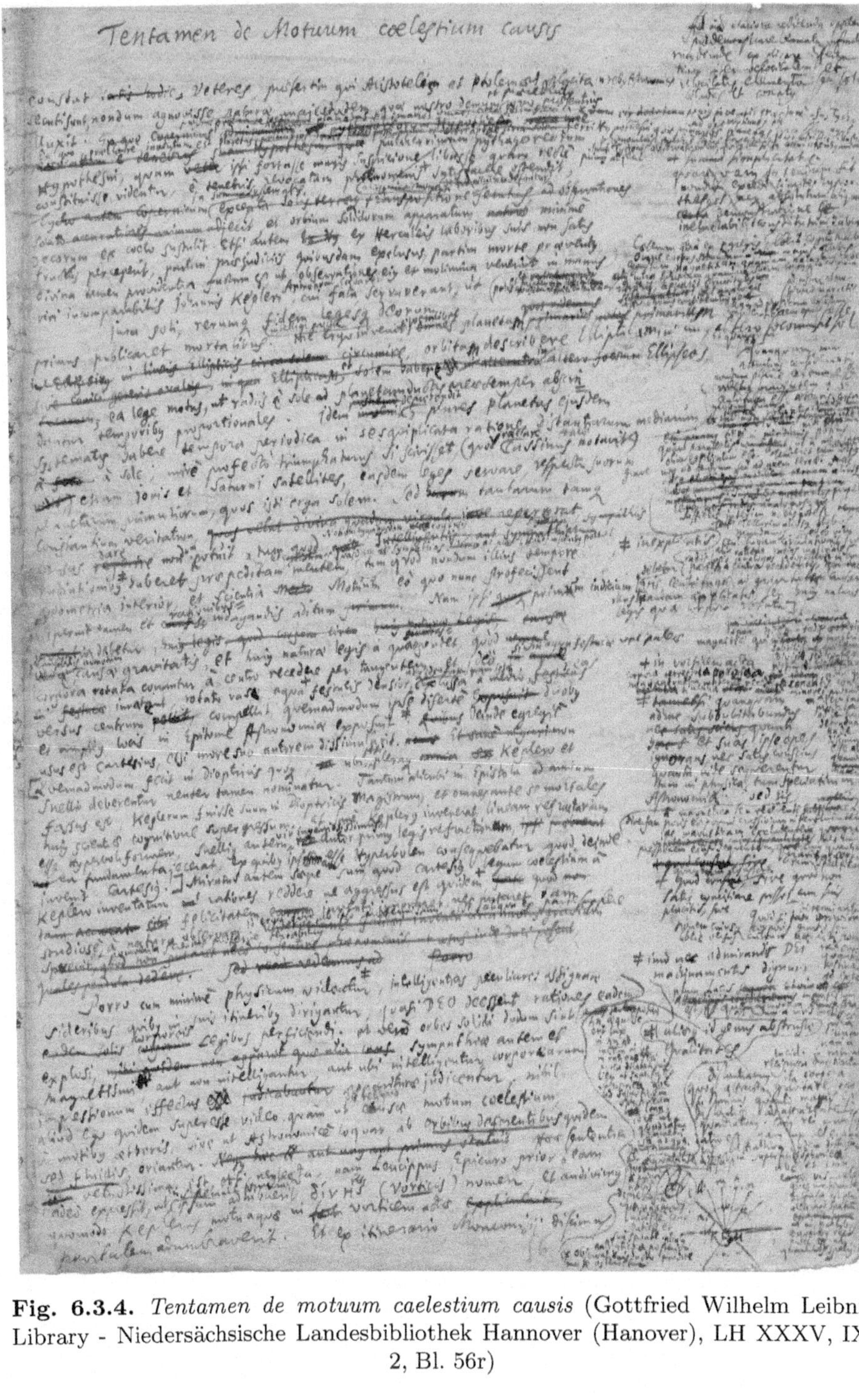

Tentamen de Motuum coelestium causis

Fig. 6.3.4. *Tentamen de motuum caelestium causis* (Gottfried Wilhelm Leibniz Library - Niedersächsische Landesbibliothek Hannover (Hanover), LH XXXV, IX, 2, Bl. 56r)

$$\frac{ds}{dt} = \frac{v}{a}, \tag{6.4}$$

so that it follows from (6.3) and (6.4):

$$\frac{a - a\frac{ds}{dt}}{a} = \frac{s}{b}, \tag{6.5}$$

or

$$dt - ds = s\frac{dt}{b},$$

i.e.

$$dt = b\frac{ds}{b - s}.$$

And now integration yields the equation

$$t = b \log \frac{b}{b - s},$$

and that is the content of proposition 3. A discussion of the further contents of the *Schediasma* can be found in [Aiton 1972]. The *Schediasma* ends with a problem that Newton had not solved in the *Principia*, namely the question concerning the path of a body moving in a medium that resists with the square of the velocity. Leibniz did not succeed in solving the problem in full generality and after criticism by Huygens Leibniz admitted it. The problem was finally solved by John Bernoulli in 1719 [Guicciardini 1999, p. 149].

In the *Tentamen* which we know that Leibniz wrote after having read the *Principia* he applied his calculus consistently to problems of celestial mechanics. Like Huygens Leibniz also could not resign himself with a large-distance force; he was also a Cartesian and was an adherent of the vortex theory. Leibniz assumed that three forces acted on a planet: a transradial force acting orthogonal to the connecting line between sun and the planet, and two radial forces acting opposingly. Following Leibniz the reason for the transradial force was a 'harmonic vortex' where particles moved in circular orbits around the sun, so that their velocity was inversely proportional to the distance from the sun. As Newton had derived Kepler's second law from 'first principles' in Propositions I and II of Book I of his *Principia* [Newton 1999a, p. 59-62], so Leibniz's harmonic vortex served to satisfy the area law since the assumption of a velocity inversely proportional to the distance to the sun implied Kepler's law. The two radial forces are the 'centrifugal conatus' (conatus=striving, urge) directed outwards and the inwardly directed 'desire of gravity'. These two radial forces are responsible for the 'paracentric' motion.

Leibniz assumed the orbit of the planet being a polygon with infinitesimally small edges. The planet moves with uniform velocity along an infinitesimal edge until it reaches an endpoint of the edge. There it changed its velocity until it reached another endpoint, and so on. Using this method which is described in [Meli 1993, p. 126ff.] and summed up in [Guicciardini 1999, p. 150ff.] Leibniz was able to show that the 'desire of gravity' was inversely proportional to the square of the distance if one assumed an elliptical trajectory of the planet. Thus, with the help of his differential calculus Leibniz had solved Newton's Problem VI in Proposition XI from Book I [Newton 1999, p. 462]:

> *'Let a body revolve in an ellipse; it is required to find the law of the centripetal force tending toward a focus of the ellipse'.*

This was in fact an ultimately elegant application of his calculus.

Both works, *Schediasma* and *Tentamen*, can be considered as the first milestones on the way to a calculus-based mechanics but both works also were failures [Guicciardini 1999, p. 152]. In the *Schediasma* Leibniz delivered a false theory of the path of a body moving in a medium in which the resistance is proportional to the square of the velocity, and in the *Tentamen* Leibniz had not succeeded to derive Kepler's third law. Both works were swiftly forgotten and did not contribute to the mathematisation of mechanics.

6.3.4 Newton's Attack on Leibniz's *Tentamen*

We have some written statements by Newton concerning Leibniz's *Tentamen*, among them the draft of a critique in the possession of Keill which was published by Edleston [Edleston 1850, p. 308-315]. This critique is entitled *Ex Epistola cujusdam ad Amicum* (From a letter of a certain author to a friend) and was written about 1712. There are also Newton's notes concerning Leibniz's *Tentamen* in [Turnbull 1959–77, Vol. VI, p. 116ff.]. The contents of both statements were examined in detail in [Meli 1993] and are not of great interest to us. Newton's written comment on the *Tentamen* was not used by him to attack Leibniz but he sent it to John Keill, cp. page 403; but here we already anticipate events. We want to keep in mind that the *Principia* and the *Tentamen* were also involved in the priority dispute, but that happened only after the turn from the 17th to the 18th century.

6.3.5 The First Reaction in France

Although France was seized by an enthusiasm concerning Newton and although even books on Newton's natural philosophy for the ladies were written, the first response to the *Principia* was rather cool. A book review was published anonymously in the Journal des Sçavans of 1688. Even today it is not clear who the author of this review was, but it is pronouncedly short. It begins [Newton 1999a, p. 588]:

> *'Monsieur Newton's work is the most perfect mechanics one can consider, since it is impossible to make the proofs more rigorous and exact*

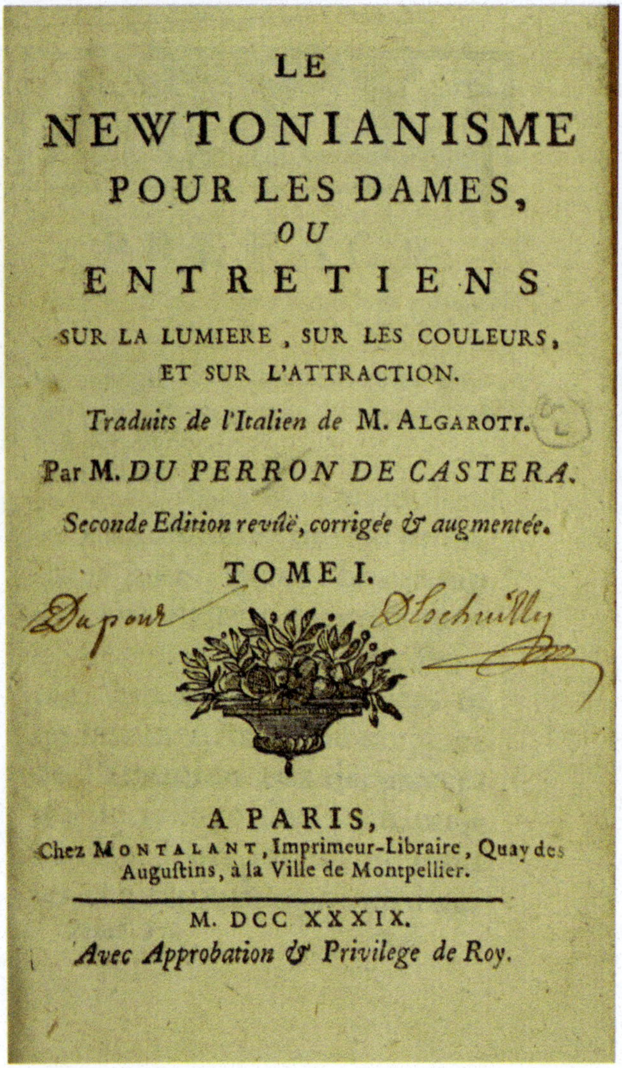

LE
NEWTONIANISME
POUR LES DAMES,
OU
ENTRETIENS
SUR LA LUMIERE, SUR LES COULEURS,
ET SUR L'ATTRACTION.
Traduits de l'Italien de M. ALGAROTI.
Par M. *DU PERRON DE CASTERA.*
Seconde Edition revûë, corrigée & augmentée.
TOME I.

A PARIS,
Chez MONTALANT, Imprimeur-Libraire, Quay des Auguſtins, à la Ville de Montpellier.

M. DCC XXXIX.
Avec Approbation & Privilege de Roy.

Fig. 6.3.5. Title page of the second edition of the second volume of Algarotti's *Le Newtonianisme Pour Les Dames* of 1739 (with kind permission by Pazzo Books, Boston). Count Francesco Algarotti (1712–1764) was a Venetian writer and art trader who popularised sciences and arts for a broader public. In the oil painting 'Die Tafelrunde von Sanssouci' by Adolph von Menzel he is depicted with Voltaire and others in illustrious company around Frederick the Great. The popular science book *Il newtonianismo per le dame ovvero dialoghi sopra la luce e i colori* (The Newtonianism for the ladies or dialogs on the light and the colours) was published for the first time in 1737 and was translated into French and German. On the marble epitaph on his grave on the Camposanto in Pisa Frederick the Great ordered to affix the inscription 'Algarotti, this successor of Ovid, this pupil of Newton, Frederick the Great'

than those which he presented in the first two books on gravity, on levity, on elasticity, on resistance of fluid bodies, and on the attracting and impulsing forces, which comprise the main foundation of physics. But one has to admit that one can take these proofs only as mechanical ones since the author himself confessed at the end of page 4 and the beginning of page 5 that he viewed their foundations not as a physicist but as a geometer.'

(Monsieur Newtons Werk ist die vollkommenste Mechanik, die man sich einfallen lassen kann, da es unmöglich ist, die Beweise noch schärfer und genauer zu machen als jene, die er in den beiden ersten Büchern zur Schwere, zur Leichtheit, zur Elastizität, zum Widerstand fluider Körper und zu den anziehenden und anstoßenden Kräften angibt, die das hauptsächliche Fundament der Physik ist. Aber man muß einräumen, daß man diese Beweise nur für mechanische halten kann, da ja der Autor selbst am Ende von Seite 4 und am Anfang von Seite 5 bekennt, daß er ihre Grundlagen nicht als Physiker, sondern bloß als Geometer betrachtet hat.)

Then Book III is discussed. The reviewer complains that all depends on Newton's hypothesis of a large-distance force of gravitation; a hypothesis being not provable so that every proof depending on it could only be a 'mechanical' one. Quite obviously, the reviewer was a Cartesian who could not be content with a vortex-free motion of the planets. The book review ended with a request to Newton [Newton 1999a, p. 588]:

'To create such a work being perfect as possible Mr. Newton has only to give us a physics which is as exact as mechanics. He will have given it to us when he will have replaced true motions for the one he has considered.'

(Um also ein so vollkommenes Werk, wie nur irgend möglich, zu erschaffen, braucht Mr. Newton uns nur eine ebenso exakte Physik zu geben, wie es die Mechanik ist. Er wird sie dann gegeben haben, wenn er wahre Bewegungen an Stelle derjenigen gesetzt haben wird, die er angenommen hat.)

In fact Leibniz had tried in the *Tentamen* to follow such a program of mechanistic explanations of the motion of planets.

6.4 The Scholium Dedicated to Leibniz

When the *Principia* was published for the first time in the year 1687 – and even in the second edition 1713 – a scholium could be found in the second chapter of the second book directly after Lemma II, in which the product rule was proven, cp. pages 259 and 460 [Newton 1999, p. 649, footnote dd]:

> *'In correspondence which I carried on ten years ago with the very able geometer G.W. Leibniz, I indicated that I was in possession of a method of determining maxima and minima, drawing tangents, and performing similar operations, and that the method worked for surd as well as rational terms. I concealed this method under an anagram*[7] *comprising the sentence: 'Given an equation involving any number of fluent quantities, to find the fluxions, and vice versa.' The distinguished gentleman wrote back that he too had come upon a method of this kind, and he communicated his method, which hardly differed from mine except in the forms of words and notations. The foundation of both methods is contained in this lemma.'*

When the third edition of the *Principia* was published in 1726 Leibniz was already dead for 10 years. The priority dispute was at its climax around 1710 and the friendly relationship expressed in Leibniz's and Newton's mutual letters had ceased. It is no wonder therefore that Newton completely revised the scholium for the third edition and removed the name of his opponent [Newton 1999, p. 649f.]:

> *'In a certain letter written to our fellow Englishman Mr. J. Collins on 10 December 1672, when I had described a method of tangents that I suspected to be the same as Sluse's method, which at that time had not yet been made public, I added: 'This is one particular, or rather a corollary, of a general method, which extends, without any troublesome calculation, not only to the drawing of tangents to all curve lines, whether geometric or mechanical or having respect in any way to straight lines or other curves, but also to resolving other more abstruse kinds of problems concerning curvatures, areas, lengths, centers of gravity of curves, ..., and is not restricted (as Hudde's method of maxima and minima is) only to those equations which are free from surd quantities. I have interwoven this method with that other by which I find the roots of equations by reducing them to infinite series.' So much for this letter. And these last words refer to the treatise that I had written on this topic in 1671. The foundation of this general method is contained in the preceding lemma.'*

It is shocking to see that on one side the old Newton was still anxious to secure his priority concerning the mathematics of infinitesimals long after the death of his opponent, and on the other side to remove the name of Leibniz from the great work. But here we have already anticipated events occurring in the development of the priority dispute.

[7] It is the first anagram in the *Epistola posterior*, cp. page 221.

7 The War is getting hot

T. Sonar, *The History of the Priority Diſpute between Newton and Leibniz*, https://doi.org/10.1007/978-3-319-72563-5_7

7.1 Newton in a Political Crisis

On 16th February 1685 the king of England, Scotland, and Ireland, Charles II passed away after having suffered a stroke. He left no legitimate heir so that his brother became his successor as James II. When James, a convinced catholic ascended the throne – he will be the last one – his declared aim was the re-catholicalisation of England. To achieve this aim he needed the control over the universities. When Edward Spence of Jesus College Cambridge gave a satirical talk on catholicism at the end of the year 1686 James struck out and forced him to revoke publicly. A month later the denunciator, Joshua Bassett of Caius College, a secret papist, was appointed master of Sidney Sussex College by the king. That was a clear attack on the protestantism at Cambridge University but what could an institution depending on the benevolence of the king do [Westfall 2006, p. 474]? The crisis reached its first hot phase only on

Fig. 7.1.1. James II with his wife Lady Anne. Their two daughters Mary (later Queen Mary II) and Anne (later Queen Anne) were inserted by Benedetto Gennari about 1680 [Pastel by HWK, Template: Painting by Peter Lely 1669, with later insertions by Benedetto Gennari]

Fig. 7.1.2. Isaac Newton ([Painting: unknown artist attributed to the English School, about 1715–1720], Detail)

16th February 1687 when the title of 'Master of Arts' was ordered to be given to the Benedictine monk Alban Francis by the king without Francis having to take any oath. Vice chancellor John Peachell withheld the king's letter for nearly two weeks because he needed to get advice [Westfall 2006, p. 475].

Meanwhile Newton was working to finish book III of his *Principia.* On 29th February 1687 he drafted a letter to an unknown correspondent [Turnbull 1959–77, Vol. II, p. 467f.]:

> *'Sr,*
>
> *Here's a strong report in ye Town yt a Mandamus has been brought to ye Vice-chancellor to admit one F. Francis a Benedictine Monck to be a Master of Arts,* [...]
>
> *For all honest men are obliged by ye Laws of God & Man to obey ye King's lawfull Commands. But if his Majesty be advised to require a Matter wch cannot be done by Law, no Man can suffer for neglect of it.* [...]
>
> *An honest Courage in these matters will secure all, having Law on our sides.'*

The senate of Cambridge University met on 4th March. It reached an agreement to support the vice chancellor in that every house of the senate sent him the opinion that it was unlawful to permit Father Francis without the required oath [Westfall 2006, p. 476]. Nothing changed. We do not know what happened on the second meeting of the senate on 21st March, only the result of the meeting is known: Isaac Newton was elected as one of the two representatives of the professorate to advise the vice chancellor that an admission of the monk was illegal. That meant that Newton must have had positioned himself clearly in this case. The king was angry and summoned the vice chancellor and eight representatives of the university to the court of the Churches Commission. The senate elected Newton as one of the eight. The delegation appeared before court four times; on 1st, 7th, 17th, and 22nd May 1687. Vice chancellor Peachell was removed from his office immediately and lost all assets and income. In this situation Newton's longstanding occupation with theological matters payed off; he began to draft an answer to the king which surely increased Newton's reputation within the university considerably due to its eloquence and passion. Only Newton himself knew of the irony of a convinced Arian carrying forward arguments of the anglican faith [Westfall 2006, p. 479].

Eventually Newton's reply was not even accepted by the Churches Commission. On 22nd May the delegates of Cambridge University gathered to receive the verdict of the commission. They were lucky: the commission identified Peachell as the one and only originator of the problems and finally decreed [Westfall 2006, p. 479]:

> '*Gentleman* [he concluded], *your best course will be a ready obedience to his majesty's command for the future, and by giving a good example to others, to make amends for the ill example that has been given you. Therefore I shall say to you what the scripture says, and rather because most of you are divines; Go your way, and sin no more, lest a worse thing come unto you.*'

Father Francis did not get the master's degree! Not least because Newton remained steadfast and obviously did not fear consequences.

When William of Orange landed in England in November 1688 the time of James II was over. Newton had become a celebrity not only in the circles of scientists and it does not wonder that he was elected as one of two members representing the university in parliament in London at the beginning of the year 1689. He had arrived in politics and in London, the heart of power.

7.2 A Friend Appeared

John Locke (1632–1704) was the son of a country lawyer in the county of Somerset and grew up in wealthy conditions. He attended the Royal Westminster School from 1647 in London's city and won a stipend for Oxford where he studied Classics at Christ Church College from 1652. He got a bachelor's degree in 1656 and a master's degree in 1658. Locke remained faithful to Oxford.

Fig. 7.1.3. Charles II of England [Painting: Philippe de Champaigne 1653], Centre: James II of England [Painting: Benedetto Gennari junior, 1685] and William III of Orange ([Painting: Godfrey Kneller] Verein der Freunde und Förder des Siegerland-museums e.V.)

He became a lecturer of Greek in 1660 and of rhetoric and ethics in 1662 [Thiel 1990, p. 7-23].

After the early death of his father in 1661 Locke inherited land and became financially independent. Through Francis Bacon and the latter's 'New Learning' Locke's interest in medicine and empirical methods grew and unofficially he listened to medical lectures. He maintained close contact with Robert Boyle and although he did not graduate he was eventually awarded a bachelor's degree in medicine in 1675 [Thiel 1990, p. 35]. When he met Sir Anthony Ashley Cooper, 1st Earl of Shaftesbury, who sought relief from a liver disease in Oxford, he became the latter's personal physician. In the year 1668 Locke conducted a risky surgery to remove a purulent ulcer of the Earl's liver which probably saved Ashley Cooper's life. Hence, Locke won a mighty mentor who eventually rose in government circles. Shaftesbury provided him with a minor governmental position which helped Locke to wealth and reputation.

Under the reign of James II Shaftesbury fell into disgrace; Locke travelled France from 1675 until 1679 and went to exile in Holland in 1683 where he stayed until 1688, the year of the 'Glorious Revolution'. Only in 1689 was he again offered a governmental position but he declined for reasons of failing health. In the year 1690 his arguably most important work *An Essay Concerning Human Understanding* appeared [Locke 1995], [Euchner, p. 25ff.].

Isaac Newton and John Locke probably met for the first time in 1689 [Westfall 2006, p. 488]. John Locke had seen the *Principia* already while in exile in Holland and had immediately tried to understand its contents, but he failed. He had therefore asked Christiaan Huygens whether he could trust the

Fig. 7.2.1. John Locke, so-called 'Kit-Kat portrait' by Sir Godfrey Kneller, since it was painted in the Kit-Kat Club in London (National Portrait Gallery London [Photo: Stephendickson 2014]) and painting by [Godfrey Kneller 1697] (Hermitage St. Petersburg, ГЭ-1345)

mathematical results and when Huygens answered positively Locke restricted himself reading the pure text only [Westfall 2006, p. 470]. Locke saw immediately that Newton was an intellectual giant of his times and became a lifelong close party man of Newton. In *An Essay Concerning Human Understandig* Locke wrote [Locke 1995, Book IV, Chapter VII, Sp 511]:

> *'Mr. Newton, in his never-enough-to-be-admired book, has demonstrated several propositions which are so many new truths, before unknown to the world, and are farther advanced in mathematical knowledge: ...'*

We know that Newton and Locke corresponded before the autumn of 1690 and there are hints that they knew each other half a year earlier [Westfall 2006, p. 488]. We now witness a new Newton; no longer the young man annoyedly keeping all correspondence as small as possible or even breaking off correspondence, but the eager correspondent who is aware of his growing fame and his importance after having finished his *Principia*. Both men mutually estimated each other much. Newton even prepared a personal copy of his *Principia* for Locke in which he noted all corrections hitherto known. The topics of conversation in the letters did not remain just mutual esteem but Newton felt that the philosopher was just the right correspondence partner to discuss his religious convictions. And indeed Newton was so open with Locke as he never was before with any other person. Both men obviously shared similar, but

Fig. 7.2.2. Jean le Clerc [Engraving of 1657] and Anthony Ashley-Cooper, 1st Earl of Shaftesbury [Painting: John Greenhill 1672–73]

unspeakable theological views [Westfall 2006, p. 490]. Newton even sent an Arian manifest to Locke that the latter was asked to send to the Netherlands to be anonimously published there. About a year later it must have become clear to Newton what an enormous risk he was taking. He requested that Locke should stop the publication under all circumstances. On 26th February 1692 Newton wrote [Turnbull 1959–77, Vol. III, p. 195]:

> *'Let me entreat you to stop their translation & impression so soon as you can for I designe to suppress them. If your friend hath been at any pains & charge I will repay it & gratify him.'*

Newton's fear was justified! Jean le Clerc (1657–1736) was Locke's correspondent in Amsterdam and he apparently knew exactly who the author was. Fifty years later the manuscript was found in the Remonstrant Library in Amsterdam where Le Clerc had deposited it, and the manuscript bore the name of Isaac Newton. If the manuscript would have been published in 1692 Newton would not only have been banned from Cambridge University but also from the English society.

Another important theme in the correspondence between Newton and Locke concerned alchemy. Unfortunaely, most of the letters dealing with alchemy are lost [Westfall 2006, p. 491ff.].

7.3 Isaac Newton and his Monkey

7.3.1 A Strange Couple

Approximately at the time when Newton picked up with John Locke Newton also met Fatio. The first encounter must have taken place at the meeting of the Royal Society on 22nd June 1689 at the latest, when Christiaan Huygens reported on light and gravitation [Westfall 2006, p. 493].

Nicolas Fatio de Duillier was born in Basel on 26th February 1664 and died on 12th May 1753 in Maddersfield close to Worcester. He was the seventh of 14 children and moved in 1672 with his parents to Duillier in what nowadays is the district Nyon of the Canton of Waadt. That explains the name affix 'De Duillier'.

Aged 18 Fatio travelled to Paris in 1682 for astronomical studies with Giovanni Domenico Cassini (1625–1712) at the observatory in Paris. Two years later Fatio succeeded in explaining the phenomenon of the zodiacal light discovered by Cassini which was a great success and made his name known in scientific circles. Thereby he became an accepted brilliant Swiss mathematician. In the year 1686 Fatio met Jacob Bernoulli and Christiaan Huygens. A close collaboration on mathematical topics concerning the computation of tangents

Fig. 7.3.1. Nicolas Fatio de Duillier [unknown painter about 1700] and Giovanni Domenico Cassini [Painter: Durangel 1879]

Fig. 7.3.2. Hampton Court, western frontage [Photo: Duncan Harris, 2012]

developed with Huygens. In spring 1687 Fatio travelled to London and experienced closely the fever in which the learned circles waited for Newton's great work which one assumed would revolutionise natural philosophy [Westfall 2006, p. 469]. In London he picked up with some scientists, among them Wallis, and was elected a member of the Royal Society by one of the founding members and by its temporary president Sir John Hoskyns (1634–1705).

Both men – Newton as well as Fatio – felt exceeding mutual attraction. In June 1689 Newton was already 46 years old, Fatio just 25. On 19th July Newton visited the two brothers Huygens at their abode in Hampton Court.

On the next day Newton, Christiaan Huygens, and Fatio drove to London since Newton was to be recommended to the king for an executive position at a college in Cambridge through the member of parliament John Hampden (1653–1696) [Westfall 2006, p. 488, p. 493]. On the 20th October Newton asked Fatio for a free room where Fatio lived. Newton wrote [Turnbull 1959–77, p. 45]:

> *'Sr*
>
> *I am extreamly glad that you .... friend & thank you most heartily for your kindness to me in designing to bring me acquainted with him. I intend to be in London ye next week & should be very glad to be in ye same lodgings with you.*
>
> *[...] Pray let me know by a line or two whether you can have lodgings for us both in ye same house at present or whether you would have me take some other lodgings for a time till ....'*

The places marked with '..........' were cut out for reasons not known. Both men were hence very closely associated at this time. But why? Frank E. Manuel has suspected a homosexual relationship [Manuel 1968, p. 191ff.] but there is no evidence. It belongs to the myth that Newton died untouched as a male virgin and we have two witnesses for this. In advanced age he confided to a relative that he never violated chastity and this relative later reported this to the poet Thomas Maude. The other witness is the physician Dr. Richard Mead (1673–1754) whose patient Isaac Newton was [Manuel 1968, p. 191]. Apparently also the great Voltaire spoke to Dr. Mead since in the *Letters concerning the English Nation* we read [Voltaire 2011, p. 64]:

> *'One very singular Difference in the Lives of these two great Men* [Descartes and Newton] *is, that Sir Isaac, during the long Course of Years he enjoy'd was never sensible to any Passion, was not subject to the common Frailties of Mankind, nor ever had any Commerce with Women; a Circumstance which was assur'd me by the Physician and Surgeon who attended him in his last Moments.'*

Be it as it may: whether Newton had a homosexual relationship with Fatio, or whether he recognised himself in the young man and was therefore attracted, or whether he saw the son in him he never had – in any case Fatio changed from a Cartesian to a dedicated Newtonian in no time. It seems that Fatio even kept Newton in leading-strings in some sense[1], since Fatio wrote to Huygens that Newton would be glad to receive comments on certain results in the *Principia* [Westfall 2006, p. 495]:

> *'I have found him ready to correct his book on the matters that I have told him about so many times that I cannot admire his facility too much ...'*

It sounds as if Fatio would understand the *Principia* even better than Newton himself and perhaps this was really what Fatio believed. In the *Principia* Newton had wrongly modelled the leaking of a fluid through a hole in a vessel and Fatio wrote in the margins of his copy of the *Principia* [Westfall 2006, p. 495]:

> *'I could scarcely free our friend Newton from this mistake, and that only after making the experiment with the help of a vessel which I took care to have prepared.'*

For some time he even planned to issue a second edition but this did not happen. Only at the beginning of the year 1692 Fatio's arrogant tone changed after he was admitted to see some of Newton's manuscripts. He then wrote to Huygens [Westfall 2006, p. 495]:

[1]Newton would have never acquiesced such a behaviour from any other person which probably shows the extraordinary bond between the two men.

'I was frozen stiff when I saw what Mr. Newton has accomplished ...'

Fatio also felt able to comment on Newton's theory of gravitation. In a memorandum by David Gregory dating 7th January 1692 one finds a remark which would not have delighted Fatio had he ever read it [Turnbull 1959–77, p. 191]:

'Mr Newton and Mr Hally laugh at Mr Fatios manner of explaining gravity.'

Newton and Fatio spent almost one month in London together between March and April 1690 [Westfall 2006, p. 496]. The plans to get a leading position at a college in Cambridge (it was King's College) for Newton were dismantled by John Hampden's descent in the king's favour. Hence Fatio contacted John Locke asking to organise an appropriate post for Newton in London. Newton wanted to leave Cambridge, also because his Trinity College was so financially spent that dividends were no longer distributed. Only in 1696 came the opportunity to finally leave Cambridge, but at that time Fatio did not play a role any longer. Concerning Newton biographers Fatio could never win sympathy. Frank E. Manuel called him 'The Ape of Newton' [Manuel 1968, p. 191] and this ape will start the war against Leibniz in 1699.

7.3.2 A New Crisis

At the beginning of the 1690s Newton's name is inseparably connected with the name of Fatio in the letters of Huygens, de l'Hospital and Leibniz [Manuel 1968, p. 195]. Since October 1690 Huygens, Leibniz and Fatio correspond with each other on two mathematical problems which Huygens introduced to the other two correspondends [Turnbull 1959–77, Vol. III, footnote (1), p. 149]. Newton had certainly introduced the young man to further works by him, in particular to his theological and alchemistical considerations. On 7th November 1690 Newton wrote to John Locke [Turnbull 1959–77, Vol. III, p. 79]:

'I suppose Mr Fatio is in Holland for I have heard nothing from him ye half year.'

In autumn 1691 Fatio was again in London and wrote to Huygens on 18 September that he expected to see Newton within the next few days [Turnbull 1959–77, Vol. III, p. 168]. With date of 20th October 1692 David Gregory informs Newton that Fatio was in London [Turnbull 1959–77, Vol. III, p. 170]. There were no letters of Fatio to Newton until the 27th September 1692, but then Fatio wrote [Turnbull 1959–77, Vol. III, p. 229f.]:

'I have Sir allmost no hopes of seeing you again. With coming from Cambridge I got a grievous cold, which is fallen upon my lungs. Yesterday I had such a sudden sense as might probably have been caused

upon my $\left\{\begin{array}{c}\textit{midriff}\\ \textit{diaphragm}\end{array}\right\}$ *by a breaking of an ulcer, or vomica, in the undermost part of the left lobe of my lungs. For about that place of my* $\left\{\begin{array}{c}\textit{midriff}\\ \textit{diaphragm}\end{array}\right\}$ *I felt a momentaneous sense of something bigger than my fist moving and acting powerfully. That sense was distinct in all that region, but not troublesome to me, tho' my surprise caused my body to bend forwards, as I was sitting by the fire. What I felt next was only a gentle and easie sense of a natural heat in that region. My pulse was good this morning; It is now (at 6. afternoon) feaverish and hath been so most part of the day. I thank God my soul is extreamly quiet, in which you have had the chief hand. My head is something out of order, and I suspect will grow worse and worse.*

[...] *If I am to depart this life I could wish my eldest brother, a man of an extraordinary integrity, could succeed in Your friendship. As yet I have had no Doctor. Perhaps wth a paracenthesis they may save my life, which I am not yet certain is in any danger.'*

Newton is deeply concernd about the state Fatio was in and wrote back on 31st September [Turnbull 1959–77, Vol. III, S. 231]:

'Sr

I have ye book[2] *& last night received your letter wth wch how much I was affected I cannot express. Pray procure ye advice & assistance of Physitians before it be too late & if you want any money I will supply you.'*

Fatio was probably indeed suffering from influenza but had excessively exaggerated the symptoms. He was much better already when Newton's letter arrived in London [Westfall 2006, p. 532] but the convalescence took longer. When the Swiss theologian Jean-Alphonse Turrettini (1671–1737) who was situated in England at this time reported to Newton on the slow recovery of Fatio in January 1693 Newton wrote to Fatio [Turnbull 1959–77, Vol. III, p. 241]:

'Sr

I understand by Mr Turretine who brings you this that your cold has not yet left you as I hoped it would have done long since. I feare ye London air conduces to your indisposition & therefore wish you would remove hither so soon as ye weather will give you leave to take a journey. For I beleive this air will agree with you better. Mr Turretine tells me you are considering whether you should return this year into your

[2] In the preceeding letter of Fatio to Newton Fatio reported that the bookseller Lea had forgotten to send a book to Newton.

own country. Whatever your resolutions may prove yet I see not how you can stirr wthout health & therefore to promote your recovery & save charges til you can recover I [am] very desirous you should return hither. When you are well you will then know better what measures to take about returning home or staying here.'

Fatio confirmed that he wanted to go back to Switzerland because his mother had died. With her inheritance, he wrote, he could live for some years in England *'chiefly in Cambridge'* [Turnbull 1959–77, Vol. III, p. 242]. The correspondence continued over the whole winter. It is about Fatio's health, about money, and about the question whether Fatio could move to Cambridge. Newton offered generous financial support although full dividends from his college had ceased for seven years [Westfall 2006, p. 532]. In May 1693 Fatio wrote again about alchemy. He had met a man knowing a certain process in which gold united with mercury would grow. Two weeks later, on 28th May, he wrote that his new friend had obtained a medicine from mercury which had fully cured him [Turnbull 1959–77, Vol. III, p. 267ff.]. The new friend had offered him a partnership concerning the new medicine. Fatio wanted to acquire a medical degree to save thousands of lives with this medicine which was cheap to produce. However, there was one obstacle: he needed between 100 and 150 pound Sterling per year for the next four years and he asked Newton for help. Newton left Cambridge on 9th June for a week; without doubt he travelled to London. At the beginning of July he is there again for another week. On 9th June Newton drafted a letter to the editor of the Acta Eruditorum, Otto Mencke [Turnbull 1959–77, Vol. III, p. 270], which he did not finish and left ignored for six months. He again conducted experiments in his laboratory. More we do not know [Westfall 2006, p. 534]! Almost four months of his life lie in darkness. It came to a life crisis the case of which is also not clear. On 23rd September 1693 an apparently mad Newton wrote to Samuel Pepys [Turnbull 1959–77, Vol. III, p. 279]:

'Sir,

Some time after Mr Millington had delivered your message, he pressed me to see you the next time I went to London. I was averse; but upon his pressing consented, before I considered what I did, for I am extremely troubled at the embroilment I am in, and have neither ate nor slept well this twelve month, nor have my former consistency of mind. I never desigend to get anything by your interest, nor by King James's favour, but am now sensible that I must withdraw from your acquaintance, and see neither you nor the rest of my friends any more, if I may but leave them quietly. I beg your pardon for saying I would see you again, and rest your most humble and most obedient servant,

Is. Newton'

On 26th September 1693 Newton was in London in the inn 'Bull' and wrote to his friend John Locke [Turnbull 1959–77, p. 280]:

'Sr

Being of opinion that you endeavoured to embroil me wth woemen & by other means I was so much affected with it as that when one told me you were sickly & would not live I answered twere better if you were dead. I desire you to forgive me this uncharitableness. For I am now satisfied that what you have done is just & I beg your pardon for my having hard thoughts of you for it & for representing that you struck at ye root of morality in a principle you laid down in your book of Ideas & designed to pursue in another book & that I took you for a Hobbist. I beg your pardon also for saying or thinking that there was a designe to sell me an office, or to embroile me, I am

your most humble & most

unfortunate Servant

Is. Newton'

What had happened and what did Pepys and Locke do? Both must have been shocked! Had Newton indeed become mad? And if so, why?

For a start Pepys enquired about the case and asked the physician Thomas Millington (1628–1704) whether he knew more details. He wrote [Brewster 1855, Vol. II, p. 143]:

'For I was loth at first dash to tell you that I had lately received a letter from him so surprising to me for the inconsistency of every part of it, as to be put into great disorder by it, from the concernment I have for him, lest it should arise from that which of all mankind I should least dread from him and most lament for, – I mean a discomposure in head, or mind, or both. Let me, therefore, beg you, Sir, having now told you the true ground of the trouble I lately gave you, to let me know the very truth of the matter, as far at least as comes within your knowledge.'

Millington wrote back on 10th October [Brewster 1855, Vol. II, p. 144f.]:

'... he [Newton] *told me that he had writt to you a very odd letter, at which he was much concerned; added, that it was in a distemper that much seized his head, and that kept him awake for above five nights together, which upon occasion he desired I would represent to you, and beg your pardon, he being very much ashamed he should be so rude to a person for whom he hath so great an honour. He is now very well, and, though I fear he is under some small degree of melancholy, yet I think there is no reason to suspect it hath at all touched his understanding, and I hope never will; and so I am sure all ought to wish that love learning or the honour of our nation, ...'*

The crisis seemed over but Newton never again in his life came back to *creative* scientific work. Mind you, he had not lost his intellectual capacities but he never again dared to tackle new scientific theories. There were numerous attempts to explain the crisis which probably began already at the beginning of 1692 when Newton wrote a somewhat paranoic letter to John Locke in which he accused the ever loyal Charles Montague of falsehood [Westfall 2006, p. 534f.]. An acute mercury poisoning was held responsible [Johnson/Wolbarsht 1979], [Spargo/Pounds 1979] which Newton could have suffered from due to his work in the laboratory. Others suggested a fit of manic depression [Lieb/Hershman 1983] or a breakdown due to the death of his mother [Spargo/Pounds 1979, p. 15]. A simpler possibility could lie in the permanent overload which was no longer bearable after years of work on the *Principia.* Often told is the story of a fire in Newton's laboratory caused by his pet dog in which he should have lost numerous manuscripts, but the dating of the event is difficult as we discussed on page 245.

Frank E. Manuel linked the breakdown of Newton directly with the disappearance of Fatio from Newton's life [Manuel 1968, p. 219]: *'The affective relationship to Fatio had approached a climax and the plans to have Fatio reside with him in Cambridge fell through.'* Fatio also suffered a phase of personal and religious tensions at that time [Westfall 2006, p. 538]. In fact, Newton's breakdown coincides with the abrupt end of the relationship with Fatio and although we will probably never know what happened between the two men the breaking off of the relationship might be a possible trigger to ignite the crisis. But perhaps more than one factor was involved and the breaking off was only the last drop causing the barrel to overflow.

7.3.3 Leibniz Back in Newton's Thoughts

At the beginning of the 1690s a younger generation of mathematicians and natural philosophers had evolved who seeked Newton's support in acquiring a professorship. One of these young men was David Gregory whom Newton had recommended successfully in 1691 for the Savilian chair of astronomy at Oxford. The Scottish mathematician John Craig (1663–1731) had copied two quadratures from Newton's papers and had taken the copies with him to Scotland. David Gregory had succeeded in discovering the binomial theorem which was behind the quadratures and allowed his friend Archibald Pitcairne (1652–1713) to publish this discovery as 'Gregory's discovery' in the year 1688 without even mentioning Newton's name. Now Gregory intended to publish a manuscript on general quadratures but did not want to raise Newton against him. Gregory submissively wrote twice to Newton that he would publish the manuscript as 'Letter to Newton' and he also mentioned Newton's priority concerning the binomial theorem. Newton's answer clearly focused on the way his binomial theorem took via Craig to Gregory, and Gregory felt not

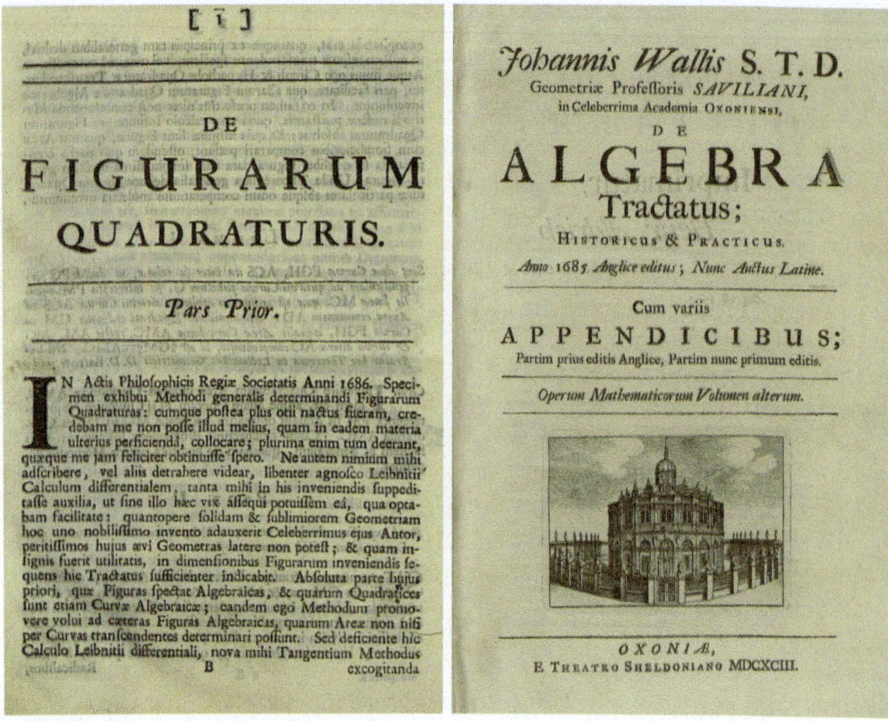

Fig. 7.3.3. John Craig's *De Figurarum Quadraturis* of 1693 (Bayerische Staatsbibliothek München (Munich), Signatur: 853766 4 Math. p. 90) and the title page of John Wallis's *Opera mathematica*, Volume 2, 1693 (Ghent University Library, BIB.MA.000006)

encouraged to publish his manuscript. However, he sent it to Huygens. It was published only in 1693 in the *Opera mathematica* of John Wallis and without any hint to Craig [Westfall 2006, p. 514].

In a letter to Gregory [Turnbull 1959–77, Vol. III, p. 182] there is a very interesting paragraph in view of Newton's attitude towards Leibniz:

> *'But since you ask me, with your usual courtesy, for my series in return, it is necessary for me first to explain some points that concern it. For when the very distinguished Mr G. G. Leibniz fifteen years ago conducted a correspondence with me as arranged by Mr Oldenburg, and I took the opportunity to expound my method of infinite series, in the second of my letters, dated 24 October 1676, I described this series.'*

Actually, Newton's wrath should have gone off on Gregory, but that did not happen! Instead, Leibniz served as a kind of lightning rod [Westfall 2006, p. 514]. This Leibniz, with whom Newton had not corresponded since 1676,

ISAACI NEWTONI Equ. Aur.
TRACTATUS
DE
QUADRATURA
CURVARUM
IN
USUM STUDIOSÆ JUVENTUTIS MATHEMATICÆ
EXPLICATIONIBUS ILLUSTRATUS
A
DANIELE MELANDER.

UPSALIÆ, MDCCLXI.

Fig. 7.3.4. Newton's *Quadratura Curvarum* (Bayerische Staatsbibliothek München (Munich), Signatur: 1584193 4Math.p.253 b, S.5)

is still in his mind in the year 1691. Just at the beginning of the letter to Gregory Newton started to mention Leibniz and one gets the feeling that he immediately took a defensive attitude in expressing his priority over Leibniz concerning his mathematical achievements, but he did not mention Gregory who justifiably should have been the actual target.

Fatio had just returned from the Netherlands and Newton had spent a week in London with him. Huygens had always informed Fatio about the recent works of Leibniz. Had Fatio told Newton about the publications of Leibniz? Had he talked about the progress made on the continent by means of the new calculus without even mentioning Newton's name? We do not know but it is not unlikely. We know however, that Fatio did not like Leibniz. Did Fatio play the role of an agent provocateur already at this time, shortly before the letter of Gregory?

The draft of the letter to Gregory in any case inspired Newton to write a summarising manuscript on his calculus of fluxions, *De quadratura curvarum* (On the quadrature of curves), which (of course!) started with a passage on the exchange with Leibniz in 1676 and which resolved both of the anagrams in the *Epistola posterior*. The manuscript ought to reject Gregory's claims but contains a breathtaking expansion of quadratures which made Gregory's manuscript look pale [Westfall 2006, p. 515]. *De quadratura* also included a new, systematic notation for the fluxions in the form $\dot{x}, \dot{y}$ we use today. In the case of quadrature Newton experimented with the letter Q serving as a replacement for Leibniz's symbol $\int$. We also find in this manuscript the Taylor expansion of a function which was rediscovered by Brook Taylor twenty years later and which now bears his name.

At the end of the year 1691 *De quadratura* was known in the circle of Newton's young admirers. It would have been a good time to publish it, but Newton let the opportunity pass. Even worse, he did not even finish the work on *De quadratura*. Fatio reported to Huygens that Newton shied from the confusion which he assumed would follow the publication [Westfall 2006, p. 516]. Only in the middle of the 1690s did Newton finish a shortened version of *De quadratura* in which the references to Leibniz were deleted. This shortened version is the one which was finally published in 1704 as an appendix of Newton's second famous book, *Opticks*.

7.3.4 The Monkey Bites

On 28th December 1691 Fatio wrote a first scathing letter to Christiaan Huygens [Westfall 2006, p. 516f.], [Turnbull 1959–77, Vol. III, p. 186f.]:

> *'It seems to me from everything that I have been able to see so far, among which I include papers written many years ago, that Mr. Newton is beyond question the first Author of the differential calculus and that he knew it as well or better than Mr. Leibniz yet knows it before the latter had even the idea of it, which idea itself came to him, it seems, only on the occasion of what Mr. Newton wrote to him on the subject. (Please Sir look at page 235 of Mr. Newton's book). Furthermore, I cannot be sufficiently surprised that Mr. Leibniz indicates nothing about this in the Leipsig Acta*[3]*.'*

For the first time here was a clear accusation of plagiarism against Leibniz openly expressed! And if that was not already enough Fatio wrote in a letter to Huygens in February 1692 [Westfall 2006, p. 517]:

> *'The letters that Mr. Newton wrote to Mr. Leibniz 15 or 16 years ago speak much more positively than the place that I cited to you from the Principles which nevertheless is clear enough especially when the*

[3]Fatio refers to the works of Leibniz on the differential calculus which were published in the Acta Eruditorum.

letters explicate it. I have no doubt that they would do some injury to Mr. Leibniz if they were printed, since it was only a considerable time after them that he gave the Rules of his Differential Calculus to the Public, and that without rendering to Mr. Newton the justice he owed him. And the way in which he presented it is so far removed from what Mr. Newton has on the subject that in comparing these things I cannot prevent myself from feeling very strongly that their difference is like that of a perfected original and a botched and very imperfect copy. It is true Sir as you have guessed that Mr. Newton has everything that Mr. Leibniz seemed to have and everything that I myself had and that Mr. Leibniz did not have. But he has gone infinitely farther than we have, both in regard to quadratures, and in regard to the property of the curve when one must find it from the property of the tangent.'

The cat was out of the bag: the differential calculus of Leibniz should be nothing more than a messed up and poor copy of the original calculus as derived by Newton! The accusation is only addressed in a private letter to Huygens but thereby it is put into the world.

In summer 1692 John Wallis offered that Newton could publish what he wanted in Wallis's *Opera* which was due to be published shortly. After all, Gregory's method of quadrature should be published in the first volume and Newton could have secured his priority only by also publishing something in Wallis's *Opera*. Newton sent Wallis a summary of *De quadratura* and thereupon Wallis began to pester Newton incessantly about Leibniz [Westfall 2006, p. 517]. Needless to say, Newton targeted Leibniz even in his summary. Both anagrams were decoded and Newton made clear that he already in 1676 had methods at his command which were not published previously. Newton asked Wallis to publish his series for the arithmetical quadrature of the circle [Turnbull 1959–77, Vol. III, p. 219] since Wallis wanted also to publish Leibniz's series. The second volume of Wallis's *Opera* came out in 1693, two years before the first volume was published in 1695. One thing is quite clear: even in 1692 Newton had meticulously in mind the question of priority concerning Leibniz! But Leibniz also got alert concerning Newton.

Huygens had made the parts of Fatio's letters concerning the progress of Newton accessible to his pupil, but not the devious parts concerning the priority question. We may assume that all mathematicians on the continent (and there were only a few) quickly received the news concerning Newton's plan to publish details on his calculus of fluxions. On 17th March 1693 Leibniz wrote again to Newton [Turnbull 1959–77, p. 257ff.]:

'To the celebrated Isaac Newton
Gottfried Wilhelm Leibniz cordial greetings

How great I think the debt owed to you, by our knowledge of mathematics and of all nature, I have acknowledged in public also when

> *occasion offered. You had given an astonishing development to geometry by your series; but when you published your work, the Principia, you showed that even what is not subject to the received analysis is an open book to you. I too have tried by the application of convenient symbols which exhibit differences and sums, to submit that geometry, which I call 'transcendent', in some sense to analysis, and the attempt did not go badly. But to put the last touches I am still looking for something big from you,* [...]
>
> [Here Leibniz got more specific and asked Newton to present details concerning the 'inverse method of tangents', i.e. methods for the solution of differential equations.]
>
> [...] *My fellow-countryman Heinson on his return assured me of your friendly feelings towards me. But of my veneration for you not only he can testify but Stepney too, once your fellow resident in the same college,* [...]
>
> [...] *I write this rather that you should understand my devotion to you, a devotion that has lost nothing by the silence of so many years, than that with empty, and worse than empty, letters I should interrupt the devoted studies by which you increase the patrimony of mankind. Farewell.'*

It is an odd letter. On one hand one can feel Leibniz's hope to renew his correspondence with Newton, on the other hand the letter is a bit too submissive, even regarding the customs of the time. Has Leibniz become nervous as Westfall [Westfall 2006, p. 518] has suspected? But why should he?

Only on the 26th October 1693 when the breakdown, the great crisis in Newton's life, seems halfway overcome, did Newton answer Leibniz's letter. He apologised for the delay with the following words [Turnbull 1959–77, Vol. III; p. 285ff.]:

> *'As I did not reply at once on receipt of your letter, it slipped from my hands and was long mislaid among my papers, and I could not lay hands on it until yesterday. This vexed me since I value your friendship very highly and have for many years back considered you as one of the leading geometers of this century, as I have also acknowledged on every occasion that offered. I was however afraid that our friendship might be diminished by silence, and at the very moment too when our friend Wallis has inserted into his imminent new edition of his History of Algebra some new points from letters which I once wrote to you by the hand of Mr Oldenburg and so has given me a handle to write to you on that question also.* [...]
>
> [Here Newton explains one of his anagrams from the correspondence in 1676]

> [...]*I hope indeed that I have written nothing to displease you, and if there is anything that you think deserves censure, please let me know of it by letter, since I value friends more highly than mathematical discoveries.'*

Then Newton addressed the mathematical questions of Leibniz and finished with the words [Turnbull 1959–77, Vol. III, p. 287]:

> *'My aim in these pages has been to give proof that I am your most sincere friend and that I value your friendship very highly. Farewell.*
> [...]
> *I wish you would publish the rectification of the hyperbola that you have referred to as a long-standing discovery of yours.'*

Westfall has criticised that Newton gave only a very short answer and did not address all of Leibniz's questions openly [Westfall 2006, p. 519]. In fact Leibniz did not answer Newton's letter. Had Leibniz understood the hint concerning the forthcoming publication in Wallis's book as a threat because he was himself hinting on the question of priority? In any case, Leibniz had not yet seen Wallis's book in June 1694 and wrote impatiently to Huygens to send him the book as soon as possible. When he received the book in autumn 1694 he was disappointed that it contained so little material on the inverse method of tangents, but Westfall has interpreted this disappointment as relief [Westfall 2006, p. 519]. Newton did not publish more than what he knew already in 1676 and Leibniz could now be sure that he himself had clearly further advanced. In this sense Newton's results were also understood by John Bernoulli who suspected that Newton had looted Leibniz's publications to come up with the method he published only now [Westfall 2006, p. 519]. We will come back to this on page 341.

Many assessments of this last correspondence between Newton and Leibniz are uncertain. One can take the manifestations of friendships of both men at face value; then the monkey Fatio had only bitten because he did not like Leibniz and because he thought himself the better mathematician in his hubris. In this case we have to assume that Fatio acted on his own account. On the other hand we can suspect that the friendly words of both men – in particular Newton's – were meant to conceal the subliminally boiling feelings in the cold war. In that case Fatio would have bitten with Newton's approval. For Newton this would have had the invaluable advantage that he could take cover – a position he always preferred to a public attack. If we try to read the letters without prejudice then they are nothing but extraordinary manifestations of mutual appreciation. Nothing points to a conflict.

The war seemed avoided, but in truth it got hot on both sides.

7.4 Wallis, Flamsteed and the Way Into the Mint

7.4.1 Wallis Jars on Newton's Nerves

John Wallis, always anxious to promote the glory of English sciences, now twisted Newton's arm. On 20th April 1695 he wrote to Newton [Turnbull 1959–77, Vol. IV, p. 100f.]:

> '*I understand (from Mr Caswell) you have finished a Treatise about Light, Refraction, & Colours*[4]*; which I should be glad to see abroad. 'Tis pitty it was not out long since. If it be in English (as I hear it is) let it, however, come out as it is; & let those who desire to read it, learn English. I wish you would also print the two large Letters of June & August 1676*[5]*. I had intimation from Holland, as desired there by your friends, that somewhat of that kind were done; because your Notions (of Fluxions) pass there with great applause, by the name of Leibnitz's Calculus Differentialis. [...] You are not so kind to your Reputation (& that of the Nation) as you might be, when you let things of worthly by you so long, till others carry away the Reputation that is due to you.* [...]
>
> [...] *I understand you are now about adjusting the Moons Motions;* [...]'

When Newton did not answer this letter Wallis tried to create facts. On 9th June 1695 he wrote again to Newton [Turnbull 1959–77, Vol. IV, p. 129f.]:

> '*Sir*
>
> *I have taken the pains to transcribe a fair copy of your two letters, which I wish were printed. I send it you with this, because I suspect there may be some little mistakes either in ye Calculation or Transcribing in some places, which therefore I desire you will please carefully to peruse, & correct to your own mind, & then (if you please) remit to me. I would have subjoined them (with your good leave) to the second volume of my Opera Math: if I had thought of it a little sooner, before that had been sent abroad; but 'tis now, I think, too late.* [...]
>
> [Wallis wants to find a printer in Oxford.]
>
> *But I find that these letters do refer to two Letters of Leibnitz, which I have never seen. If you have copies of them by you, it would be proper to print those with these.*'

[4]The book *Opticks: or, a Treatise of the Reflexions, Inflexions and Colours of Light* will not be published before 1704.

[5]Meant are the two *Epistolae*, the *Epistola prior* and the *Epistola posterior*

Fig. 7.4.1. John Wallis

Once again Newton did not react. He only reacted after Wallis wrote a third, pressing letter on 13th July 1695 [Turnbull 1959–77, Vol. IV, p. 139]. Should Newton not agree to have both his letters published he should at least correct Wallis's transcription so that Wallis could have deposited them in the Savilian Library in Oxford,

> *'which will be no dishonour to you, but confirm to you the reputation of your having discovered these notions so long ago.'*

Eventually Newton responded in a letter dated July 1695 [Turnbull 1959–77, p. 140f.]:

> *'Sr*
>
> *I am very much obliged to you for your pains in transcribing my two Letters of 1676 & much more for your kind concern of right being done me by publishing them. I have perused your transcripts of them & examined ye calculations & corrected some few places wch were amiss.'*

To express his own worry concerning priority Newton cited a passage from a letter by Collins from 28th June 1673 showing that Newton had his method of tangents even *before* Sluse had his.

The pressing questions concerning a publication of the book on optics with which Wallis probably got on Newton's nerves were not heard. The reason was no other than – Robert Hooke! Newton had had enough trouble with Hooke on questions of priority in the case of the law of gravitation just after

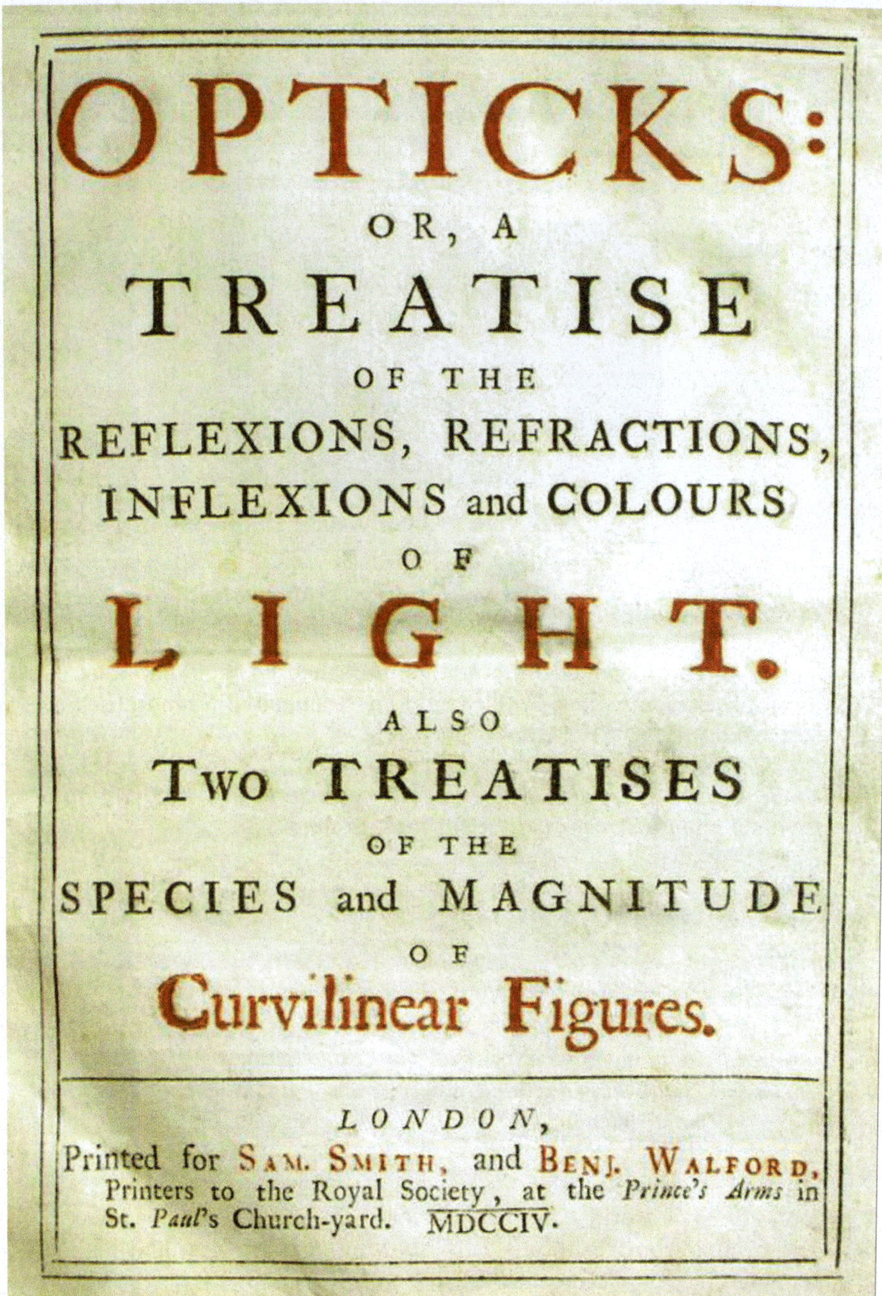

OPTICKS:

OR, A

TREATISE

OF THE

REFLEXIONS, REFRACTIONS, INFLEXIONS and COLOURS

OF

LIGHT.

ALSO

Two TREATISES

OF THE

SPECIES and MAGNITUDE

OF

Curvilinear Figures.

LONDON,

Printed for SAM. SMITH, and BENJ. WALFORD, Printers to the Royal Society, at the *Prince's Arms* in St. *Paul's* Church-yard. MDCCIV.

Fig. 7.4.2. Title page of *Opticks* published in 1704

the *Principia* were published and he did not need to pick a second quarrel on questions of optics. Indeed, Newton waited for the death of his opponent in March 1703; he felt so much damaged by Hooke. *Opticks* came out in 1704. Already at the very beginning Newton outlined that his research on optics was based on a manuscript which he had sent to the Royal Society in 1675 and that the theory was completed 12 years later in 1687. Even the reason why the book did not see an earlier publication was named [Newton 1979, p. cxxi]:

> *'To avoid being engaged in Disputes about these Matters, I have hitherto delayed the printing, and should still have delayed it, had not the Importunity of Friends prevailes upon me.'*

The following remark is of particular importance to us [Newton 1979, p. cxxii]:

> *'In a Letter written to Mr. Leibnitz in the year 1679*[6]*, and published by Dr. Wallis, I mention'd a Method by which I had found some general Theorems about squaring Curvilinear Figures, or comparing them with the Conic Sections, or other the simplest Figures with which they may be compared. And some Years ago I lent out a Manuscript containing such Theorems, and having since met with some things copied out of it*[7]*. I have on this Occasion made it publick, prefixing to it an Introduction, and subjoining a Scholium concerning that Method. And I have joined with it another small Tract concerning the Curvilinear Figures of the Second Kind, which was also written many Years ago, and made known to some Friends, who have solicited the making it publick.'*

Here, in a book on optics, the world finally saw the first publication of Newton's calculus of fluxions! It was the *Tractatus de quadratura curvarum* (Treatise on the quadrature of curves), a shortened version of the manuscript *De quadratura*. The second work appended is *Enumeratio linearum tertii ordinis* (Enumeration of the lines of third order) on the classification of cubic curves which went back to Newton's work of the late 1660s or the early 1670s [Guicciardini 2009, p. 109]. Guicciardini has discussed this work as well as the associated earlier works in [Guicciardini 2009, p. 109ff.].

[6]This is a printing error in the 4th edition. Of course, the year must have been 1676 and in the first edition (http://www.rarebookroom.org/Control/nwtopt/index.html) this correct date is given.

[7]Newton refers to George Cheyne's book *Fluxionum methodus inversa* of 1703 which we will discuss in chapter 8.

Fig. 7.4.3. Bust of John Flamsteed in the museum of the Royal Observatory Greenwich [Photo: Klaus-Dieter Keller 2006] and 'The Royal Observatory' [Thomas Hosmer Shepherd, 1824]

7.4.2 Newton's Quarrel with Flamsteed

Not only had Newton's nerves in the 1690s to suffer from Wallis's permanent urge but also from a conflict with the first Astronomer Royal, John Flamsteed. The history of this conflict has been vividly described by D.H. Clark and S.P.H. Clark in [Clark/Clark 2000]. However, this book is completely useless from a scientific point of view, since not one source used is mentioned. Therefore we are going to follow the statements by Iliffe [Iliffe 2006, Vol. 1, p. xix ff.] and Westfall [Westfall 2006, p. 541ff.].

John Flamsteed was born on 19th August 1646 and schooled at home for some time due to an illness. As an autodidact he occupied himself with astronomy and in 1669 he sent the (correct) prediction of a solar eclipse to the Royal Society which much impressed Collins and Oldenburg and quickly made him a well-known astronomer. In autumn 1670 Flamsteed had enrolled as a student at Jesus College, Cambridge, and he even attended lectures by Isaac Newton in 1674. At Easter 1675 he was ordained to the clergy, and in February 1676 he became a member of the Royal Society. Due to the need for more accurate star tables for navigational purposes the English King Charles II ordered a Royal Observatory to be built at Greenwich. He also created the title and the position of an 'Astronomer Royal' and he employed Flamsteed to be the first in 1675. Flamsteed lived in Greenwich until the year 1684 in which he became the priest of the parish of Burstow in Surrey where he eventually was buried after his death in 1719. He held both positions until his death, hence was Astronomer Royal and priest of Burstow in personal union.

Fig. 7.4.4. left: Tycho Brahe and his mural instrument (based on [Astronomiae Instauratare Mechanica, woodcut 1598] later coloured, unknown artist), right: English mural instrument by John Bird 1713 in the Museum of the History of Science in Oxford [Photo: Heinz-Josef Lücking 2006]

Already in the 1670s Flamsteed had designed and built exact pendulum clocks, quadrants, and sextants to follow the stars [Iliffe 2006, Vol. 1, p. xx]; from 1689 he disposed of a large mural instrument and some refractor telescopes in Greenwich, which could be exactly adjusted by means of micrometer screws. For their purchase he had used mainly his own money.

The 'heroes' in Flamsteed's life were the astronomers Jeremiah Horrocks (also known as Horrox) (1619–1641) and Tycho de Brahe (1546–1601), whom he tried to emulate. His aim was the creation of a comprehensive star catalogue which in respect to accuracy should surpass everything known before. The delays which were caused by these high demands constituted not least the struggle with Newton.

At the end of the year 1680 Flamsteed observed the so-called 'Great Comet' C/1680 V1 (also known under the names of 'Kirch's Comet' and 'Newton's Comet'), which disappeared at the end of November behind the sun and reappeared at the beginning of December (cp. page 254). Flamsteed wrote to Newton in January 1681 that he had calculated the return of the comet and he would be sure that it was just *one* comet and not two different ones, and that this comet was redirected by 'magnetic repulsion' of the sun. Newton answered via Crompton in a friendly manner on 28th February 1681 [Turnbull 1959–77, Vol. II, p. 340ff.], but then started his criticisms. For one thing the comet's trajectories were not consistent before and after orbiting the sun

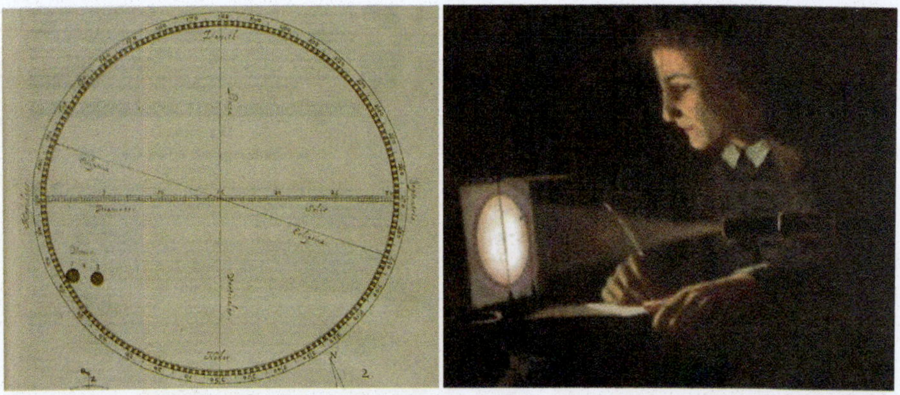

Fig. 7.4.5. Drawing of the Venus transit of 1639 [Drawing: Jan Hevelius 1662] and Jeremiah Horrocks observing the transit of Venus in front of the sun for the first time ([Painting: William Richard Lavender 1903], Collection of Ashley Hall Museum and Art Gallery, Chorley Council)

in the sense that the comet had increased and decreased its velocity several times. The trajectories therefore could barely belong to the same heavenly body. Furthermore, a magnetic repulsion of the sun was hardly explainable, since the sun was known to attract bodies. And finally one knew that magnets would lose their magnetic properties in great heat. So where should this repulsive force come from? Newton stuck to it: there were but *two* comets. His opinion only held unto autumn 1685, however; then Newton had found the true source of the attracting force of the sun – gravitation, not magnetism! Therefore, Rob Iliffe holds the correspondence of Flamsteed and Newton as important for the development of the ideas in the *Principia* as the correspondence between Hooke and Newton in the years 1679-80 [Iliffe 2006, Vol. 1, p. xxi]. The fact that Newton never even mentioned Flamsteed's role in this correspondence or the data of the trajectory he delivered has certainly laid the seed of the coming conflict.

Thereby also Flamsteed was surely no easy man to deal with; he seems to have been quite like Newton [Manuel 1968, p. 293]. Like Newton he had to cope with a great loss in his childhood: the death of his mother. Like Newton suffered from the separation from his mother, he suffered from his childhood impressions and became depressed. When his relations with Newton were still intact he had already broken off with Hooke and Halley. Hooke had humiliated Flamsteed in that he stated that he could grind much better lenses and could build much better instruments as the Astronomer Royal, but he never gave evidence for his claims. Halley was taught by Flamsteed in the year 1675 and both men remained in friendship for some time, but after the publication of the *Principia* Flamsteed accused him of plagiarism and detected an unhealthy hunger for acclaim in him [Iliffe 2006, Vol. 1, p. xxii].

Fig. 7.4.6. The Great Comet of 1680 above Rotterdam [Painting: Lieve Verschuier as from 1680]

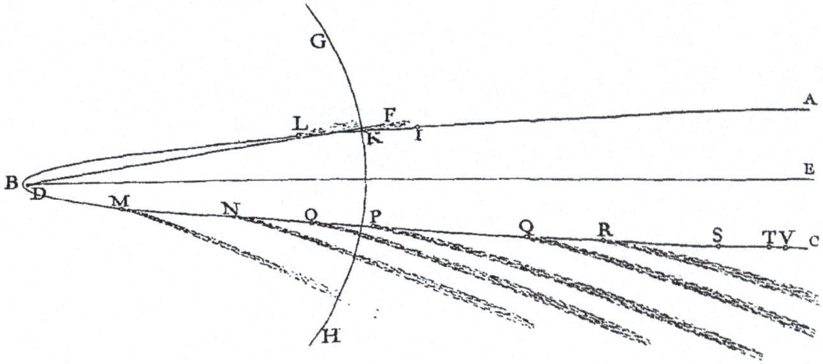

Fig. 7.4.7. The trajectory of the Great Comet of 1680 modelled by a parabola in Newton's *Principia*

Fig. 7.4.8. A part of the Royal Observatory in Greenwich today left: [Photo: ChrisO], right: [Photo: Peter Smyly 2009]

Flamsteed had delivered important data for the *Principia* in an extensive correspondence between 1685 and 1686 which were evaluated by Newton and used in the *Principia*. Flamsteed was somehow upset because his 'ingredients' to the *Principia* found little reference, while Halley, being the editor of Newton's oeuvre, was highly praised. However, the relationship between Flamsteed and Newton was still intact. In his records Flamsteed had noted [Iliffe 2006, Vol. 1, p. 15]:

> *'1687. his principles published, little notice taken of her Maties Observatory.'*

The break occurred only when Newton started in 1694 to work on improving his theory of the moon and took up a correspondence with Flamsteed in autumn. We may assume that Flamsteed was aware of his significance when he sent 50 new measurements of the moon's position to Newton [Iliffe 2006, Vol. 1, p. xxii, p. 15],

> *'drawn up in 3 large Synopses & on his request gave him copys of them he promising me not to impart or communicate them to anybody.'*

Under the same conditions, namely *not* to pass on the data, Flamsteed sent additional one hundred observation data during the following months, while continuing his work on his great star catalogue. Already shortly after, Flamsteed learned that Newton had not complied with his conditions, since Edmond Halley as well as David Gregory, Savilian professor of astronomy in Oxford, were in the possession of the data. Flamsteed felt betrayed and was shocked; superficially because the measurements were taken using a sextant and did not satisfy Flamsteed's high demands on accuracy. Actually, Flamsteed wanted Newton to wait for the great star catalogue, since positions of the moon are given relative to fixed stars. But Newton did not understand why the catalogue took so long to be finished and he started in 1695 to complain about Flamsteed. The tone of Newton's letters became abrasive; Flamsteed's answers grew increasingly moralising.

Newton had treated the three body problem in his *Principia* but he was not far advanced. Now he wanted to take up again the calculations to tackle the problem of which no analytical solution existed. The calculations were demanding. A nagging problem in the observations of star positions was the atmospheric refraction, since there was no theory available to correct it. In November 1694 Newton sent a first table with computations to Flamsteed who immediately doubted their accuracy [Westfall 2006, p. 544]; this had to be accepted by Newton. Newton sent a corrected table to Flamsteed only in March 1695. Newton's computational techniques lay beyond the empirical approach of Flamsteed who could barely understand most of it; some of it he did not understand at all. Newton's harsh tone in the letters eventually led to Flamsteed's deliberate withholding of data, which in turn made Newton even more impatient. Flamsteed should simply send raw data and should not care processing them which Newton would do himself, so Newton wrote. When Flamsteed wrote that he had heard gossip from London that the second edition of the *Principia* would not contain a new theory of the moon because Flamsteed would withhold data, Newton exploded in a letter of 19th July 1695 [Turnbull 1959–77, Vol. IV, p. 143]. Flamsteed answered on 28th July 1695 [Turnbull 1959–77, Vol. IV, p. 150f.]:

> '*I have just cause to complaine of the stile & expressions of your last letter, they are not freindly but that you may know me not to be of yt quarrelsome humor I am represented by ye Clerk of ye Society* [Halley]. *I shall wave all save this expression* that what you communicated to me was of more valew yn many observations. *I grant it. as ye Wier is of more worth then ye gould from which twas drawne: I gathered ye gould melted refined & presented it to you sometimes unasked I hope you valew not my paines ye less because they became yours so easily.*'

Although Flamsteed later tried to get closer again Newton did not react, so this last letter by Flamsteed marked the end to a new theory of the moon [Westfall 2006, p. 548]. Newton had reached the age of 53 with his birthday at the beginning of 1696. He had written the *Principia*, the manuscript of the *Opticks* was finished, and two mathematical works which he appended in 1704 to his *Opticks* which were supposed to show his importance as a mathematician. Anyway, he had failed on the theory of the moon[8]. Surely he blamed Flamsteed for his failure, but he must also have experienced his forces fading and that he was no longer capable of achieving a true breakthrough in this area. It was too late. Hence, it is not surprising that he decided to leave Cambridge in 1696.

[8]The motion of the moon shows anomalies like the *evection* or the *second anomaly*, a periodic perturbation of the orbit. Anomalies of this kind were known already in antiquity but are not easy to model (cp. [Verdun 2015, p. 118ff.]).

7.4.3 Newton and the Mint

It seems that Newton became tired of his reclusive life in Cambridge. He had looked for an appointment in London already at the beginning of the 1690s. With the rise of his patron and friend Charles Montague these plans could finally be realised.

Montague enrolled at Trinity College and was elected fellow in 1683. He was always on good terms with Newton and when he was made Chancellor of the Exchequer in 1694 there quickly were rumours that an occupation in the Royal Mint should be offered to Newton [Turnbull 1959–77, Vol. IV, p. 188]. On the 29th March 1696 Montague wrote to Newton [Turnbull 1959–77, Vol. IV, p. 195] to offer him a position as Warden of the Mint, and Newton immediately seized the opportunity. Less than a month later Newton took up his service but for the time being he kept his chair and fellowship at Trinity College.

The mint and the whole financial sector faced a deep crisis. In the year 1689 William of Orange was crowned to become the English King William III who had freed England from the reign of the hated catholic James II. Besides the Jacobites, the followers of James II, who would not accept William's reign, there was a war with France which ate up enormous amounts of money. A national bankruptcy would have led to unforeseeable consequences; the Glorious Revolution would have been doomed to failure at last. It was not clear

Fig. 7.4.9. Charles Montague (1661-1715), first Earl of Halifax [Painting: Michael Dahl, about 1700]

Fig. 7.4.10. Medieval mint; the coins were minted with a hammer by hand ([Hans Burgmair] probably 16th c)

in 1696 whether the country would finally be able to cover the costs but in this regard the new Warden of the Royal Mint could not help. Newton was rather dealing with another problem which also caused much trouble. Coins were minted by hand until 1662 and they had no defined edges. Hence the problem of clipping became epidemic, i.e. small parts were cut off the edges of coins which led to serious currency decline. Furthermore, the old coins were easily counterfeited since each coin minted by hand appeared to look different from another. The business of the coiners and clippers was very successful. Under the reign of Charles II a decree was issued that coins must be minted only by means of machines. The most important innovation was the knurling

Fig. 7.4.11. A gold coin minted by hand from the time of James I, and to the right a new one, minted with a machine and with knurled edge with reverse side from the reign of William III
[Classical Numismatic Group Inc., U.K]

of the edges which made clipping impossible because it became clearly visible. The larger coins additionally got the edge lettering 'Decus et Tutamen' (decoration and protection) which can still be seen on the English pound coins. A Frenchman called Peter Blondeau had developed the new machines at the mint and he knurled the coins himself until his death in 1672. The process of knurling stayed the secret of only a few men. Newton had to swear an oath that he would never reveal this secret; however, one could find all the details of the process in French sources already [Westfall 2006, p. 553].

When Newton became Warden of the Mint, the mint being installed in the Tower, both currencies existed side by side; the old, minted by hand, and the new. On the market another problem had occurred: the silver trade from England to Amsterdam and Paris [Levenson 2009, p. 110]. With the same amount of silver coins one could purchase more gold in Paris than in London. Hence, smart traders collected silver coins in London to melt them down and then bought gold on the continent with which they purchased silver coins in London – a kind of financial perpetual motion. Due to this trade large amounts of silver left the country and the new coins having trustable weight got short. Thereby the old currency got prevalent again and many counterfeit workshops flourished again. A vicious circle was established.

Apparently, Newton saw very clear from the beginning what had to be done. To eliminate the parallelism of the two currencies new coins had to be minted. That meant that the old silver coins had to be pulled off the market, melted down, and re-issued as new coins. This step alone would have solved the problem of clipping. To resolve the problems of exchange rates the ratio between the mere value of the metal and the nominal value had to be changed to stop the drain of silver leaving the country – we would call it depreciation today. In this way it could be guaranteed that the value of a silver coin in the country complied with its value abroad [Levenson 2009, p. 118]. In the summer of 1698 the whole process of new coinage was already completed – Newton had presented himself as the natural born organiser and had established a firm

Fig. 7.4.12. The Tower in London in which the Royal Mint was located ([Engraving by Wenzel Hollar, between 1637 and 1677] Thomas Fischer Rare Book Library, Plate No P908)

organisation of the mint against all odds [Westfall 2006, p. 566]. He had not acted like a Warden but like the highest officer of the mint, the Master of the Mint, Thomas Neale. When Neale died in 1699 Newton indeed became the new Master of the Mint.

7.5 Leibniz, his Fellow Campaigners, and the First Great Achievements of the Calculus

7.5.1 Leibniz again in Hanover

After Leibniz's return to Hanover in June 1690 from a long journey (two years and eight month) to Southern Europe it did not take long for him to feel cut off again from the scientific world.

In Hanover there were only Duchess consort Sophia (1630–1714) and her daughter Sophia Charlotte of Hanover (1668–1705) with whom he liked to conduct talks on scientific and philosophical issues. He held contact even after the marriage of Sophia Charlotte with Frederick III, Elector of Brandenburg and Duke in Prussia. When Frederick William, Elector of Brandenburg died, Frederick III became Duke of Prussia and on 18th January 1701 he ascended the throne to become the first King in Prussia as Frederick I. Leibniz corresponded further on with numerous persons in all of Europe about a confusingly large number of topics. In 1691 he became librarian in Wolfenbüttel

Fig. 7.5.1. Princess Sophia Charlotte 1668 ([Photo: James Steakley 2007] City museum Berlin) and the Dukes Rudolph Augustus ([Painting: H. Hinrich Rundt about 1700], Duke August Library Wolfenbüttel) and Antony Ulrich of Brunswick and Wolfenbüttel ([Painting: Christoph B. Franke] Duke Anton-Ulrich-Museum Brunswick)

while retaining his occupation in Hanover. In Wolfenbüttel he worked for the Dukes Anthony Ulrich (1633–1714) and Rudolf Augustus (1627–1704) who ruled mutually. These two Dukes were very art-minded and Leibniz felt that he was much better understood by these two than by his Duke Ernest Augustus (1629–1698) in Hanover. Between 1692 and 1696 he again worked on the mining problems in the Harz Mountains.

Together with Sophia Charlotte Leibniz adhered to a common plan of founding an academy in Berlin-Brandenburg; he also tried to convince the Tsar Peter I (1672–1725) of the founding of a Russian academy.

One fruit of his journey to Southern Europe began to ripen slowly and lead to a posthumous publication: the *Protogaea* [Leibniz 1949], the prehistory of the earth. Leibniz's interest in geology was already aroused by his works in the Harz Mountains [Waschkies 1999]. On his journeys he had seen fossils and he now went to develop a theory of the emergence and development of the earth. Only an announcement containing some excerpts found its way into the Acta Eruditorum in 1693. Not until 1749 was a Latin edition printed in Göttingen and a German edition appeared in Leipzig. The *Protogaea* was actually meant as an introduction to the history of the House of Welfs, but finally that did not materialise.

The first part of his work *Specimen Dynamicum pro admirandis naturae legibus circa corporum vires et mutuas actiones detegendis et ad suas causas revocandis* (Dynamic specimen to expose the admirable laws of nature with regard of the forces and mutual actions of bodies and to the reduction [of the laws of nature] to their causes) appeared in print in 1695 in the Acta Eruditorum. In it he set out his ideas concerning the science of dynamics.

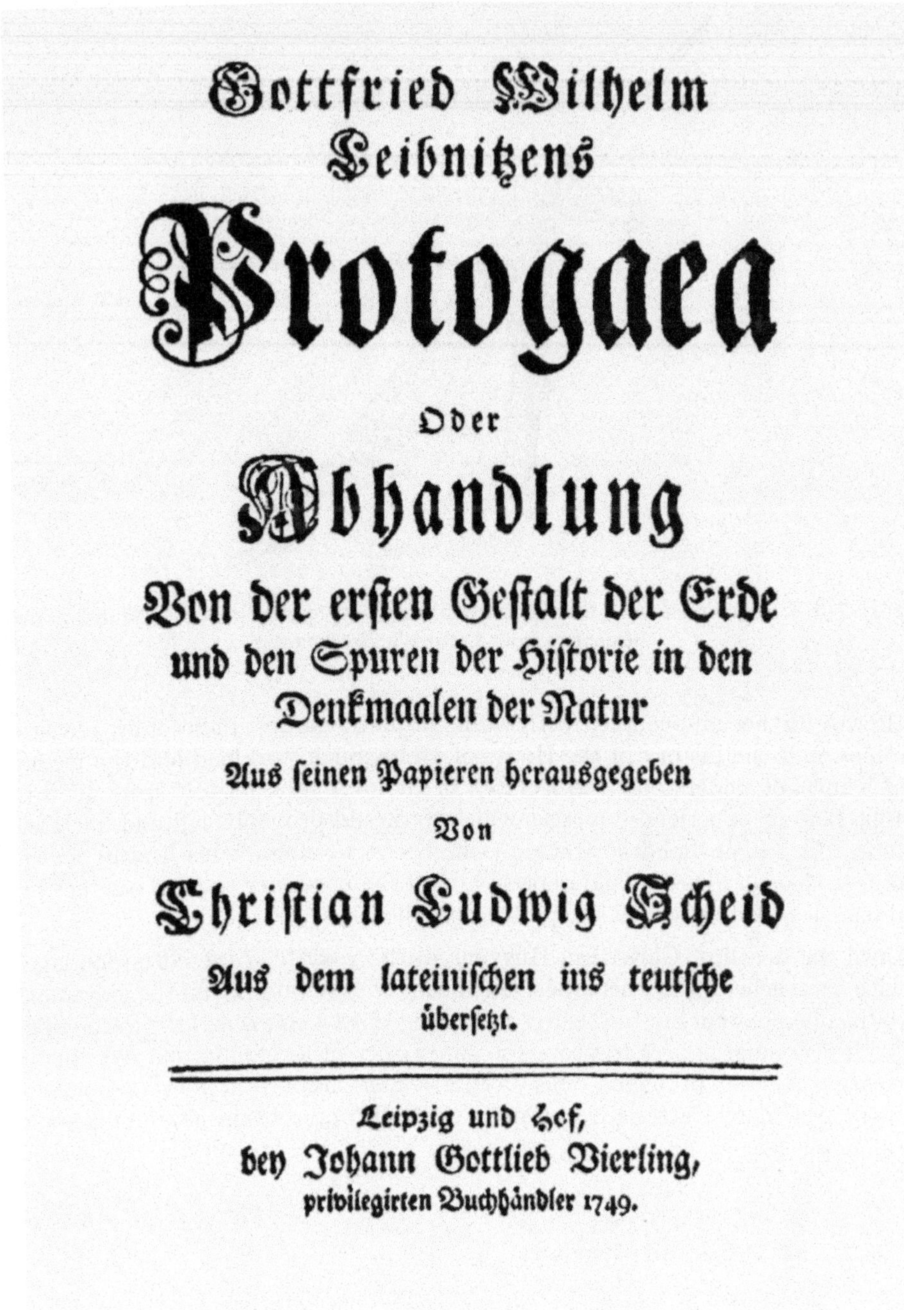

Gottfried Wilhelm
Leibnitzens
Protogaea
Oder
Abhandlung
Von der ersten Gestalt der Erde
und den Spuren der Historie in den
Denkmaalen der Natur
Aus seinen Papieren herausgegeben
Von
Christian Ludwig Scheid
Aus dem lateinischen ins teutsche
übersetzt.

Leipzig und Hof,
bey Johann Gottlieb Vierling,
privilegirten Buchhändler 1749.

Fig. 7.5.2. Title page of the German edition of the *Protogea* of 1749 (The Library Curtis Schuh's Bibliography of Mineralogy, Tucson, Arizona)

Fig. 7.5.3. A 'reconstruction' of a unicorn based on fossile finds (the Quedlinburg unicorn) from Leibniz's *Protogaea*

He was further on occupied with works in mathematics, philosophy, physics, ethics, and the history of the House of Welfs. Such workload and the feeling of being not understood and isolated in Hanover took its toll: from 1693 to 1696 Leibniz experienced a crisis which we would probably call mid-life crisis today. In a hypochondriacal manner he began to monitor his health; always in fear that an illness would deprive him of the necessary time to conduct his researches [Antognazza 2009, p. 322], [Görlich 1987, p. 117f.].

Until the death of Christiaan Huygens in the year 1695 the correspondence with his teacher on mathematics, the theory of gravitation, and the movement of the planets went on. In the first half of the 1690s Leibniz and the Bernoullies published some papers to prove the superiority of the Leibnizian calculus in treating concrete problems. This was particularly important in France because there were many Cartesians who stood critically opposite the new infinitesimal calculus.

7.5.2 The Bernoulli Brothers

The true triumph of Leibniz's differential and integral calculus on the continent began not least with the works of three of Leibniz's followers, the brothers Jacob and John Bernoulli, and the French aristocrat Guillaume François Antoine, Marquis de l'Hospital. As practised before by Pascal it became custom-

Fig. 7.5.4. Page from the manuscript *Specimen Dynamicum* (Gottfried Wilhelm Leibniz Library - Niedersächsische Landesbibliothek Hannover (Hanover), LH XXXV IX 4 Bl 1r)

Fig. 7.5.5. Relief of Jacob I Bernoulli (Medaillon [Photo: Mattes 2012] Historisches Museum Basel) and his brother John Bernoulli [Mezzotint by Johann J. Haid 1742, after a painting by Johann Rudolf Huber]

ary again from 1690 on to publicly challenge the mathematicians of Europe with mathematical exercises. This led to an unimaginable boost for the new calculus of Leibniz and a new mathematical science – the calculus of variations – emerged. However, in this kind of competition also lay the seed of conflicts.

Like his brother John (1667–1748) Jacob Bernoulli (1654–1705) came from a trading family. The grandfather, also a Jacob, was a chemist and trader of spices from Amsterdam who became a citizen of Basel by marriage.

The father of John and Jacob Bernoulli, Nicolaus Bernoulli, took over the successful business and became a member of the Grand Council; the mother Margaretha Schönauer was the daughter of a banker who was a member of the city council [DSB 1971, Vol. II, p. 46f., J.O. Fleckenstein]. The family was quite wealthy.

Jacob studied at the University of Basel and got a Magister Artium in philosophy in 1671 and in 1676 the theological licentiate *lic. theol.*; he thereby followed the wish of his father. However, against the wish of his father he studied mathematics and astronomy on the side. He went to Geneva as a tutor in 1676 and then spent two years in France where he became acquainted with the methods and science of Descartes's. Between 1681 and 1682 he undertook two educational trips to Holland and England. In Holland he met Jan Hudde, in England Robert Boyle and Robert Hooke. The result of this journey was a (defective) theory of comets (1682) and a theory of gravitation (1683) which were highly praised by his contemporaries [DSB 1971, Vol. II, p. 46, J.O. Fleckenstein]. In 1684 Jacob married Judith Stupanus who was the daughter of a wealthy pharmacist. Their son Nicolaus became a city councillor and head of the guild of artists in Basel.

In the year 1685 Jacob gave lectures on the mechanics of solid bodies and fluids in his hometown. He published in the Journal des Sçavans and the Acta Eruditorum and worked through the *Geometria* by René Descartes. He also worked on logic and wrote a treatise on probability theory. Through the study of works by Wallis and Barrow he was introduced to infinitesimal methods. Jacob became a master of calculus and pushed the development of Leibniz's differential and integral calculus forward; together with his brother John, but sometimes even in bitter dispute with him. Jacob was a slower, but deeper thinker than his younger brother.

In the meantime the twelve years younger brother John turned out to be useless concerning a career in his father's business. Grudgingly the father in 1683 gave him his permission to enrol at the University of Basel. The tenth child of his father finally acquired the degree of a Magister Artium after a disputation conducted by his brother.

Thereafter John devoted his time to study medicine while at the same time studying mathematics under his brother. In 1687 Jacob became professor of mathematics at the University of Basel. At that time the brothers worked through the publications of Leibniz and Tschirnhaus which had appeared in the Acta Eruditorum. Although Leibniz's first publication concerning differential calculus, *Nova methodus pro maximis et minimis*, published in 1684, made very hard reading, John later remarked in his autobiography [Hall 1980, p. 81]: *'It was for us only a matter of a few days to unravel all its secrets'*. John spent most of 1691 in Geneva where he taught the basics of the differential calculus to the brother of Fatio and where he enlarged his own knowledge of mathematics [DSB 1971, Vol. II, p. 52, E.A. Fellmann, J.O. Fleckenstein]. In the autumn of 1691 John went to Paris and got access to the scientific circle around Nicolas Malebranche (1638–1715). There he met the young aristocrat Guillaume François Antoine, Marquis de l'Hospital (1661–1704), a gifted mathematician who employed John as his tutor in mathematics. He also met Pierre de Varignon (1654–1722) in 1692 and made friends with him. In the following year John started the correspondence with Leibniz. The solution of the problem of the catenary from the year 1691 which we will discuss from page 328 on was John's first own work and catapulted him immediately to the level of Huygens, Leibniz, Newton, and Jacob Bernoulli.

In the year 1695 John became professor of mathematics at the University of Groningen. There was also the offer of a professorship at the University of Halle but Huygens took care that he, his wife Dorothea Falkner, and their seven months old son Nicolaus I[9] came to Groningen [DSB 1971, Vol. II, p. 53, E.A. Fellmann, J.O. Fleckenstein]. At that time the relationship with his brother Jacob deteriorated. Jacob was sensitive, easily irritable; he liked to criticise others and was addicted to recognition. John was also – perhaps even

[9]The scientists of the Bernoulli family are so numerous that one has to give numbers to the several family members with the names Jacob, John, and Nicolaus.

Fig. 7.5.6. Daniel Bernoulli [Painting by J. Haid after R. Huber, 18th c] and title page of his *Hydrodynamica* of 1738

more than his brother – argumentative and combative. In Jacob's eyes John was his pupil, and the pupil could only repeat what he had learned from his teacher. But John was an independent spirit. He was surely the more gifted of the two brothers concerning mathematical intuition, and he was the faster thinker. The calculus of variations[10] is very significantly the product of the heated conflict between the two brothers.

The rivalry began around the year 1692 when Jacob introduced his brother John to an interesting problem: In what form will the sail of a ship swell when the wind moves in? Jacob had solved the problem and had shown that the form of the sail will be that of a cycloid. John also found a solution which he wholeheartedly published in the Journal des Sçavans. This upset his older brother. In 1696, now on a par with his older brother, John proposed the problem of the brachistochrone curve which we will look at shortly in some detail. Jacob could solve the problem, but proposed yet another under the title *'Solution of the problem of my brother whom I offer another one'*. John apparently underestimated the problem and published a wrong solution [DSB 1971, Vol. II, p. 53, E.A. Fellmann, J.O. Fleckenstein], on which Jacob showered him with bitter scorn. Therewith the conflict escalated and ended only with Jacob's death.

[10]The name calculus of variations stems from the fact that a whole suite of functions has to be 'varied' in order to find one with some extremal properties.

When Jacob died in 1705 at age 50 John followed him on the professorship in Basel. After Newton's death in 1727 John Bernoulli became the unchallenged leading mathematician in Europe. His famous pupil, the great Leonhard Euler (1707–1783), significantly advanced Leibniz's analysis in the 18th century and penetrated the science of mechanics of solid and fluid bodies completely.

The brothers Jacob and John Bernoulli became the founding fathers of a whole dynasty of natural scientists and mathematicians which reached well into our age. Particular celebrity was acquired by one of John's sons, Daniel Bernoulli (1700–1782).

Daniel studied medicine in Basel and Heidelberg and was appointed to the Russian Academy of Sciences in St Petersburg in 1725 together with his brother Nicolaus I but he did not like it there. Back in Basel he held the chairs of anatomy and medicine, but since his heart was with mathematics and the natural sciences and since he had already written outstanding works in these areas he took hold of the chair of physics when it was offered him in 1750. His main oevre, the *Hydrodynamica*, was already published in 1738, and it throws a characteristic light on his father John that he plagiarized his son's book under the title *Hydraulica* and redated it seven years back [Szabó 1996, p. 165ff.]. Daniel Bernoulli was a close friend of Leonhard Euler.

7.5.3 Marquis de l'Hospital

De l'Hospital (1661–1704) originated from an esteemed family of French nobility and his mathematical talent was recognised early on. He solved a problem by Pascal concerning the cycloid when he was just 15 years old. However, his status required him to become an officer and indeed he became an officer of the cavalry. This was a profession for which he was not suited due to his severe myopia. Hence, he gave up the career in the army to devote himself to mathematical studies. About the year 1690 he belonged to the circle around Nicolas Malebranche in Paris where he met John Bernoulli in 1691. Fascinated by Leibniz's differential calculus which was presented to the Paris circle by Bernoulli he payed to get private tutorship by John Bernoulli in Paris and at de l'Hospital's chateau Oucques in what is now the canton of Marchenoir.

Thereafter de l'Hospital and John Bernoulli stayed in correspondence and the Marquis payed Bernoulli for sending him new mathematical results. The well-known formula named after de l'Hospital actually came from John Bernoulli. In the year 1693 de l'Hospital was accepted as a member of the French Acadêmie des Sciences. Since he could not have been accepted without problems as he came from aristocracy he was made vice president. He corresponded with Leibniz, Huygens, and the Bernoulli brothers, and published the first textbook[11] on Leibniz's differential calculus, *Analyse des Infiniment Petits*

[11]This statement is not quite correct historically since Bernard Nieuwentijt had published such a textbook before which had but no impact (cp. p. 443).

Fig. 7.5.7. Guillaume François Antoine, Marquis de l'Hospital and Louis XIV at a visit of the Academy of Sciences 1671 [Engraving: S. Le Clerc]

pour l'Intelligence des Lignes Courbes (Analysis of the infintely small [quantities] for the understanding of curved lines) [Bradley et al. 2015].

The book was comprised of some 200 pages; it contained the differential calculus only and was published anonymously, although everybody knew that the Marquis de l'Hospital was the author [Truesdell 1958, p. 59]. It contained original material by John Bernoulli, saw many editions, and remained the defining text until the works of Euler were published. John Bernoulli wrote to Leibniz, Varignon, and others that he was actually the author of the book and that de l'Hospital had freely used the first part of the introduction to the differential and integral calculus which Bernoulli had lent to the Marquis in Paris. That had been part of the financial arrangement between Bernoulli and the Marquis, but in his autobiography John Bernoulli wrote less positively and with a growing worry concerning the contents of the book. In the year 1742 John Bernoulli published part II, the integral calculus, but no part I, because, as he remarked, this part was already published by de l'Hospital.

De l'Hospital expressed his commitment to John Bernoulli clearly in the preface of the *Analyse des Infiniment Petits* and remarked, that the book contained the results of different persons. In the text half a dozen persons were named but not John Bernoulli [Truesdell 1958, p. 59].

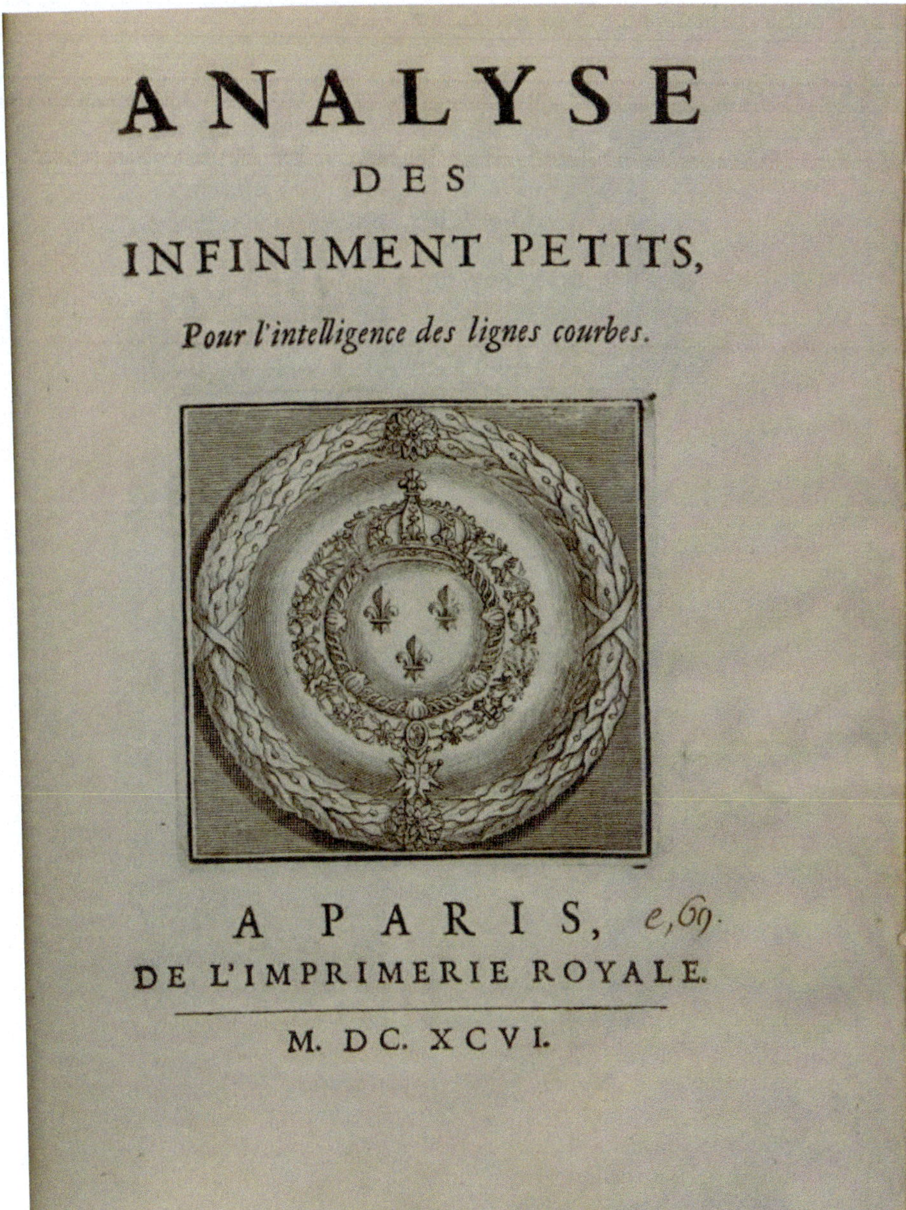

ANALYSE
DES
INFINIMENT PETITS,
Pour l'intelligence des lignes courbes.

A PARIS, e,69.
DE L'IMPRIMERIE ROYALE.
M. DC. XCVI.

Fig. 7.5.8. Title page of *Analyse des Infiniment Petits* by Guillaume François Antoine, Marquis de l'Hospital, 1696 (Courtesy of Sophia Rare Books, Copenhagen)

The Problem of the Isochrone

Already in 1659 Christiaan Huygens had shown that the cycloid was the tautochrone, i.e. the curve on which a point of mass moves frictionless from any starting point to a fixed endpoint. As we have reported in section 2.7, Huygens had used this property of the cycloid in his construction of an accurate clock. In autumn 1687 Leibniz proposed another problem in the journal 'Nouvelles de la République des lettres': find a curve $x \mapsto y(x)$ on which a point of mass slides with constant vertical velocity $dy/dt = -b$ [Hairer/Wanner 2000, p. 134f.]. We employ the negative sign here because we draw the y-axis upward as in figure 7.5.9.

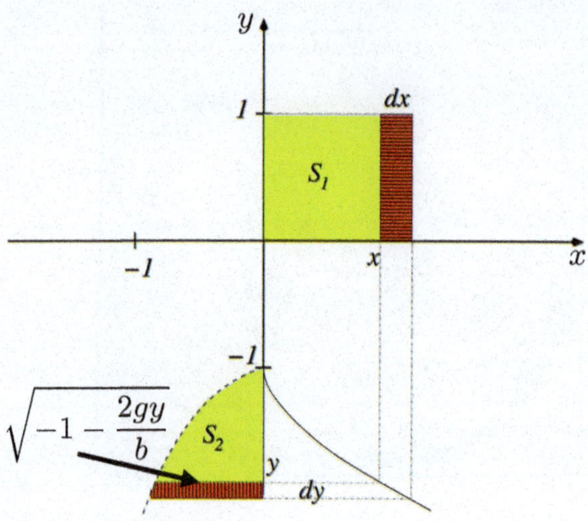

Fig. 7.5.9. Concerning the construction of Leibniz's isochrone

Already after one month a solution by Huygens arrived which was unsatisfactory however, since Huygens guessed the solution and then proved that the guessed function in fact solved the problem [Hairer/Wanner 2000, p. 134]. A further solution obtained by means of a general method of solution was published by Jacob Bernoulli in the Acta Eruditorum: *Analysis problematis ante hac propositi, de inventione lineae descensus a corpore gravi percurrendae uniformiter, sic ut temporibus aequalibus aequales altitudines emetiatur: & alterius cujusdam Problematis Propositio* (Analysis of a formerly posed problem to find the uniformly passing line of a heavy body, so that equal heights are traversed in equal times, and proposal of a certain other problem). We follow Jacob Bernoulli's solution approach.

Galilei had shown that a body in the gravitational field moves along the y-axis with velocity

$$v = \sqrt{-2gy},$$

where $g = 9.81 \frac{m}{s^2}$ denotes the gravitational acceleration. If s denotes the arc length of the sought curve then

$$\left(\frac{ds}{dt}\right)^2 = -2gy$$

has to hold, and since after Pythagoras's theorem it holds $ds^2 = dx^2 + dy^2$, it follows

$$\frac{dx^2 + dy^2}{dt^2} = -2gy.$$

We know that $dy/dt = -b$ has to hold, and dividing the above equation by $(dy/dt)^2 = b^2$ yields

$$\frac{(dx^2 + dy^2)dt^2}{dy^2\, dt^2} = \left(\frac{dx}{dy}\right)^2 + 1 = -\frac{2gy}{b^2},$$

hence

$$dx = -\sqrt{-1 - \frac{2gy}{b^2}}\, dy.$$

Bernoulli interpreted this equation as the equality of two rectangles. The area $1 \cdot dx$ is always exactly as large as the area $-\sqrt{-1 - \frac{2gy}{b^2}} \cdot dy$ and that is the equality of areas S_1 and S_2 in figure 7.5.9. Bernoulli had solved the differential equation for the sought curve by the method of separation of variables and needed only to integrate. He wrote:

> *'Hence also their integrals will be made equal'*
>
> (Ergo & horum Integralia aequantur)

and this is the first time that the word 'integral' appeared in mathematics. By the way the solution is Neile's parabola

$$x = \frac{b^2}{3g}\left(-1 - \frac{2gy}{b^2}\right)^{\frac{3}{2}}.$$

The Problem of the Tractrix

Already in his time in Paris Leibniz learned of a problem of the anatomist, physician, and architect Claude Perrault (1613–1688): which curve is the trajectory of a pocket watch if it is pulled at the fob chain with constant length along a horizontal line? The sought curve is the so-called minimum turning curve or tractrix.

With the notations from figure 7.5.10 we have

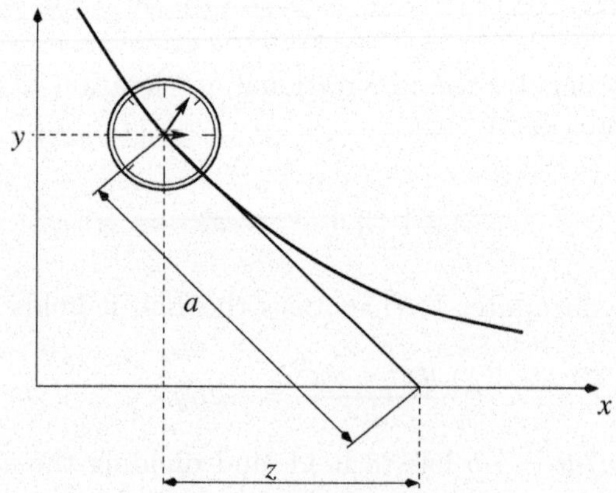

Fig. 7.5.10. The problem of Perrault

$$\frac{dy}{dx} = -\frac{y}{z}$$

and from the theorem of Pythagoras it follows

$$z^2 + y^2 = a^2.$$

Therewith we arrive at the differential equation

$$y' = \frac{dy}{dx} = -\frac{y}{\sqrt{a^2 - y^2}},$$

which Leibniz had solved in *Supplementum geometriae dimensioriae, seu generalissima omnium Tetragonismorum effectio per motum: Similiterque multiplex constructio lineae ex data tangentium conditione* (A supplement of the gauging geometry or the most general realisation of all tetragonisms by means of movement and the polymorphic construction of a line from a given tangent condition effectuated by a similar kind) in the Acta Eruditorum by means of separation of variables in 1693.

The solution is given by the function

$$x = -\sqrt{a^2 - y^2} - a \ln \frac{a - \sqrt{a^2 - y^2}}{y}.$$

The Problem of the Catenary

In the same work of 1690 in which Jacob Bernoulli solved the problem of the isochrone of Leibniz Bernoulli posed yet another problem:

What form is taken by an idealised chain under its own weight if it is fixed at the ends at the same height?

Galileo Galilei had described the following experiment in his *Discorsi* [Galilei 1989, p. 143]:

> *'The other way to draw on the prism the line we seek is to fix two nails in a wall in a horizontal line, separated by double the width of the rectangle in which we wish to draw the semiparabola. From these two nails hang a fine chain, of such length that its curve* [sacca] *will extend over the length of the prism. This chain curves in parabolic shape, so that if we mark points on the wall along the path of the chain, we shall have drawn a full parabola.'*

In [Galilei 1973, p. 123] the German tranlsation reads:

> *'This chain has the shape of a parabola, ...'*
>
> (Diese Kette hat die Gestalt einer Parabel, ...)

The editor of this old German translation of the *Discorsi*, Arthur von Oettingen, has added a remark at the end of the last sentence cited [Galilei 1973, Anmerkung 23, p. 139]:

> *'It is well known that this is a mistake since the chain takes the form of the so-called catenary which has only superficial similarity with a parabola.'*
>
> (Bekanntlich ist das ein Irrthum, da die Kette die Form der sogenannten Kettenlinie bildet, welche nur äussere Aehnlichkeit mit der Parabel hat.)

Knobloch in his work on Galilei and Leibniz [Knobloch 2012, p. 12] has criticised the translation of Arthur von Oettingen which stems from the 19th century. Galileo did not write *'this chain is a parabola'*, but *'This chain curves in parabolic shape'*[12] and Knobloch has rightly remarked that a human being who has the appearance of a child has not necessarily to be a child.

Although Leibniz had written in *Communicatio suae pariter, duarumque alienarum ad edendum sibi primum a Dn. Jo. Bernoullio, deinde a Dn. Marchione Hospitalio communicatarum solutionum problematis curvae celerrimi descensus a Dn. Jo. Bernoullio Geometris publice propositi, una cum solutione sua problematis alterius ab eodem postea propositi* (Communication of both his own solution as well as solutions of two other men, communicated to him first by Mr John Bernoulli, then by Marquis de l'Hospital, to be published as communicated solutions of the problem of the curve of fastest decent which Mr John Bernoulli had presented publicly to the geometers, together with his solution of a second problem, presented by the very same) [Leibniz 2011, p. 297ff.] in the Acta Eruditorum of May 1697 (to which we will come back later on) [Knobloch 2012, p. 11]:

[12] In German: Galileo did not write *'Diese Kette hat die Gestalt einer Parabel'* but *'Dieses Kettchen biegt sich in parabolischer Gestalt'*.

Fig. 7.5.11. The 'Gateway Arch' in St. Louis, USA, the 'Gateway to the West', was built in the shape of a reversed catenary
[Photo: Bev Sykes from Davis, CA, USA, 2005]

> *'Galilei was in fact a highly ingenious and discriminating man. But since in his day the analytical art was not yet sufficiently developed and its higher or infinitesimal part lay still in the dark, he could not hope for such solutions. He conjectured however that the catenary was a parabola and the line of fastest descent a circle. But he went grossly astray.'*
>
> (Galilei war in der Tat ein höchst geistreicher und urteilsfähiger Mann. Aber da zu seiner Zeit die analytische Kunst noch nicht hinreichend fortentwickelt war, ihr höherer oder infinitesimaler Teil noch im Dunkeln lag, durfte er derartige Lösungen nicht erhoffen. Er vermutete freilich, dass die Kettenlinie eine Parabel und die Linie des kürzesten Abstiegs ein Kreis ist. Aber er ging im höchsten Maße fehl.),

but this could not have been deduced from Galileo!

In the discussion of the movement of a thrown body and the trajectory of a thrown mass, respectively, Galileo came back again to the catenary [Galilei 1989, p. 256 f.]:

> *'But I wish to cause you wonder and delight together by telling you that the cord thus hung, whether much or little stretched, bends in a*

line that is very close to parabolic. The similarity is so great that if you draw a parabolic line in a vertical plane surface but upside down – that is, with the vertex down and the base parallel to the horizontal – and then hang a little chain from the extremities of the base of the parabola thus drawn, you will see by slackening the little chain now more and now less, that it curves and adapts itself to the parabola; and the agreement will be closer, the less curved and the more extended the parabola drawn shall be. In parabolas with an elevation of less than 45°, the chain will go almost exactly along the parabola.'

The German translation by Knobloch in [Knobloch 2012, S. 12], staying close to the original text, is very close to Stillman Drake's translation cited above.

Hence, Galileo had by no means claimed that the catenary *is* a parabola. István Szabó [Szabó 1996, p. 486] had seen a contradiction in the two citations [Galilei 1989, p. 143] and [Galilei 1989, p. 256 f.] (in the old German translation) and tried to resolve this contradition by conjecturing that the latter part was probably written later than the first and that a correction of the first part remained undone. With Knobloch and Drake we have to state here that such contradictions simply do not exist in Galileo's writings.

The proof that the catenary can not be a parabola had been found already by the French Jesuit priest Ignatius Gaston Pardies (1636–1673) (who had pestered Newton in 1671/72 with his critique of the theory of light) in the 1660s, and even earlier by the Lübeck born Joachim Jungius (1587–1657) [Jahnke 2003a, p. 109]. It was also clear to the young Huygens that the solution could not be a parabola. Solutions to this problem were given by Huygens, Leibniz, and Jacob's brother John Bernoulli in the Acta Eruditorum in 1691. The catenary or catenoid proved to be a hyperbolic function, namely the hyperbolic cosine, which results as a solution of a certain differential equation.

The solution of Bernoulli is based on an idea by Pardies. The forces at two arbitrary points A and B of the chain are the same of those of a light thread (i.e. the weight can be neglected) tangential with respect to the chain in A and B and loaded with the weight of the cord D as shown in figure 7.5.12.

Considering only one half of the catenary and shifting the point C into the apex B, the weight of the chain is located as shown in figure 7.5.13 at point E. Using the notations in figure 7.5.13 the curve BA is the sought one. Let their arc length be denoted by s. Furthermore, let $Gg = dy$ and $Ha = dx$. It follows from the parallelogram of forces, that

$$w : H = EL : AL = AH : Ha = Gg : Ha = dy : dx,$$

and since w is proportional to the arc length s and since the horizontal force H has to be constant, it follows

$$c\frac{dy}{dx} = s,$$

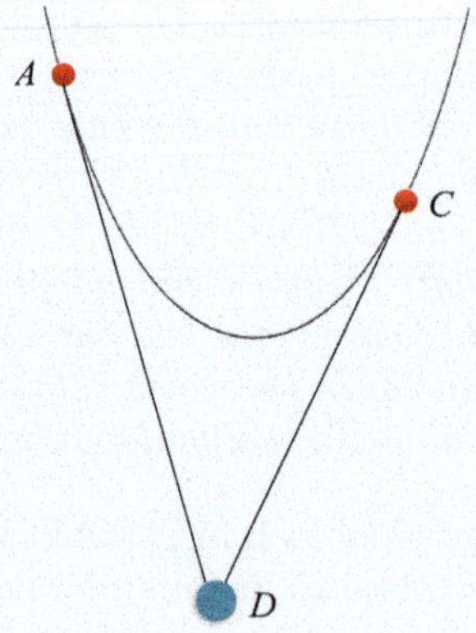

Fig. 7.5.12. Catenary with a thread tangential to the chain at the points A and C

where c contains the constant of proportionality in w and the constant of the horizontal force. The slope of the sought curve has thus to be proportional to the arc length.

At this point the computations of Bernoulli get confusing [Hairer/Wanner 2000, p. 137] since he employed differentials of second order. The resulting solution is

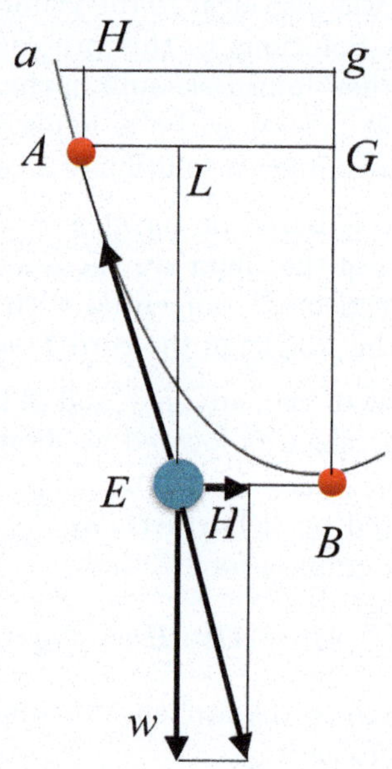

Fig. 7.5.13. Catenary with a thread tangential in the points A and B

$$y = K + c \cosh \frac{x}{c}$$

with K being a constant of integration.

The Problem of the Brachistochrone Curve

We have already cited Leibniz on page 330 who wrote:

> *'Galilei was in fact a highly ingenious and discriminating man. [...] He conjectured however that the catenary was a parabola and the line of fastest descent a circle. But he went grossly astray.'*

That Galileo had by no means taken the catenary for a parabola was already mentioned, but Leibniz was also wrong in the question of the brachistochrone. Never had Galileo posed the problem of the brachistochrone but rather shown that the movement on a quarter circle was always faster than that on any inscribed polygonial course. This is a completely different problem (which Galileo solved correctly!). Compare the statements by Szabó [Szabó 1996, p. 490ff.] and Knobloch [Knobloch 2012, p. 14ff.].

In June 1696 John Bernoulli published his invitation *Problema novum ad cuius solutionem mathematici invitantur* (New problem to the solution of which the mathematicians are invited) in the Acta Eruditorum, and he repeated it in his announcement *Programma, editum Groningae anno 1697*. In the invitation he wrote [Stäckel 1976, p. 3]:

> *'If in a vertical plane two points A and B are given, one should assign a path AMB to the moveable point M on which it can get from A to B only by its own gravity in the shortest time.*
>
> [...] *To counter a premature judgement it may be remarked that although the straight line AB is the shortest between A and B it will not be traversed in shortest time. Arguably the curve AMB is well known to the geometers and I will specify it if no one else will do it in the course of this year.'*
>
> (Wenn in einer verticalen Ebene zwei Punkte A und B gegeben sind, soll man dem beweglichen Punkte M eine Bahn AMB anweisen, auf welcher er von A ausgehend vermöge seiner eigenen Schwere in kürzester Zeit nach B gelangt.
>
> [...] Um einem voreiligen Urtheile entgegenzutreten, möge noch bemerkt werden, dass die gerade Linie AB zwar die kürzeste zwischen A und B ist, jedoch nicht in kürzester Zeit durchlaufen wird. Wohl aber ist die Curve AMB eine den Geometern sehr bekannte, die ich angeben werde, wenn sie nach Verlauf dieses Jahres kein anderer genannt hat.)

Hence, the sought curve is the brachistochrone (brachistos = shortest, chronos = time).

In January 1697 the invitation [Stäckel 1976, p. 3f.] appeared as a flyer in Groningen:

> *'John Bernoulli, official professor of mathematics, sends greetings to the most acute mathematicians on the globe.*
>
> *Since experience shows that noble minds are driven to work on the increase of knowledge by nothing more than by presenting to them difficult and at the same time useful tasks so that they can make a name for themselves by solving them and to appear as everlasting monument to posterity.* [...]
>
> *Now I have posed such a problem half a year ago in the Leipzig Acta Eruditorum,* [...] *A limit of six months was granted the geometers from the day of the publication* [of the problem], *and if there should not have been a solution sent in within this period of time I promised to communicate mine. This period of time has elapsed and not a trace of a solution has appeared. Only the famous Leibniz, who has served higher geometry so well, informed me by letter that he happily had resolved the knot of this, as he wrote, very beautiful and hitherto unheard-of problem. He asked me to extend the deadline until next Easter so that the problem may meanwhile be published in France and Italy so that nobody would have reason to lament on a too tight determination of the time period.'*
>
> (Die scharfsinnigsten Mathematiker des ganzen Erdkreises grüsst Johann Bernoulli, öffentlicher Professor der Mathematik.

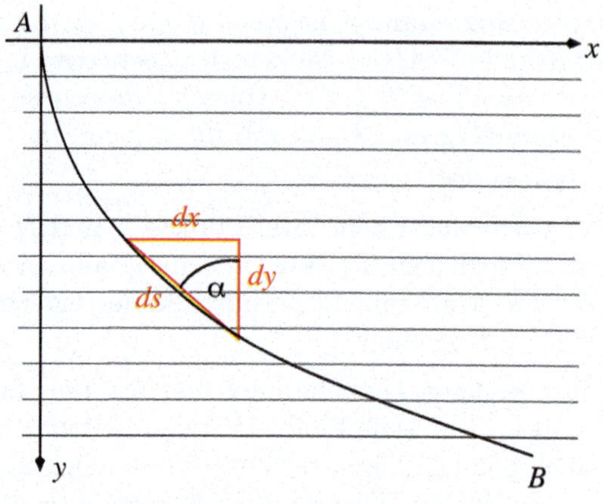

Fig. 7.5.14. The solution approach of John Bernoulli via the optical law of refraction. The characteristic triangle is much enlarged

Da die Erfahrung zeigt, dass edle Geister zur Arbeit an der Vermehrung des Wissens durch nichts mehr angetrieben werden, als wenn man ihnen schwierige und zugleich nützliche Aufgaben vorlegt, durch deren Lösung sie einen berühmten Namen erlangen und sich der Nachwelt ein ewiges Denkmal setzen, [...]

Nun habe ich vor einem halben Jahre im Junihefte der Leipziger Acta Eruditorum eine solche Aufgabe vorgelegt, [...] Sechs Monate Frist vom Tage der Veröffentlichung ab wurde den Geometern gewährt, und wenn bis dahin keine Lösung eingelaufen wäre, versprach ich die meinige mitzutheilen. Verflossen ist dieser Zeitraum, und keine Spur einer Lösung ist erschienen. Nur der berühmte, um die höhere Geometrie so verdiente Leibniz theilte mir brieflich mit, dass er den Knoten dieses, wie er sich ausdrückte, sehr schönen und bis jetzt unerhörten Problems glücklich aufgelöst habe, und bat mich freundlich, die Frist bis zum nächsten Osterfeste ausdehnen zu wollen, damit die Aufgabe inzwischen in Frankreich und Italien veröffentlicht werden könne, und Niemand Veranlassung hätte sich über eine zu enge Bemessung des Zeitraums zu beklagen.)

John Bernoulli then announced the extension of the time period and repeated again the task for all those who did not have access to the Acta Eruditorum. He sent the announcement also to London to the Philosophical Transactions and to Wallis and Newton [Westfall 2006, p. 582].

The problem of the brachistochrone marked the actual starting point of the calculus of variations[13]. It is an application of the differential calculus to *families* of functions in order to compute functions with certain extremal properties. Before we go on in our description it is worthwhile to take a look at John Bernoulli's very elegant solution.

Bernoulli divided a virtual medium in infinitesimally small horizontal strips as in figure 7.5.14 and imagined that a ray of light with speed $v = \sqrt{2gy}$ (analogously to Galileo's velocity of fall!) would be refracted in each strip according to Fermat's principle

$$\frac{v}{\sin\alpha} = K,$$

where K denotes some constant. If s denotes the arc length of the sought curve which we can write in the form

$$ds = \sqrt{dx^2 + dy^2}$$

due to Phythagoras's theorem, it holds that $\sin\alpha = dx/ds$. Hence it follows

[13]Before that Newton in the *Principia* had already computed the body of revolution experiencing the least resistance but he used wrong assumptions on the fluid. However, this solution had no influence on the development of the calculus of variations.

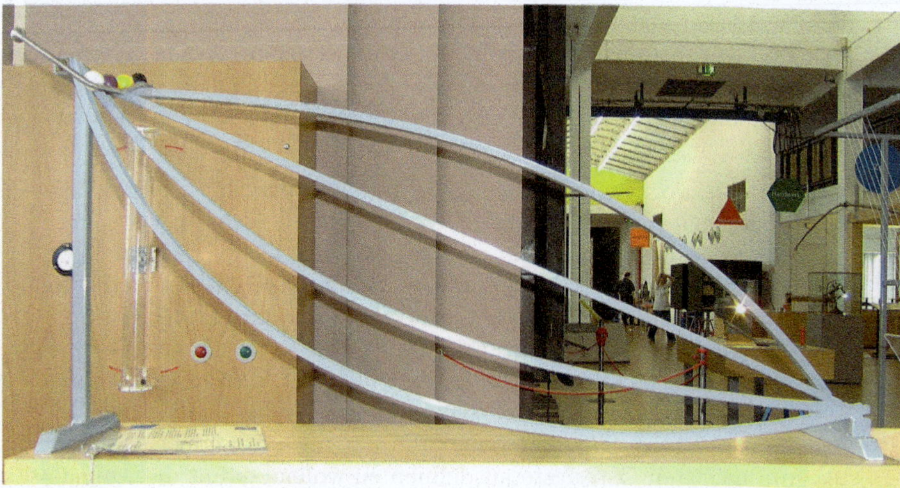

Fig. 7.5.15. An installation of four paths in the Landesmuseum für Technik und Arbeit in Mannheim. All four spheres can be started at the time; the sphere on the brachistochrone (third path from above) is the fastest

$$\frac{v}{\sin\alpha} = \frac{v\,ds}{dx} = \frac{v\sqrt{dx^2+dy^2}}{dx} = \frac{v\,dx\sqrt{1+\frac{dy^2}{dx^2}}}{dx} = v\sqrt{1+\frac{dy^2}{dx^2}} = K,$$

i.e.

$$\sqrt{1+\frac{dy^2}{dx^2}}\sqrt{2gy} = K.$$

Taking squares this yields

$$y + y\frac{dy^2}{dx^2} = \frac{K^2}{2g} =: c,$$

so that

$$dx = \sqrt{\frac{y}{c-y}}\,dy$$

follows, where we have again separated variables and hence can integrate directly. With the help of the substitution $y = c\sin^2 u = \frac{c}{2} - \frac{c}{2}\cos 2u$ we arrive at the solution

$$x - x_0 = cu - \frac{c}{2}\sin 2u$$

and this is the common cycloid, where x_0 denotes the constant of integration. The brachistochrone hence is the tautochrone which surely had delighted Bernoulli's contemporaries.

The relationship between the Bernoulli brothers was already ruined at that time. Jacob's solution of the problem of the brachistochrone was much more general than that of his brother and carried within it already more than only

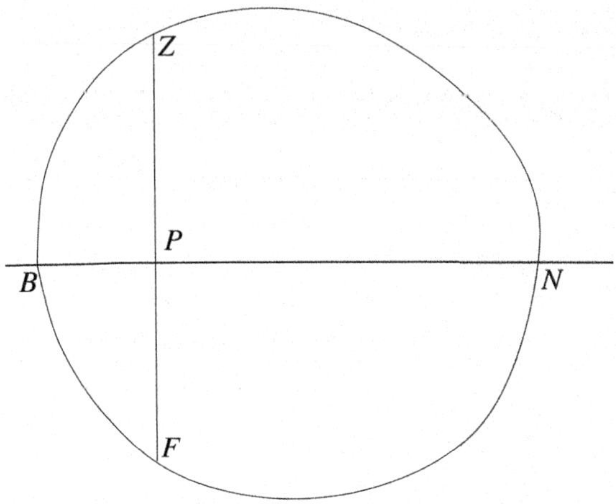

Fig. 7.5.16. The isoperimetric problem with which Jacob challenged his brother John

the seed of generalisation to other problems. The title which Jacob gave his solution in the Acta Eruditorum of May 1697 is symptomatic: *Solutio problematum fraternorum cum propositione reciproca aliorum* (Solution of the brotherly problems with a reversed proposal of others [Lösung der brüderlichen Probleme mit einem umgekehrten Vorschlag anderer]) [Stäckel 1976, p. 14ff.]. The new problem established a whole class of variational problems, the so-called 'isoperimetric problems' (isoperimetric = with the same circumference). In the classical version (Dido's problem) one asks for a closed curve with given length so that the enclosed area is maximal. Numerous generalisations of this problems are also called isoperimetric problems. Jacob gave the following description of these problems [Stäckel 1976, p. 19], cp. figure 7.5.16:

> *'Among all isoperimetric figures above a common basis BN the curve BFN should be determined which not itself encloses the largest area, but another curve BZN, the ordinate PZ of which is proportional to some power or root of the line PF or of the arc BF.'*
>
> (Unter allen isoperimetrischen Figuren über der gemeinsamen Basis *BN* soll die Curve *BFN* bestimmt werden, welche zwar nicht selbst den grössten Flächeninhalt hat, aber bewirkt, dass es eine andere Curve *BZN* thut, deren Ordinate *PZ* irgend einer Potenz oder Wurzel der Strecke *PF* oder des Bogens *BF* proportional ist.)

7.6 The Monkey Attacks

7.6.1 Wallis's Algebra of 1693

In the year 1693 the second volume of Wallis's *Opera mathematica* was published including a new edition of the *Algebra*. Thereby the second volume appeared in print two years *before* the first volume was published. The first edition of *Treatise on Algebra. Both historical and practical* was published in English in 1685 and now a new edition in Latin was to follow. All what could be read in Wallis's second volume concerning fluxions – and it was little enough – came from Newton's quill [Hall 1980, p. 96]. At the end of a few pages on fluxions Wallis wrote:

> *'Analogous to this method is the differential method of Leibniz and that other method, older than either, which Barrow expounded in his Geometrical Lectures; and this is acknowledged in the Leipzig Transactions (January 1691) by a writer making use of a method similar to that of Leibniz ...'*

Of course, the 'Leipzig Transactions' are the Acta Eruditorum and 'a writer making use of a method similar to that of Leibniz's' is Jacob Bernoulli with his work *Specimen calculi differentialis in dimensione Parabolae helicoidis, ubi de flexuris curvarum in genere; earundem evolutionibus* (Specimen of the differential calculus for the measuring of the helical parabola, where curvature of curves are concerned in general, their evolutes, and others) published in January 1691. Concerning the theory of infinite series Wallis wrote unfair and wrongly that it was [Hall 1980, p. 93]

> *'long ago introduced by Mr. Isaac Newton and pursued by Nicolas Mercator, Mr. Leibniz and others.'*

The comparisons between the method of fluxions and the differential calculus which follow came certainly from Newton since they include identical sentences from a manuscript [Turnbull 1959–77, Vol. III, p. 222] of Newton of autumn 1692. Here is the text from the *Algebra* [Hall 1980, S. 94]:

> *'By fluents Newton understands indeterminate quantities which in the generation of a curve by the local motion are perpetually increased or diminished, and by their fluxions he understands the swiftness of their increase or decrease. And although at first glance fluents and their fluxions seem difficult to grasp, since it is usually a hard matter to understand new ideas; yet he thinks the notion of them quickly becomes more familiar than does the notion of moments or least parts or infinitely little differences;* [...] *Although he does not neglect the use of such parts but uses them only when by their means the work is to*

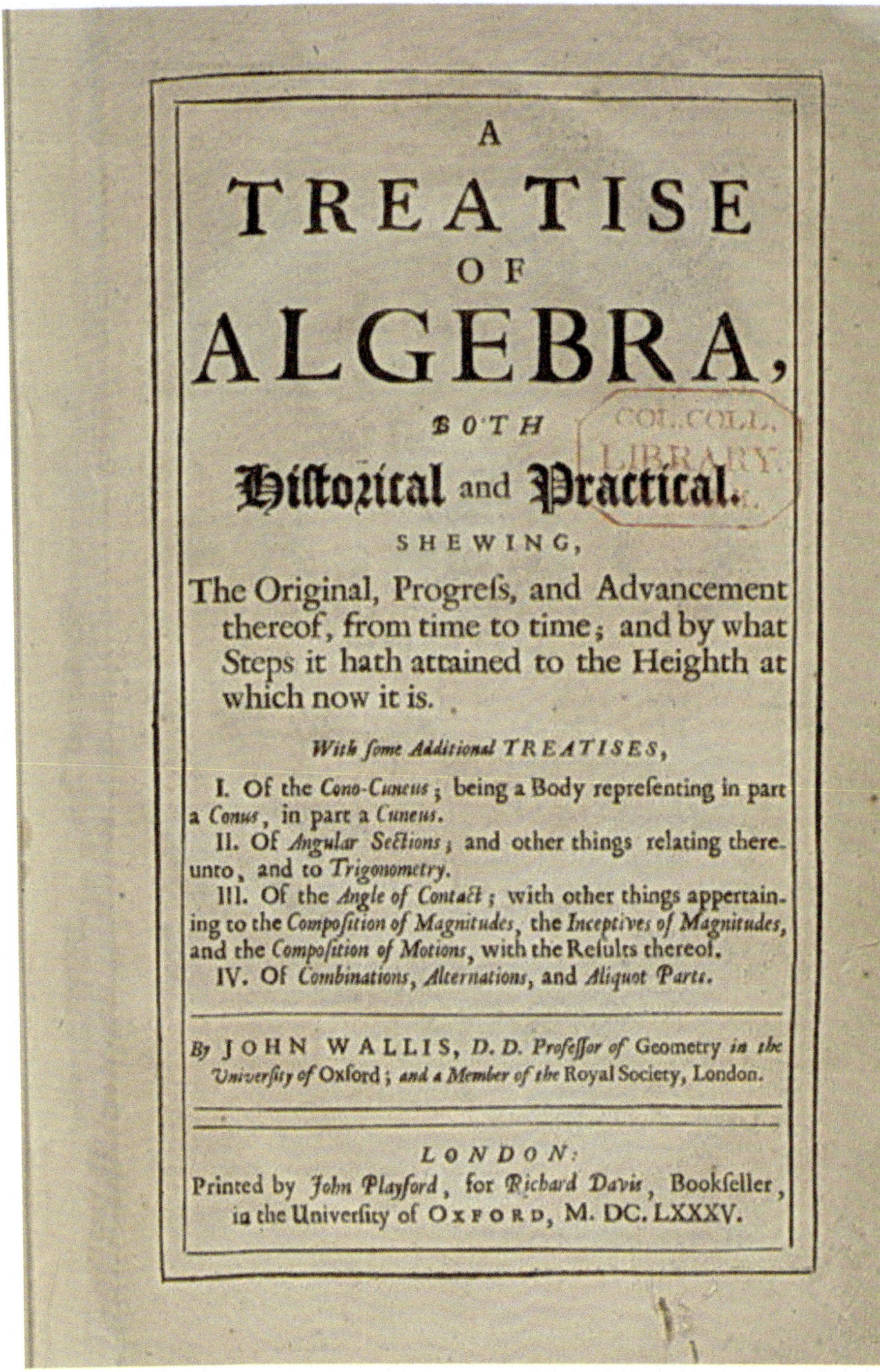

A

TREATISE

OF

ALGEBRA,

BOTH

Historical and Practical.

SHEWING,

The Original, Progreſs, and Advancement thereof, from time to time; and by what Steps it hath attained to the Heighth at which now it is.

With ſome Additional TREATISES,

I. Of the *Cono-Cuneus*; being a Body repreſenting in part a *Conus*, in part a *Cuneus*.

II. Of *Angular Sections*; and other things relating thereunto, and to *Trigonometry*.

III. Of the *Angle of Contact*; with other things appertaining to the *Compoſition of Magnitudes*, the *Inceptives of Magnitudes*, and the *Compoſition of Motions*, with the Reſults thereof.

IV. Of *Combinations*, *Alternations*, and *Aliquot Parts*.

By JOHN WALLIS, *D. D. Profeſſor of* Geometry *in the Univerſity of* Oxford; *and a Member of the* Royal Society, London.

LONDON:

Printed by *John Playford*, for *Richard Davis*, Bookſeller, in the Univerſity of OXFORD, M. DC. LXXXV.

Fig. 7.6.1. Title page of the English Algebra of the year 1685 [John Wallis, 1685] (Oxford University Press, 2002)

Fig. 7.6.2. Title page of the first volume of the *Opera mathematica* by John Wallis, 1695. The second volume contains the Latin *Algebra* (Sotheby's Picture Library, London)

> *be done more briefly and clearly, or leads to the discovery of the ratios of the fluxions.'*

All in all Wallis's description in the Latin *Algebra* of 1693 is nevertheless very fair. For the first time he used Newton's dot notation $\dot{x}$ in print and stated that all things concerning fluxions are made public for the first time. It is nowhere asserted or suggested that Leibniz was communicated anything concerning the calculus of fluxions in October 1676. There is not even a hint that Leibniz had taken anything from Newton or that Newton had helped him in some way [Hall 1980, p. 94]. In the preface of the first volume of the *Opera mathematica*, published in 1695, Wallis wrote [Hall 1980, p. 95]:

> *'Here* [in the second volume, i.e. the Latin *Algebra*] *is set out Newton's method of fluxions, to give it his name, which is of a similar nature with the differential calculus of Leibniz, to use his name for it, as anyone comparing the two methods will observe well enough though they employ different notations ...'*

But then in this first volume Wallis started to take sides in favour of Newton. He wrote that Leibniz had taken Newton's method from the two famous letters of the year 1676 [Hall 1980, p. 95], *'which were then communicated to Leibniz in almost the same words, where he explains this method to Leibniz, having been worked out by him more than ten years previously.'*

As we have already noticed Newton had *never* communicated his method to Leibniz – except in the anagrams there was nowhere even a word on fluxions and fluents – and Wallis got the material for his *Algebra* surely not from the *Epistolae* alone.

The most important British correspondent of Leibniz in the 1690s was the Scotsman Thomas Burnet (1656–1729), son of the Royal Physician in Scotland[14] who received a degree at the University of Leiden. Leibniz wrote to Burnet [Hall 1980, p. 95]:

'I am very satisfied with Mr. Newton, but not with Mr. Wallis who treats me a little coldly in his last works in Latin, through an amusing affectation of attributing everything to his own nation.'

Wallis always was a fighter for all matters English. He had defended the writings of Thomas Harriot's against Descartes and had fought in favour of Neile's priority over Heuraet. Now that he sensed that Newton's achievements ran into danger to be overlooked he fought for Newton.

Leibniz's review of volumes I and II of Wallis's *Opera mathematica* was published in the Acta Eruditorum VI of the year 1696 [Hofmann 1973, p. 259, footnote 86]. Concerning the second volume Leibniz wrote politely but determinedly, that he had been in possession of the infinitesimal calculus for 20 years [Hofmann 1973, p. 260]. Answering this review Wallis wrote still in 1696 that he knew long since that letters had been exchanged between Newton and Leibniz concerning the mathematics of infinitesimals but that he did not know any details. He asked Leibniz to send him copies. Leibniz could not comply to Wallis's request since some copies were lost and others he could not find in the disorder of his papers. Wallis emphasised yet again that Leibniz's differential calculus coincided strongly with Newton's calculus of fluxions. Leibniz protested and wrote that there were substantial differences [Hofmann 1973, p. 268].

Even before John Bernoulli saw Leibniz's review he remarked in a letter to Leibniz that the latter's calculus was not mentioned enough by Wallis (cp. page 301).

[14]We have to be very cautious when discussing people with this family name since there existed at least seven Scottish lineages with the name of Burnet(t) which produced numerous men with the name Thomas. The most famous of them is arguably the theologian Thomas Burnet (about 1635–1715). Our Thomas is a 'Burnett of Kemnay' (or Kemney). A short genealogy of the Burnets of Kemnay can be found on the internet: http://www.burnett.uk.com/kemnay.htm.

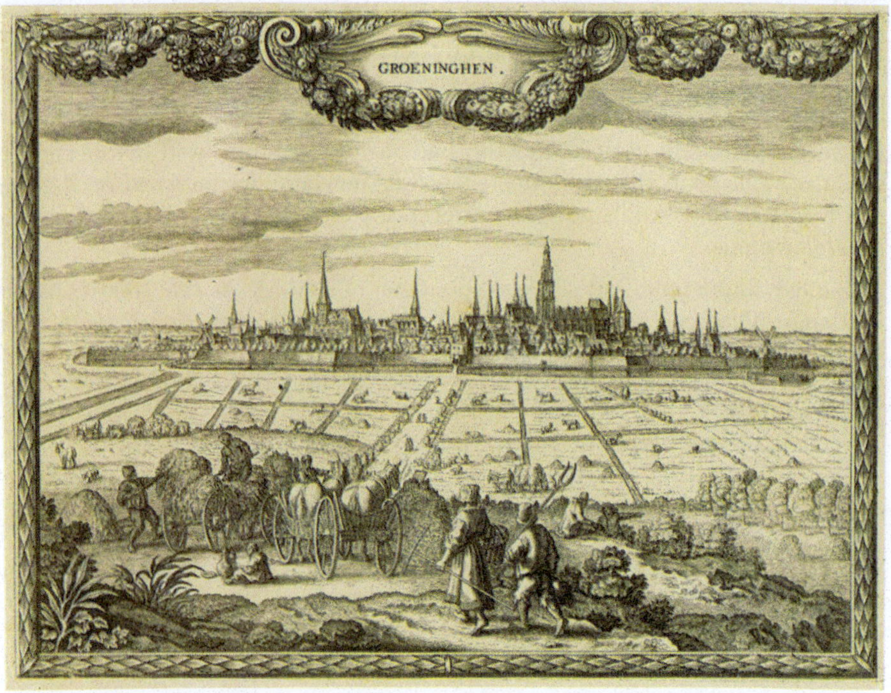

Fig. 7.6.3. The town of Groningen in John Bernoulli's days (Copperplate, collection RHC Groninger Archieven, 1536–3779)

Leibniz answered very calmly and that must have upset the younger Bernoulli completely. In his reply letter [Leibniz 2004, Vol. II, p. 295ff.] it burst out of him that the calculus fluxions was nothing than the differential calculus. Leibniz's differential, so Bernoulli wrote, was Newton's fluxion[15], Leibniz's sum was Newton's fluent. And then he arrived at an outrageous accusation [Hall 1980, p. 117]:

> *'so that I do not know whether or not Newton contrived his own method after having seen your calculus, especially as I see that you imparted your calculus to him, before he had published his method.'*

Here we therefore find an accusation of plagiarism against Newton which Leibniz did not follow. Such thoughts were obviously completely alien to him at this time.

Newton's *Epistolae* had still not appeared in print in the very wording and Wallis now planned to publish a third volume of his *Opera mathematica.*

[15]This is a technical mistake of Bernoulli since the fluxion $\dot{x}$ is *not* identical to Leibniz's dx. It is rather the *moment* $x'(t)\,dt$ that corresponds to the differential dx. (Newton wrote $\dot{x}o$, where o denotes an infinitely small interval of time) [Guicciardini 1989, p. 3].

In that third volume not only the *Epistolae* but also Leibniz's letter should be printed. Wallis asked Leibniz and Newton for detailed accounts of their respective methods in order to be able to compare them. Additionally he asked to publish the texts unabridged and Leibniz agreed [Hofmann 1973, p. 269].

Wallis and David Gregory now started to examine the correspondence between Leibniz and Oldenburg in the 1670s in the documents of the Royal Society and they made a find. In an undated memorandum in the estate of David Gregory, probably written about the 22nd September 1697, one finds the information that Wallis had received four letters from Newton in September 1697, but strictly speaking there were five [Scriba 1969, p. 74f.]; two of them were letters by Leibniz. On 24th September 1697 David Gregory noted [Scriba 1969, p. 75]:

> '*By Libnitz's letter to Mr Oldenburg dated 27 August 1676* [6. September 1676] *it is plain that Libnitz then knew nothing of his after differential Methode: For in that long letter, tho there be great occasion given for it, he speaks nothing that can be interpreted that way.*
>
> *Which is yet plainer by Monsr Leibnitz's letter to Mr Oldenburg of the 18 Novr 1676* [28. November 1676] *from Amsterdam after he had been in London in Octr before, a part of which Mr Collins inserts in his letter from London to Mr Newton at Cambridge dated the 5. March 1676* [15. März 1677] *wherein he proposes Calculating by Mr Bakers tables for the tangents of Curves*[16]. *Now he speaks here so sillily on this purpose that it is plain to any body who reads his letter, that then he knew nothing at all of the differential methode or method of fluxions.*

Gregory continued that Leibniz finally received Newton's *Epistola posterior* of 24th October 1676 (hence 3rd November in the Gregorian calender). One could extract from Leibniz's answer to Oldenburg dated 21. Juni (Julian) that he then had learned the *'differential method'* [Scriba 1969, p. 75]. But on the other hand Collins's letter to Newton of 18th June 1673 (Julian) implied that Newton was already in the possession of his method at that time. Further on in David Gregory's note it is said [Scriba 1969, p. 75]:

> '*These letters are to be printed in the folio that Dr Wallis is now aprinting, in the order of their dates, without any notes or Commentaries or reflections: But let the letters themselves speake.*'

We cannot help to postulate that David Gregory and John Wallis were convinced that Leibniz was a plagiarist. In contrast, however, stands the tone in

[16] This proposal of Leibniz could indeed raise a strange feeling since Leibniz had not yet mentioned his differential calculus although its discovery dated one year previously [Scriba 1969, S. 76].

Fig. 7.6.4. The brothers Jacob and John Bernoulli at work on mathematical problems [Engraving from *Encyclopedia Britannica Online.* Published in: L. Figuier: Vie des savants illustres du XVIIIe siécle. Paris 1870]
(© Collection: Photos.com/Thinkstock)

the letters of the year 1698 between Wallis and Leibniz [Hall 1980, p. 98f.]. The patriotic Wallis, meanwhile 82 years old, seemed to have accepted Leibniz's high esteem on the continent.

The third volume of Wallis's *Opera mathematica* carried the imprimatur of 13th March 1699 (Julian), i.e. 23rd March 1699, but it was not immediately available on the continent. De l'Hospital received it in July 1699 [Hofmann 1973, p. 270], probably together with a manuscript by Fatio which we will have to discuss later on. De l'Hospital immediately recognised the biased choice of letters in this volume and wrote to Leibniz that the English apparently absolutely wanted to claim the discovery of the infinitesimal methods for themselves [Antognazza 2009, p. 428][17]

> *'Wallis has published a third volume of his mathematical works in which he has inserted some of your letters to Mr Newton and others, and this, I believe, with the intention of attributing to the latter the invention of your differential calculus, which Newton calls 'fluxions'. It seems to me that the English are using every means possible to attribute the glory of this invention to their nation.'*

[17]The French original can be found in [Leibniz 2004, Vol. I, p. 336].

Leibniz replied [Leibniz 2004, Vol. I, p. 337f.] that he had consented to the publication of his letters to Oldenburg at that time and that he could not imagine that Newton was aware of the attacks of Fatio since Newton would know the truth [Hofmann 1973, p. 270f., footnote 143]. Even when John Bernoulli wrote to Leibniz in December 1699, repeated de l'Hospital's observation and identified Wallis's behaviour as that of a *'valient champion of English glory'* [Antognazza 2009, p. 429], Leibniz replied calmly [Antognazza 2009, p. 429][18]:

> *'That Wallis, as you have said, is a valient champion of English glory is grounds for praise rather than blame. I sometimes reproach my fellow countrymen that they are not sufficiently valiant champions of German glory. Competition amongst nations, while it should not lead us to speak ill of others, will nevertheless have the virtue of making us strive to equal or surpass others. The fruit of such competition comes to everyone; its praise to those who deserve it.'*

The worst was still to come, however. Newton's monkey Fatio entered the battle ground in connection with the problem of the brachistochrone and opened a new stage in the controversy. The war turned hot.

7.6.2 The Aftermath of the Problem of the Brachistochrone: The Monkey is Insulted

In the work *Communicatio suae pariter, duarumque alienarum ad edendum sibi primum a Dn. Jo. Bernoullio, deinde a Dn. Marchione Hospitalio communicatarum solutionum problematis curvae celerrimi descensus a Dn. Jo. Bernoullio Geometris publice propositi, una cum solutione sua problematis alterius ab eodem postea propositi* (Communication of both, his own as also two external solutions of the problem of the curve of fastest descent first communicated to him for publication by Mr John Bernoulli, then by Marquis de l'Hospital, which Mr John Bernoulli had publicly presented to the geometers, together with his solution of a second problem, presented by the very same) [Leibniz 2011, p. 297ff.] which appeared in May 1697 in the Acta Eruditorum Leibniz presented the problem of the brachistochrone and its solutions in a kind of introduction. Subsequently the five further solutions were published in the May edition of the Acta Eruditorum which were sent in, namely the ones of John Bernoulli, Jacob Bernoulli, de l'Hospital, Tschirnhaus, and Newton.

Newton had published his solution anonymously in January 1697 in Volume XIX, p. 384-389, of the Philosophical Transactions of the Royal Society. John Conduitt (1688–1737), Newton's successor as Master of the Mint who married

[18] The Latin original can be found in [Leibniz 2004, Vol. III, p. 620ff.].

Newton's niece Catherine Barton and became Newton's first biographer[19], wrote [Iliffe 2006, Vol. I, p. 181]:

> *'When the problem in 1697 was sent by Bernoulli – Sr I. N. was in the midst of the hurry of the great recoinage and did not come home till four from the Tower very much tired, but did not sleep till he had solved it wch was by 4 in the morning.'*

Newton sent his solution to his friend and mentor Charles Montague, then President of the Royal Society, who looked after the anonymous publication. Nevertheless, John Bernoulli immediately recognised the author *'tanquam ex ungue leonem'* (the lion by the claw). Apparently it was clear to Newton that the problem came from Leibniz and was meant to challenge him. Leibniz wrote to the Royal Society and emphasised that he was *not* the author of the problem. However much Newton must have been annoyed, poor Flamsteed had to experience what Newton wrote on 16th January 1699 concerning his theory of the moon [Turnbull 1959–77, Vol. IV, p. 296]:

> *'I do not love to be printed upon every occasion much less to be dunned & teezed by forreigners about Mathematical things or to be thought by our own people to be trifling away my time about them when I should be about ye Kings business*[20]

The remark crucial for the outbreak of war can be found in Leibniz's *Communicatio suae pariter* [Leibniz 2011, p. 302]:

> *'And in fact it is not inappropriate to remark that only those have solved the problem whom I had assumed that they could solve it; hence only those who are sufficiently* [far] *penetrated into the secrets of our differential calculus. And when I predicted such for the brother of the poser of the problem and for the Marquis de l'Hospital in France, I added that in my opinion Mr Huygens, if he were* [still] *alive, Mr Hudde, if he had not given up this studies, and Mr Newton, if he would take the effort, would be up to the task; ...'*
>
> (Und in der Tat ist es nicht unangemessen anzumerken, dass nur diejenigen das Problem gelöst haben, von denen ich angenommen hatte, dass sie es lösen könnten, also nur jene, die in die Geheimnisse unseres differentialen Kalküls ausreichend [weit] eingedrungen waren.

[19] The famous *Éloge de M. Neuton* for Newton [Fontenelle 1989, p. 326-348] of the year 1727 and from the quill of the secretary of the Academie des Sciences in Paris, Bernard le Bovier de Fontenelle (1657–1757), drew heavily on material which Conduitt had communicated

[20] The 'King's business' was the order to Newton to employ detectives and informants to send coiners and clippers to their doom. This episode from the life of Newton seemed so thrilling to a novelist that he wrote a crime novel thereon [Kerr 2002].

> Und als ich solches außer für den Herrn Bruder des Problemstellers für den Marquis de l'Hospital in Frankreich vorhergesagt hatte, hatte ich [noch] obendrein hinzugefügt, dass meines Erachtens Herr Huygens, wenn er denn [noch] lebte, Herr Hudde, wenn er diese Studien nicht längst aufgegeben hätte, und Herr Newton, wenn er diese Mühe auf sich nehmen sollte, der Aufgabe gewachsen wären; ...)

Here Newton's monkey, living in England again since 1698, got his teeth in! There it was in cold print: Leibniz did not see him, the great Fatio, capable of solving the problem of the brachistochrone! Fatio de Duillier hit back openly in his work *Lineae brevissimi descensus investigatio geometrica duplex* (Twofold geometric investigation of the line of steepest descent). On page 18 of the 20-pages long writing we read [Hess 2005, p. 65]:

> *'I am forced to accept by the evidence of the case that Newton was the first and – with many years in advance – oldest inventor of this kind of calculation. I do not want to decide whether Leibniz, the second inventor, has taken something from him, but leave that decision to those who have seen Newton's letters and his other writings.'*
>
> (Ich bin durch die Evidenz der Sachlage gezwungen anzuerkennen, dass Newton der erste und – mit vielen Jahren Vorsprung – älteste Erfinder dieser Rechnungsart ist. Ob Leibniz, der zweite Erfinder, von ihm etwas übernommen hat, möchte ich weniger selbst entscheiden als dem Urteil derjenigen überlassen, die Newtons Briefe und seine anderen Handschriften gesehen haben.)

And further on [Fleckenstein 1956, p. 23]:

> *'No one who has thoroughly studied what I myself have revealed will be deceived by Newton's modest silence or Leibniz's pressing bustle.'*
>
> (Niemanden, der durchstudiert, was ich selber an Dokumenten aufgerollt habe, wird das Schweigen des allzu bescheidenen Newton oder Leibnizens vordringliche Geschäftigkeit täuschen.)

Here was an open blow against Leibniz's scientific integrity and Leibniz was heart stricken. Fatio's disrespectful remarks in the private letters to Huygens were one thing, but the *Lineae brevissimi* carried the imprimatur of the Royal Society of which Leibniz was a member.

Leibniz addressed Wallis in a letter of 6th August 1699 and expressed his displeasure. On 8th September Wallis replied. He assured Leibniz that neither he nor the Royal Society would agree with Fatio's attack. The imprimatur of the Royal Society was granted by the Vice President Sir John Hoskyns (1634–1705) who apparently thought that it was a purely mathematical work [Antognazza 2009, p. 429]. Obviously, Fatio had obtained the imprimatur

Fig. 7.6.5. Otto Mencke [Engraving: Martin Bernigeroth before 1712] and Sir John Hoskyns [unknown artist about 1800]

by fraud [Hofmann 1973, p. 271] what Leibniz had already suspected in his letter to Wallis [Wahl 2012, p. 280]. The secretary of the Royal Society, Sir Hans Sloane (1660-1753), was informed by Wallis and reacted with a letter of apology to Leibniz via Wallis [Antognazza 2009, p. 429].

Now Leibniz prepared Otto Mencke that a response to Fatio's publication would be necessary in the Acta Eruditorum. Mencke replied [Wahl 2012, p. 282]:

> *'Whatever my highly esteemed patron will publish against him we will be pleased to print it in the Acta, if only it is not too long. It will be necessary, however, that the book by Mr Fatio should be unbiasedly reviewed before, so that we do not appear indecent.'*
>
> (Was mein Hochgeehrtester Patron wieder ihn publiciren wil, wollen wir hertzlich gern, wen es nur nicht gar zu groß, denen Actis inserieren. Es wird aber nötig seyn, daß des H^{n} Fatio buch, damit wir nicht anstoßen, zugleich, aber doch priore loco gantz unparteyisch recensiert werde.)

Mencke proposed to Leibniz to ask John Bernoulli to write the review and some days later Leibniz sent Fatio's publication and his thoughts in a response to Bernoulli [Wahl 2012, p. 282]. Bernoulli proposed some aggravating changes and reported that he had heard from a transient person named Manneville that Fatio had worked day and night on the problem of the brachistochrone without success, and that he tried again one year and a half later. Leibniz proposed to publish the review together with a response and excerpts of Bernoulli's letter and the anonymous review appeared in print in the Acta Eruditorum XI of the year 1699. Thereupon it also followed an excerpt from John Bernoulli's letter to Leibniz of 17th August 1699 [Leibniz 2004, Vol. III, p. 602ff.] which was defused and hence mitigated by the editor Otto Mencke – much to Bernoulli's anger. But a response by Leibniz was not published! Bernoulli now must have had the fear that Fatio's wrath would discharge wholly upon him. He requested that Leibniz should really publish the announced response [Wahl 2012, p. 283].

The reply of Fatio concerning the excerpts of the letter of Bernoulli which Mencke received was so aggressive that he only published an excerpt which was limited to technical issues in the Acta Eruditorum III of 1701. This excerpt was redacted by Leibniz, but Mencke refused to publish any further comments in the Acta [Hofmann 1973, p. 271, footnote 145]. Eventually Leibniz published his response to Fatio's accusations under his name in the Acta Eruditorum of May 1700 under the title *Responsio ad Dn. Nic. Fatii Duillerii imputationes. Accessit nova Artis Analyticae promotio specimine indicata; dum Designatione per Numeros assumtitios loco literarum, Algebra ex Combinatoria Arte lucem capit* (Reply to the accusations of Mr Nicolas Fatio de Duillier. It follows a new promotion of the analytic art, demonstrated by means of a specimen, through which the algebra gains clarification through the combinatorial art due to the notation with fictive numbers instead of letters) [Leibniz 2011, p. 308-343]. Since Leibniz was aware that Mencke surely would not publish purely private remarks concerning the conflict in the Acta he added a method for the resolution of implicit functions by means of formal power series.

We have to note that Leibniz did not act honestly in the case of Fatio! He had sent forward John Bernoulli to publish a polemic against Fatio and only then had published his moderate response. Quite obviously Leibniz wanted to appear as the lenient grand seigneur of the republic of scholars. He had exploited the quick-tempered Bernoulli to act as his polemic spearhead.

Leibniz remained very moderate in his response since he assumed that Fatio had no real support in England. He wrote [Leibniz 2011, p. 328f.]:

> '*When the recently published treatise by Mr Nicolas Fatio de Duillier on the curve of fastest descent and on the body with least resistance (in a medium) published in London reached me I was a touch surprised that this man, whom I never had offended, appeared so negative to me.* [...]
>
> [Leibniz reports that he had thought of probably not answering, but then he feared that Fatio could interpret his silence as disregard. He

Fig. 7.6.6. John Bernoulli's four volume work *'Opera Omnia, tam antea sparsim edita'* in the first edition of 1742. Below the opened inner title with a portrait of John Bernoulli. (Courtesy of Sophia Rare Books, Copenhagen)

admonishes all scholars to gradually drop their attitude to mutual attacks with biting words. He also mentions the letter of the secretary of the Royal Society, Hans Sloane.]

[...] *Perhaps some will suspect me having done something so that Mr Duillier would justly be excited. If such has happened by injudiciousness an admonition would have sufficed; actually it is my opinion that I still can improve. But the very words of the man show that his supposed insult is only that he was not named as one of the men I thought easily able to solve the problem of the line of fastest descent posed by Mr John Bernoulli, if only they would concentrate upon it. But how could he be named since in this paper he insists that he was not* **willing** *to release something of what he possessed concerning these investigations.* [...]

[Leibniz cites a sentence of Fatio from *Lineae brevissimi* that the latter was never wiling to reveal his solutions of mathematical problems in public writings.]

[...] *Therefore our ignorance concerning his progress was excusable.'*

(Als mich die neulich in London veröffentlichte Abhandlung von Herrn Nicolas Fatio de Duillier über die Kurve des schnellsten Abstiegs und über den Körper mit geringstem Widerstand (in einem Medium) erreichte, war ich nicht wenig verwundert, dass dieser Mann, den ich niemals gekränkt hatte, sich mir gegenüber so negativ eingestellt zeigte. [...]

[...] Vielleicht werden einige argwöhnen, ich hätte etwas getan, worüber Herr Duillier mit Recht erregt sein würde. Wenn solches freilich durch Unüberlegtheit vorgekommen sein sollte, hätte es lediglich einer Ermahnung bedurft; ich bin nämlich der Überzeugung, mich noch vervollkommnen zu können. Aber gerade die Worte des Mannes zeigen, dass seine vermeintliche Kränkung allein darin besteht, dass er nicht unter denen genannt wurde, die eine Lösung des von Herrn Johann Bernoulli aufgeworfenen Problems der Linie des schnellsten Abstiegs geliefert oder durch Proben ähnlicher Herleitungen erreicht hatten, dass man ihnen [die Lösung] leicht zutrauen könnte, wenn sie ihre Gedanken [nur] darauf gerichtet hätten. Aber wie hätte er genannt werden können, da er doch selbst in dieser Schrift daran festhält, dass er nicht **gewillt** war, etwas von dem herauszugeben, was er hierzu an Untersuchungen besaß.[...]

[...] Deshalb war uns unsere Unkenntnis über seine Fortschritte zu verzeihen.)

Leibniz then suspected that Fatio wanted to pursue an open trial and he started to concentrate particularly on two of Fatio's accusations: the 'rampant' kind to pose problems, and 'to assign appreciation and rank to individual geometers as coming down from a mathematical throne'. Concerning the second accusation he remarked that he had confidence that also other excellent mathematicians could have solved the problem. He named Wallis, Hooke, Halley, Craig, but also Ole Rømer from Denmark, Tschirnhaus, Pierre de Varignon, and Philippe de La Hire from France.

Leibniz furthermore remarked that there were still other mathematicians who could solve problems different from the ones of the infinitesimal calculus. But in the further course of the work Fatio yet received some flogging. Leibniz wrote [Leibniz 2011, p. 333f.]:

> *'Mr Duillier says that he had found the general foundations and most of the rules of calculus which we call differential, by himself in the year 1687. We want to believe that this is true (at least in part, since I think that he not even now understands all basics of this calculus sufficiently ...'*
>
> (Herr Duillier sagt, er hätte im Jahr 1687 aus eigenem Vermögen die allgemeinen Grundlagen und die meisten Regeln des Kalküls, den wir differential nennen, gefunden. Wir wollen glauben, dass dem so ist (wenigstens zum Teil, denn ich denke, dass ihm nicht einmal jetzt alle Grundlagen dieses Kalküls hinreichend bekannt sind ...))

Then Leibniz came to mention Newton [Leibniz 2011, p. 335]:

> *'Up to know Mr Duillier pursued his personal and in part, as he thought, a public issue: but now that he advocates the interests, so to speak, of the excellent geometer Isaac Newton and others against me he will forgive me that I do not reply to anything until he will produce a mandate to represent in particular Mr Newton to whom my relation was not strained. In any case the admirable Sir has always expressed a high opinion of me in quite a number of conversations with my friends and has never, as far as I know, expressed complaints; in public he has treated me in a manner that I would be unjust if I would complain. But I have taken every opportunity to praise his exceptional merits and he himself knows that better than others. When he published his* Principia *in the year 1687 he also pointed out publicly that in the case of certain new geometrical discoveries which he as well as I had achieved, no one had received something enlightening from each other, but that each one of us had owned* [these discoveries] *to his own considerations, and that I had communicated these already ten years before*[21]*. In any case, when I published my calculus in the year 1684 I did not even know more* [from hearsay] *about his discoveries in this field than he himself communicated once in letters, ...'*
>
> (Bisher betrieb Herr Duillier teils sein persönliches, teils, wie er

[21]Leibniz refers here to the Scholium after Book II, Section II, Proposition VII.

Fig. 7.6.7. Ole Rømer (1644-1710) was a Danish astronomer who gave the first evidence on the finiteness of the speed of light. During his stay in Paris between 1672 and 1681 he made the acquaintance of Leibniz who highly esteemed him. From 1681 he was professor of mathematics at the University of Copenhagen and introduced a national uniform system of lengths and weights. In the year 1705 he became mayor of Copenhagen and did a great service by restructuring the water supply and the sewer system as well as by the introduction of a street lighting. Hans Sloane (1660-1753), secretary of the Royal Society from 1693 on, their president since 1727. Sloane even served as Royal Physician to three English monarchs: Anne, George I, and George II. After his death the endowment of his natural history collection served as a foundation gift for the British Museum which opened with an exhibition of Sloane's collection in 1759 [Engraving after a portrait by T. Murray, probably 18th c]

glaubte, ein öffentliches Anliegen: nun aber, da er gleichsam auch die Interessen des hervorragenden Geometers Isaac Newton und anderer gegen mich vertritt, wird er mir nachsehen, wenn ich [so lange] nicht auf alles antworte, bis er sein Vertretungsmandat, sei es für die anderen, sei es insbesondere für Herrn Newton vorweist, zu dem mein Verhältnis nicht gespannt war. Jedenfalls hat der vortreffliche Herr in etlichen Gesprächen mit meinen Freunden anscheinend immer eine gute Meinung von mir gehabt und niemals, soweit ich weiß, Klagen vorgebracht; öffentlich aber hat er mich so behandelt, dass ich ungerecht wäre, wenn ich klagen würde. Ich aber habe bei sich bietender Gelegenheit gerne seine außerordentlichen Verdienste gerühmt, und er selbst weiß das von allen am besten. Auch hat er, als er im Jahr 1687 seine Mathematischen Prinzipien der Natur publizierte, in aller Öffentlichkeit hinlänglich darauf hingewiesen, dass bei gewissen neuen geometrischen Entdeckungen, die er ebenso wie ich gemacht hat, keiner vom anderen Erhellendes erhalten habe, sondern dass jeder [diese Entdeckungen] seinen eigenen Überlegungen verdanke, und dass ich [selbige] bereits zehn Jahre zuvor mitgeteilt hätte. Jedenfalls war mir, als ich meinen Kalkül im Jahr 1684 herausgegeben habe, nicht einmal [vom Hörensagen] mehr von seinen Erfindungen auf diesem Gebiet bekannt, als was er selbst einst in Briefen mitgeteilt hatte, ...)

And further on Leibniz wrote [Leibniz 2011, p. 336]:

'There is only one thing left where I [yet] *see some need for justification. When Mr John Bernoulli sent the invocation inviting the geometers to find the line of fastest descent* [also] *in particular to Mr Newton, several voices raised in England saying that I had challenged Newton*[22]*; also Mr Duillier seems to be of the opinion that I had proposed, even arranged, to send* [this invocation]. *However, that* [this] *was done completely without my knowledge will be testified by Mr Bernoulli himself.'*

(Nur eine Sache steht aus, in der ich für mich [noch] etwas Rechtfertigungsbedarf sehe. Als Herr Johann Bernoulli den Aufruf, mit dem die Geometer eingeladen wurden, die Linie des schnellsten Abstiegs zu suchen, insbesondere [auch] an Herrn Newton geschickt hatte, sind etliche Stimmen in England laut geworden, die sagten, Newton sei von mir herausgefordert worden; dieser Meinung scheint auch Herr Duilier zu sein, als ob ich angeregt, ja veranlasst hätte, [diesen Aufruf] zu versenden. Dass [dies] jedoch völlig ohne mein Wissen geschehen ist, wird Herr Bernoulli selbst bezeugen.)

As a reaction to Leibniz's reply a letter by Fatio was sent to Mencke. In this letter Fatio admitted that Newton did not agree with his action [Wahl 2012, p. 285]. That was enough for Leibniz to declare the end of this conflict.

7.7 Fatio's Fate

Fatio de Duillier did never again appear in this confrontation. He stayed in London as a teacher and worked together with the French clockmaker brothers Peter and Jacob de Beaufré on the use of perforated rubies for the bearings in clocks. They received an English patent in 1704.

Shortly afterwards Fatio joined the Camisards, huguenots from the Cévennes in the south east part of the French Massiv Central. These protestants being descendants of the Waldensians emigrated in large number to protestant

[22]Thomas Burnett wrote in his letter of 14th May 1697 that Newton had solved the problem of the brachistochrone within two hours. There Burnett emphasized that Newton had highly esteemed Leibniz

countries due to their constant oppression in France. In London a community of 'French Prophets' was founded and Fatio quickly advanced to the leader of this group which seemed suspicious to the governing class in London. Fatio was suspected to plan a political conspiracy and eventually he and two of his followers had to stand trial and were sentenced.

As was suspected by Domson, Fatio's orientation towards prophetic religious ideas began already in the years 1692/93; a time where he could have absorbed Newton's theological ideas [Domson 1981, Chapter II]. However, the opinions concerning religious mysticism drifted apart between Newton and Fatio [Domson 1981, p. 64ff.], as happened with their opinions concerning gravitation before. Thereby it may be possible that Fatio's interest in alchemy, in the exegesis of the Holy Scripture and the Kabbalah, affected his mechanical interpretations as similar interests had affected Newton's work [Heyd 1995, p. 257f.].

On 2nd December 1707 Fatio stood on the pillory at Charing Cross in London. On his hat was an inscription describing him as a propagator of dangerous and wrong prophecies. Only by the intervention of the Duke of Ormonde, whose brother had been taught by Fatio, was he protected from the mob.

Following these events Fatio wanted to convert the world by missionary work; the active Camisard Elie Marion (1678–1713) had come to England in 1706 and with him Fatio now travelled through the German lands to Turkey, where Marion died aged 35. In 1712 Fatio was back in England again and settled in Worcester. He was still active in science and proud of his membership in the Royal Society. He died in spring 1753 and was buried close to the church of St Nicolas in Worcester.

However, Leibniz's position regarding Fatio after 1700 is remarkable. When Thomas Burnet started his second journey to the continent Leibniz referred to the two Fatio brothers in Basel and called them excellent mathematicians. When in 1708 he heard of Nicolas Fatio de Duillier's relationship with the Camisards and the negative consequences for Fatio Leibniz wrote that he was in grief [Hall 1980, p. 100],

> *'because of my love for Mr. Fatio, for he is a man excellent in mathematics, and I do not understand how he could have got involved in such an affair.'*

There is no doubt that it is a sign of Leibniz's humanity to stand behind Fatio *after* Fatio's shabby behaviour towards Leibniz.

7.8 The 'Case Leibniz'

We have already referred at a few places that Leibniz behaviour was not always flawless. That was the case when he stated that he did not know the *Principia* before he published his three papers on physics in the Acta Eruditorum (cp. page 265) and when he pushed forward John Bernoulli in the conflict with Fatio. Charlotte Wahl has presented a detailed analysis of Leibniz's behaviour in the affairs of the republic of scholars which we will summarise at this point.

We were already concerned time and again with attempted cheating and tactical manoeuvres which were typical for the 17th century. These include the use of anagrams, anonymous publishing, and the public challenge of other mathematicians. Why did even the best utilise such tricks? The distribution of new results and applications of methods took very long. Neither the Acta Eruditorum nor the Transactions of the Royal Society nor French science journals were available everywhere so that it was quite possible that an English scientist had made a discovery that a French had published years earlier – or vice versa. It was simply not verifiable whether the colleague coming years too late had arrived at a discovery by his own or whether he had received wind of the original publication. Under these conditions posing problems to a circle of scientists as large as possible was a method to show ones own reflections in all of Europe. If one hid scientific knowledge in anagrams one made it difficult (if not impossible) for potential competitors to publish ones results as their own. If yet a colleague had the same idea one could always resolve the anagram and secure ones priority. Anonymous publication was always the method of choice when delicate issues like attacks on colleagues were concerned. Leibniz was a true master in all issues of tactics.

In public Leibniz liked to characterise himself as being a lenient and open scientist who was actually always distracted by other works (which is certainly the truth). That he acted differently behind the scene was shown already in connection with Fatio's attack. He saw himself surely not as being equal among equals, but as the *spiritus rector* of the new infinitesimal mathematics; in other words, as the 'doyen' of European mathematics.

Hence, in his work *Communicatio* concerning the problem of the brachistochrone in May 1697 he did *not* publish his own solution but reflected only on the solutions of others. The reason for this can be found in the fact that Leibniz clearly saw that the problem of the brachistochrone belonged to a really new class of problems for which his calculus had not yet delivered an algorithm which was generally applicable [Leibniz 2011, p. 303]:

> '*These kind of problems posed in such a way concerning maxima and minima show something unusual which surpasses the common questions concerning maxima and minima* [of a curve] *by far.*'

Fig. 7.8.1. Stamp of the German Federal Mail (Deutsche Bundespost) on occasion of the 250 th anniversary of the death of Leibniz 1966 (left) and a special issue stamp of 1980

> (Diese Art auf solche Weise vorgelegter Probleme über Maxima und Minima hat aber etwas Ungewöhnliches, das die gängigen Fragen nach den Maxima und Minima [einer Kurve] bei weitem übersteigt.)

Hence Leibniz clearly saw the difference between calculating extreme values for a *single* curve and choosing one curve with extremal properties from a whole flock of curves. He did not want to give away the key to the solutions of such problems [Wahl 2012, p. 275]. John Bernoulli had sent two different solutions of the problem of the brachistochrone to Leibniz. Leibniz wanted to obtain permission from Bernoulli to skip one solution in order to increase the excitement of the other scientists working on their solutions, but Bernoulli did not agree, since the solutions anyway would not be published immediately. Now Leibniz argued [Wahl 2012, p. 276]:

> *'If I would be in your place I would postpone this second method a little bit, also because it reaches farther, so that the sources may not be shown to those who would suppress the sources or would redirect them into their streams.'*

> (Jene zweite Methode würde ich noch etwas aufschieben, wenn ich an Ihrer Stelle wäre, auch deshalb, weil sie weiter reicht, damit nämlich die Quellen nicht gleich jenen angezeigt werden, die später die angezeigten [Quellen] unterdrücken oder in ihre Bäche umlenken.)

Bernoulli could endorse much of this agenda; however, he offered publicly to send his second solution to everyone interested in it. John's brother Jacob

Fig. 7.8.2. Stamp of the German Democratic Republic of 1950 celebrating the 250th anniversary of the German Academy of Sciences in Berlin (left), and beside it a stamp of 1927

was asked to keep his solution (which Leibniz had not seen then) secret as well. Apparently, Leibniz did ask for secrecy under the impression that the publication of Bernard Nieuwentijt's *Analysis Infinitorum* (cp. section 10.1), in which Nieuwentijt had developed an alternative infinitesimal calculus and had claimed that the calculus of Leibniz could easily be derived from his own. Leibniz wanted to prevent such a 'cold takeover' of his new mathematics in the future. But Jacob had sent his solution already to Otto Mencke for publication, so that the plan to keep it secret could not be realised.

Addressing the solution of the isoperimetric problems which Jacob had posed to his brother John, Leibniz advised reluctance, too. In case of a problem concerning orthogonal trajectories where John Bernoulli had failed but Leibniz had sent him the solution, Leibniz urged to keep it secret. He asked John Bernoulli to cease posing problems publicly which could set others on the track [Wahl 2012, p. 277]. John was further asked to distract his brother by means of other problems so that Jacob would not find the solution and would be kept from publishing it. In the correspondence with John Bernoulli Leibniz also discussed what should be revealed to their common correspondent de l'Hospital. The problem was posed originally by John Bernoulli who in 1694 led Leibniz to the following question: Given a family of curves, find the curve which intersects all given curves of the family in a right angle [Engelsman 1984, p. 60ff.]. If a trajectory intersects a curve in a right angle, then

$$\left.\frac{dx}{dy}\right|_{\text{Curve}} = -\left.\frac{dy}{dx}\right|_{\text{Trajectory}}$$

has to hold. If the family of curves is given by $S(x, y; a) = 0$ where a denotes the parameter of the family, differentiating with respect to x results in

$$S_x(x, y; a) + S_y(x, y; a)y' = 0,$$

where the subscripts denote the variables of differentiation[23]. Hence it holds

$$S_x(x, y; a)\,dx + S_y(x, y; a)\,dy = 0.$$

At the point of intersection we get from the first equation $dx|_{\text{Curve}} = -dy|_{\text{Trajectory}}$ or $dy|_{\text{Curve}} = -dx|_{\text{Trajectory}}$, hence it finally follows

$$S_x(x, y; a)\,dy - S_y(x, y; a)\,dx = 0,$$

and that is a differential equation which can be solved if one can solve $S(x, y; a) = 0$ explicitly for the parameter a and then insert this a. This method fails for 'transcendent' curves, e.g. if the family of curves is given by an integral of the form

$$y(x) = \int_{x_0}^{x} p(z; a)\,dz.$$

The solvability for the parameter a can not be guaranteed in general in these cases. Leibniz did work on 'differentiation across the curves of a family' and finally succeeded at the beginning of 1697 when he discovered the exchange of the derivative with respect to a parameter with integration [Engelsman 1984, p. 67], i.e. the rule

$$\frac{d}{da}\int_{x_0}^{x} p(z; a)\,dz = \int_{x_0}^{x} \frac{\partial}{\partial a} p(z; a)\,dz.$$

Thereby John Bernoulli was able to compute orthogonal trajectories even in the case of transcendent functions. Later Leibniz used this problem as a weapon when he challenged Fatio in 1700 and then all English mathematicians in 1715 [Wahl 2012, p. 278] in order to show the superiority of his calculus.

In the end, Leibniz's policy of publication did not pay off. The calculus of variations was not developed into a general theory within his lifetime and the English were able to solve the problem of orthogonal trajectories with their own methods, too [Wahl 2012, p. 278], [Engelsman 1984, p. 71ff.]. It remains a hollow aftertaste and a shadow falls on Leibniz concerning his acting behind the scenes. Even in the conflict between the brothers Bernoulli he did not always put his cards on the table [Wahl 2012, p. 287ff.].

Additionally, Leibniz had made himself unpopular with the English with his work *Animadversio ad Davidis Gregorii Schediasma de Catenaria, quod habetur in Actis Eruditorum A. 1698. p. 305. seqq.* (Remarks concerning the contribution of David Gregory on the catenary, which can be found in the

[23] These are *partial* first derivatives for which we now use the notation $\partial/\partial x$. We see here the first steps to apply the differential calculus to functions of multiple variables. This historical development was examined thoroughly by Engelsman [Engelsman 1984].

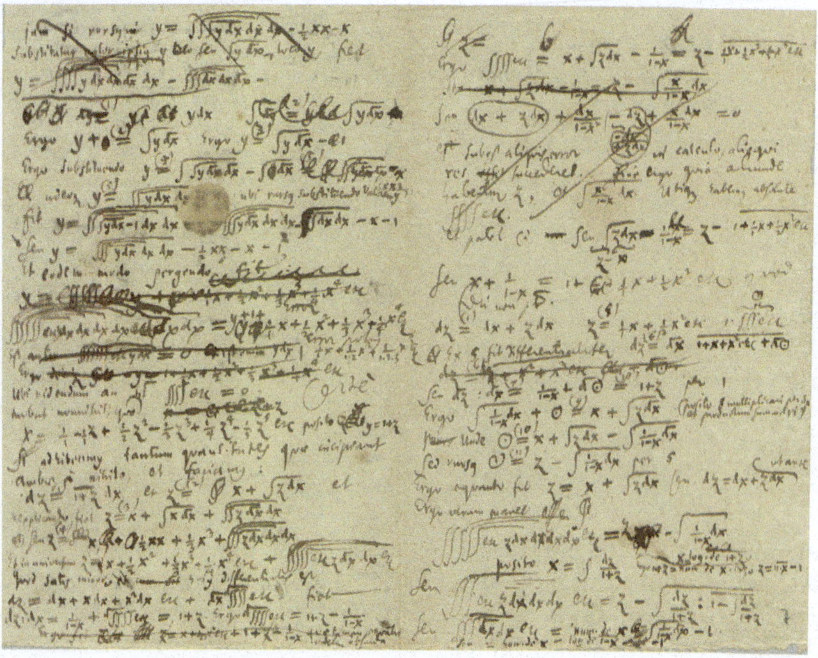

Fig. 7.8.3. A kind of rough paper by Leibniz gives evidence of his advanced mathematics (Gottfried Wilhelm Leibniz Library - Niedersächsische Landesbibliothek Hannover (Hanover), LH XXXV VIII 9 Bl. bv 7r)

Acta Eruditorum of the year 1698 on p. 305ff.) [Leibniz 2011, p. 319ff.] in the Acta Eruditorum of February 1699 which was published anonymously (!), although its authorship was immediately clear to all involved. David Gregory had published a solution to the problem of the catenary in the Philosophical Transactions of August 1697, after it was solved by others seven years before. A reprint of this work appeared in the Acta Eruditorum in the year 1698 and Leibniz immediately sent critical remarks to John Bernoulli who recommended a publication in the Acta. Thereupon Leibniz asked Bernoulli (!) to send his remarks anonymously to Mencke. The remarks are mainly concerned with an improper use of differential quantities and somewhat vague relations to the physical foundations. Leibniz dismembered Gregory's work; he critisised wrong calculations with fluxions and the use of obscure mechanical principles. At the end he wrote [Leibniz 2011, p. 325]:

> *'It is to be assumed that the learned Gregory himself will admit this openly after further reconsideration, if he at least will invite the advice of the famous Newton, since he follows the method of the latter as he explicitly stated.'*
>
> (Es ist anzunehmen, dass der gelehrte Gregory selbst dies nach nochmaligem Überdenken offen zugeben wird, wenn er zumindest den Rat des

berühmten Newton eingeholt hat, nach dessen Methode er erklärtermaßen vorgeht.)

Hence Leibniz implied that the trouble that Gregory had prevented him from arriving at his goal (although the result was correct. Leibniz talks about the annulation of errors) was rooted in Newton's calculus of fluxions. Mr Leibniz had laid another fuse!

Additionally, an appendix to Johann Groenig's book *Historia cycloeidis* (History of the cycloid) appeared in 1701 containing a list of errors in Newton's *Principia* which was issued as being a work of Christiaan Huygens. This suggested, of course, that numerous errors in Newton's magnus opus were found by Huygens, but the list of errors came from Newton himself who had given it to Fatio in 1691 to pass it to Huygens. Among Newton's friends it was out of the question that Leibniz was involved in the publication. But was he?

7.9 Summary and the Front Line in the Year 1699

The position entrenched with Fatio's appearance in Newton's life. Fatio's nasty remarks concerning Leibniz in his letters to Huygens at the end of 1691 and the beginning of 1692 clearly expressed an accusation of plagiarism against Leibniz, but the accusation was still concealed in private correspondence. We have no evidence that Newton stood behind Fatio and may well assume that Fatio here did *not* express Newton's opinion.

The printing of both of the *Epistolae* which Newton sent to Leibniz in 1676 as well as some of the letters of Leibniz in Wallis's third volume of his *Opera mathematica* exacerbated the situation considerably, but only on the side of continental Europe. John Bernoulli, always willing to fierce and polemic dispute, wrote in a letter of 1696 to Leibniz and critisised that Leibniz's calculus was not appreciated in volumes I and II of the *Opera mathematica*, but Leibniz stayed unaffected. Thereupon Bernoulli raised an accusation of plagiarism against Newton in his correspondence with Leibniz, but Leibniz could not subscribe to this allegation. Wallis's tendency in volume I, to maintain the priority of his fellow countryman Newton over Leibniz, elicited from Leibniz only the remark that he was treated in a coolish manner by Wallis. During the preparation of volume III David Gregory and John Wallis found transcriptions of letters by Leibniz in the letter books of the Royal Society. They concluded that Leibniz could only have developed his calculus after receiving the *Epistolae*. We have already reported that Oldenburg did not copy letters in full in the letter books and that the activities of Collins contributed more to confusion than to transparency. Hence it is perhaps not surprising that Wallis and Gregory were led on this trail.

After the publication of the third volume Leibniz received a letter of de l'Hospital in which the latter pointed out the favourable treatment of Newton in that volume, but also included a copy of Fatio's work *Lineae brevissimi* in which serious offences against Leibniz were included and which carried the imprimatur of the Royal Society. Here is the public declaration of war and Leibniz reacted with a letter to Wallis, which really showed concern and wished for an apology of the Royal Society, since Fatio had obtained the imprimator by fraud. Leibniz also reacted publicly with his work *Responsio ad Dn. Nic. Fatii Duillerii imputationes*, but only *after* he had exploited John Bernoulli who had agreed to the printing of excerpts of his letters in the Acta Eruditorum. Fatio was forced to admit that Newton did not agree with his approach and Leibniz thereupon declared the conflict terminated.

We have not the trace of evidence that Newton had any doubts concerning the integrity of Leibniz in 1699; Leibniz clearly showed his appreciation for Newton.

But Fatio's declaration of war upon Leibniz made sure that the fellows and partisans on both sides, Newton's as well as Leibniz's, began to arm and took a stand. David Gregory, John Wallis, and Fatio, of course, were convinced that Leibniz was a plagiarist. Fatio did not play any role further on. John and Jacob Bernoulli, as well as the Marquis de l'Hospital, were convinced that 'the English' would claim a discovery which actually belonged to Leibniz. Gregory felt particularly injured by Leibniz's anonymous remarks in the Acta Eruditorum concerning his computation of the catenary, to which he reacted in the Philosophical Transactions of the year 1700. Wallis also assisted his fellow countryman and defended him in September 1699 in a letter to Leibniz.

There was still not one of Newton's manuscripts on the calculus of fluxions in print! The Continent knew about the calculus of fluxions only from Wallis's *Opera mathematica*. In contrast, Leibniz had published on his calculus since 1684 and the triumph of the calculus was a fact due to the works of the brothers Bernoulli and the text book by de l'Hospital. Perhaps Newton had perceived this fact too late. Only in 1704, as an appendix to his *Opticks*, the manuscript *De quadratura* concerning his calculus of fluxions became publicly known.

8 War of Extermination

T. Sonar, *The History of the Priority Diſpute between Newton and Leibniz*, https://doi.org/10.1007/978-3-319-72563-5_8

Wallis's *Opera mathematica* blatantly showed that even the great old man of English mathematics did not know anything about Newton's mathematical achievements before 1676 – how then could Leibniz have known? Wallis's nationalist pride carried him away in his *Opera* to clearly take Newton's side. In the preface to volume I Wallis wrote that Newton had explained his method to Leibniz (*'methodum hanc Leibnitio exponit'*), which was obviously false, but Wallis wrote under the control of Newton [Hall 1980, p. 129]. There were no longer witnesses of the development of Leibniz's calculus between 1675 and 1684, nor were there witnesses of Newton's developments between 1666 and 1685. John Collins would have had to say something, but he had been dead for 20 years in 1703. Now the events overturned.

8.1 The Mood Shifts

8.1.1 George Cheyne and His Effect on Newton

In the year 1703, one year before Newton's *Opticks*, the Scottish physician George Cheyne (1671–1743) published a book with the title *Fluxionum methodus inversa* (The inverse method of fluxions).

In a remark [Newton 1979, p. cxxii] in his *Opticks* (cp. page 305) Newton wrote: *'And some Years ago I lent out a Manuscript containing such Theorems, and having since met with some Things copied out of it, I have on this Occasion made it publick, ...'*. This remark was directly referring to Cheyne's

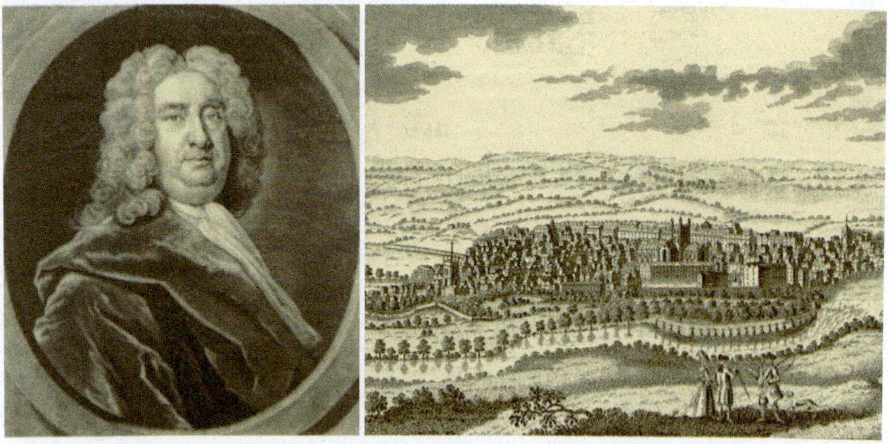

Fig. 8.1.1. George Cheyne practised as a physician from 1702 on in Bath. He developed groundbreaking ideas concerning social psychology, psychology, and vegetarian nutrition [Painting by John Faber jr., 1732]. Bath in the year 1772 [Engraving of an unknown artist]

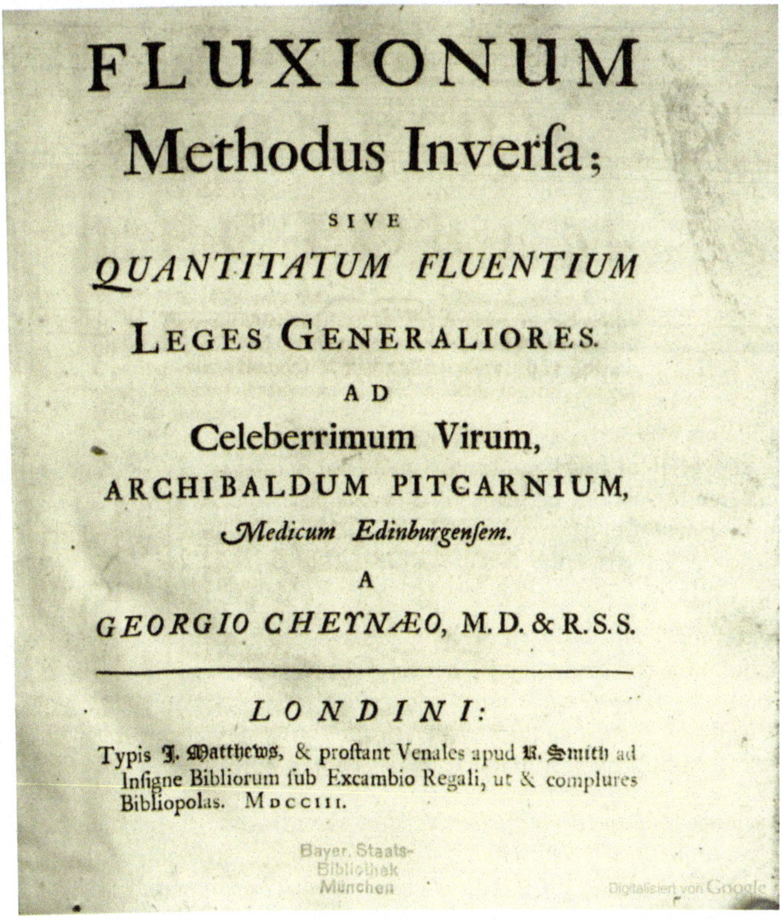

FLUXIONUM
Methodus Inverſa;
SIVE
QUANTITATUM FLUENTIUM
LEGES GENERALIORES.
AD
Celeberrimum Virum,
ARCHIBALDUM PITCARNIUM,
Medicum Edinburgenſem.
A
GEORGIO CHEYNÆO, M.D. & R.S.S.

LONDINI:
Typis J. Matthews, & proſtant Venales apud R. Smith ad Inſigne Bibliorum ſub Excambio Regali, ut & complures Bibliopolas. MDCCIII.

Fig. 8.1.2. Title page of *Fluxionum methodus inversa* by George Cheyne (Bayerische Staatsbibliothek München (Munich), Signatur: 845532 4 Math. p.75, Title page)

book [Westfall 2006, p. 639]. In a memorandum by David Gregory of 11th March 1704 Newton's reaction to the publication of Cheyne's book is described as follows [Westfall 2006, p. 639]:

> *'Mr. Newton was provoked by Dr. Cheyns book to publish his Quadratures, and with it, his Light & Colours, &c.'*

In the collection of anecdotes by John Conduitt the following episode [Iliffe 2006, Vol. I, p. 177][1] is told. When Cheyne came to England from Scotland, he was introduced to Newton by Dr. Arbuthnot to whom he reported having written a book, but no money to print it. Newton offered him *'a bag of*

[1] Cp. also [Iliffe 2006, Vol. I., p. 189].

money' but Cheyne declined the offer and both men were embarrassed. Newton refused to see Cheyne afterwards. Cheyne later wrote that the manuscript which he showed Newton seemed not to be intolerable in the eyes of the latter (*'thought it not intolerable'* [Westfall 2006, p. 639]).

Cheyne's book is not a masterpiece; he did not present new mathematics but wanted to explain and expound. There were, however, many errors left. Although quite noting these errors John Bernoulli wrote to Leibniz, that the book seemed to comprise [Hall 1980, p. 131][2]:

> *'a most remarkable little book, stuffed with very clever discoveries; I know of no one in Britain since Newton who has penetrated so far into these deeper levels of geometry.'*

Now Cheyne was a dyed-in-the-wool Newtonian. Although he had discussed some discoveries of mathematicians from the Continent he attributed everything to Newton whom he praised to the skies. Naturally, this did not remain hidden to Bernoulli. He reported to Leibniz that Cheyne would try to turn them into *'Newton's apes, uselessly retracting his steps of long before'* [Hall 1980, p. 132].

When Leibniz eventually received Cheyne's book he reacted like being an Anglophobe for the first time – even stronger than Bernoulli – and wrote [Hall 1980, p. 132]: *'Whoever has once understood our work can easily put together such a book; he furnishes no new series nor an elegant theorem'.* In the same letter to Bernoulli he continued [Hall 1980, p. 132f.][3]:

> *'He tries ineptly to claim for Newton the method of series employing assumed arbitrary coefficients, determined by comparison of the terms, for I published that* [in 1693], *when it was not apparent to me or to any one else (at least, in the public domain) that Newton too possessed such a thing. Nor did he attribute it to himself, rather than to me. Which of us two had it first, I have not declared. I already displayed it in my ancient treatise on the arithmetic circle-quadrature, which Huygens and Tschirnhaus read in Paris* [in 1675]*.'*

Then Cheyne got reprimanded [Hall 1980, p. 133]:

> *'*[...] *it may be the case that just as Mr Newton discovered some things before I did, so I discovered others before him. Certainly I have encountered no indication that the differential calculus or an equivalent to it was known to him before it was known to me.'*

Here Leibniz contradicted his own reply letter to the *Epistola posterior* in which he showed himself satisfied with Newton having found an analogous

[2]Latin source see [Leibniz 2004, Vol. III, p. 724].

[3]Latin source see [Leibniz 2004, Vol. III, p. 725ff.].

Fig. 8.1.3. Matthew Stuart [Painting in the Art Collection of the University of Edinburgh], Archibald Pitcairne [Engraving by Rob Stranae] and Colin McLaurin [11e Comte de Buchan after a portrait by James Ferguson] and some other Scottish scholars emerged as Newton's supporters

method! Apparently something had happened to Leibniz who seemed to have become thin-skinned since the attack of Fatio. Now the next fighter for Newton's calculus of fluxions appeared on the scene and took the achievements of Leibniz and the Bernoulli brothers with a grand gesture for the English party! Although the sources from which Cheyne's book was fed have not been unearthed until today [Hall 1980, p. 134] we can infer from the content that Cheyne had certainly studied the works of the Leibniz school which were published in the Acta Eruditorum. In a letter to his mentor David Gregory Cheyne wrote in 1702 [Hall 1980, p. 134]:

> '*all these are but a few examples of Mr Newton's (excepting yours) Methods, and* [...] *all found within these 20 years by these or not unlike Methods are but either repetitions of, or easie corollaries from these things which he* [Newton] *has either imparted to his friends or the publick* [...]'

Concerning this paragraph Cheyne wrote that Newton would certainly have represented it differently. The remark *excepting yours* was surely meant to flatter Gregory; however, Gregory here had at least the same problem that was claimed for Leibniz: his methods were grown verifiably on Newton's ideas.

Hall has remarked [Hall 1980, p. 134] that about the year 1700 the weight of mathematical research had shifted in Great Britain from England to Scotland as it had on the Continent from France to the German lands. And in fact all enthusiastic mathematicians on Newton's side were Scots at this time: David Gregory, John Craig (or Craige), Archibald Pitcairne, George Cheyne, John and James Keill, James Stirling, Matthew Stewart, and Colin Maclaurin. Scotsmen were (not only in the past) very proud of their courage, their grim belligerence, and their pronounced unwillingness for compromise. All of these characteristics attributed to Scots could be clearly recognised on the side of the Newtonians. The only mathematician of rank dealing reasonably with the colleagues on the Continent was Brook Taylor, but he was a 'Sassenach' in the eyes of the Scots, i.e. an Englishman [Hall 1980, p. 134], to whom one could only look down.

It *could not* have suited Newton to see his ideas published by a man such as Cheyne. It appeares credible that he got the impetus to finally publish himself through Cheyne's book. After the death of Robert Hooke the opportunity opened up to present his *Opticks* in public. Bound with[4] it was *De quadratura*, a shortened version of the manuscript that Newton had begun in 1691 when he was pressed by Gregory's publication concerning the quadrature of series. Certainly, the work was and still is a masterpiece, but at the time of its publication it lagged some 20 years behind! If we accept that not later than 1690 Newton arrived mathematically at a point at which Leibniz, Bernoulli, de l'Hospital, and others had arrived in about 1700, then we understand that *De quadratura* could not impress Leibniz oder Bernoulli in 1704.

In Newton's introduction to *De quadratura* we find the paragraph [Whiteside 1967–81, Vol. VIII, p. 123]:

> *'By considering, then, that quantities increasing and begotten by increase in equal times come to be greater or lesser in accord with the greater or less speed with which they grow and are generated, I was led to seek a method of determining quantities out of the speeds of motion or increment by which they are generated; and, naming these speeds of motion or increment 'fluxions' and the quantities so born 'fluents', I fell in the year 1665 upon the method of fluxions which I have here employed in the quadrature of curves.'*

Newton repeated here his claim of priority which could be found already in Wallis's *Opera mathematica*. Hall [Hall 1980, p. 137] called this paragraph 'tactless'; but is it?

8.1.2 The Response to Newton's *Opticks*

John Bernoulli received the *Opticks* in Groningen in December 1704; some weeks later in Berlin a copy came into Leibniz's hands. Leibniz called the *Opticks* 'profound' (profundum), but he pulled *De quadratura* to pieces. The second mathematical manuscript appended, *Enumeratio linearum tertii ordinis* (Enumeration of lines of third order) would be correct and surely no innovation to geometry, but *De quadratura* comprised nothing new or difficult, as he wrote to John Bernoulli on the 25th January 1705 [Leibniz 2004, Vol.

[4]The mathematical papers were appended to the first edition only; they are missing in all further editions.

III, p. 760f.]. Apparently Newton had given up on forwarding the state of the art of the methods of quadrature as Bernoulli and himself had achieved it. In a letter to John Bernoulli from Hanover on 28th July 1705 he wrote [Leibniz 2004, Vol. III, p. 771] (original in Latin):

> *'To exceed quadrature beyond the boundaries that we have up to now and that Newton himself has left* [behind] *other arts are required in my view.'*

From this sentence we can clearly infer the esteem concerning Newton's achievements. Leibniz discussed both works appended to the *Opticks* in an anonymous review of five pages in the Acta Eruditorum of January 1705. Although this review contained great praise for Newton Leibniz burned a new fuse: In the review of *De quadratura* he discussed how differences arose from the momentary flux of a point moving along a curve. He wrote that this idea, as the idea inverse to it, namely the calculus of summation, are at the foundation of the differential calculus which was presented by its inventor Mr G.W. Leibniz and further developed by the same and others. Leibniz went on [Hall 1980, p. 138][5]:

> 'Accordingly instead of the Leibnizian differences, Mr Newton employs, and has always employed, *fluxions, which are almost the same as the increments of the fluents generated in the least equal portions of time.* He has made elegant use of these both in his *Principia Mathematica* and in other publications since, just as Honoré Fabri in his *Synopsis Geometrica* substituted the advance of movements for the method of Cavalieri.'

Leibniz then went on to explain the operations of differentiation and integration with the aid of his symbols dx, dy, and $\int$. Mr Newton, so Leibniz wrote, had worked very successfully on differentiation as well as on quadrature, but further details the reader should look up in the more recent publications of Cheyne and Craig. The fact that Leibniz referred the interested readers to the books by John Craig and George Cheyne must have shown Newton clearly that for Leibniz there was nothing in *De quadratura* what was not published already by others. But did Newton ever read this review? We may assume that he at most skimmed it, since in 1711 Newton said that he did not know it. Alarmed by John Keill the actually harmless passage [Hall 1980, p. 138]

> *'Accordingly instead of the Leibnizian differences Mr Newton employs, and has always employed, fluxions,* [...]*'*

became the stumbling block about which we will report immediately.

[5]The parts set in italics are set in italics also in the original source.

Fig. 8.1.4. Honoré Fabri (about 1608–1688). The Jesuit taught philosophy and natural sciences at the Jesuit college in Arles, then logic in Aix-en-Provence, and worked as a professor for logic and mathematics at his former school in Lyon from 1640 on. He is recognised as one of the pioneers of infinitesimal mathematics and corresponded with Leibniz. Italian title page of his opus *'Opusculum Geometricum'* of 1659 (from: 8 MATH III, 1770 (2); SUB Göttingen)

Concerning the general tone of the review one can only say that it is friendly and appreciative. Whether Leibniz had consciously set some peaks or whether he simply wrote them thoughtlessly can not be known. Was he consciously tactless? The comparison of Newton with Fabri and Cavalieri can in fact be read as a disguised attack: As (second class mathematician) Fabri had altered the methods of (first class mathematician) Cavalieri, so did Newton alter the methods of Leibniz? Was it meant in this way? I can only read it in this way. Leibniz had experienced the jealously of the English party (and I count Fatio on the English side, of course), but not in the case of Newton. He was literally pushed by John Bernoulli to insinuate illicit dealings on the side of the English. Did he become suspicious or was it as Hall [Hall 1980, p. 140f.] suspected in favour of Leibniz, namely that the review was written in great haste and that Leibniz would have chosen his words more carefully had he took more time? We have no reason to doubt Hall's assessment since Leibniz was never quarrelsome or even vindictive at any time in his life. The unstable ceasefire which settled after Fatio's first heavy attack held for another five years *after* Leibniz's review.

8.2 War is Officially Declared

8.2.1 The Last Years in Peace

Either Newton did not read the Acta Eruditorum or the explosiveness inherent in Leibniz's book review did not strike him while he browsed through its pages. After all his living conditions had completely changed. Due to his very successful work as 'Warden of the Mint' he was advanced to the 'Master of the Mint' in 1699. In the year 1701 he left Cambridge definitly behind and gave up his fellowship at Trinity College. After the death of his hated opponent Robert Hooke in the year 1703 who acted as one of two secretaries of the Royal Society Newton was elected the 11th President of the Royal Society. In 1703 Newton also became associated foreigner of the French Académie des Sciences. In April of 1705 he was even knighted by Queen Anne. Newton had turned from an academic hermit to a 'man of the world' and socialised with the political elite.

Newton's lectures on algebra, given in Cambridge a long time ago where the manuscript was deposited for more than twenty years, now became public.

Fig. 8.2.1. Two portraits of William Whiston, left: in younger years ([Painting, early 18th c], Original at Clare Collage Cambridge); in the right picture William Whiston holds a diagramm concerning his theory of planetary orbits in his hand [unknown artist of the 18th c]. He succeeded Newton to the Lucasian Chair in Cambridge in 1702. The theologian, historian, and mathematician openly made profession to Arianism, thereupon losing the Lucasian Chair in 1710 and having to leave the university. Whiston did a lot to popularise Newton's ideas but his relationship to Newton was never really close

Newton's successor on the Lucasian Chair was William Whiston (1667–1752), who eventually published the lecture under the title *Arithmetica universalis* in 1707, but Newton did not want to see his name on the book. In the year 1722 a revised edition was published by Newton himself.

The learned men of these days of course also knew immediately that the author of this anonymous publication was Newton. Although the results in the *Arithmetica universalis* were some 30 years old when published they aroused great interest in Leibniz and John Bernoulli and Leibniz wrote a very positive book review in the Acta Eruditorum.

Not a bit of an aversion of Leibniz against Newton can be felt. For Leibniz there were even no reasons for such an aversion. Leibnizians held sway over all of scientific Continental Europe; the Newtonians with their rather mystic entity 'gravity' contrasting Descartes's vortices were recognised as being rather idiosyncratic. Yes, Newton was a great mathematician about 1710 who was esteemed by Leibniz, but some kind of competition was nothing that Leibniz had to worry about. Therefore it is probably natural and not at all odd that the breach of the wobbly peace came from the side of the Newtonians 'fighting for recognition in a Cartesian universe' [Hall 1980, p. 143].

8.2.2 John Keill Appoints Himself as the Army Commander

We have already reported that most of the Newtonians were Scotsmen from 1700 on. One of these combative Scots was John Keill (1671–1721) who raised himself as the army commander of the Newton troops, started the first heavy attack, and thereby declared war. Keill studied with David Gregory at the University of Edinburgh, his hometown, and followed his master to Oxford in 1694 when the latter became Savilian Professor of Astronomy (with the support of Newton). He soon became well known in Oxford since he was an 'experimental philosopher' who enriched his lectures on Newtonian physics with experiments (the first of their kind). Such a lecture style was practised twenty years before in the Netherlands but in England this was revolutionary. So he became Lecturer for experimental philosophy, gave his lectures 1694 in rooms of Balliol College and thereafter in Hart Hall[6]. From 1699 he was the assistant of the Sedleian Professor for Natural Philosophy, Thomas Millington (1628–1704). In 1701 he published his lectures in Latin, in 1720 there followed the English translation entitled *An Introduction to Natural Philosophy, or Philosophical Lectures Read in the University of Oxford.* In the year 1703 Keill

[6]Hart Hall was founded in Oxford in 1282. During the middle ages the 'Halls' did not enjoy the same status as the colleges since they served mainly as homes for professors and students. In 1740 Hart Hall got the official title of a college. Since the founder of Hart Hall was Elias de Hertford, the college got the name Hertford College. This college is famous nowadays for its 'Bridge of Sighs'.

Fig. 8.2.2. The Bridge of Sighs near Catte Street, Oxofrd, was finished in 1914 and connects the old part of Hertford College with the new building across New College Lane [Photo: Chensiyuan 2012, Detail]

changed to Christ Church College. When Millington died in 1704 and his chair became vacant Keill refused to take the chair. He even declined the offer to take the Savilian Chair when his teacher Gregory died in 1708 and joined the civil service instead. He travelled to New England, returned to England in 1711 and worked as a cryptologist for the government of Queen Anne from 1712 until 1716. In the year 1712 he finally became Savilian Professor of Astronomy. He became a member of the Royal Society in 1700.

Keill was a very pugnacious Scotsman who had already sharply criticised the speculative cosmogonies[7] of Thomas Burnet and William Whiston in 1698. He also was in conflict with classical philologist Richard Bentley (1662–1742). Keill was a wool-dyed Newtonian, but he did not command a significant scientific talent [Hall 1980, p. 144]. We can only suspect that it was Gregory who introduced him to Newton, but Newton certainly did not appreciate Keill's agitation which the latter tried to push into the public. In Newton's private library there was just one book by Keill of 1702 which is an indication that both men were not very close. Even the later correspondence between Newton and Keill shows that Newton surely always stayed suspicious.

We may assume with some probability that it was not Newton who harnessed Keill for his purposes but that Keill made himself a servant of Newton. We

[7]Theories of the origins of the world.

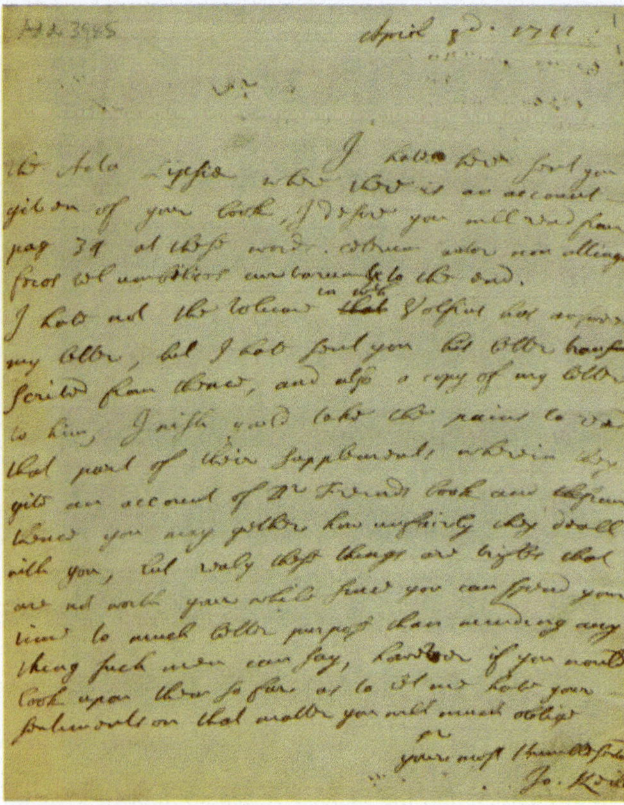

April 3d. 1711

Sr

I have here sent you the Acta Lipsiae where there is an account given of your book, I desire you will read from pag 37 at these words. [illegible] to the end.

I have not the volume in which Wolfius hath answered my letter, but I have sent you his letter transcribed from thence, and also a copy of my letter to him, I wish you would take the pains to read that part of their supplements wherein they give an account of Dr Freinds book and thence you may gather how unfairly they deal with you, but truly these things are trifles that are not worth your while since you can spend your time to much better purpose than in reading any thing such men can say, however if you would look upon them so far as to let me have your sentiments on that matter you will much oblige

Sr

your most humble servt

Jo. Keill

Fig. 8.2.3. Letter from Keill to Newton, about 1711-1718 (Cambridge University, Digital Library, ADD.3985 page 1:1r)

can also assume – like in the case of Fatio – that the first attack on Leibniz took place without Newton's knowledge [Hall 1980, p. 144].

In any case Keill's perception was apparently sharper than Newton's when he read the review of both mathematical manuscripts appended to the *Opticks* by Leibniz (cp. page 369). Here Keill saw an accusation against Newton. In Number 317 of the Philosophical Transactions of autumn 1708 Keill published a work on centrifugal forces where at the end he carried out the first strike [Westfall 2006, p. 715f.]:

> *'All of these* [propositions] *follow from the now highly celebrated Arithmetic of Fluxions which Mr. Newton, beyond all doubt, First Invented, as anyone who reads his Letters published by Wallis can easily determine; the same Arithmetic under a different name and using a different notation was later published in the Acta Eruditorum, however, by Mr. Leibniz.'*

Later Newton said that he was not aware of Keill's attack and that he was not excited, at least until Keill showed him Leibniz's review of *De quadratura* of 1705. However, Keill's work was read to the Royal Society on 14th November 1708; Newton was present as President and the members accepted the publication in the Philosophical Transactions! Hence it seems utterly unlikely that Newton had not been informed [Westfall 2006, p. 716]. Since the work read in 1708 was only published in 1710 we may also assume that the attack on Leibniz might have been a later addition [Hall 1980, p. 145]. If this would be the case Newton would have been rehabilitated. We can further assume that Keill performed the attack to please Newton, but this certainly was not all. Keill's work was concerned with forces – attracting forces, too – and the critique of Newton's concept of gravitation of which nobody knew the causes, got more and more vitriolic in the Acta Eruditorum. It is therefore more likely that Keill understood his role as a defender of Newton's ideas of the movement of the planets and that he saw the calculus as a good point of attack against the opposite side.

8.3 Leibniz Responds and Keill Fights Back

That Keill's work was published only in 1710 and that it took some time for the journals to be distributed explains why Leibniz responded to the attack by means of a letter to the Royal Society only in March 1711. Surely he remembered Fatio's attack from 1699 when he had protested successfully and when Fatio could not expect any backing from the English. But now things had changed!

8.3.1 Leibniz Asks the Royal Society for Help

On 11th March 1711 Leibniz wrote to the Secretary of the Royal Society and therefore the editor of the Transactions, Hans Sloane [Turnbull 1959–77, Vol. V, p. 97]:

> *'I could wish that an examination of the work did not compel me to make a complaint against your countrymen for the second time. Some time ago Nicholas Fatio de Duillier attacked me in a published paper for having attributed to myself another's discovery. I taught him to know better in the Acta Eruditorum of Leipzig, and you* [English] *yourselves disapproved of this* [charge] *as I learned from a letter written by the Secretary of your distinguished Society (that is, to the best of my recollection, by yourself). Newton himself, a truly excellent person, disapproved of this misplaced zeal of certain persons on behalf of your nation and himself, as I understand. And yet Mr. Keill in this*

very volume, in the [*Transactions* for] *September and October 1708, page 185, has seen fit to renew this most impertinent accusation when he writes that I have published the arithmetic of fluxions invented by Newton, after altering the name and the style of notation. Whoever has read and believed this could not but suspect that I have given out another's discovery disguised by substitute names and symbolism. But no one knows better than Newton himself how false this is; never did I hear the name calculus!of fluxions spoken nor see with these eyes the symbolism that Mr. Newton has employed before they appeared in Wallis's Works. The very letters published by Wallis prove that I had mastered the subject many years before I gave it out; how then could I have published another's work modified of which I was ignorant? However, although I do not take Mr. Keill to be a slanderer (for I think he is to be blamed rather for hastiness of judgement than for malice) yet I cannot but take that accusation which is injurious to myself as a slander. And because it is to be feared that it may be frequently repeated by imprudent or dishonest people I am driven to seek a remedy from your distinguished Royal Society. For I think you yourself will judge it equitable that Mr. Keill should testify publicly that he did not mean to charge me with that which his words seem to imply, as though I had found out something invented by another person and claimed it as my own. In this way he may give satisfaction for his injury to me, and show that he had no intention of uttering a slander, and a curb will be put on other persons who might at some time give voice to other similar* [charges]*.'*

Apparently in this case Sloane consulted Newton who approached Keill [Hall 1980, p. 169]. On 14th April 1711 Keill replied to Newton and included a copy of the review of *De quadratura* from the Acta Eruditorum. He wrote [Turnbull 1959–77, Vol. V, p. 115]:

'Sr

I have here sent you the Acta Lipsiae where there is an account given of your book, I desire you will read from pag 39 at these words. ceterum autor non attingit focos vel umbilicos curvarum &c [=Incidentally the author does not touch upon the foci or umbilics of curves][8] &c *to the end.'*

Thereupon Newton wrote in April 1711 (the particular day is not known) to Hans Sloane [Turnbull 1959–77, Vol. V, p. 117]:

[8]The full sentence in the Acta Eruditorum reads: *'Caeterum Autor non attingit focos vel umbilicos curvarum secundi generis, & multo minus generum altiorum'* meaning 'Incidentally the author does not touch upon the foci or umbilics of curves of genus two and even much less of higher genera'.

Fig. 8.3.1. Bust of Hans Sloane in the British Library [Photo: Fæ 2011]

'Sr

Upon speaking wth Mr Keill about ye complaint of Mr Leibnitz concerning what he had inserted into the Ph. Transactions, he represented to me that what he there said was to obviate the usage which I & my friends met with in the Acta Leipsica, & shewed me some passages in those Acta, to justify what he said. I had not seen those passages before, but upon reading them I found that I have more reason to complain of the collectors of ye mathematical papers in those Acta then Mr Leibnitz hath to complain of Mr Keil. For the collectors of those papers everywhere insinuate to their readers that ye method of fluxions is the differential method of Mr Leibnitz & do it in such a manner as if he was the true author & I had taken it from him, & give such an account of the Booke of Quadratures as it it was nothing else than an improvement of what had been found out before by Mr Leibnitz Dr Sheen & Mr Craig. Whereas he that compares that book with the Letters wch passed between me & Mr Leibnitz by means of Mr Oldenburg before Mr Leibniz began to discover his Knowledge of his differential method will see yt the things contained in this book were invented be-

fore the writing of those Letters. For the first Proposition is set down in those Letters enigmatically.'

Quite obviously Keill did not have to spend much time to persuade Newton of the meanness of the Leibnizian party. It seems that on both sides the barrel now overflowed which had started to be filled years ago.

8.3.2 The Royal Society Instructs Keill

On 15th April 1711 Keill defended his case before the Royal Society and this defence turned out to be very successfull [Hall 1980, p. 169]. Keill was instructed to put pen to paper and describe the accusations and the whole dispute. In particular he was asked to justify his specific reference to Leibniz since Leibniz was then a member of the Royal Society for nearly 40 years [Turnbull 1959–77, Vol. V, p. 116f., footnote (7)]. The result was a letter to Sloane to be sent to Leibniz. Its content was summarised in the Journal Book of the Royal Society and got the approval to be sent on 11th July 1711 [Turnbull 1959–77, Vol. V, p. 133ff.]. Apparently it had taken Keill a while to compose this large letter and suspicion is strong that he had arranged with Newton. This, however, is not certain. Admittedly Keill showed himself as an expert on the development of Newton's mathematics which rather suggests a contact with Newton. To me it only seems certain that Newton was relieved not to fight for himself but to put Keill in charge. This seems to fit his temperament much better than a personal and public conflict. As a most clear-cut proof of Newton's mastery of the calculus even *before* 1676 Keill made reference to a letter of Newton to Collins of 20th December 1672 [Turnbull 1959–77, Vol. I, p. 247ff.] which was found under the estate papers of Collins'.

These papers had come into the possession of the London mathematics teacher William Jones who in 1711 published a book with the title *Analysis Per Quantitatum Series, Fluxiones, ac Differentias: cum Enumeratione Linearum Tertii Ordinis* (Analysis by means of series, fluxions, and differences of quantities together with an enumeration of lines of third order) serving as a documentation of Newton's achievements in the area of mathematics since it contained Newton's work *De analysi*, cp. page 118. Hence Keill could also have been informed of Newton's mathematical works by Jones without discussing it with Newton.

Keill's letter [Turnbull 1959–77, Vol. V, p. 142] started conciliatory. He did not want to disparage Leibniz but he wrote:

'I admit that I said that the Arithmetic of Fluxions was discovered by Mr. Newton, which was published with a change of name and method of notation by Leibniz but I do not mean these words to be understood as though I were arguing that either the name which Newton gave to his

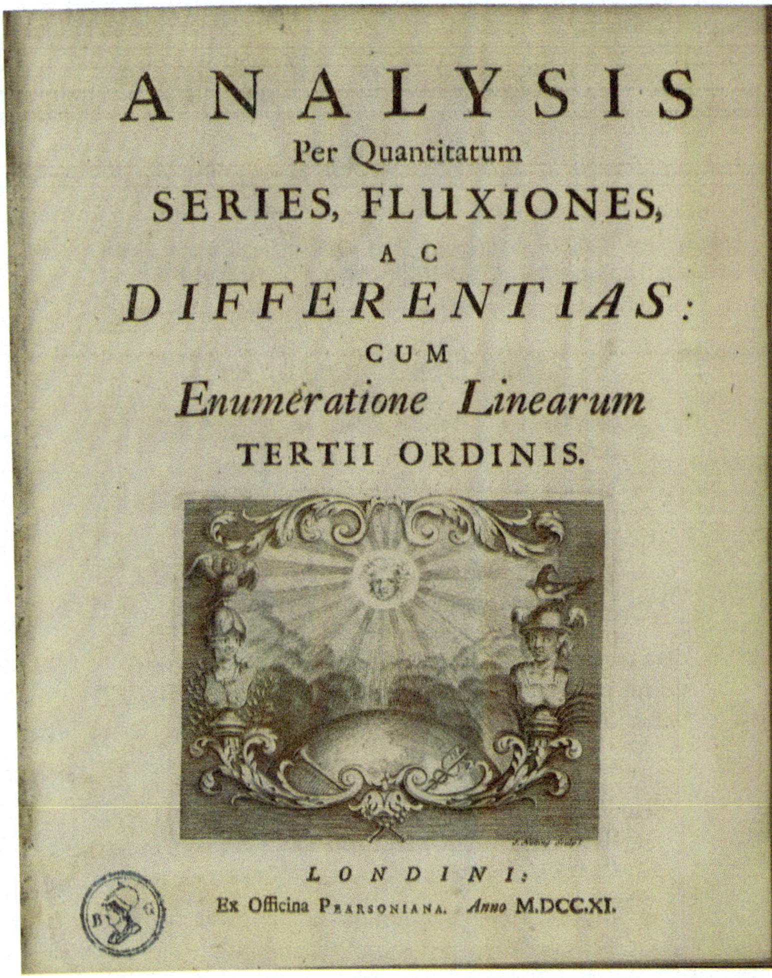

ANALYSIS
Per Quantitatum
SERIES, FLUXIONES,
AC
DIFFERENTIAS:
CUM
Enumeratione Linearum
TERTII ORDINIS.

LONDINI:
Ex Officina PEARSONIANA. *Anno* M.DCC.XI.

Fig. 8.3.2. Title page of the book by William Jones of the year 1711 serving as a documentation of Newton's achievements
(Ghent University Library, BIB.MA.000248)

method or the form of notation that he developed were known to Mr. Leibniz; I suggested only this, that Mr. Newton was the first discoverer of the Arithmetic of Fluxions or Differential Calculus; however, as he had in two letters written to Oldenburg (which the latter transmitted to Leibniz) given pretty plain indications to that man of most perceptive intelligence, whence Leibniz derived the principles of that calculus or at least could have derived them; But as that illustrious man did not need for his reasoning the form of speaking and notation which Newton had used, he imposed his own.'

The passage that Leibniz used a 'change of name' or 'form of notation' for the arithmetic of fluxions or even that Newton had communicated these to Leibniz constituted a strategic error of Keill [Hall 1980, p. 171] since he was not able to present evidence.

Then Keill noted the book review in the Acta Eruditorum of 1705 as the actual reason for his allegations. From it, he wrote, he could learn that Leibniz was the actual inventor of Newton's calculus although Newton had found his calculus at least 18 years before Leibniz. Eventually Keill cited the letter from Newton to Collins of the year 1672 and wrote [Turnbull 1959–77, Vol. V, p. 144]:

> *'It clearly appears from this letter that Mr. Newton had the method of fluxions before the year 1670, that in which Barrow's lectures were published.'*

At the end of the letter [Turnbull 1959–77, Vol. V, p. 149] Keill stated

> [...] *'that the hints and examples of Newton were sufficiently understood by Leibniz,* AT LEAST AS TO THE FIRST DIFFERENCES; FOR AS TO THE SECOND DIFFERENCES IT SEEMS THAT LEIBNIZ WAS RATHER SLOW TO COMPREHEND THE NEWTONIAN METHOD, AS PERHAPS I WILL SHOW MORE CLEARLY IN A LITTLE WHILE.*'*

The addition in small caps came from Newton's quill in Newton's copy of the letter. This addition later appeared exactly as cited here in 1712 in the Commercium epistolicum.

Hence Keill's concillatory short beginning of the letter became a farce subsequently. Ultimately it all revolved around the content of the two *Epistolae* from which Leibniz should have been able to reconstruct Newton's method. But this is also the great weakness of Keill's argumentation: It simply makes no sense to believe that Leibniz could have arrived at the very same method as Newton. If Leibniz really would have been so clever to have learned everything from the 'hints and examples' of Newton, how could he have imitated Newton's method? If he had imitated Newton's method, how could the hints and examples have served him [Hall 1980, p. 174]? Keill argued that in the *Epistola prior* Newton had explained the method of series expansions with which one could represent the increments of flowing quantities [Turnbull 1959–77, Vol. V, p. 146]:

> *'In the first Letter sent to Leibniz by Oldenburg Mr. Newton taught the method by which quantities may be reduced to infinite series, that is, by which the increments of flowing quantities may be displayed;* [...]*'*

but that was simply not true! Newton had communicated the binomial expansion to Leibniz, but by no means an example of its use for differences (increments) of fluxions. In the discussion of the contents of the *Epistola posterior* Keill claimed [Turnbull 1959–77, Vol. V, p. 148]:

> '[...], *he gave examples of the procedure from differentials to integrals.'*

This was also not in accordance with the facts. For one thing Newton did not give a single example of such operation, furthermore he did not use words like 'differential', 'integral', or 'Fluxion'.

Hence instead of an apology Leibniz received another attack on his scientific reputation, and also the accompanying letter by Sloane which was initiated by Newton [Hall 1980, p. 176] was barely polite. Leibniz must have been surprised; at least he took some time before he wrote a further letter to Sloane on 29th December [Turnbull 1959–77, Vol. V, p. 207f.]:

> *'Gottfried Wilhelm Leibniz presents a grand salute to the very celebrated Mr. Hans Sloane*
>
> *What Mr. John Keill wrote to you recently attacks my sincerity more openly than* [he did] *before; no fair-minded or sensible person will think it right that I, at my age and with such a full testimony of my life, should state an apologetic case for it, appearing like a suitor before a court of law, against a man who is learned indeed, but an upstart with little deep knowledge of what has gone before and without any authority from the person chiefly concerned.*
>
> [...]
>
> *I, too, and my friends have on several occasions made obvious our belief that the illustrious discoverer of fluxions arrived by his own efforts at basic principles similar to our own. Nor do I lay a less claim than his to the rights of the discoverer, as Huygens (who was a most clever and incorruptible judge) also acknowledged before the public – rights which I have not hastened to claim for myself but rather concealed the discovery for nine years, so that no one can claim to have forestalled me.*
>
> *Thus I throw myself upon your sense of justice,* [to determine] *whether or not such empty and unjust braying should not be suppressed, of which I believe even Newton himself would disapprove, being a distinguished person who is thoroughly acquainted with past events; and I am confident that he will freely give evidence of his opinion on this* [issue]*.'*

8.4 Newton Springs into Action

Sloane received Leibniz's letter in January 1712. At this time Keill by persistent arguing had ultimately persuaded Newton that the latter's honour was attacked. As inherent in his character Newton would not do things by halves: he not only wanted to clearly prove that he was the first inventor of the calculus but also that Leibniz had stolen it from him. At this point it was out of the question to somehow share the honour of the discovery – Leibniz needed to be dishonoured [Hall 1980, p. 177].

8.4.1 The Fast Track Leading to the *Commercium epistolicum*

Leibniz's letter was presented to the President on 11th February 1712, *'delivr'd to the President to consider of the Contents thereof'*, as it was noted in the Journal Book of the Royal Society [Whiteside 1967–81, Vol. VIII, p. 480]. A few days later the President Sir Isaac Newton started to compile a defence on his own account. This defence began as a letter to Sloane [Turnbull 1959–77, Vol. V, p. 212ff.] but developed into a speech to the *'Gentlemen'* of the Society [Whiteside 1967–81, Vol. VIII, p. 539ff.], [Turnbull 1959–77, Vol. V, p. xxivf.]:

> *'Gentlemen*
>
> *The Letter of Mr Leibnitz wch was read before you when I was last here relating to me as well as to Mr Keill I have considered it, & can acquaint you that I did not see the papers in the Acta Leipsica till the last summer & therefore had no hand in beginning this controversy. The controversy is between the author of those papers & Mr Keil. And I have as much reason to complain of that author for questioning my candor & to desire that Mr Leibnitz would set the matter right without engaging me in a dispute wth that author as Mr Leibnitz has to complain of Mr Keil for questioning his candor & to desire that I would set the matter right without engaging him in a controversy with Mr Keil. For if that author in giving an account of my book of Quadratures gave every man his own, as Mr Leibnitz affirms, he has taxed me with borrowing from other men & therebye opposed my candor as much as Mr Keil has opposed the candor of Mr Leibnitz & and so was the agressor. Mr Leibnitz & his friends allow that I was the inventor of the method of fluxions: & claim that he was the inventor of the differential method. Both may be true because the same thing is often invented by several men. For the two methods are one & ye same method variously explained & no man could invent the method of fluxions without knowing first how to work in the augmenta momentanea of fluent quantities wch augmenta Mr Leibnitz calls differences.*

Dr Barrow & Mr Gregory drew tangents by the differential method before the year 1669. I applied it to abstracted aequationes before that year & therebye made it generall. Mr Leibnits might do the like about the same time; but I heard nothing of his having the method before the year 1677. When and how he found it must come from himself. By putting the fluxions of quantities to be in the first ratios of ye augmenta momentanea I demonstrated the method & thence called it the method of fluxions: Mr Leibnitz uses it without a Demonstration. [9]'

This speech to the Royal Society sounds exceptionally defensive and fair. Leibniz was granted the right of having likewise found a calculus but Newton (justly) attached importance to the fact that he was the first inventor. In the draft, i.e. the planned letter to Sloane, Newton used a stronger language and mentioned also the *Epistolae* [Turnbull 1959–77, Vol. V, p. 213]:

'[...] *the author is the first author, & I am not yet convinced that he* [Leibniz] *was the first author of that method.*

[...]

For By the Letters wch passed between him & me in the years 1676 & 1677 he knows that I wrote a treatise of the methods of converging series & fluxions six years before I heard of his differential method.'

We do not want to follow the further course of discussions in the Journal Book of the Royal Society, but on 17th March there is the entry [Turnbull 1959–77, Vol. V, p. xxv]:

'*Upon account of Mr Leibnitz's Letter to Dr. Sloane concerning the Dispute formerly mentioned between him and Mr Keill, a Committee was appointed by the Society to inspect the Letters and Papers relating thereto; viz. Dr Arbuthnot, Mr Hill, Dr Halley, Mr Jones, Mr Machen, and Mr Burnet, who were to make their Report to the Society.*'

Already on 31st March Mr Robartes was received into the committee; Mr Bonet, the envoy of the Prussian King in London, on 7th April, and Abraham de Moivre, Francis Ashton, and Brook Taylor on 28th April. The last three could not have contributed much to the committee's report since the official report was presented to the Royal Society already one week after their election on 5th May 1712 [Turnbull 1959–77, Vol. V, p. xxv]. The time elapsed from the installation of the committee on 17th March to the final report was just fifty days! To speak about a thorough inspection of the documents is therefore out of the question.

[9]The entry in the Journal Book of the Royal Society terminates at this point.

8.4.2 The *Commercium epistolicum*

The judgement of the Royal Society concerning the priority dispute became known as the *Commercium epistolicum* and carries the full title *Commercium Epistolicum D. Johannis Collins et aliorum de analysi promota: jussu Societatis Regiae in lucem editum* (Corresponence of Mr John Collins and others on the advancement of analysis, edited on behalf of the Royal Society). The document consists of 122 printed pages which contain excerpts of letters and scientific papers. The final judgement reads as[10]:

> *'We have consulted the Letters and Letter-books in the Custody of the Royal Society, and those found among the Papers of Mr. John Collins, dated between the Years 1669 and 1677 inclusive; and shewed them to such as knew and avouched the Hands of Mr. Barrow, Mr. Collins, Mr. Oldenburg and Mr. Leibnitz; and compar'd those of Mr. Gregory with one another, and with Copies of some of them taken in the Hand of Mr. Collins; and have extracted from them what relates to the Matter referr'd to us; all which Extracts herewith deliver'd to you, we believe to be genuine and authentick: And by these Letters and Papers we find,*
>
> *I. That Mr. Leibnitz was in London in the beginning of the Year 1673, and went thence in or about March to Paris, where he kept a Correspondence with Mr. Collins by means of Mr. Oldenburg, till about September 1676, and that Mr. Collins was very free in communicating to able Mathematicians what he had receiv'd from Mr. Newton and Mr. Gregory.*
>
> *II. That when Mr. Leibnitz was the first time in London, he contended for the Invention of another Differential Method properly so call'd; and notwithstanding that he was shewn by Dr. Pell that it was Mouton's Method, persisted in maintaining it to be his own Invention, by reason that he had found it by himself, without knowing what Mouton had done before, and had much improved it. And we find no mention of his having any other Differential Method than Mouton's, before his Letter of 21st of June 1677, which was a Year after a Copy of Mr. Newton's Letter, of 10th of December 1672, had been sent to Paris to be communicated to him; and above four Years after Mr. Collins began to communicate that Letter to his Correspondents; in which Letter the Method of Fluxions was sufficiently describ'd to any intelligent person.*
>
> *III. That by Mr. Newton's Letter of the 13th of June 1676 it appears, that he had the Method of Fluxions above five Years before the writing*

[10]The *Commercium epistolicum* can be found completely on the internet, e.g. at `http://books.google.de/books/about/Commercium_epistolicum_D_Johannis_Collin.html?id=YDgPAAAAQAAJ`.

of that Letter. And by his Analysis per Aequationes numero Terminorum Infinitas, communicated by Dr. Barrow to Mr. Collins in July 1669, we find that he had invented the Method before that time.

IV. That the Differential Method is one and the same with the Method of Fluxions, expccting the Name and Mode of Notation; Mr. Leibnitz calling those Quantities Differences, which Mr. Newton calls Moments or Fluxions; and marking them with the Letter d, a Mark not used by Mr. Newton. And therefore we take the proper Question to be, not who invented this or that Method, but who was the first Inventor of the Method. And we believe that those who have reputed Mr. Leibnitz the first Inventor, knew little or nothing of his Correspondence with Mr. Collins and Mr. Oldenburg long before; nor of Mr. Newton's having that Method about Fifteen Years before Mr. Leibnitz began to publish it in the Acta Eruditorum of Leipsick.

For which Reason, we reckon Mr. Newton the first Inventor; and are of Opinion, that Mr. Keill in asserting the same, has been no ways injurious to Mr. Leibniz. And we submit to the Judgment of the Society, whether the Extract of Letters and Papers now presented to you, together with what is extant to the same purpose in Dr. Wallis's third Volume, may not deserve to be made Publick.'

There is no question that this text flowed directly from Newton's quill since his draft with his own notes still exists [Turnbull 1959–77, Vol. V, p. xxv]. Not only was Leibniz accused as a plagiarist but also as somebody who consciously had hidden his knowledge of the priority of others. Did he not conceal Newton's right as first inventor and did he not later on even denied him this right? Newton must have been intensely occupied with Leibniz from 1711 on and it took him the summer of the year 1712 to draft the *Commercium epistolicum*. There is only little to be found in his correspondence since Newton was excessively cautious [Turnbull 1959–77, Vol. V, p. xxvii].

In the Journal Book of the Royal Society one then finds [Turnbull 1959–77, Vol. V, p. xxvif.]:

'To which Report the Society agreed nemine contradicente, and ordered that the whole Matter, from the Beginning, with the Extracts of all the Letters relating thereto, and Mr Keill's and Mr Leibnitz's Letters, be published with all convenient speed may be, together withe the Report of the said Committee.

Ordered that Dr Halley, Mr Jones, and Mr Machin, be desired to take care of the said Impression (which they promised) and Mr Jones to make an Estimate of the Charges against the next Meeting.

Mr Keill said he would draw up an Answer to Mr Leibnitz's last Letter, it relating chiefly to himself; which he was also desired to do, and that it should be read at a Meeting of the Royal Society.'

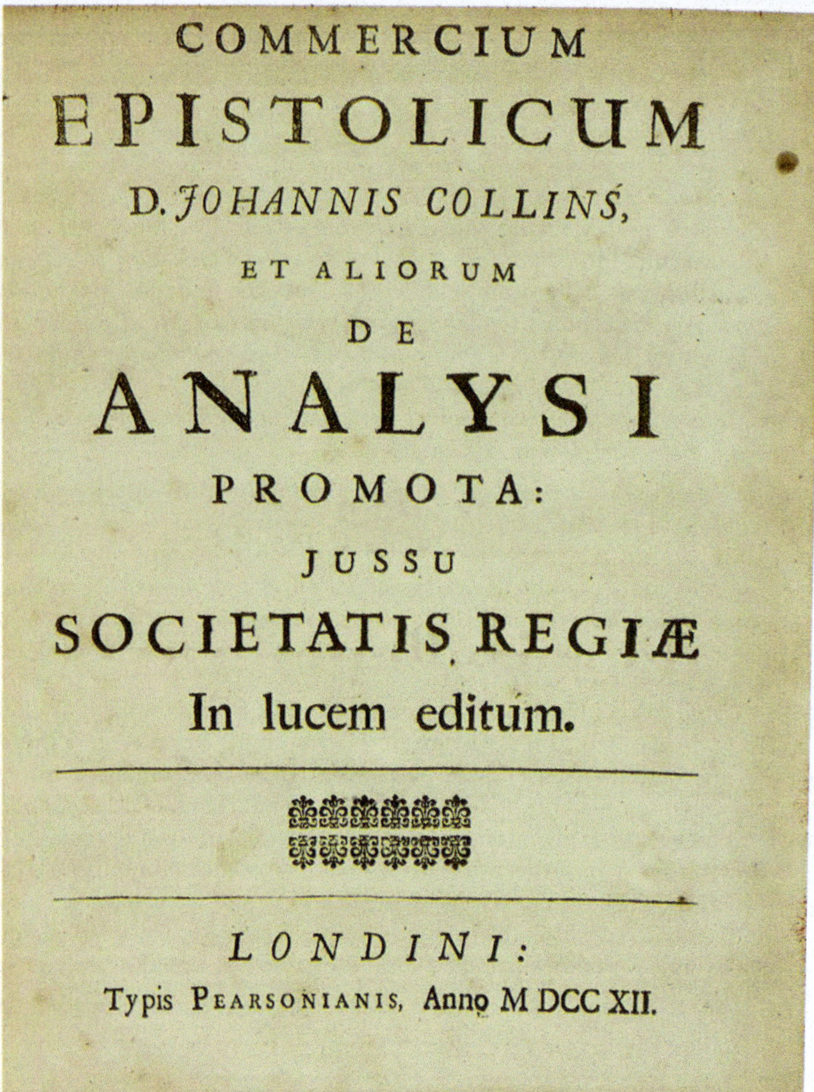

COMMERCIUM
EPISTOLICUM
D. JOHANNIS COLLINS,
ET ALIORUM
DE
ANALYSI
PROMOTA:
JUSSU
SOCIETATIS REGIÆ
In lucem editum.

LONDINI:
Typis PEARSONIANIS, Anno M DCC XII.

Fig. 8.4.1. Title page of the *Commercium epistolicum* (PBA Galleries/pbagalleries.com, San Francisco)

The *Commercium epistolicum* was published in London in January 1713[11]. Leibniz resided at this time again in Vienna and was informed by John Bernoulli only some months later [Antognazza 2009, p. 488].

[11]Due to the Julian calander the printing date on the title is 1712.

8.5 Escalation

8.5.1 An Error in the *Principia*

In a letter of 7th July to Leibniz, John Bernoulli wrote from Basel [Turnbull 1959–77, Vol. VI, p. 1] where he meanwhile became the successor of his brother Jacob [Turnbull 1959–77, Vol. VI, p. 3] [Antognazza 2009, p. 495]:

> *'My nephew brought from Paris a single copy of the Commercium Epistolicum Collinsii et aliorum de analysi promota, which the Abbé Bignon had handed to him, having received a number of copies sent from London for distribution to the learned. I have read it, not without a fair amount of attention. This hardly civilized way of doing things displeases me particularly; you are at once accused before a tribunal consisting, as it seems, of the participants and witnesses themselves, as if charged of plagiary, then documents against you are produced, sentence is passed; you lose the case, you are condemned.*
>
> [...]
>
> *But I am driven to break off for the present; I do indeed beg you to use what I now write properly and not to involve me with Newton and his people, for I am reluctant to be involved in these disputes or to appear ungrateful to Newton who has heaped many testimonies of his goodwill upon me. More another time; for now, farewell, etc.'*

Fig. 8.5.1. Isaac Newton [Oil painting about 1715–1720, unknown painter, attributed to the 'English School'], and Gottfried Wilhelm Leibniz [Painting by Johann Friedrich Wentzel the Elder, about 1700]

Fig. 8.5.2. Abbé Jean-Paul Bignon (1662–1743) of the Congregation of the Oratory of Saint Philip Neri was a French cleric and statesman and served Louis XIV. He was a member of the Académie française (left: [Painting by Hyacinthe Rigand, 1693 or 1707] Palace of Versailles; right: [Engraving by Edelinck, 1700])

The nephew was Nicolaus I Bernoulli (1687–1759), a son of the painter and councillor of Basel, Nicolaus Bernoulli. He studied mathematics with his uncles Jacob and John and got a degree from the University of Basel under Jacob in 1704. He earned his doctorate of law in 1709; in his thesis he studied the applications of probability theory to problems of law. He travelled to Holland, England and France in 1712, became member of the Berlin-Brandenburg Academy of Sciences in 1713, member of the Royal Society in 1724, and member of the Academy of Bologna (Istituto delle Scienze di Bologna) in 1724. He became the successor of Jacob Hermann as professor of mathematics in Padua in 1716 but left Italy in 1722 to take the chair for logic at the university of his hometown which he exchanged for a professorship of law in 1731. He stood on the side of his uncle John in the priority dispute and played an important role concerning the critique of some of the results in the *Principia* [DSB 1971, Vol. II, p. 56f., J.O. Fleckenstein].

In the letter above it is interesting to note the clear backdown of John Bernoulli with respect to Leibniz. John was received into the Royal Society in 1712 and was certainly worried that Leibniz would again exploit him as a battering ram on his own account as he already did in the conflict with Fatio in 1699.

To John Bernoulli Leibniz wrote on 17th June 1713 [Turnbull 1959–77, Vol. VI, p. 8]:

'I have not yet seen the little English book directed against me; those idiotic arguments which (as I gather from your letter) they have brought forward deserve to be lashed by satirical wit. They would maintain Newton in the possession of his own invented calculus and yet it appears that he no more knew our calculus than Apollonius knew the algebraic calculus of Viète and Descartes. He knew fluxions, but not the calculus of fluxions which (as you rightly judge) he put together at a later stage after our own was already published. Thus I have myself done him more than justice, and this is the price I pay for my kindness.'

Now also Leibniz's mood had shifted; as John Bernoulli he seemed to be convinced that actually Newton was the plagiarist. However, we have to keep in mind that this letter was written in a highly tense mood. In a letter to Leibniz of 18th July 1713 [Turnbull 1959–77, Vol. VI, p. 12f.] John Bernoulli piled the fire up against the English. John and the mathematician, philosopher, and Leibnitian Christian Wolff (1679–1754) urged Leibniz to publish his own statement concerning the history of the development of calculus and Leibniz was not averse [Hall 1980, p. 193]. He considered writing his own *Commercium epistolicum* but eventually dropped the idea. In 1714 he began to write down his recollections concerning the developments in the work *Historia et origo calculi differentialis* (History and origin of the differential calculus) but the manuscript remained fragmentary. An answer to the *Commercium epistolicum* based on historical facts was never published.

It is hard to say whether the reading of the *Commercium epistolicums* ever changed anyones opinion – the English behind Newton had their clear convictions as the Europeans on the Continent on Leibniz's side had their own. The most important effect of the *Commercium epistolicums* was the escalation of the dispute it caused since now the conflict extended to other areas. One can identify clearly two main lines in the development of the escalation [Hall 1980, p. 193]. The first main line was aimed at questioning Newton's capacity as a mathematician. In this way one could question that a scientist making mistakes could be the inventor of the calculus. Generalissimo[12] for this main line was John Bernoulli who launched serious attacks but liked to entrench himself behind Leibniz. After Leibniz's death John Bernoulli openly continued the fight against the English. The second main line resulted from the critique of Leibniz and others (Huygens, for example) on Newton's theory of gravitation in the *Principia*. Leibniz and Huygens were seeking for a mechanical explanation of the phenomenon of gravitation and were always closer to Descartes's mechanistic vortex theory than to the 'occult' nature of Newton's gravity. Even in his philosophical writings Leibniz showed his aversion against a force with long-distance effects; e.g. in his *Theodizee* (Theodicy) which was

[12] I do use this outdated notion for the 'General of the Generals' consciously. It was invented for Albrecht von Wallenstein in the year 1625.

Fig. 8.5.3. René Descartes ([Painting by Frans Hals about 1649] understood as being a study for the painting in the Louvre, Statens Museum for Kunst, Kopenhagen); the illustration of Descartes's vortices ([Descartes, Principia Philosophiae 1656, p. 72] Bayerische Landesbibliothek, München (Munich), Sign.: 858338 4 Ph.u. 44 858338 4 Ph.u. 44)

published in 1710. Leibniz had set out the foundations of his dynamics shortly after the *Principia* had been published and now it was claimed that the actual, true progress lay in Leibniz's dynamics. The Generalissimo of that main line was Leibniz himself.

John Bernoulli started combat already at a time when the *Commercium epistolicum* was in the making. In the *Principia* in Book II, Proposition 10, Newton had treated the following problem: Suppose a body moves along a trajectory under the influence of a constant gravitation and a resisting force proportional to the product of density of the medium and the square of the velocity. Given the trajectory, find the density of the medium and the velocity at any point of the path [Newton 1999a, p. 264]. John had great interest in such mechanic problems and was surprised when he arrived at a result differing from Newton's in considering a specific case [Hall 1980, p. 194]. He started checking Newton's calculations and found a consequence that hinted to a real error of Newton. John informed Leibniz in 1710 about this error and sent a corresponding paper to the French Academy in Paris in the following year. However, this paper was published only in 1714. When John's nephew Nicolaus I stayed in London in September 1712 Newton learned that this error had been uncovered. Newton and John Bernoulli had always kept a friendly correspondence with each other, but Newton knew exactly that John was a fierce partisan and close associate of Leibniz. It is very likely that Nicolaus I did not

ESSAIS
DE
THEODICÉE
SUR LA
BONTE' DE DIEU,
LA
LIBERTE' DE L'HOMME,
ET
L'ORIGINE DU MAL.
PAR M. LEIBNITZ.
NOUVELLE EDITION,
Augmentée de l'Hiſtoire de la Vie & des
Ouvrages de l'Auteur,
PAR M. L. DE NEUFVILLE.
TOME PREMIER.
A AMSTERDAM,
Chez FRANÇOIS CHANGUION.
MDCCXXXIV.

Fig. 8.5.4. Title page of an Amsterdam edition of the *Theodizee* of 1734 (UCB Lausanne)

say anything about the publication which was submitted for publication to the Academy which included some contemptuous remarks concerning Newton as a mathematician. As it turned out the error proved to be not as serious as John had hoped and Newton could easily remedy it; but after all Newton was corrected by John Bernoulli in this instance. A detailed examination of Proposition 10 can be found in [Guicciardini 1999, p. 233ff.]. The reason of Newton's error was explained by Nicolaus I in a postscript to the paper of his uncle. Nicolaus I wrote [Hall 1980, p. 197]:

> '[It was Newton's] *method of changing indeterminate, variable quantities into converging series, and making the terms of these series serve as successive differentials, which has led Mr Newton into error.'*

In contrast to Newton's correct explanation of successive fluxions (i.e. higher derivatives) he failed in the proof of Proposition 10: If z^n is the fluent changing in an infinitesimal amount of time from z to $(z+0)$ then it follows from the binomial theorem the series expansion

$$z^n + noz^{n-1} + \frac{(n^2-n)o^2z^{n-2}}{2} + \frac{(n^3-3n^2+2n)o^3z^{n-3}}{6} + \dots$$

Newton then wrote without justification that the second summand would be the first increment or the difference of the fluent being proportional to the first fluxion. The third summand is the increment of second order or the difference being proportional to the second fluxion, and so on. This, wrote Nicolaus I, is correct only for the first summand following z^n, since in the second summand the increment is only $(n^2-n)o^2z^{n-2}$, in the third $(n^3-3n^2+2n)o^3z^{n-3}$, and so on. Nicolaus I pointed out that if one would multiply the summands with 1, 2, 6, and so on, and would then follow Newton's argumentation, one would arrive at the correct result. For Generalissimo John Bernoulli this was a clear sign that Newton could not even differentiate correctly [Hall 1980, p. 198]!

The correction by Bernoulli had entirely practical effects since this part of the *Principia* was already printed in the second edition. Therefore we should look more closely to the events happening in connection with this second edition.

8.5.2 The Second Edition of the *Principia* Becomes Due

Newton had started to be concerned with thoughts and notes concerning a new edition soon after the *Principia* was publsihed. About the year 1709 he went seriously to work. This was also the time of his first clash with the Astronomer Royal.

Flamsteed Reloaded

In the year 1711 the results of Flamsteed had still not been published in form of the planned great work *Historia coelestis* (History of the heavens). Queen

Fig. 8.5.5. Personal physicians to Queen Anne, left: Richard Mead (1673–1754) cared for Queen Anne on her deathbed and became personal physician to King George II in 1727. He was witness of the meeting between Newton and Flamsteed in Crane Court [Mezzotint by R. Houston after A. Ramsey, 1st half of 18th c]; right: John Arbuthnot (1667–1735) became a member of the Royal Society in 1704 [Painting by Godfrey Kneller 1723]

Anne had ordered (influenced by Newton?) to tie the Greenwich observatory to the leash of the Royal Society and therefore to place Flamsteed under supervision. Who was behind this order is not certain; Flamsteed thought it certain that it was Newton. Based on this order Flamsteed received on 25th March 1711 a letter by John Arbuthnot, personal physician to the Queen, informing him that the Queen had ordered him to finish the *Historia coelestis.*

Although Flamsteed did not react initially his past data were already at the Royal Society. On 5th April 1711 the Astronomer Royal learned that the printing of the *Historia coelestis* was already in progress [Westfall 2006, p. 689]. Now Flamsteed wrote to Arbuthnot that he was pleased that the printing had finally started but that he had no new results. He would need help with the evaluation and he asked Arbuthnot for a meeting to discuss the further steps. For Newton this was the last straw. He wrote an unambigiuous letter to Flamsteed ordering him to immediately send the missing data [Turnbull 1959–77, Vol. V, p. 102]. Whether the letter was ever sent is questionable since Flamsteed nowhere gives a reference to it. A few days later he saw the first pages of the printing and found changes and errors which were inserted by his archenemy Halley. Although Arbuthnot tried hard to pour oil on the troubled waters Flamsteed stopped being accessible to rational arguments or discussions. He saw his life's work in the hands of his enemies but had no option to withdraw his data. Newton and Halley alone now cared for the publication of the *Historia coelestis* in the year 1712.

Fig. 8.5.6. George I (1660-1727), King of Great Britain and Ireland, about 1715 ([Painting by Joachim Kayser] Residenzmuseum Celler Schloss (Celle Castle))

In the preface it was noted that Flamsteed refused to release his observations and that only an order by Prince George had solved that problem. Halley appeared as the one having corrected the errors of Flamsteed. Even missing charts of star constellations were attributed to Flamsteed. Newton cared that Halley received £150 for his work; £25 more that one would have paid Flamsteed.

But Newton was not satisfied yet. He went through the first edition of the *Principia* and erased Flamsteed's name at 15 positions [Westfall 2006, p. 693]. That was not all. As overseer of the observatory he ordered Flamsteed to attend a meeting at Crane Court to where the Royal Society had moved in 1710.

Fig. 8.5.7. Residence of the Royal Society in Crane Court. Newton as the president had made sure that the Royal Society bought this house and moved in 1710 from the former residence in Gresham College. Crane Court was the residence of the Royal Society until 1780 [Walter Thornbury, 1873]

The meeting took place at 6th November in the presence of Hans Sloane and Richard Mead. The 64 years old Flamsteed was suffering from gout and could only climb the stairs with some assistance. Newton wanted to lay his hands on Flamsteed's instruments to secure further observations.

Now Flamsteed could triumph since he could unequivocally prove that all instruments in the Royal Observatory of Greenwich were his property and his property alone. Newton went berserk! As Flamsteed wrote to Abraham Sharp who worked with him in the observatory [Westfall 2006, p. 693]:

> *'Y^e^ P^r^ ran himself into a great heat & very indecent passion.'*

Newton shouted in anger: *'as good have no Observatory as no Instruments'.* Flamsteed complained about the publication of his star catalogue and that one had robbed him of the fruits of his labour. To Sharp Flamsteed wrote [Westfall 2006, p. 693]:

> *'at this he fired & cald me all the ill names Puppy &c. that he could think of. All I returned was I put him in mind of his passion desired him to govern it & keep his temper, this made him rage worse, & he told me how much I had receaved from ye Goverm^t^ in 36 years I had served.'*

Hence the meeting at Crane Court turned out to be a failure for Newton. Eventually Flamsteed even got afloat when Queen Anne died in 1714 and a government of Whigs replaced the Tories. This change came with House of Hanover taking over the reign. In 1714 Leibniz's Hanoverian Duke George Louis, son of Ernest Augustus and Sophia, became King of Great Britain and Ireland as George I.

Not later than 1715 the close contact between Newton and the new government had ceased while Flamsteed's contact was excellent due to the Duke of Bolton who acted as Lord Chamberlain [Westfall 2006, p. 694]. This Duke offered to send Flamsteed all available copies of the *Historia coelestis* in order to remove them from the market and to enable Flamsteed to publish his star catalogue exactly as he had planned it long ago. Flamsteed agreed and Newton received an order from Bolton to send the remaining 300 copies to the Astronomer Royal. After further quarrels Flamsteed received the copies on 8th April 1716.

Flamsteed had nearly finished his *Historia coelestis* when he died in 1719. His two former assistants Joseph Crothwait and Abraham Sharp cared for the completion of the work which finally was published posthumously in 1725 as *Historia coelestis britannica.*

Principia reloaded

In 1709 Newton seriously started work on the second edition of the *Principia.* From the publishing day of the first edition in 1687 Newton nurtured the plan to publish a new edition. He had written lists of errors and omissions and had a special copy bound with empty pages between the printed ones. In this copy he could write the changes for the new edition [Westfall 2006, p. 698], [Cohen 1971, p. 200ff.]. Already Fatio once wanted to care for a second edition and David Gregory later on, but this did not work out, probably because Newton intentionally delayed it. With Newton's departure from Cambridge and the time consuming work at the Mint in London a long time passed until Newton could finally embark. Apparently he was pushed into it because the Master of Trinity College, Richard Bentley, wanted to arrange an edition for the

Fig. 8.5.8. Bust of Roger Cotes, posthumously created by Peter Scheemakers in 1758; right: Richard Bentley (1662–1742), he was a classic scholar who was elected Master of Trinity College in 1700; he tried several times to reform the College from which he greedily drew large sums of money at the same time (Bust by Louis-François Roubiliac [Photo: Andrew Dunn 2004] in the Wren Library of the Trinity College in Cambridge)

University Press. In mid-1708 Newton saw the first pages of this edition but did not react. Only with the appearance of a brilliant young man in the year 1709 the project could really start. This man was Roger Cotes (1682–1716) [Gowing 1983].

Cotes was a supporter of Bentley in Cambridge concerning the latter's efforts to free the College from its scientific lethargy and to breath new, young life into it. Cotes came to Trinity College in 1699 and Bentley soon took notice of him, so that he became the first Professor of Astronomy on the newly created Plume Chair[13] in 1705, even before he had earned a Master's degree. Cotes got to know Newton via Bentley and when Bentley ordered the building of an observatory for the Plume Professor, Cotes travelled to London and visited Newton, who promised the construction of a pendulum clock for the new observatory. In fact Cotes got the clock – but only seven years later [Westfall 2006, p. 703].

The brilliant young man came into play apparently because Bentley saw that he would neither have the time, nor the mathematical abilities to care for a second edition of the *Principia*. Furthermore Bentley felt that Cotes was obliged beyond all measure to satisfy his patron. Hence in the autumn 1709 a fascinating correspondence between the 27 years old Cotes and the 66 years old Newton began. On 22nd October 1709 – Cotes had written to Newton to ask for the copies concerning the corrections and to correct two errors in

[13]The chair was endowed in 1704 by Thomas Plume (1630–1704) who was a member of Christ College, Cambridge, and Archdeacon of Rochester.

Newton's *De quadratura* at the same time – Newton replied [Gowing 1983, p. 15]:

> *'I would not have you be at the trouble of examining all the Demonstrations in the Principia. Its impossible to print the book w'out some Faults & if you print by the copy sent you, correcting only such faults as appear in reading over the sheets to correct them as they are printed off, you will have labour more than is fit to give you.'*

Cotes had no interest in just correcting obvious error and he turned out to be a godsend for Newton. He was not only a brilliant mathematician who really understood the *Principia*, he was also not afraid to mention obscure calculations and wrong formulae and to make proposals concerning phrases used by Newton which he thought failed. Newton accepted some of Cotes's changes but some he did not. Cotes stayed patient throughout and showed such a great admiration for Newton that Newton finally accepted his work. When the printing press had to be stopped because Cotes had detected a problem in Book II in Section VI Newton initially was reluctant, but then the old physicist by and by relished the scientific exchange with Cotes. On 12th July 1710 Newton wrote to Cotes [Westfall 2006, p. 707], [Turnbull 1959–77, Vol. V, p. 707]:

> *'I am wth my humble service to your Master & many thanks to your self for your trouble in correcting this edition*
>
> *S^{r}*
>
> *Your most humble servant*
>
> *Is. Newton'*

The hardest piece of work was Section VII in Book II. Here the resistance experienced by a body moving in a medium was concerned. We should remind ourselves that a satisfactory theory of friction became possible not before the mid-19th century; hence Newton actually fought on a lost front line. Book II was in fact the weakest book of the *Principia* and contained many errors. As István Szabó mercilessly judged [Szabó 1996, p. 152]:

> *'The second book concerning the fluids is however nearly completely independent and almost entirely wrong.'*
>
> (Das zweite Buch, welches die Flüssigkeiten behandelt, ist hingegen fast vollkommen eigenständig und beinahe ganz falsch.)

Already Fatio had pointed out an error in Proposition 36, concerning the outflow of a fluid from a vessel, to Newton (cp. page 290). Cotes did not like Newton's newly revised Section VII which he got forwarded in autumn 1710. He particularly mistrusted Proposition 36 since he had seen an experiment in a book by Edme Mariotte (about 1620–1684) with which Newton's theory was not compatible. Although Newton played down the issue Cotes bothered on, but had to wait long for an answer. The year 1711 marked a burdensome time in Newton's life. Not only was he concerned with the second edition of the *Principia*, but also with the activities within the Mint, since the government wanted to issue a new silver coin [Westfall 2006, p. 710f.]. To make matters worse Leibniz's complaint letter concerning Keill arrived in 1711.

In any case the printing press stood still for half a year before Cotes was satisfied with the changes Newton made in Section VII, but he soon found new problems in Section VIII [Westfall 2006, p. 729], so that the work only slowly advanced further to Book III. In the autumn of 1712 Nicolaus I Bernoulli resided in London and told Newton about an error which his uncle John had found. On 25th October Newton wrote to Cotes [Turnbull 1959–77, Vol. V, p. 347]:

> *'There is an error in the tenth Proposition of the second Book, Prob. III, wch will require the reprinting of about a sheet & an half. I was told of it since I wrote to you & am correcting it. I will pay the charge of reprinting it, & send it to you as soon as I can make it ready.'*

In January 1713 Newton sent an advance notice to Cotes that the *Principia* would end with a *General Scholium.* This scholium begins with the statement that Descartes's vortex theory would have some problems [Newton 1999, p. 939]:

> *'The hypothesis of vortices is beset with many difficulties.'*

Then remarks concerning the motion of the planets follow and then – God! Newton designed a concept of God and deduced the existence of absolute space and absolute time as consequences of God's infinite extension and his infinite duration. And although God now could have served as the cause of gravitation, Newton [Newton 1999, p. 943] wrote:

> *'Thus far I have explained the phenomena of the heavens and of our sea by the force of gravity, but I have not yet assigned a cause to gravity.*
>
> [...]
>
> *I have not as yet been able to deduce from phenomena the reason for these properties of gravity, and I do not feign hypotheses*[14]*.'*

Meanwhile Cotes brooded over a preface in which he could take Leibniz and his *Tentamen* – written shortly after the first publication of the *Principia* –

[14]This is probably the most cited statement of Newton: *'Hypotheses non fingo'* - I do not feign hypotheses.

Fig. 8.5.9. Samuel Clarke (1675–1729) ([Engraving by Nixon after T. Gibson, after 1702] Wellcome Images/Wellcome Trust, London) was an English philosopher and theologian and an intimate of Newton. Clarke became known in essence through his correspondence with Leibniz in the years 1715 and 1716 in which he defended Newton's philosophy against the objections of Leibniz and against the accusation of atheism. The correspondence terminated with Leibniz's death, but Clarke published the correspondence in 1717 in London

polemically to court [Westfall 2006, p. 749]. Cotes tried to win either Newton or Bentley to write such a preface, but both declined. Bentley was of the opinion that Leibniz's name should not be mentioned at all and Newton even refused to read Cotes's preface because he feared to be made responsible for it. Cotes finally could persuade Samuel Clarke to read the preface and to remark on it.

As wished by Bentley Leibniz's name was not mentioned, but the preface contained an acid reply to Newton's critics. Hence, the addressee of this preface –

Leibniz – was apparent to every informed reader. It goes without saying that Leibniz understood the preface as a renewed provocation.

The second edition of the *Principia* was eventually finished at the beginning of July 1713; Newton got six copies of which he sent one to Abbé Bignon, Fontenelle and Varignon, respectively. Apparently he wanted to impress the members of the French Academy. John Bernoulli did not receive a copy. Newton wrote as an apology that he had got only a few copies. There were in fact 700 copies printed[15] at an overall cost of £117. At the end of 1715 there were only 71 copies left in the possession of Bentley who made a fortune of nearly £200. The demand on the Continent was so large that a reprint was published already in 1714 in Amsterdam and a second one in 1723 [Westfall 2006, p. 750].

Cotes neither received any money nor any words of thanks for his work. In a draft of the preface by Newton from autumn 1712 Cotes received thanks, but this part was deleted by Newton. This probably happened because Newton was dissatisfied with Cotes in connection with the error found by John Bernoulli and communicated by Nicolaus I, which required the reprint of some of the printing sheets [Gowing 1983, p. 16]. Perhaps Newton was disappointed that Cotes had not detected this error. Out of the blue, Cotes suddenly received an additional list of errors, change requests and additions, of which Newton apparently expected that Cotes would attach it to the already finished book. Cotes answered somewhat coldly on 2nd January 1714 [Gowing 1983, p. 17f.]:

> *'I observe You have put down about 20 Errata besides those in my Table*[16]*. I am glad to find they are not of any moment, such I mean as can give the reader any trouble. I had myself observ'd several of them, but I confess to You I was asham'd to put 'em in the Table, lest I should appear to be too diligent in trifles. Such Errata the Reader expects to meet with, and they cannot well be avoided. After You have now Your self examined the Book & found these 20, I beleive You will not be surpriz's if I tell You I can send You 20 more as considerable, which I have casually observ'd, & which seem to have escap'd You: & I am far from thinking these forty are all that may be found out, notwithstanding that I think the Edition to be very correct. I am sure it is much more so than the former, which was carefully enough printed; for besides Your own corrections & those I acquainted You with whilst the Book was printing, I may venture to say I made some Hundreds* [corrections]*, with which I never acquainted You.'*

The correspondence of the two men which began so warm-hearted and led to a fruitful collaboration over four years terminated at that point. Cotes died

[15] Gowing speaks of 750 copies [Gowing 1983, p. 16].
[16] Cotes had already attached a table of printing errors he had found.

33 years old at the beginning of the year 1716 of a violent fever. His cousin Robert Smith (1689–1768) published Cotes's mathematical works in 1722 and hoped that Newton would help him, but Newton did not stir a finger [Westfall 2006, p. 751].

8.5.3 A Flyer is Published

In order to answer the *Commercium epistolicum* which was freely distributed in Europe Leibniz came up with the idea of an anonymous pamphlet which became known under the Latin name *Charta volans* (flyer) and carried the date of the 29th July 1713 [Turnbull 1959–77, Vol. VI, p. 15ff.]. In the *Charta volans* Leibniz processed Bernoulli's letter of 7th June (cp. page 387) in which also Newton's error in the *Principia* was treated. Since he was in too much of a hurry to deal with the accusations of the English, Leibniz referred to an 'eminent mathematician' without naming Bernoulli. Leibniz wrote that it would follow from the explanations of this mathematician that Newton, when he claimed the invention of the differential calculus for himself, [Turnbull 1959–77, Vol. VI, p. 19],

> [...] *'he was too much influenced by flatterers ignorant of the earlier course of events and by a desire for renown; having undeservedly obtained a partial share in this, through the kindness of a stranger* [namely Leibniz], *he longed to have deserved the whole – a sign of a mind neither fair nor honest. Of this Hooke too has complained, in relation to the hypothesis of the planets* [the law of gravitation], *and Flamsteed because of the use of* [his] *observations.'*

Here finally Leibniz openly flung the gauntlet and accused Newton of plagiarism. He was also not even afraid to turn the quarrels of Newton with Hooke and Flamsteed, which he knew of only by hearsay, against Newton. More conciliatory Leibniz finished the letter [Turnbull 1959–77, Vol. VI, p. 19] as follows:

> *'Moreover, there is no doubt that in England many distinguished persons deplore this vanity and injustice among Newton's disciples; and the bad conduct of the few should not be imputed to the whole nation.'*

We know from Newton's own writings that he received the *Charta volans* already in autumn 1713 from John Chamberlayne (1666–1723) [Hall 1980, p. 202]. Chamberlayne was a writer, translator, and journalist, who became a member of the Royal Society in 1702 and corresponded with Leibniz from 1710 on. On the 10th March 1714 Chamberlayne wrote to Leibniz as if he had heard about the conflict between Leibniz and Newton just recently and offered himself to act as a mediator [Turnbull 1959–77, Vol. VI, p. 71]. In any case, Newton did not react at once to the *Charta volans*; perhaps he did not attribute much importance to an anonymous pamphlet. The *Charta volans* was reprinted in a French translation together with Leibniz's remarks concerning the dispute in the November/December edition of the 'Journal Littéraire de la Haye' which appeared every second month. In the May/June edition of 1713 of this journal Keill had already published a long history of the differential calculus from Newton's viewpoint. Thereby he had effectively popularised the *Commercium epistolicum* in the French language. It seems certain that Keill had not pursued this publication without having the backing of Newton, and it takes no wonder that Leibniz also replied in this journal. The remarks of Leibniz were – as could be expected – not very flattering for the Newtonians.

Perhaps Newton had hoped that he could silence Leibniz with the *Commercium epistolicum*, but this hope was in vain. Newton wanted to react and turned to Keill in a letter of 13th April 1714 [Turnbull 1959–77, Vol. VI, p. 79f.]

> *'Mr Leibnitz in August last, by one of his correspondents published a paper in German conteining the judgement of a nameless mathematician in opposition to the judgment of the Committee of the Royal Society, with many reflexions annexed.* [...] *And the whole is printed in the journal Littéraire pag. 445. And now it is made so publick I think it requires an Answer.*
>
> [Newton requests from Keill to draft this answer and wants to send him his thoughts with later mail.]
>
> *You need not set your name to it.'*

Now Newton himself compiled a reply to the *Charta volans* and some accompanying remarks for publication in the Journal Littéraire and this reply is preserved in both, an English as well as in a French version [Turnbull 1959–77, Vol. VI, p. 80ff.]. This reply, planned to be published in the Journal Littéraire was never published. However, Keill's reply was printed in the Journal Littéraire de la Haye in summer 1714 [Hall 1980, p. 207f.]. Keill referred to the alleged error made by Newton in the *Principia* and remarked that Newton also had found errors in Leibniz's *Tentamen*. Newton's remarks on the *Tentamen* are preserved [Turnbull 1959–77, Vol. VI, p. 116f.]. It seems strange that Keill exploited here the weapons of the Leibnizians, obviously wanting to repay in the same coin.

Keill's reply to the *Charta volans* had hardly been published when rumours came up that Newton had a finger in the pie. On the other side of the Channel John Bernoulli felt more and more uneasy. That Leibniz had exploited his letter anonymously in the *Charta volans* was one thing, but had he not criticised the *Principia* and pointed to the error which was communicated to Newton by Nicolaus I? In spring 1714 he got the uncomfortable feeling that Newton

Fig. 8.5.10. Willem Jacob 's Gravesande had studied law at the University of Leiden and became a Doctor of Law in 1707. On the occasion of George's I coronation he visited London and resided there for one year. He became a member of the Royal Society and got acquainted with Newton and Keill, before he became Professor of Astronomy and Mathematics at the University of Leiden. By means of experiments concerning the depth of penetration of brass globes in soft clay he showed that the formula $E = mv^2$ of kinetic energy must be correct and that Newton's formula $E = mv$ had to be wrong. He communicated his results to Madame du Châtelet (1706–1749) who translated Newton's *Principia* into French and cared for a popularisation of Newton's physics in France together with Voltaire [Painting by Hendrik van Limborch, 18th c]. Bernard le Bovier de Fontenelle (1657–1757) is today recognised as one of the most important personalities of the early French enlightenment. He was a member of the Académie des Sciences from 1697 on and acted as their secretary until 1740. His numerous laudatory writings concerning scientists are famous, which he composed in an extraordinaryly elegant style ([Painting by Louis Galloche], Palace of Versailles)

thought badly of him. He also believed that Newton would exclude him from the Royal Society [Hall 1980, p. 211f.], although he was calmed by Abraham de Moivre. Nonetheless Bernoulli demanded further counter-attacks against the accusations of the English from Leibniz. But somehow a state of standstill had been reached, a stalemate, and Leibniz did not take the opportunity to reply again to Keill's reply in the Journal Littéraire. In the German lands the *Acta Eruditorum* were completely on Leibniz's side; in Holland contributions of either the English or the German side were printed, although the Journal Littéraire de la Haye was inclined to the English side since one of its editors, Willem Jacob 's Gravesande (1688–1742), had arguably become the greatest Newtonian on the Continent.

In France not much attention was payed to the priority dispute, probably due to the War of the Spanish Succession in which Germany, Holland, and England were allied against France [Hall 1980, p. 213f.]. After all, this war

Fig. 8.5.11. Caroline, Princess of Wales, an educated and beautiful woman, was married in the year 1705 as Wilhelmina Charlotte Caroline of Brandenburg-Ansbach to George Augustus, son of George I and his wife Sophia Dorothea of Celle. She lived as Princess of Wales in England since 1714 and became Queen of Great Britain after the death of George I in 1727 as wife of George II. She corresponded with Leibniz and was a great supporter of Händel who dedicated his Water Music to her. She supported Voltaire during his English exile from 1726 until 1729 who thankfully dedicated his *La Henriade* to her. ([Painting by Michael Dahl, about 1730], Shire Hall, Warwick)

raged until 1714. The Academy in Paris was inclined to the side of Leibniz; their mathematician Pierre Varignon had learned mathematics from John Bernoulli and corresponded busily with Leibniz. The Academy's Secretary, Bernard le Bovier de Fontenelle, was an adherent of the ideas of Descartes and therefore initially not inclined to Newton's side, but that changed when Newton's *Opticks* became known in France.

After the War of the Spanish Succession was over relations between France and England improved. Some Englishmen moved to Paris to live there, and some Frenchmen moved to London hoping for the acquaintance with Newton or with his remarkably beautiful and charming niece (actually half-niece), Catherine Barton (1679–1739), who kept house for Newton and was a close friend of Jonathan Swift. Young French scientists eagerly absorbed Newton's theories and really wanted to see the experiments with light, colour, and the vacuum which they hitherto had only read about [Hall 1980, p. 214f.]. All earned esteem for John Bernoulli and Gottfried Wilhelm Leibniz: The action was in London and in Paris, and not in Basel and not at all in Hanover.

Flamsteed had profited from the take-over of the Hanoverian George I; this can not be said in case of Leibniz. For some time the combatants around Newton must have been worried that George I, after all Leibniz's Duke, would prefer Leibniz's side in the priority dispute, but all concerns quickly appeared unfounded. Leibniz had lost the trust of the Court already under Duke Ernest Augustus since he was not advancing fast enough in the history of the Guelfs. Prospects remained bleak under George. The Londoners had Newton! Yes, that was a great man! And what had the Hanoverians? An elderly philosopher working much too slow, taking too much freedoms, and travelling for most of the time.

John Bernoulli went on persuading Leibniz to react reasonably to the *Commercium epistolicum.* In a letter of 6th February 1715 he wrote to Leibniz [Turnbull 1959–77, Vol. VI, p. 204]:

> '*You would do well to publicize some* [problems] *where Newton would, as you know, find himself in difficulties. Doubtless there are many of them to hand, which were once discussed between us, and which are not easily dealt with by the ordinary differential method:* [...]'

Here Bernoulli referred to the new problems of the calculus of variations. In fact, the English did not look too well concerning the problems which now were publicly posed by Leibniz and Bernoulli; but the course of the priority dispute could not be changed. Merely the front lines were hardened.

At the end of the year 1715 Leibniz wrote to Abbé Antonio Schinella Conti (1677–1749). The letter contained a problem concerning the calculation of normals to a family of curves which we will have to discuss shortly. Conti, known as Abbé Conti, was an Italian mathematician, philosopher, and physician from Padua, who played a somewhat unfortunate role in the priority dispute. He had lived in France and became an admirer of Malebranche. Malebranche was a member of the Congregation of the Oratory of Saint Philip Neri, hence an Oratorian. Since he was eager to impress others by his acquaintance with great men he also wrote to Leibniz in the spring of 1715. He wrote that he would go to England to support Leibniz as he had already done in Paris [Hall 1980, p. 217]. In London Conti apparently felt comfortable. He was amicably received by Leibniz's acquaintance Wilhelmina Charlotte Caroline of Brandenburg-Ansbach, known as Caroline of Ansbach, who lived in England from 1714 as Princess of Wales after she had married George, a son of George I and later King George II.

He also made acquaintance with Newton and found him very reasonable and cordial. In fact, Newton treated him as if he was a famous guest and Conti was tremendously flattered. Conti gave Leibniz's letter concerning the normals to families of curves to Newton and this made Leibniz feel uncomfortable [Hall 1980, p. 216f]. He wrote to Caroline and complained about Conti's disloyalty.

8.6 The Leibniz-Clarke Controversy

Leibniz exploited Conti and the interest of Caroline in his philosophical correspondence with Samuel Clarke to start another attack on Newton. In a letter to Conti of 6th December 1715 he attacked substantial parts of Newton's natural philosophy [Turnbull 1959–77, Vol. VI., p. 250ff.]: If every body is heavy, then gravitation would be an occult or at least supernatural quality. If one says that God had created this law of nature, the law itself is nowhere near to be obvious if it violates the nature of creation. God is not a 'world soul' and does not require to use the material world as an organ for sensory impressions; a nerve system, so to speak. Phenomena like universal gravitation, atoms, and the void could neither be proved by Newton nor by observations of experimental sciences. Newton's method of induction which started from the phenomena is excellent, so Leibniz wrote, but if the initial data would be faulty it seemed permissible to formulate hypotheses and to trust them before better facts would be known. It is a pity, Leibniz remarked inamicably, that Newton did not possess able disciples, while he himself was so much happier.

These thoughts had ripened in Leibniz for quite some time and his *Theodizee* of 1710 had at least indirectly given the impetus for the controversy between him and Samuel Clarke [Hall 1980, p. 218f.]. Caroline asked Clarke to translate the *Theodizee* into English but he declined, since the ideas expressed therein were too different from his own. When Caroline showed Clarke a letter by Leibniz in which he expressed concerns about Newton's philosophy Clarke replied to Leibniz and a correspondence began which only ended with Leibniz's death in November 1716 [Clarke 1956]. Clarke must have been a fascinating personality himself. Although a brilliant thinker his Arian views had prevented a career in the Church of England. Voltaire wrote that Clarke had deserved the Chair of the Archbishop of Canterbury had he not been an Arian [Voltaire 2011, Letter VII, p. 32]. Nevertheless Clarke managed not to be excluded from church and to retain his position. Clarke provided a Latin translation of Newton's *Opticks* in 1706 where probably also Abraham de Moivre had a finger in the pie. Following a legend Newton rewarded him with the enormous sum of £500 [Hall 1980, p. 219]. Newton and Clarke certainly knew each other long before 1715. Since they were neighbours there exists no correspondence. There is no hint that Newton played an active role in the sense that he drafted parts of the correspondence himself in the Leibniz-Clarke controversy, but it seems certain that Clarke was well briefed by Newton, talked to Newton, and made the letters which Leibniz wrote to Caroline available to Newton. That Newton ultimately spoke through Clarke can be clearly seen from the fact that Clarke's letters comprise the most detailed representation of Newton's natural philosophy together with its epistemological, mathematical, physical, and theological foundations and conclusions not to be found in any of Newton's works in this clarity, cp. the Introduction in [Clarke 1956, p. ixff.]

Fig. 8.6.1. Gottfried Wilhelm Leibniz [Painting: © Historisches Museum Hannover (Hanover)] and Samuel Clarke (1675–1729), ([Engraving by J. Goldar after T. Gibson], Wellcome Library London, Wellcome Images/Wellcome Trust)

Leibniz wrote to Conti (and to Clarke) via the Princess of Wales, so that also the Court was informed about the contents of the letters. We know that Newton was requested to react since he wrote [Hall 1980, p. 220f.]:

> *'When Mr l'Abbé Conti had received a letter from Mr Leibnitz with a large Postscript against me full of accusations forreign to the Question, & the Postscript was shewed to the King, & I was Pressed for an answer to be also shewed to his Majesty,* [...] *the same was afterwards sent to Mr Leibnitz* [...]*'*

Apparently Newton suspected that Leibniz had exploited his contacts to the Court to set him under pressure. The letter which Newton had to draw up exists in an official version, later given to the printer by Newton himself, and in a much longer draft. Essentially the letter was concerned with the mathematical developments of the 1670s and repeated the topics already known from the *Commercium epistolicum*. Only in passing did he mention Leibniz's critique concerning Newton's natural philosophy; it is only in the draft that we find a significantly longer discussion [Hall 1980, p. 221] in which Newton for his part attacks Leibniz's opinions of space and time.

The connection between God and the infinitely extended space postulated by Newton in the *Principia* and even more so in the *Opticks* is fairly close to the analogy of body and soul in men. This *must* have attracted Leibniz's critique. In Leibniz's metaphysics God is necessary since he created the laws of nature and got the universe going, but from the point of creation this perfect clockwork continues. Newton was inclined to explain the forces of nature by an intervening God. Following Leibniz's belief Newton thereby viewed God as a bad or unfit or not perfect clockmaker who had to impinge on the course of the universe from time to time. Although Clarke declared that Leibniz had not rightly understood Newton's point we can hardly misinterpret or misunderstand Newton's remarks concerning this matter. In the 'Queries' which Newton had added to his *Opticks* the numbers 28 and 31 are of particular interest. In Query 28 Newton discussed the problems of the ether, of the resistance in fluids, and the movement of the planets. He then came to *'the first cause'* and asked for the causes of gravity, why we are able to move by our own will, and where the instinct of animals came from. Then he wrote [Newton 1979, p. 370]:

> *'And these things being rightly dispatch'd, does it not appear from Phaenomena that there is a Being incorporeal, living, intelligent, omnipresent, who in infinite Space, as it were in his Sensory, sees the things themselves intimately, and throughly perceives them, and comprehends them wholly by their immediate presence to himself:* [...]*'*

In Query 31 we read [Newton 1979, p. 403]:

> *'Also the first Contrivance of those very artificial Parts of Animals, the Eyes, Ears, Brain, Muscles, Heart, Lungs, Midriff, Glands, Larynx, Hands, Wings, swimming Bladders, natural Spectacles, and other Organs of Sense and Motion; and the Instinct of Brutes and Insects, can be the effect of nothing else than the Wisdom and Skill of a powerful ever-living Agent, who being in all Places, is more able by his Will to move the Bodies within his boundless uniform Sensorium, and thereby to form and reform the Parts of the Universe, than we are by our Will to move the Parts of our own Bodies.'*

That is the propellant which fuelled the Leibniz-Clarke controversy. In order to document Leibniz's reaction concerning Newton's statements in the *Opticks* we cite from a letter of Leibniz to John Bernoulli of spring 1715 [Hall 1980, p. 222]:

> *'When I was told that Newton says something extraordinary about God in the Latin edition of his Opticks, which until then I had not seen, I examined it and laughed at the idea that space is the sensorium of God, as if God from whom every thing comes, should have need of a sensorium.* [...] *And so this man has little success with metaphysics.'*

8.7 Newton's *Account* and Raphson's *History of Fluxions*

The Leibniz-Clarke controversy took a large part in Leibniz's last year of his life. During this time two publications appeared in England, one of them was *An Account of the Book entitled Commercium Epistolicum Collinii & aliorum* [Hall 1980, p. 263ff.], a product of Newton's quill. The *Account* was published anonymously in the Philosophical Transactions of the year 1715 and was later added to the new edition of the *Commercium epistolicum* in a Latin translation in 1722. It is the only long, coherent history of the priority dispute from Newton's own hand. The *Account* emerged from the reply which Newton planned to publish in the Journal Littéraire de la Haye in 1714 and which was revised several times [Hall 1980, p. 226].

The *Account* is structured roughly into five sections. In the first Newton showed that he had his method of fluxions at his disposal already in the year 1669 in that he not only gave *De analysi* as evidence but also letters which were exchanged at this time. The second section was supposed to show that Leibniz could not have had developed his calculus before 1677 and that he betrayed himself only after he had received the *Epistola posterior* in the reply letter to Oldenburg of 1st July 1677 (cp. page 223), writing there that he had learned from the *Epistola posterior*. Like a good historian Newton deduced from the dates and contents of letters that Leibniz always was at least one step behind him and that all of which Leibniz boasted came from Newton or Gregory. Leibniz's method to determine tangents would actually stem from Barrow. Newton wrote [Hall 1980, p. 286] that Leibniz

> ' [...] *has followed this Method of Tangents exactly, excepting that he has changed the Letters a and e of Dr. Barrow into dx and dy.*'

The third section of the *Account* is devoted to the difference between the fluxions and the differentials. Newton played down the importance of a symbolism to which Leibniz always attached great emphasis. He tried to show that the calculus of fluxions was inherently superior to the differential calculus. The method of fluxions was based on geometry while Leibniz's infinitely small differentials neither existed in geometry nor in nature. Leibniz's method would therefore be suitable only for analysis; in contrast Newton's would also allow for proofs (synthesis) [Hall 1980, p. 296]:

> *'Mr. Newton's Method is also of greater Use and Certainty, being adapted either to the ready finding out of a Proposition by such Approximations as will create no Error in the Conclusion, or to the demonstrating it exactly: Mr. Leibnitz's is only for finding it out.'*

In a fourth section Newton wrote that he had treated fluxions of second and higher order already from the beginning of the development of the calculus of fluxions. These explanations can certainly be understood as a reaction to the critique of Bernoulli. Then it followed an interesting and famous passage concerning the role of the calculus of fluxions in writing the *Principia* [Hall 1980, p. 296]:

> *'By the help of the new Analysis Mr. Newton found out most of the Propositions in his Principia Philosophiae: but because the Ancients for making things certain admitted nothing into Geometry before it was demonstrated synthetically, he* [Newton] *demonstrated the Propositions synthetically, that the Systeme of the Heavens might be founded upon good Geometry. And this makes it now difficult for unskilful Men to see the Analysis by which those Propositions were found out.'*

This section is certainly a work of fiction since there is not even a single reference to a 'fluxional' development of the *Principia* in Newton's manuscripts.

In the fifth and last part Newton defended his natural philosophy against the 'mechanical philosophy' of Leibniz.

The *Account* was presumably read only in England and the Englishmen were already on Newton's side. In November 1715 a French translation was published in the Journal Littéraire de la Haye with assistance by John Keill. Newton's hope that the *Account* would hit the Continent like a bomb was not met even with the French version. Leibniz replied with a few disparaging sentences in the Nouvelles Littéraires that the Continent knew since 1684 that Leibniz was the actual inventor of the calculus until the Englishmen had surprised the world with an opposite assertion. Then Leibniz again cited the judgement of Bernoulli [Hall 1980, p. 231]. The *Account* did not change anything; it even hardened the front lines further.

Also in the year 1715 Joseph Raphson's book *De historia fluxionum. The history of fluxions, showing in a compendious manner the first rise of, and various improvements made in that incomparable method* was published posthumously. We know little of Raphson (1648–1715)[17]; he attended Jesus College Cambridge and left with the title of an MA in 1692. He became a member of the Royal Society already in 1689 and a year later published the book *Analysis aequationum universalis, seu ad aequationes algebraicas resolvendas methodus generalis, et expedita, ex nova infinitarum serierum doctrina, deducta ac demonstrata* (Universal analysis of equations or general and easy methods to solve algebraic equations based on the new doctrine of infinite series derived and proven) containing the approximative method for roots which we today know as the Newton-Raphson method [Thomas/Smith 1990, p. 151]. Raphson met Newton in Cambridge in 1691; Roger Cotes was acquainted with him and he knew already in 1711 that Raphson planned to publish on fluxions. *The history of fluxions* is much less a history of fluxions than rather an

[17]Even the years of his birth and death are uncertain, cp. [Thomas/Smith 1990, p. 159].

apologia in favour of Newton [Hall 1980, p. 224]. Some of the statements in Raphsons book of which a Latin translation appeared yet in 1715 are simply wrong and there were even deficiencies concerning his presentation of the calculus of fluxions [Cajori 1919, p. 49f.]. Newton always maintained that he had no knowledge about the book before it was published or had even prevented its publication for three or four years; this is hard to believe [Hall 1980, p. 225]. After Leibniz's death in 1716 Newton took care in 1717 or 1718 that Raphson's book was reissued with a list of errors, the old title page, and an added last chapter on the *Commercium epistolicum.*

9 Beyond the Grave

T. Sonar, *The History of the Priority Diʃpute between Newton and Leibniz*, https://doi.org/10.1007/978-3-319-72563-5_9

The priority dispute could have been terminated finally with Gottfried Wilhelm Leibniz's death in November 1716, but neither on the English nor on the Continental European side was one inclined to show piety or forbearance.

9.1 Poor Abbé Conti

9.1.1 An Addendum to a Letter of 6th December 1715

John Bernoulli had urged Leibniz to put pressure on the English by means of new mathematical problems concerning the differential calculus. The letter to Conti of 6th December 1715 contained such a problem [Westfall 2006, p. 773]: *'To test the pulse of our English Analysts ...'*. Here Leibniz posed the problem of finding a curve intersecting a whole family of curves under a right angle. The problem came from John Bernoulli and in order to illustrate it Leibniz gave as an example a family of hyperbolas. This had consequences since the English understood the problem as if they should solve the problem for hyperbolas only. Solutions arrived quickly; from Halley, Keill, Pemberton, Taylor, and even from an Oxford student by the name of James Stirling. Newton also attempted a solution which was published anonymously in the Philosophical Transactions.

By then Newton was approaching his 75th birthday and he was no longer able to solve the actual problem. Leibniz wrote four times to England to explain the problem and to clarify that it was concerned with families of arbitrary curves and not only with families of hyperbolas. He also asked John Bernoulli for help who provided him with a more general formulation of the problem, which Leibniz sent to Conti, therebye revealing Bernoulli as being the originator of the problem. In England one said contemptuously that Leibniz simply stated a new problem after one had solved the previous one successfully. One also remarked that John Bernoulli had again confronted Newton [Westfall 2006, p. 774].

9.1.2 Newton Becomes Active Again

One of the friends and supporters of Leibniz in England was a John Arnold. In February 1716 Arnold wrote to Leibniz to inform him about some noteworthy events [Turnbull 1959–77, Vol. VI, p. 274f.]. On the occasion of a visit to Conti Arnold had met Newton and the talk that developed was concerned with the priority dispute. Upon Newton's request Arnold admitted that he would possess a copy of the *Charta volans* which he wanted to show Newton. So Conti and Arnold visited Newton who blamed Leibniz and Keill entirely for the priority dispute. He had prevented Keill from writing anything, so Newton said, until Leibniz finally accused him for plagiarism. Keill had written much too

Fig. 9.1.1. Gravestone on the grave of James Stirling (1692–1770) in Greyfriar's Kirkyard, Edinburgh. The Scotsman Stirling went to Oxford as an 18-year-old to study at Balliol College but was expelled from university due to his contacts to Jacobites. In Venice he earned his living as a Professor of Mathematics and published on Newton in England. With the help of Newton he returned to London in 1725. He published important works concerning mathematics and the question of the true shape of the earth [Photo: Roegel 2012]

ruggedly, but would have been annoyed about the problem posed which he saw as a challenge to the whole nation. Then Newton courted Arnold and invited him and Conti to stay for dinner. Newton presented to Arnold the French edition of the *Account* for edification. Arnold remarked that similar courtesies had pulled Conti to Newton's side. Newton would now have begun to gain the attention of the Hanoverian court. That way he had arranged to demonstrate optical experiments to the mistress of the King, Baroness of Kielmannsegg (1675–1725). Arnold additionally informed Leibniz that Newton planned to draw the Diplomatic corps into the priority dispute.

Indeed, as we know from a letter which Conti wrote to Taylor five years after the events, Newton organised a reception for ambassadors at the end of February or the beginning of March 1716 in the rooms of the Royal Society in Crane Court. It was planned to present the originals of the letters which were cited in the *Commercium epistolicum* to the ambassadors. Almost surely Newton had expected that the ambassadors would carry the conviction on the correctness of the assertions in the *Commercium epistolicum* into other countries, but this plan went awry completely [Westfall 2006, p. 775]. The Baron Johann

Adolf von Kielmansegg, husband of Baronesse von Kielmansegg, thought this approach not sufficient. He postulated that the priority dispute could only be settled by a direct correspondence between Newton and Leibniz. All ambassadors present agreed and later even the King showed his sympathy towards such a solution. Surely Newton was unhappy concerning these course of events. It is likely that he never wanted to reply to Leibniz's letter to Conti, but due to the intervention of the King he could not escape. Within a few days in February 1716 he put a letter to paper which pleased George I. We do not find anything new in it; it is nothing but a condensed version of the *Account* [Hall 1980, p. 233]. The letter to Leibniz via Conti was written on 8th March 1716 and shows an irreconcilable Newton who reproached Leibniz for not having answered personally to the *Commercium epistolicum* but that he instead had pushed forward an 'eminent mathematician' [Turnbull 1959–77, Vol. VI, p. 285]:

> *'Mr Leibnitz has hitherto avoiding returning an Answer to the same; for the Book* [it is the *Commercium epistolicum*] *is matter of fact & uncapable of an Answer. To avoid answering it he pretended the first year that he had not seen this Book nor had leasure to examin it, but had desired an eminent Mathematician to examin it. And the Answer of the Mathematician (or pretended Mathematician) dated 7 June 1713, was inserted into a defamatory Letter* [that is the *Charta volans*] *dated 29 July following & published in Germany without the name of the Author or Printer or City where it was printed.'*

Of course Newton was aware of the identity of the 'eminent mathematician', hence he added the little nasty remark *'or pretended Mathematician'* consciously as a thorn against John Bernoulli whom he meanwhile held for a traitor. It is no wonder that Leibniz passed on the letter directly to Bernoulli to incite him even more [Westfall 2006, p. 776].

Abbé Conti treated the letter as if it would be completely public. The letter circulated at the Court for one month before it was sent off. The French refugee, theologian and book printer Pierre Coste (1668–1747), translated it into French since the King of England could neither read and speak, nor understand English. Coste later translated the *Opticks* into French but also corresponded with Leibniz.

Conti's role was more than unfortunate. In his effort to shine through the contact with great men he appeared miserable and had rekindled the priority dispute again unconsciously. A small glimpse into Leibniz's character can be picked up from a letter he wrote to Baroness von Kielmansegg [Hall 1980, p. 237]. There he told the story of a shoemaker from Leiden who listened to Latin disputations although he could not understand Latin. The question how he could ever judge who stayed winner in the disputes he answered that the one becoming angry was the one who ran out of arguments. This little story shows

Fig. 9.2.1. Andreas Gottlieb von Bernstorff (1649–1726) was Chancellor in the service of the principality Brunswick-Lüneburg. He went with George Louis of Hanover to London, but remained leading minister in Hanover [unknown painter, 18th c]. George I as Knight of the Order of the Garter 1701 ([Detail, after a painting by Godfrey Kneller], Residenzmuseum Celler Schloss (Celle Castle))

that Leibniz had not lost his good humour since it was obviously Newton who became angry first.

Newton's letter to Leibniz via Conti was significantly longer as the one Leibniz had written to Conti. Now Leibniz wrote a reply double in length whereupon Newton replied with a letter of the same length. Most likely this correspondence would have built up if Leibniz had not died.

9.2 Leibniz Dies

The last years of Gottfried Wilhelm Leibniz were overshadowed by grief and illness. From the point of view of the Hanoverian Court he did not proceed fast enough with the history of the Guelfs because he pursued his own objectives. When the Hanoverian sovereign ascended to the throne of England as George I in the year 1714 Leibniz must have cherished hopes to be allowed to accompany his ruler to London. On the contrary, he got a strict travel ban with a hint to the completion of the history of the Guelfs. The payment of his salary was stopped and his secretary Eckhart was commissioned by the Hanoverian minister Andreas Gottlieb von Bernstorff (1649–1726) to report on the progress of Leibniz's work.

In the year 1716 Leibniz was a 70 year old man who spent most of his life seated, either at a desk or in coaches on longer journeys. He suffered from gout and an open leg sore had developed which he tried to heal with curious means by himself. Eckhart wrote that he [Eckhart 1982, p. 197f.]:

'*believes that it came from this* [from sitting] *that a fluxion or open damage developed on the right leg. This incommoded him in walking, he thus tried to heal it, in fact with nothing else other than with blotting paper;* [...] *To alleviate the pain and to make the nerves impalpable he ordered wooden vices to be made and screwed them on where he felt pain.*'

(glaube, daß es daher [vom Sitzen] kam, dass sich am rechten Beine eine Fluxion oder offener Schaden formierte. Dieses incommodirte ihn im Gehen, er suchte es also zuzuheilen, und zwar mit nichts anders, als darauf gelegtem Löschpapier; [...] Die Schmerzen aber zu verhindern, und die Nerven unfühlbar zu machen, ließ er hölzerne Schraubstöcke machen, und dieselbe überall, wo er Schmerzen fühlte, anschrauben.)

On the 3rd November the 'chiagra', gout in the hands, accompanied the 'podagra', gout in the feet, and Leibniz had to stop writing on 6th November. Abdominal colics joined in. Eckhart reported [Eckhart 1982, p. 190f.]:

[...] '*was he violently attacked by the gout in his shoulders.* [...] *great stone pains joined in which caused convulsions and death within an hour. He did not think he had already to die and discussed until shortly before his end how the well known Furtenbach had transformed half of an iron nail into gold. As he was so weak and his servants reminded him to take the Holy Communion he replied: they should leave him in peace; he never did anybody harm and had nothing to confess. He died 14th November 1716 and I have taken all precautions to bury him in dignity in the earth.*'

([...] wurde er von der Gicht, so ihm in die Schultern trat, heftig angegriffen. [...] es kamen große Steinschmerzen darzu, welche ihm Convulsionen und den Tod in einer Stunde Zeit verursachten. Er meinte nicht, daß er schon sterben müßte, und discourirte noch kurz vor seinem Ende, wie der bekannte Furtenbach einen eisernen Nagel halb in Gold verwandelt. Wie er so schwach war, und ihm seine Diener erinnert, ob er nicht das heil. Abendmahl nehmen wolte, hat er geantwortet: sie sollen ihn zufrieden lassen; er habe niemand etwas zu leyde getan; habe nichts zu beichten. Er starb den 14. Novemb. 1716. und habe ich alle Sorge getragen, ihn ehrlich zur Erde zu bestatten.)

We have all reason to mistrust Eckhart who was envious. Other descriptions, as for example the one of the dying of Leibniz by Ludovici [Ludovici 1966, Vol. I, p. 242f.], are also to be treated with great caution, since they romaticised strongly. They are all essentially based on Fontenelle's *Eloge* [Fontenelle 1989, p. 289ff.], which in turn depended heavily on Eckhart's communications. In truth Leibniz's death was extremely painful. The circumstances concerning the death and the burial were thoroughly researched in [Sonar 2008]. Most

Fig. 9.2.2. The copperplate 'Leibniz dies' showing a romanticised view of the dying of Leibniz, was added to the biography by Eckhart [Eberhard/Eckhart 1982]

credibility must be paid to the descriptions of Leibniz's amanuensis Johann Hermann Vogler and his coachman Henrich.

In his last hours Leibniz was visited by the physician Johann Philipp Seip for whom Leibniz had sent. Later Seip tried to tell the world the story that Leibniz could not anymore take his prescribed medicine. This does not comply with the truth according to the information given by Vogler.

The date of the burial in the Neustädter Church (St John's Church of the Court and the City in the New Town at Hanover) in Hanover was the 14th December 1716, hence exactly one month after Leibniz's death. Graeven wrote [Graeven 1902, p. 569]:

> *'The long period is not surprising since it took some time to brick up the tomb in the Neustädter Church which was assigned as the resting place.'*
>
> (Die lange Frist ist nicht verwunderlich, denn es bedurfte einige Zeit, um die zur Ruhestätte bestimmte Gruft in der Neustädter Kirche auszumauern.)

Vogler wrote that Leibniz's corpse was stored in a vault of the Neustädter Church for one month [Ritter 1916, p. 251f.]:

> *'The Master counsellor Eckhart had to come to the room of the privy councillor* [Leibniz] *at noon on Sunday* [it is the day after Leibniz's death] *since he was ordered to lay the corpse of the gentle privy councillor only in a fir coffin so that he could be brought to the Neustädter Church already in the evening, since order had been given that at this time the King's horses and service cart would collect him, and so it was. Master Erythropel's servants and Master Henning's Johann walked in front with lanterns; myself and Henrich besides the vehicle on which was the coffin; Master counsellor Eckhart's and Monseigneur Göbel's servants behind the coffin: on which Master counsellor Eckhart's vehicle followed in which he sat with Monseigneur Göbel. 4 grooms have lifted the corpse on and off which was set in the sand of the vault only until further order comes from the King or until the heirs will be here.'*
>
> (Der H. Rath Eckhart muste am Sonntag Mittag in die Geh. Raths Stube kommen, da man Ihm befohlen, den Cörper des seel. H. Geh. Raths nur in ein Tannen Sarg zu legen, damit er noch den Abend nach der Neustädter Kirche könte gebracht werden, weil schon Befehl gegeben, daß gegen die Zeit des Königs Pferde und Rüstwagen Ihn abholen sollten, welches auch geschehen. H. Erythropels Diener und H. Hennings Johann gingen forn mit Laternen; Ich und Henrich neben dem Wagen, auf welchem der Sarg; H. Rath Eckharts und Mons. Göbels Diener hinter dem Sarg: worauf H. Rath Eckharts Wagen folgete, darinn er mit Mons. Göbeln saß. 4. Königl. Stall-Knechte haben die Leiche auf und abgehoben; welche nur so lange in ein Gewölbe ins Sand gesetzet, biß weitere Ordre vom König kömmt, oder die Erben selber hie sind.)

It was an honour to be buried in the church of the court, but a state funeral as was granted later to Newton was refused. An ecclesiastical ceremony took

Fig. 9.2.3. The upright positioned gravestone in the Neustädter Church (Hanover) together with the stone sarcophagus containing Leibniz's bones
[Photo: Anne Gottwald 2007]

place, however, on the occasion of which 'Oberhofprediger' (chief chaplain of the court) Erythropel sang the collection. The bells also rang but no member of the court was present – privy councillor Leibniz was much too awkward in the eyes of the Hanoverians. The tomb slab contained no markings so that visitors in the 18th century could not find the tomb. Only in 1790 with the growing number of visitors a brass inscription 'Ossa Leibnitii' (bones of Leibniz) was applied. However, for a long time it was not certain whether the inscription had been assigned to the right tomb. Only an excavation of the tomb and an examination of the remains in it by the pathologist Krause from Berlin at the beginning of the 20th century brought some certainty that it really was the tomb of Leibniz. After further reconstructions in recent times the Ossa Leibnitii are now contained in a box in a side alcove of the sanctuary.

9.3 The Dogs of War

Before we turn to Newton's last years we have to take a look at the two warriors which Hall called the *'dogs of war'* [Hall 1980, p. 235] and at their behaviour shortly before and after Leibniz's death: John Bernoulli and John Keill.

9.3.1 Bernoulli versus Keill, but in Favour of Newton

When Leibniz reformulated the problem of finding a curve orthogonal to a given family of curves in greater generality the English had to give up. Bernoulli now turned back to the attack on him by Keill in the Philosophical Transactions of summer 1714. Already in 1710 Bernoulli had claimed that Newton had not presented sufficient reasons in his *Principia* that only conic sections were possible as orbits for planets if the inverse square law of gravitation was assumed. Keill had struck back in 1714 and claimed that Bernoulli's calculations were just a reworked version of Newton's calculations concerning Proposition 41 in Book I of the *Principia* [Hall 1980, p. 236]. Perhaps Bernoulli was too proud to reply directly to Keill; he did that in a letter to Christian Wolff who promptly published it anonymously in the Acta Eruditorum of July 1716. Anonymity was pointless – Bernoulli's spiteful quill was clearly identified by all involved. Keill replied sharply; Bernoulli set up one of his pupils to reply under his name. In that way it went back and forth.

After Leibniz's death further mathematicians were drawn into the quarrel. On the English side Brook Taylor; on the French side Pierre Rémond de Montmort (1678–1719). Now it also became widely known that Bernoulli was the 'eminent mathematician' to which the *Charta volans* referred. Leibniz had actually published the French translation of the *Charta volans* in the Nouvelles Literaires under the title *Lettre de M. Jean Bernoulli de Bâle, du 7 de Juin 1713* (letter by Mr John Bernoulli from Basel, 7th June 1713) and that he did against the will of Bernoulli. Interestingly enough, Bernoulli claimed firmly that he was *not* this 'eminent mathematician' until the death of Newton.

John Bernoulli undisputedly became the most important mathematician on the Continent after Leibniz had died, but he was also isolated. He fought on the biting quarrel with Keill until the death of Keill in the year 1721, but he wanted to make peace with Newton. That was not to happen; the bad smell of a traitor stuck to him too strongly in Newton's thoughts. In the spring after the death of Leibniz Bernoulli approached Rémond de Montmort because he knew that the latter had contact to Newtonians. Montmort was asked to convince Newton that Bernoulli had played only a passive role in 'testing the pulse of our English Analysts' (cp. page 414). He confessed [Hall 1980, p. 238]:

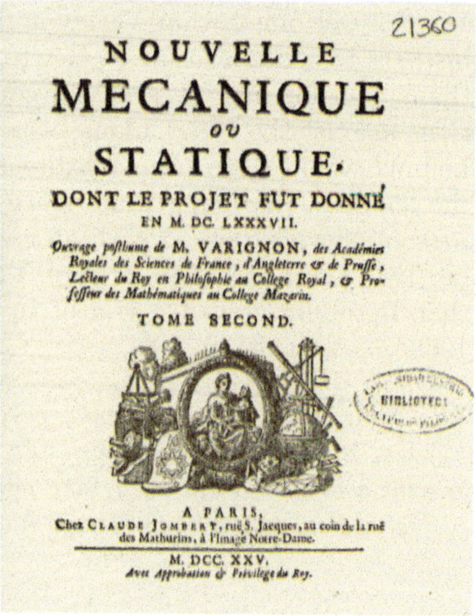
NOUVELLE
MECANIQUE
OU
STATIQUE.
DONT LE PROJET FUT DONNÉ
EN M. DC. LXXXVII.
Ouvrage posthume de M. VARIGNON, des Académies Royales des Sciences de France, d'Angleterre & de Prusse, Lecteur du Roy en Philosophie au College Royal, & Professeur des Mathématiques au College Mazarin.
TOME SECOND.
A PARIS,
Chez CLAUDE JOMBERT, ruë S. Jacques, au coin de la ruë des Mathurins, à l'Image Notre-Dame.
M. DCC. XXV.
Avec Approbation & Privilege du Roy.

Fig. 9.3.1. Pierre Varignon (1654–1722) was a French mathematician and physicist who acted as the director of the Academy in Paris from 1712 to 1719. He was befriended by Leibniz, Newton, and the Bernoullis and is recognised as a great supporter of Leibniz's calculus in France which he applied successfully to problems concerning mechanics in Newton's *Principia*; Title page of the second volume concerning mechanics by Varignon; the first volume appeared in 1725 (Biblioteca Histórica. Universidad Complutense de Madrid. BH FLL 21360)

> *'I desire nothing so much as to live in good fellowship with him, and to find an opportunity of showing him how much I value his rare merits, indeed I never speak of him save with much praise* [...]*'*

The situation between Bernoulli und Newton got better only when Pierre Varignon stepped in. When in 1717 the second English edition of the *Opticks* was published and Newton sent three copies to the Academy in 1718, Varignon sent a copy to John Bernoulli. Varignon informed Newton in November 1718 [Hall 1980, p. 239], [Turnbull 1959–77, Vol. VII, p. 16]:

> *'I add fresh gratitude to the thanks just offered because you have correctly judged me wholly innocent of acting against you in the Leibnizian controversy: quite the contrary, I have taken so little part in it that I have rather always kept silent about that controversy in my letters to Mr Leibniz and Mr Bernoulli, only lamenting to myself and in private that such great men are troubled by it, whom, if I had any influence, I would have restored to their former cordiality; and this was the sole object that I had in mind when I sent to John Bernoulli from yourself the new edition of your book on colours.'*

Newton welcomed what Varignon did and Bernoulli took the opportunity for his part to write a letter to Newton [Turnbull 1959–77, Vol. VII, p. 42ff.] to thank him for 'his' present. Of course, the letter was not free from accusations against the Newtonians swearing against every foreigner, guilty or not. Leibniz would have been misdirected in assigning his, Bernoulli's, name to the 'eminent mathematician' since he would under no circumstances be this person. Newton replied with an extraordinaryly friendly letter of 10th October 1719 [Turnbull 1959–77, Vol. VII, p. 69f.], although we may safely assume that Bernoulli stayed a traitor in Newton's head. Newton wrote [Turnbull 1959–77, Vol. VII, p. 70]:

> *'Now that I am old I take very little pleasure in mathematical studies, nor have I ever taken the trouble of spreading opinions through the world, but rather I take care not to allow them to involve me in wrangles. For I have always hated disputes.'*

9.3.2 Pierre des Maizeaux and His *Recueil*

The French refugee Pierre des Maizeaux (also: Desmaizeaux) (1672 or 1673–1745) was an huguenot and became known as the translator and biographer of Pierre Bayle[1]. He prepared a book on the mathematical and also the philosophical quarrel between Leibniz and Newton. To this end he had received the first four letters of the Leibniz-Clarke controversy from Leibniz already in August 1716 [Hall 1980, p. 242]. In the background was Abbé Conti [Whiteside 1967–81, Vol. VIII, p. 520]. Des Maizeaux had thought of writing a book concerning the correspondence between Leibniz and Clarke already since 1716[2]. Now the priority dispute joined in and it was agreed upon to publish the correspondence in the second volume of des Maizeaux's *Recueil de Diverses Pièces sur la Philosophie ...* (Collection of different pieces on philosophy ...). Des Maizeaux now asked Newton for permission to also use his documents and letters to Conti and Newton delivered unpublished manuscripts and drafts. When des Maizeaux received the galley proofs he forwarded them immediately to Newton to read over it. Newton was not satisfied and drafted a whole new structure of the book so that the documents could be brought into an intelligible order. Des Maizeaux accepted without complaint, but that was

[1]Pierre Bayle (1647–1706) was a French writer and philosopher who is recognised as a central figure of the French enlightenment together with Fontenelle. In 1675 he became professor of philosophy at the protestant Academy Sedan in the Dukedom of Lorraine but went to Rotterdam as a secondary school teacher after the Academy was closed in 1681. In the time period from 1684 to 1687 he acted as the editor of the *Nouvelles de la République des Lettres*. In 1691 he lost his position and worked on his *Dictionnaire historique et critique* which saw more than 10 editions until 1760.

[2]Clarke forestalled him in 1717.

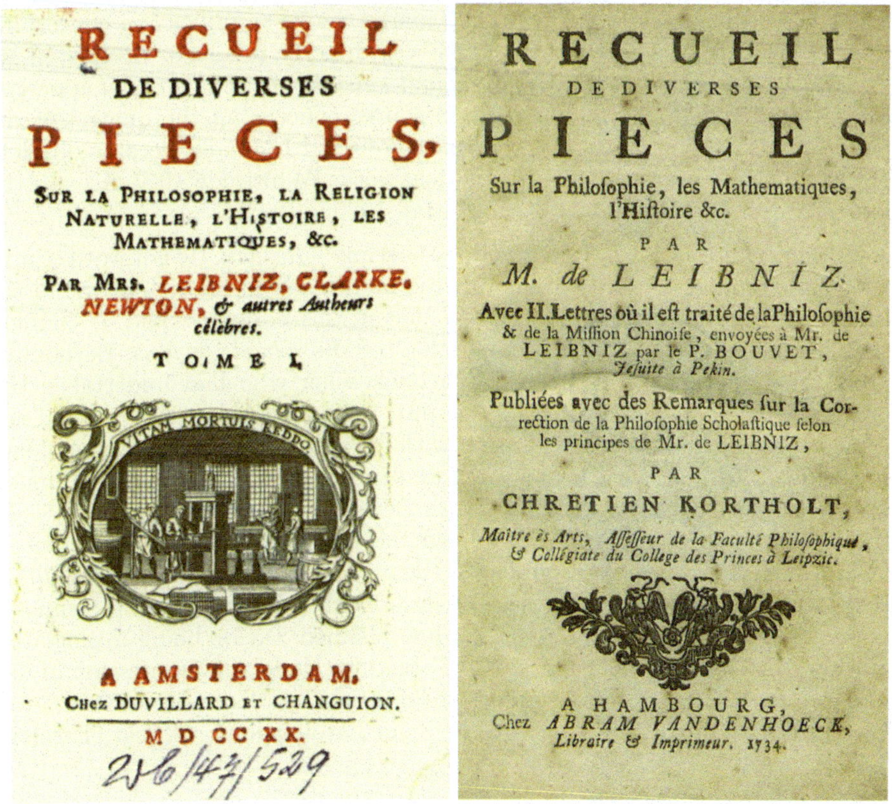
RECUEIL
DE DIVERSES
PIECES,
SUR LA PHILOSOPHIE, LA RELIGION NATURELLE, L'HISTOIRE, LES MATHEMATIQUES, &c.
PAR MRS. LEIBNIZ, CLARKE, NEWTON, & autres Autheurs célèbres.
TOME I.
VITAM MORTUIS REDDO
A AMSTERDAM,
CHEZ DUVILLARD ET CHANGUION.
M D CC XX.

RECUEIL
DE DIVERSES
PIECES
Sur la Philosophie, les Mathematiques, l'Histoire &c.
PAR
M. de LEIBNIZ.
Avec II. Lettres où il est traité de la Philosophie & de la Mission Chinoise, envoyées à Mr. de LEIBNIZ par le P. BOUVET, Jesuite à Pekin.
Publiées avec des Remarques sur la Correction de la Philosophie Scholastique selon les principes de Mr. de LEIBNIZ,
PAR
CHRETIEN KORTHOLT,
Maitre ès Arts, Assesseur de la Faculté Philosophique, & Collégiate du College des Princes à Leipzic.
A HAMBOURG,
Chez ABRAM VANDENHOECK, Libraire & Imprimeur. 1734.

Fig. 9.3.2. Title page of the *Recueil De Diverses Pieces* of 1720 (Amsterdam, Detail; Bayerische Staatsbibliothek München (Munich), Sign.: 5901843 Var. 250-1 5901843 Var. 250-1) and of 1734 (A Hambourg, Vandenhoeck, Detail; ULB Sachsen Anhalt, Halle, VD18 10779299)

not all! Newton sent many pages of *Corrigenda*, however with partly wrong dates [Whiteside 1967–81, Vol. VIII, p. 522f.]. Newton remarked [Whiteside 1967–81, Vol. VIII, p. 523]:

> *'And ever since I wrote that Book* [the *Principia*] *I have been forgetting the Methods by which I wrote it.'*

Pure irony? In the accompanying letter to des Maizeaux he suddenly changed from wholly harmless remarks to a harsh tone and stated [Whiteside 1967–81, Vol. VIII, p. 523]:

> *'In all this controversy the true Question has been whether Mr Leibnitz or I were the first inventor* [&] *not who invented this or that method.'*

To Varignon Newton absolutely insisted that he would not have known anything about des Maizeaux's *Recueil*.

The edition of des Maizeaux's *Recueil* marks the last publication concerning the priority dispute in which Newton was directly involved, at least if we neglect the re-edition of the *Commercium epistolicum*. The *Recueil* appeared in 1720 and remained the most useful source concerning the quarrel of the philosophers. It saw reprints in the years 1740 and 1749 and became obsolete only with the edition of the mathematical works [Whiteside 1967–81] and the letters of Newton [Turnbull 1959–77] [Hall 1980, p. 242].

John Bernoulli was not happy with des Maizeaux's *Recueil*. He got word of it in 1719 and complained of the tone. Newton had informed him about the forthcoming publication and told Bernoulli that in this book it would be claimed that he was the 'eminent mathematician' of the *Charta volans*. Bernoulli's protest led Newton to pay the Dutch book seller who had undertaken the print in order to delay the publication until 1720. Apparently Newton did not want to jeopardise the recent ceasefire with Bernoulli. We also know that Newton planned to buy the whole edition and to pay off des Maizeaux; so much was he interested in keeping peace with Bernoulli [Westfall 2006, p. 788]. In the year 1719 Brook Taylor had joined the English front and based on an initiative of Newton he had published the work *Apologia contra Bernoullium* (Defence against Bernoulli) in the Philosophical Transactions. The 'eminent mathematician' of the *Charta volans* had claimed Taylor being a plagiarist and the *Apologia* was clearly a kind of reaction. Under the impression of des Maizeaux's *Recueil* Bernoulli asked Newton on 21st December 1719 to stop the attacks of Taylor and Keill, but he immediately added a threat [Turnbull 1959–77, Vol. VII, S. 78]:

> *'There are letters to me from certain learned men of nations having no part in this national strife; if I laid these before the public I know not whether those of your countrymen who quarreled so warmly with me as to become insulting would gain from them great reason for boasting. I have amongst other authentic documents a copy* [of a letter] *from Mr Montmort, a mathematician not long dead who was, as you know, learned and attached to neither party while he lived, since he was French; I have, I say, a copy (sent from him to me) of a certain letter which he had written to Mr Taylor on 18 December 1718, and which alone would certainly dispel a large part of the dispute; but not to the taste of Taylor and the rest of his following. However, I will willingly refrain from making these things public, if only your party will cease to try our patience, which I desire for the sake of peace.'*

Newton understood the letter as an attack and requested Keill to let loose his reply to Bernoulli which Keill had prepared long ago but was hitherto hindered to publish by Newton. Newton himself drafted a frosty reply to Bernoulli [Turnbull 1959–77, Vol. VII, p. 80] which probably was never sent [Westfall 2006, p. 789]. The ceasefire, mediated by Varignon, did not last long.

However, Varignon tried further to bring about peace until he finally died in 1722. Letters between Newton and Varignon on one side, and Bernoulli and Varignon on the other side went back and forth. In des Maizeaux's *Receuil* one could find letters by Newton in which the 'eminent mathematician' of the *Charta volans* was called 'pretended mathematician', 'novice', and 'knight errand'. Bernoulli was insulted. Varignon decided to show Newton the letters of Bernoulli which he had received. All went via de Moivre who sent the replies of Newton again to Varignon. Death redeemed Varignon from this ungrateful task and de Moivre refused to be drawn further into the struggle. But Newton was not at his wits' end.

9.3.3 *Commercium epistolicum* Reloaded

Confronted with endless complaints and claims by Bernoulli, Newton published a re-issue of the *Commercium epistolicum* in 1722. He added the anonymous letter of the 'eminent mathematician' of the *Charta volans* and a very detailed explanation from Newton's quill. Bernoulli's name was not mentioned. While the first edition of the *Commercium epistolicum* was freely distributed by the Royal Society the second edition was offered for sale.

But Bernoulli was also not at the end of his wits. In a letter to Newton of 6th February 1723 [Turnbull 1959–77, Vol. VII, p. 218ff.] he reported indignantly about a book by the Dutch Nicolas Hartsoeker (1656–1725) in which Newton was accused of plagiarism in the *Opticks*. Hartsoeker was a mathematician and physicist who was a dedicated Cartesian and who had corresponded with Leibniz. In his book *Recueil de Plusieurs Pièces de Physique où l'on fait principalement voir l'Invalidité du Système de Mr. Newton* (Collection of different pieces concerning physics, where primarily the invalidity of the system of Mr Newton is made visible) he violently attacked Newton and claimed that he, Hartsoeker, had published the theory of light already in 1694. This alone might not have flustered Bernoulli but Hartsoeker had also attacked him in that he had twisted Newton's remarks concerning the *Charta volans* in des Maizeaux's *Recueil.* Newton took note of Bernoulli's letter but did not reply. Thereby the priority dispute finally ceased [Westfall 2006, p. 792].

9.4 Newton's Last Years

Perhaps influenced by the priority dispute Newton in his last years looked after new editions of his works. The second Latin edition of his *Opticks* appeared in 1719; the third English edition in 1721. On the Continent the first edition of the *Opticks* did not make a great impact which was partly owed to Edme Mariotte. Mariotte had – as others also did – repeated Newton's optical experiments included in the *Opticks* and he was not successful. In 1715 a French

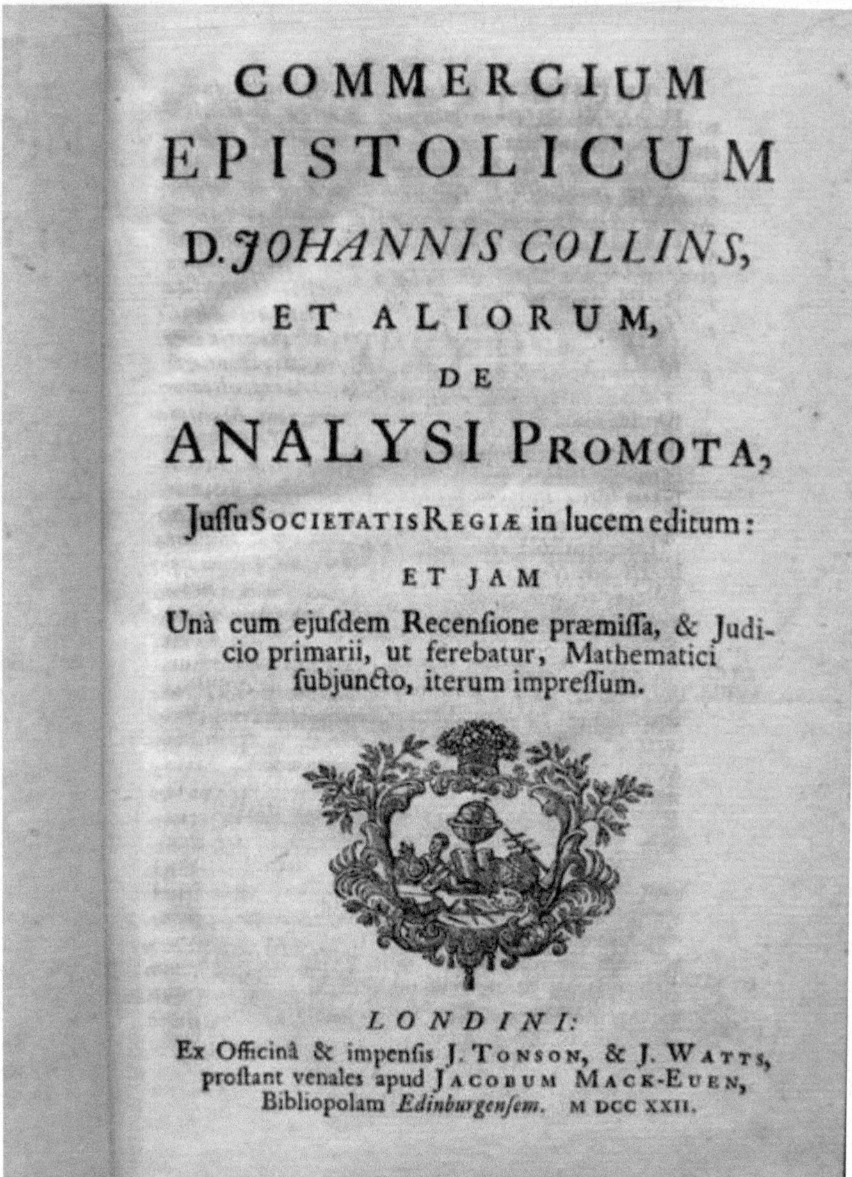
COMMERCIUM
EPISTOLICUM
D. *JOHANNIS COLLINS*,
ET ALIORUM,
DE
ANALYSI PROMOTA,
Juſſu SOCIETATIS REGIÆ in lucem editum:
ET JAM
Unà cum ejuſdem Recenſione præmiſſa, & Judicio primarii, ut ferebatur, Mathematici ſubjuncto, iterum impreſſum.

LONDINI:
Ex Officinâ & impenſis J. TONSON, & J. WATTS, proſtant venales apud JACOBUM MACK-EUEN, Bibliopolam *Edinburgenſem*. M DCC XXII.

Fig. 9.3.3. Title page of the new edition of the *Commercium epistolicum* of 1722 (Courtesy of the Trusties of the Edward Worth Library, Dublin, Online Exhibition)

delegation led by Montmort came to England to observe an eclipse of the sun. Newton took the opportunity to demonstrate some of his optical experiments to the delegation and this demonstration served to popularise Newton's ideas outside England as did the second Latin edition of the *Opticks*. It was this edition which encouraged the French Black Friar Sébastien Truchet[3] (1657–1729) to repeat the experiments and this time he succeeded. The interest in the *Opticks* grew strongly. In 1720 Pierre Coste published a French translation of the *Opticks* in Amsterdam which was reviewed by Varignon who tended a second edition in 1721 in Paris [Westfall 2006, p. 795].

The *Principia* also should be re-issued in a third edition [Cohen 1971, p. 265ff.]. Perhaps it was the reprint of the second edition in Amsterdam in 1723 which gave the impetus; perhaps it was a serious illness which Newton contracted in 1722 which made it clear to him that he would not have much time at his disposal. Newton ensured the help of young Henry Pemberton (1694–1771) who had returned from medical studies in Leiden in 1715 where he also had read mathematics. For further education he took up work at St Thomas Hospital and did his doctorate in 1719 in Leiden under Herman Boerhaave (1668–1738). Because he was of poor health he practised only little but wrote medical articles as a fellow of the Royal Society. He became professor of medicine at Gresham College in London in 1728. Richard Mead showed Newton a paper by Pemberton in which Leibniz's idea of forces was criticised and the 'great Isaac Newton' was mentioned. Newton was flattered and arranged for the publication of the paper in the Philosophical Transactions. Since Pemberton was also an able mathematician and had already excelled in the popularisation of Newton's natural philosophy, Newton chose the young man to look after the third edition of his *Prinipia*. The 80 year old Newton was no longer capable of a critical discourse as had taken place with Roger Cotes in preparation of the second edition, although Pemberton tried to discuss scientific questions in the *Principia* with Newton.

Already in 1719 Newton had got news of the critique of Nicolaus I Bernoulli concerning the motion of pendulae in resisting media from James Stirling, but Newton had not replied. The theory was left as it was. Brook Taylor pointed out to Pemberton that the calculation of the precession contained an error – this calculation was also left unchanged [Westfall 2006, p. 799]. Molyneux informed John Conduitt, husband of Newton's niece, that a nutation of the earth's axis had been observed which could perhaps undermine the whole natural philosophy. Molyneux very carefully wrote to Newton but received the dry answer that this might be so. All this shows a certain indifference of Newton. Now Pemberton wrote to Newton, but Newton did not reply. We know, however, that Pemberton also talked to Newton during visits in Newton's house. All in all the third edition which was published eventually in 1726 was not substantially different from the second. There were 1250 copies printed; 50 thereof on finest paper. In contrast to Cotes Pemberton at least got some words of thank in the preface.

[3]Truchet was a typographer, mathematician, inventor, and physicist and an expert on hydraulics. He also was successful as a clockmaker and inventor of mechanisms for the transfer of trees in the Garden of Versailles.

Fig. 9.4.1. Third edition of the *Principia* of 1726 ([Photo: Paul Hermans 2009] John Ryland Library, Manchester)

Pemberton had also planned to publish an English translation of the *Principia* but Andrew Motte (1693–1783) anticipated him since the latter's translation [Newton 1995] appeared in 1729. In the year 1728 Pemberton published the book *View of Sir Isaac Newton's philosophy.*

Newton also tended new editions of his mathematical works; probably also in retrospect of the priority dispute. In the year 1720 Raphson's English translation of the *Arithmetic universalis* was published as *Universal Arithmetick* and two years later a second edition of the Latin version. The book by William Jones of the year 1711 (cp. page 378) which contained Newton's *De quadratura* was published as a reprint in 1723 in the Netherlands.

Newton's active interest in scientific questions had evaporated with age, but he was still intensively interested in theological questions. Theology became Newton's main occupation in his last years and the chronology and prophecies were in the centre of his interest. Since 1690 Newton had occupied himself with theological chronology and when Abbé Conti came in closer contact to Newton and discussed Newton's new principles of chronology with the Princess of Wales in 1716 she asked for a copy of this work. As always Newton was not willing to give his work freely away to others[4]. He compiled an 'Abstract'

[4]Westfall suspected that the original work contained so much heretical issues that Newton might have been removed from his position at the Mint [Westfall 2006, p. 805].

and delivered it to the Princess within a few days. Conti also got a copy and the 'Abstract' quickly spread. When Conti returned to France in 1716 the 'Abstract' went with him and quickened interest, although it contained barely more than a table of datings of biblical events. Parisian scholars and book printers started to pester Newton for sending more and Newton fully blamed Conti for these disturbances [Westfall 2006, p. 811]. Only after Newton's death was the whole work published by Conduitt in 1728 under the title *The Chronology of Ancient Kingdoms Amended.* Apparently Newton had exploited astronomical calculations besides theological speculations to assemble his chronology. Newton's *Observations upon the Prophecies of Daniel, and the Apocalypse of St. John*, going back to the early years of the 18th century, was also published posthumously in 1733. There Newton 'proved' that Judgement Day was not to come earlier than 2060 [Westfall 2006, p. 816]. Other theological manuscripts of Newton's later years were not published. They were bought by economist John Maynard Keynes (1883–1946) at an auction at Sothebys in London on 13th July 1936. The whole history of the manuscripts is thoroughly researched in [Dry 2014] and reads as thrilling as a detective story.

9.5 Newton Dies

Up to his end Newton stayed President of the Royal Society. Since 1713 the Society was completely under Newton's control. When the Society lost £600 in the South Sea Bubble of 1720 Newton offered to pay this £600 out of his own pocket. It is said that he himself had lost £20000 private investment [Westfall 2006, p. 861].

The South Sea Company was founded in 1711 as a trading company and got the monopoly to trade with South America and all of the yet undiscovered territories. When the company assumed debts of the English government it got the right to issue additional new shares. By and by the company assumed more and more government debts and increased the number of shares further. In 1720 this speculative bubble burst parallel to the Mississippi Bubble in France. The debts were assumed by the East India Company and the Bank of England. Thereby the Chancellor of the Exchequer, Robert Walpole (1676–1745), could increase his power considerably. He became Prime Minister under King George II. The South Sea Company continued to exist until it was dissolved in 1850.

In 1722 Newton gave an original manuscript of Tycho Brahe as a gift to the Royal Society. By this time he had become a supporter of a young generation of scientists on a grand scale. James Stirling, Colin Maclaurin, and the astronomer James Pound were, among others, financially supported by him or recommended to professorships. The famous President of the Royal Society had also become a kind of tourist attraction and he received many visitors. We have a detailed description of such a visit in the summer of 1725 by Abbé Alari, the teacher of Louis XV [Westfall 2006, p. 833]. The two men talked

Fig. 9.5.1. The 'South Sea Bubble' in a painting by Edward Matthew Ward (1816-1879) in the style of Hogarth (Tate Britain, London)

about ancient history and when the Abbé could show that he was an educated man Newton invited him for lunch which turned out to be miserable not only because cheap wine was served. In the afternoon Newton took the Abbé to attend a meeting of the Royal Society where he was allowed to sit to the right of the president and where Newton soon fell asleep. After the meeting Newton took Alari with him homewards where he entertained him until nine o'clock.

In 1718 Newton made the acquaintance of William Stukeley who practised in London as a physician. The two men, Stukeley being 31 and Newton being 76 years old, made friends and when Stukeley moved to Grantham he started to collect information on Newton's life as Conduitt had done. The memoirs of Stukeley and Conduitt comprise the most authentic documents concerning the life of the old Newton. However, they carry features of hero worship. Both men report carefully on Newton's fading memory and signs of senility.

The last years were overshadowed by illness. We have already reported on health problems in 1722; in 1723 he was so seriously ill that he delivered himself to the care of the physicians Richard Mead and William Cheselden (1688–1752). A kidney stone went through the urinal tract without much pain [Westfall 2006, p. 866] but bladder weakness became a great problem. Incontinence joined in and Newton gave up driving in his coach since its movements worsened his problem. He also stopped going out for lunch and renounced meat completely in his diet. In January 1725 Newton developed a violent cough which turned into pneumonia. The Conduitts persuaded him

Fig. 9.5.2. William Stukely (1687–1765) was an antiquarian from Lincolnshire who became a member of the Royal Society in 1717. In 1752 he published a biography of Newton under the title *Memoirs of Sir Isaac Newton's life*. He also examined Stonehenge which he held for a Roman racecourse for horses. (left: Wellcome Images/Wellcome Trust, London [Mezzotint by J. Smith, 1721, after a painting by Godfrey Kneller]), right: [Painting, English School, 1740]

to move to a house in Kensington and the better air did him good. Since Newton felt his end nearing he began to sell his estates. He generously supported the large family of his relatives. Conduitt reported that he also burned manuscripts. After the summer break of 1726 Newton attended only four more meetings of the Royal Society. On the 13th March 1727 he presided over the Society for the last time. A few days before his death Newton received the visit of Zachary Pearce (1690–1774), vicar of St Martin-in-the-Fields in London.

Pearce reported on the visit [Westfall 2006, p. 869]:

> *'I found him writing over his Chronology of Ancient Kingdoms, without the help of spectacles, at the greatest distance of the room from the windows, and with a parcel of books on the table, casting a shade upon the paper. Seeing this, on my entering the room, I said to him, 'Sir, you seem to be writing in a place where you cannot so well see.' His answer was, 'A little light serves me.' He then told me that he was preparing his Chronology for the press, and that he had written the greatest part of it over again for that purpose. He read to me two or three sheets of what he had written, (about the middle, I think, of the work) on occasion of some points in Chronology, which had been mentioned in our conversation. I believe, that he Continued reading to*

Fig. 9.5.3. Zachary Pearce was a student and later fellow of Trinity College, Cambridge. He became the personal chaplain of Lord Chancellor Thomas Parker, 1st Earl of Macclesfield. After a career as a cleric in different parishes he became Bishop of Bangor in 1748, Bishop of Rochester in 1756, declined becoming the Bishop of London, and became Dean of Westminster from 1756 until 1768 [Artist: John Faber jr. after Thomas Hudson]; the main entrance of Trinity College Cambridge [Photo: Andrew Dunn 2004]

> *me, and talking about what he had read, for near an hour, before the dinner was brought up.'*

The visits and the meeting of 13th March had tried Newton and the violent cough returned. John Conduitt sent for Mead and Cheselden who diagnosed a bladder stone. Newton suffered great pain and sweat ran down his face [Westfall 2006, p. 869]. Stukeley noted in his *Memoirs* [Iliffe 2006, Vol. I, p. 302]:

> *'it* [his suffering] *rose to such a height, that the bed under him, & the very room, shook with his agonys, to the wonder of those that were present, such a struggle had his great soul to quit its earthly tabernacle!'*

As Leibniz also Newton refused to receive the sacraments which was consistent with view on his almost lifelong Arian conviction. On 26th March Newton seemed better off but shortly after he went downhill. He lost consciousness on Sunday, 30th March, and died the following morning at one o'clock without

Fig. 9.5.4. The Jerusalem Chamber in Westminster Abbey on a photo taken between 1870 and 1900 [unknown photographer, 1914]

further suffering. An ugly tug of war for his not insignificant assets began between his relatives [Westfall 2006, p. 870ff.]. John Conduitt could secure the now available position of Master of the Mint for himself. Catherine Conduitt, née Barton, died in 1739, two years after her husband. Her daughter, also christened Catherine, married John Wallop, Viscount Lymington, in 1740 and their son became the second Earl of Portsmouth. Through his daughter Catherine, Newton's papers and manuscripts came into the Portsmouth family and went from there to the University Library of Cambridge.

Newton's corpse was laid out on 8th April 1727 in the Jerusalem Chamber of Westminster Abbey.

A tomb monument was erected in Westminster Abbey in 1731 which Westfall called a *'baroque monstrosity'* [Westfall 2006, p. 874]. Newton is seen in relaxed, reclining pose, ornamented by angles. On top Astronomy is shown as a woman crying above a globe. The inscription reads[5]:

> *'Here is buried Isaac Newton, Knight, who by a strength of mind almost divine, and mathematical principles peculiarly his own, explored the course and figures of the planets, the paths of comets, the tides of the sea, the dissimilarities in rays of light, and, what no other scholar has previously imagined, the properties of the colours thus produced. Diligent, sagacious and faithful, in his expositions of nature, antiquity and the holy Scriptures, he vindicated by his philosophy the majesty of God mighty and good, and expressed the simplicity of the Gospel in*

[5]http://www.westminster-abbey.org/our-history/people/sir-isaac-newton

Fig. 9.5.5. The cenotaph for Isaac Newton as planned by Boullée [E.-Louis Boullée: Cénotaphe à Newton, 1784]

> *his manners. Mortals rejoice that there has existed such and so great an ornament of the human race! He was born on 25th December 1642, and died on 20th March 1726.'*
>
> (H. S. E. ISAACUS NEWTON Eques Auratus, / Qui, animi vi prope divinâ, / Planetarum Motus, Figuras, / Cometarum semitas, Oceanique Aestus. Suâ Mathesi facem praeferente / Primus demonstravit: / Radiorum Lucis dissimilitudines, / Colorumque inde nascentium proprietates, / Quas nemo antea vel suspicatus erat, pervestigavit. / Naturae, Antiquitatis, S. Scripturae, / Sedulus, sagax, fidus Interpres / Dei O. M. Majestatem Philosophiâ asseruit, / Evangelij Simplicitatem Moribus expressit. / Sibi gratulentur Mortales, / Tale tantumque exstitisse / HUMANI GENERIS DECUS. / NAT. XXV DEC. A.D. MDCXLII. OBIIT. XX. MAR. MDCCXXVI)

In the 18th century Newton had become a celebrity of the first rank also in France. The admiration of the French was so grand that the revolution architect Étienne-Louis Boullée (1728–1799) planned to build a cenotaph as a tribute to Newton. This cenotaph was planned to be a sphere of 150 meter in diameter resting on a ring-shaped plinth which was planted with cypresses. The interior of the sphere which was supposed to symbolise the universe showed the starry sky due to perforations in the hull of the sphere through which light could enter.

Fig. 9.5.6. The tomb monument for Newton in Westminster Abbey
[Photo: Klaus-Dieter Keller 2006]

10 The Early Challenges

T. Sonar, *The History of the Priority Dispute between Newton and Leibniz*, https://doi.org/10.1007/978-3-319-72563-5_10

Newton's calculus of fluxions and the differential calculus of Leibniz had hardly seen the light of day when they were already criticised. This critique could not stop the triumph of the new infinitesimal mathematics, of course; too big were the successes at which one arrived at in consequently applying this new kind of mathematics to problems of physics. As we have seen already the brothers Jacob und John Bernoulli established the calculus of variations on the Continent. Besides many other problems of optimisation, with this type of calculus one was able to solve the problem of the brachistochrone, hence the problem of finding the curve on which a point of mass moves without friction in the least period of time from one given point to another under the influence of gravity. John's son Daniel Bernoulli (1700–1782) established hydrodynamics and John's great pupil Leonhard Euler (1707–1783) was called 'analysis incarnate' already by his contemporaries. He founded all of mechanics on Leibniz's differential and integral calculus and expanded it considerably.

In England the French Abraham de Moivre (1667–1754) worked successfully with fluxions. He generalised Newton's binomial theorem resulting in the multinomial theorem. In 1707 he found the famous formula $(\cos x + i \sin x)^n = \cos(nx) + i \sin(nx)$ which now bears his name, and he made great progress in probability theory and astronomy. In 1715 Brook Taylor (1685–1731) published his main work *Methodus incrementorum directa et inversa* (Method of direct and inverse incrementation) in which the calculus of finite differences was examined in relation to fluxions. The famous power series now bearing his name can also be found in this book. Roger Cotes (1682–1716) worked closely with Newton on the new edition of the *Principia* and refined the numerical quadrature rules which we know under the names of Newton and Cotes. Colin Maclaurin (1698–1746) authored the work *A treatise of fluxions* in 1742 which comprised one of the first systematic presentations of Newton's infinitesimal mathematics. A power series was named after him as was the famous Euler-Maclaurin formula. Further English mathematicians of the 18th century were Thomas Bayes (about 1701–1761), Thomas Simpson (1710–1761), Matthew Stewart (1717–1785), John Landen (1719–1790), Robert Woodhouse (1773–1827), and Edward Waring (1736–1798).

How could criticism have come up at all if the infinitesimal mathematics allowed the successful treatment of hitherto unsolvable problems? Of course it was the use of the infinitely small which caused the trouble. Carl Benjamin Boyer (1906–1976) in his classical book on the history of differential and integral calculus suspected that the neglect of the metaphysical rationalism, hence the philosophical superstructure of infinitely small quantities, by Leibniz's successors triggered critics to come to light [Boyer 1959, p. 224]. Infinitesimals were freely understood as being truely infinitely small quantities or simply nil, as for example Euler did. Following Boyer it was the missing clarity in the discussion of infinitely small quantities in Newton's works and his inconsistent notation which encouraged replies. Boyer therefore called the 18th century the 'period of indecision'; a notion which we should in no case follow! It is more the century of storm and stress concerning the new mathematics of infinitesimals.

In this chapter we want to examine two early challenges in more detail since at least one of them had some impact. The archetype of an early critic of Newton's calculus of fluxions is the Irishman George Berkeley (1685–1753), a theologian and philosopher who became Bishop of Cloyne in 1734. Even earlier than Berkely Bernard Nieuwentijt (1654–1718) had criticised the foundation of Leibniz's calculus.

10.1 Bernard Nieuwentijt and Leibniz's Calculus

10.1.1 A Life in North Holland

Bernard Nieuwentijt was a philosopher and mathematician from the Netherlands who was the very first to criticise Leibniz's infinitesimal calculus.

Born in West-Graftdijk in the north of Holland Bernard was the son of the cleric Emmanuel Nieuwentijt and his wife Sara d'Imbleville [DSB 1971, Vol. X, p. 120f., H. Freudenthal]. Although he was assumed to enter an ecclesiastical office he studied natural sciences. He enrolled for medicine early in 1675 in Leiden; later in the year 1675 he enrolled at the University of Utrecht where he studied medicine and law. In Utrecht he defended his medical PhD thesis

Fig. 10.1.1. George Berkely ([Engraving by W. Holl], Wellcome Images/Wellcome Trust, London) and Bernard Nieuwentijt [unknown engraver]

in 1676 and opened a practice in Purmerend some 14 kilometers east of West-Graftdijk. In November 1684 he married the widow of a captain, Eva Moens. After the death of his wife he married in March 1699 for the second time; the wife was Elisabeth Lams, daughter of the mayor of the city of Wormer which was a neighbouring municipality of Purmerend and West-Graftdijkt.

10.1.2 Nieuwentijt's Short Philosophical Fame

He wrote two extensive works for which he became famous in his home country. The first appeared 1714 in the Dutch language in Amsterdam and was translated in several other languages. In the English translation by Chamberlayne of 1718 the title is *The Religious Philosopher, or the Right Use of Contemplating the Works of the Creator: (I) In the Wonderful Structure of Animal Bodies, (II) In the Formation of the Elements, (III) In the Structure of the Heavens, Designed for the Conviction of Atheists.* The second greater work was published posthumously in 1720 also in the Dutch language. In this work Nieuwentijt undertook a methodical attack on rationalism which would lead to 'Spinocism' and other kinds of atheism in his opinion. The notion of Spinocism was used in the 18th century as a philosophical battle cry to depreciate the philosophy of Baruch de Spinoza in which God was just a substance. Mind and matter are not separate substances as was taught by the Cartesians but two attributes of the substance God. Despite the fame he received in his days Nieuwentijt's influence on the development of philosophy was negligible [DSB 1971, Vol. X, p. 121, H. Freudenthal]. However, as a methodologist his work was unique well up to our times due to his proximity to mathematics and the natural sciences.

10.1.3 The Fight Against Leibniz's Infinitesimals of Higher Order

Nieuwentijt assimilated the mathematics and natural philosophy of his time as a young man and we have to see him as an educated mathematician. According to the fashion of the time he became a Cartesian since the doctrines of René Descartes were most popular in the Netherlands. The period where Nieuwentijt engaged himself in a dispute with Leibniz is most interesting to us; it is the period between 1695 and 1700 [DSB 1971, Vol. X, p. 120, H. Freudenthal].

Nieuwentijt's writings concerning the infinitesimal mathematics were all published in the two years 1694 and 1695. Following Rienk H. Vermij (b 1957) they were not sufficiently studied for their own significance [Vermij 1989, p. 69]. We actually know nothing concerning the mathematical education of the young Nieuwentijt. Vermij cites Nieuwentijt from the treatise *Considerationes circa analyseos ad quantitates infinite parvas applicatae principia, et calculi differentialis usum in resolvendis problematibus geometricis* (Considerations on the principles of the applied analysis based on infinitely small quantities and on the use of the differential calculus in solving geometric problems) [Vermij 1989, S. 70]:

> *'until I finally decided, because I was to exhibit the infinitesimal calculus for the use of a young man to whom I was related by marriage, to work over my dispersed annotations into an elementary treatise.'*

This 'young man' could only have been Nieuwentijt's stepson, Hendrik Munnik, who was the son of Eva Moers from her first marriage. Hence the date must have been some time after 1684. The 'elementary treatise' mentioned was *Analysis infinitorum seu curvilineorum proprietates ex polygonorum natura deductae* (Analysis of the infinite or the properties of curved objects, deduced from the nature of polygons) which appeared in Amsterdam in 1695. The book has a didactical character in essence and is, as Nieuwentijt wrote in the preface, 'a little work, written by a beginner for beginners' [Vermij 1989, p. 70]. It is actually the very first elementary textbook on infinitesimal mathematics since the *Analyse des infiniments petits* (Analysis of the infinitely small) by Marquis de l'Hospital appeared but one year later.

In the preface Nieuwentijt reported that he would derive all results from only one principle, namely the interpretation of curves as traverses. And in fact the *Analysis infinitorum* in this respect was a most systematic work. To be sure, Nieuwentijt did in no way circumvent infinitely small quantities. As did Barrow he computed the tangent to curves, but where Barrow omitted infinitesimal quantities ('for these terms have no value', see section 2.4) Nieuwentijt explained *why* he omitted these terms. An infinitesimal quantity a is, geometrically speaking, the infinitesimal part of a given quantity h. In other words: If the quantity h will be divided by an infinitely large m the resulting infinitesimal is a, hence $a = h/m$. If the infinitesimal quantity a is multiplied with the infinitely large quantity m the given finite quantity h is recovered again. Infinitesimal quantities like a must *not* simply be omitted. But if $e = k/m$ is also an infinitesimal quantity and if the product $a \cdot e$ is considered, then this is nothing but $(h/m) \cdot (k/m) = hk/m^2$. If one multiplies with m then there remains hk/m and since m is infinitely large such terms *are* to be omitted since they *are* nil.

The most important consequence of this more than questionable argumentation by Nieuwentijt is the nonexistence of higher differentials, and that is exactly the reason for the controversy with Leibniz.

Surely the basis for Nieuwentijt's *Analysis infinitorum* was provided by English mathematicians like Barrow, Wallis, Gregory, and Newton [Vermij 1989, p. 75]. Also mathematicians well known in the Netherlands were cited, of course; Huygens, Descartes, de Sluse, Hudde, and Gerhard Mercator. Although Leibniz had published his calculus in 1684, Nieuwentijt had read this publication obviously fairly late while working on his *Analysis infinitorum*. The name of Leibniz appeared only on page 274 in connection with the computation of maxima and minima. In the work *Considerationes circa analyseos*, published in 1694, Nieuwentijt described his work on the *Analysis infinitorum* as follows [Vermij 1989, p. 75]:

> *'until finally I learned, about four years ago, after I had provided myself with that most useful work of the Acta eruditorum, not only that great men considered this very subject, but also that Leibniz had treated long before an algorithm of the differential calculus and exhibited, just as I have done, the sums, products and quotients of infinitely small quantities and of differentials (which I had deduced from another algorithm, more fundamental than Leibniz's). So I conjectured, not without some joy, that I had found the same principle as so great a man.'*

We hence may safely assume that Nieuwentijt had discovered Leibniz's infinitesimal calculus only during the year 1690 when he already had finished the first six chapters of the *Analysis infinitorum*. From chapter seven on Nieuwentijt referred to Leibniz. In the eighth chapter he recorded that Leibniz's differentials of higher order as $(dx)^2$, $dx\,dy$, and so on could not exist since they were nil following his arguments, and the calculation with nils would be pointless. In the *Considerationes* Nieuwentijt got even more explicit. This work, written after the *Analysis infinitorum* but published earlier, was initiated by Nieuwentijt's reading of a paper by John Bernoulli in the Acta Eruditorum of the year 1694. Leibniz got both works of Nieuwentijt in the year 1695 but apparently he took it lightly [Vermij 1989, p. 77]. Already in 1695 Leibniz published the paper *Responsio ad nonnullas difficultates a Dn. Bernardo Niewentijt circa methodum differentialem seu infinitesimalem motas* (Reply to some objections put forward against the differential or infinitesimal method by Mr Bernard Nieuwentijt) [Leibniz 2011, p. 271 ff.] in the Acta Eruditorum. Here he concentrated on three questions:

1. When may infinitesimal quantities be omitted?
2. Can infinitesimal quantities be applied to exponential equations?
3. Do higher differentials satisfy the same definitions as 'quantities'?

The second question, also posed by Nieuwentijt, is concerned with his statement that the differential quotient of an exponential function

$$y^x = z$$

can not be computed. Nieuwentijt reasoned as follows [Nagel 2008, p. 202]. The quantity dz is the infinitesimal difference

$$dz = (y + dy)^{x+dx} - y^x = y^{x+dx} + xy^{x+dx-1}dy - y^x,$$

if all terms are omitted which contain products of infinitely small quantities. If one now omits all infinitely small quantities still present, then one gets

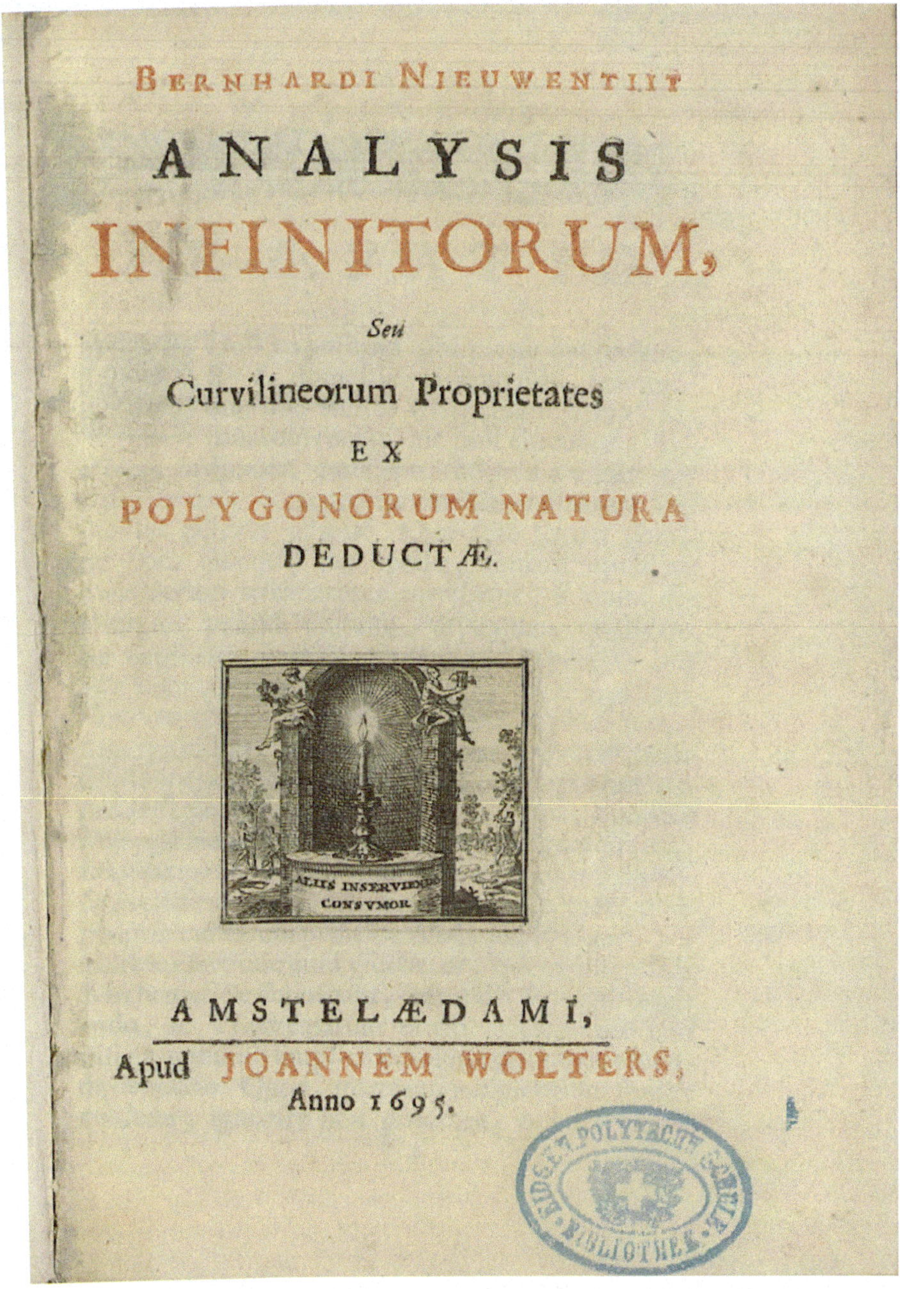
BERNHARDI NIEUWENTIJT

ANALYSIS

INFINITORUM,

Seu

Curvilineorum Proprietates

EX

POLYGONORUM NATURA

DEDUCTÆ.

ALIIS INSERVIENDO CONSUMOR

AMSTELÆDAMI,

Apud JOANNEM WOLTERS,

Anno 1695.

Fig. 10.1.2. Title page of Nieuwentijt's *Analysis Infinitorum* of 1695 (ETH-Bibliothek Zürich, Alte und Seltene Drucke, Rar 5204:1)

$$0 = y^x - y^x,$$

which is obviously correct but not very useful.

Leibniz admitted that he also had problems in treating exponential functions, but he referred to Jacob Bernoulli and gave the following example [Nagel 2008, p. 204f.]. If $x^v = y$, then it follows

$$v \log x = \log y.$$

Since

$$\log x = \int \frac{dx}{x} \quad \text{and} \quad \log y = \int \frac{dy}{y}$$

it follows

$$v \int \frac{dx}{x} = \int \frac{dy}{y},$$

what, after taking the derivative, leads to

$$\frac{dy}{y} = dv \log x + v \frac{dx}{x}.$$

Hence since $y = x^v$ one gets

$$dy = d(x^v) = x^v dv \log x + v x^{v-1} dx.$$

Unfortunately there were printing errors in the *Responsio* so that Leibniz's derivation was wrong. But since at the end he made a proof and arrived at the correct results only typos could be blamed for the error and these typos were simply overlooked by Leibniz.

Leibniz tried to answer the first question with pointing to his notion of equality: Two quantities are not only equal if their difference is nil, but also if their difference is infinitely small. Concerning question 3 Leibniz again gave an example to clarify the use of higher differentials. One month after the response to Nieuwentijt's objections Leibniz published a supplement with the title *Addenda ad schediasma Responsio* (Supplementary answer to the short report *Responsio*) in the Acta Eruditorum.

All this was not enough for Nieuwentijt. In 1696 Nieuwentijt's riposte *Considerationes secundae circa calculi differentialis principia et responsio ad Virum Nobilissimum G. G. Leibnitium* (Second considerations on the principles of the differential calculus and answer to the most noble Master G. W. Leibniz) was published. Nieuwentijt emphasised again that he was not fighting against the infinitesimal calculus but that he wanted to free the foundations of the calculus from errors [Nagel 2008, p. 205]. Apparently Leibniz did not like to reply to this riposte but that he did not need to do! He found a worthy representative in this quarrel in a pupil of Jacob Bernoulli, Jacob Hermann (1678–1733).

Fig. 10.1.3. The Swiss mathematician Jacob Hermann in two paintings [Painter unknown, prob. 18th c]

In 1700 Hermann's 62 pages long booklet bearing the title *Responsio ad Clarissimi Viri Bernh. Nieuwentiit Considerationes Secundas circa calculi differentialis principia* (Answer to the second considerations of the most famous Master Bernhard Nieuwentijt on the principles of the differential calculus) was published. The first five of six chapters were devoted to Nieuwentijt's objections against the infinitesimal calculus; in the sixth chapter Hermann presented the relevant propositions of the differential calculus [Nagel 2008, p. 206ff.]. The argumentation ran exactly as to suit Leibniz and the Bernoullis. A quantity as b/m^2, so Hermann wrote in the fourth chapter, must not only be multiplied by m, but with mm instead. In doing so one would arrive again at the given finite quantity b without contradiction and had secured the existence of higher differentials. In the fifth chapter Hermann pointed to the errors in Leibniz's *Responsio* and corrected the calculations for the derivative of $x^v = y$. Hermann recommended studying Newton's *Principia*, but also the works of his teacher Jacob Bernoulli in the Acta Eruditorum and de l'Hospital's *Analyse des infiniments petit.* In case of Hermann this response to Nieuwentijt developed into a key publication of his career. Recommended by Jacob and John Bernoulli, Leibniz promoted his election to the Brandenburg Academy and provided him with the professorship in Padua which Galileo Galilei once occupied. In 1725 Hermann went to the Chair of Higher Mathematics at the Academy of Sciences in St Petersburg and in 1731 took the Chair of Ethics, Natural-, and International Law at the University of Basel. Hermann's *Responsio ad Clarissimi Viri Bernh. Nieuwentiit* marked the very end of the Leibniz-Nieuwentijt controversy.

10.2 Bishop Berkeley and Newton's Fluxions

At least in my home country Germany George Berkeley led a shadowy existence for too long. Only in 1987 a first biography [Kulenkampff 1987] was published by Arend Kulenkampff in the German language; in 1989 Wolfgang Breidert's biography [Breidert 1989] appeared which thoroughly covered Berkeley's mathematical work. Breidert also is the translator and editor of Berkeley's work [Berkeley 1985]. More light was shed recently by Jesseph in his important work [Jesseph 1993].

10.2.1 A Life between Theology, Science, and Tar Water

George Berkeley was born on the 12th March 1685 in Kilcrene in Southern Ireland. The paternal grandfather had come from England but George saw himself already as a genuine Irishman. His father William Berkeley was a wealthy man and could send three of his sons to university; one of them was George who received his school education at Kilkenny College, which had a reputation as being the 'Eton Irelands' [Breidert 1989, S. 13].

It was the same College where the famous Irish satirist and writer of *Gulliver's Travels*, Jonathan Swift, was educated. At Kilkenny College Berkeley made

Fig. 10.2.1. Two portraits of George Berkeley, left as bishop ([Painting by John Smybert about 1727], National Portrait Gallery Washington, NPG.89.25) and (right) as reverend ([Engraving, about 1760], Wellcome Images/Wellcome Trust, London)

lifelong friendship with Thomas Prior (1680–1751) who in 1731 founded the 'Dublin Society for the Promotion of Agriculture, Manufactures, Arts, and Sciences'.

In 1700, as a 15-year-old, Berkeley became a student at Trinity College Dublin where he studied mathematics, logic, languages, philosophy, and theology. He got the degree of a Baccalaureus Artium in 1704 and started to wait for a free fellowship at Trinity College. This fellowship, however, was available not before autumn 1706. Apparently he used the spare time to work as a private tutor and to educate himself further. In spring 1707 his first work *Arithmetica absque Algebra aut Euclide demonstrata cui accesserunt cogitata nonnulla de radicibus surdis, de aestu aeris, de cono aequilatero et cylindro eidem sphaerae circumscriptis, de ludo algebraico et paranetica quaedam ad studium matheseos, praesertim algebrae* (Arithmetic proven without algebra or Euclid [=geometry]. Appended are some thoughts on irrational roots, on the heat of the air, on cylinder and equilateral cone, which are circumscribed to the same sphere, on an algebraic game, and certain admonitions concerning the study of mathematics, in particular of algebra).

It is highly likely that this work was composed to cut a figure in view of the candidacy for the fellowship. The work clearly shows Berkeley's youthful enthusiasm for mathematics, but no evidence of a refusal of infinitesimal mathematics can be found. In fact, infinitesimal mathematics did not even appear in the work. At the time of publication though, Berkeley kept a philosophical diary, the 'Philosophial Commentaries' (also known as 'Commonplace Book') [Berkeley 1979], in which he expressed himself already quite critically about mathematicians working with infinitesimals [Breidert 1989, p. 15].

Of interest to us are the great thinkers which Berkeley referred to in his first work.

They are Isaac Newton in physics, John Locke in epistemology, and the mathematicians André Tacquet (1612–1660), John Wallis, and Bernard Lamy (1640–1715). The Jesuit Tacquet was a brilliant mathematician from Brabant who is counted as one of the precursors in the pre-history of differential and integral calculus. His work *Cylindricorum et annularium liber quintus; una cum dissertatione physico-mathematica de circularium volutatione per planum* (Fifth book on cylindrical and ring-shape things together with a physical-mathematical dissertation on the revolution of some circular by a plane) was published in 1651 and influenced Pascal and probably some others of his contemporaries [Boyer 1959, p. 140]. Bernard Lamy belonged to the Oratori of Saint Philip Neri and was a theologian and mathematician. In the year 1679 he published the *Traité de Mécanique* (Treatise on mechanics), in which the parallelogram of forces could be found. In 1685 the publication of *Les éléments de géométrie* (The elements of geometry) followed.

Fig. 10.2.2. Bernard Lamy ([Engraving, 18th c, unknown artist] Revue historique et archéologique du Maine, 1894) and Jonathan Swift [unknown engraver, The International Magazine 1850]

The *Arithmetica* was widely ignored, but with the help of it and after an examination on the 9th June 1707 Berkeley became fellow of Trinity College [Berkeley 1985, p. 12]. On 19th November 1707 Berkeley gave a lecture on *Of Infinities* to the newly founded Dublin Philosophical Society. It was already here that young Berkeley criticised the weaknesses of the infinitesimal calculus and we have to inspect this talk in more detail later.

In January 1713 Berkeley travelled to London and got back to Ireland only for short periods of times during the following years. The journey pursued several targets. For one it was an educational tour; secondly Berkeley wanted to improve his health. Finally it was a missionary journey since Berkeley wanted to make his thoughts accessible to English scholars [Breidert 1989, p. 31]. Due to his kindness Berkeley went down well with his English contemporaries; he had already published four books and carried the manuscript of a fifth one with him. His friend Jonathan Swift (1667–1745) paved the ways to culturally interested circles in London. Berkeley was a determined immaterialist. He did not assign objective existence to the spatial outside world. The existence comes only through its appearance to us. We can imagine that such an ontological inclination led to interesting discussions in London. In such discussions Berkeley always stood fast and defended his philosophy, but he also stayed sociable and friendly [Breidert 1989, p. 34]. In Berkeley's *Philosophial Commentaries* we find the famous ontological principle [Berkeley 1871, p. 422]:

> *'Existence is percipi, or percipere, [or velle, i.e. agere]. The horse is in the stable, the books are in the study as before.'*

In London Berkeley met publisher Richard Steele (1672–1729) who planned to publish a new journal, the 'Guardian'. Steele could persuade Berkeley to participate. Through Steele Berkeley also got acquainted with the poet Alexander Pope (1688–1744) whose famous couplet on Newton:

Nature and nature's law lay hid in night;
God said 'Let Newton be!' and all was light.

played a decisive role even in the Hollywood movie *The Da Vinci Code* starring Tom Hanks in the leading role.

The 'Grand Tour', a common journey for educated men through the Continent down to Italy, was still missing. The opportunity came in autumn 1713 when Berkeley could travel to Italy as the spiritual advisor to Lord Peterborough. In 1714 Berkeley travelled back to London on the land route where he arrived about the day of the death of Queen Anne on 1st August [Breidert 1989, p. 36]. In autumn 1716 another opportunity to travel to Italy came up; this time as travelling companion and tutor of the sickly son of George Ashe, Bischop of Clogher, whom Berkeley knew since his days as a student at Trinity College when the director of the College was Ashe. This second trip to Italy lasted for four years and Berkeley could even experience an eruption of the Vesuvius.

It is not known where Berkeley stayed after his return from Italy between autumn 1720 until autumn of the following year. Breidert suspected that he stayed in London [Breidert 1989, p. 46]. Then Berkeley returned to Trinity College Dublin where he obtained the degree of Doctor of Theology in November 1721. To supplement his meagre income from the College Berkeley sought for a position as Dean without official duties. In fact there was such a position in Dromore but a quarrel on the occupation of the position started with the Bishop of the diocese, which ended in a settlement. Berkeley only received the title of a Dean but no income. A new position had to be found.

In the early winter of 1722 Berkeley travelled to London to remind important people of his existence, following the advice of a friend employed by the Earl of Grafton. This friend had already supported him in the application concerning the position in Dromore. Meanwhile Berkeley's thoughts concerning a 'Bermuda project' had ripened. He wanted to do 'good things for humanity' [Breidert 1989, p. 48] in that he wanted to look after education and religious upbringing of the young people in a College or Seminary, even for 'a number of young American savages'. Before the idea of a Bermuda project could manifest Berkeley got the Deanery of Londonderry in May 1724 with the support of his Trinity College. That meant that he had to leave Trinity.

Those were the days of enthusiastic missionary awakening and an expansion of politics towards the west [Breidert 1989, p. 53]. Perhaps the Bermuda project would never have been started had Berkeley not received a surprising inheritance. His friend Jonathan Swift had relationships with two women at that time [Breidert 1989, p. 50f.]. Swift lived with his housekeeper Esther Johnson in concubinage but dallied since 1708 with a much younger woman, Esther van Homrigh, who jealously looked at Swift's relationship with Esther Johnson. When rumours spread concerning a secret marriage of Swift with Esther Johnson, Esther van Homrigh demanded clarification. However, Swift let her down and she died shortly after. Shortly before her death she had changed her last will and testament; not Swift inherited, but George Berkeley and a man named Robert Marshall. Berkeley interpreted the inheritance as divine intervention in favour of the Bermuda project.

Berkeley threw himself into planning the project and even found some supporters, for example his friend Thomas Prior and some other persons connected to Trinity College. Well known characters could be won over as financiers, the Archbishop of Canterbury and the Bishop of London, so that the Bermuda project became a public matter. Finally even the King agreed with the foundation of a College under the name of St Paul's College and Berkeley was supposed to be its first president. However, there was still not enough money so that Berkeley successfully applied for the sale of landed property on one of the Lesser Antilles. Berkeley had to wait long for the money, though. In the first week of September 1728 he secretly sailed to America with a small group of people. On board was also his young wife, Anne Berkeley, who was the daughter of Chief Justice John Forster from Dublin. Berkeley had married her one month before departure. He said [Berkeley 1901, p. liii]:

> *'I chose her for her qualities of mind and her unaffected inclination to books. She goes with a great thankfulness, to live a plain farmer's life, and wear stuff of her own spinning. I have presented her with a spinning-wheel.'*

Strangely enough this secret journey did not lead to Bermuda but to Newport, Rhode Island. There Berkeley bought a house in Middletown and waited for the money from England. It had probably become clear to him that Rhode Island was the better place for the planned College, but the purchase of land certainly unsettled some of the supporters of the project in England. Also his secret departure upset some people and in clerical circles one was justifiably annoyed that Berkeley put to rest his deanship for so long without releasing his position. The promised money at least never flew; Prime Minister Walpole refused to pay the money and the Bermuda project finally failed.

After years of waiting Berkeley arrived back in England in October 1731. He made his house in America, the land, and eight boxes of books a gift to the Seminary of Yale; to Trinity Church, in which he had preached, he endowed a new organ, and to the library of Harvard he sent a box of Latin classics from England. Berkeley stayed in London for the time being; to return to Londonderry would have been politically unwise. He was interested in the deanery of Down, but the Archbishop of Dublin intervened and prevented the vocation because Berkeley was a *'maniac'* [Breidert 1989, p. 64]. After

Fig. 10.2.3. Bischop Berkeley's house, Newport/Rhode Island ([Photo: Joshua Appleby Williams, 1859–1885], New York Public Library). This is a stereographic photo which gives rise to a 3-d impression when viewed with a particular optical device (and that already in the second half of the 19th century!)

all Berkeley was promised the next available bishopric and he returned to Ireland. In the beginning of 1734 he was offered a position in Cloyne – far enough from Dublin – and Berkeley was consecrated Bishop of Cloyne on the 19th May 1734 in Dublin.

It was clear to Berkeley that only the 'freethinkers' could be blamed for the failure of the Bermuda project, i.e. those who dealt with atheistic thoughts and stood far from the Anglican Church. These convictions led Berkeley to write the philosophical-theological dialogue *Alciphron or the Minute Philosopher* [Berkeley 1871, Vol. II, p. 13ff.] which was published shortly after his return from America 1732 in London, and which was thought to challenge the freethinkers. But Berkeley was still not satisfied, for he automatically included mathematicians who dealt with infinitesimal methods in his notion of freethinkers. In a letter of 7th January 1734 Berkeley wrote to his friend Thomas Prior that he was occupied with some mathematical things in the morning hours [Breidert 1989, p. 69]. The fruit of this occupation was the book *The Analyst* [Berkeley 1992, p. 159 ff.] which was published in the last days of March 1734. In the Introduction the behaviour of an infidel mathematician was criticised as having misled unwary persons in religious matters. This mathematician is thought to be no other than Edmond Halley [Breidert 1989, p. 69].

Cloyne lies in some 40 km distance from the East-Irish seaport of Cork. In his capacity as Bishop of Cloyne Berkeley had to look after widely scattered parishes where the protestants comprised a minority and the catholics were

Fig. 10.2.4. 20 Authors [Engraving by J. W. Cool, 1825]. George Berkeley as one of 20 authors in Grabb's Historical Dictonary. Berkeley is seen left in the second last row (Wellcome Images/Wellcome Trust, London)

about eight times as much in the diocese [Breidert 1989, p. 71]. Berkeley was constantly eager to improve the living conditions of the poor population. He wrote on national and financial economy; he founded a school for spinning

Fig. 10.2.5. Berkeley's travel party ([Detail of a painting by John Smybert], Yale University Art Gallery)

for the children and a workhouse for vagrants. With his wife he had seven children of which four survived infancy. Berkeley taught Greek and Latin to his children himself, for the music he employed an Italian musician.

An episode in Berkeley's later life has brought him much ridicule and mockery in recent times, and this episode is known as his 'tar water period'. Medicine was in a regrettable state in the 18th century; there was no public health care, for example. After the extremely harsh winter 1739/40 a dysentery epidemic began and Berkeley started to experiment with medications which would help the people. He had some success with finely grated resin in broth, but from America he already knew tar water with which the Indians tried to prevent outbreaks of smallpox. After having studied several books on chemistry Berkeley succeeded in producing a clear extract of a tar infusion which he applied first to himself, then to his family, and finally to the population of Cloyne. In 1744 Berkeley published a book on his tar water under the title *Siris*. The book became an instant success, saw many editions and parts of it were translated into other languages [Breidert 1989, p. 76]. The book led to a tar water 'addiction'; pharmacists in London did not sell anything else than tar water and a contemporary remarked with some sense of humour that one could not even write a letter without having tar water in the ink. Of course, there was resistance of the licensed orthodox medical practitioners against whom Berkeley wrote satirical poems. After all, he thought his tar water was a universal remedy, which it certainly was not. Aqueous extracts of tar certainly have a slightly antiseptic effect due to the phenols they contain, but we know today that they also have side effects like nodular irritation and can even cause cancer.

In old age Berkeley led a secluded life on his rural Bishop's see. In March 1751 his 14 year old son William who was very close to his heart died; in October of the same year he was in grief about the loss of his friend Thomas Prior. When Berkeley's oldest son George went to study in Oxford Berkeley accompanied him with his wife and daughter Julia to oversee the son, who tended to a luxurious lifestyle. Shortly prior to departure he made his last will and testament. One should bury him exactly where he would die. The costs of the burial should be bounded by a maximum of £20; the poor should be abundantly given to. His wife was to inherit before the children. Wholly captured in the primal fear of the time to be buried alive, Berkeley disposed that his corpse should lay unaffected for five days before burying it. On Sunday, 14th January, the family had tea and read the Bible. When Berkeley's daughter Julia wanted to serve her father a cup of tea she noticed that he had passed away. A stroke had killed him secretly.

10.2.2 Berkeley's Philosophy of Mathematics

Before we have to discuss Berkeley's critique concerning the infinitesimal mathematics it is worthwhile to take a look at Berkeley's general attitude towards mathematics.

We may safely assume that Berkeley – as all other educated men of his time – had a solid background in classical ancient philosophy. Plato (428/427 BC–348/347 BC) had located mathematical ideas in the heaven of ideas; the ideal circle, the ideal straight line, and so on. He saw the real objects, the drawn circle, the drawn straight line, and so on, as mere images or approximations of the ideal shapes.

Plato's pupil Aristotle (384–322 BC) would not believe in his teacher's heaven of ideas. There was a problem, however. The mathematical objects could not depend on the structure of the real world since exact circles could simply not be found in the real world. On the other hand they must not come from a metaphysical heaven of ideas. Aristotle solved this problem in a most elegant way: Yes, mathematical objects are born of our experiences in the real world, but their mathematical specification takes place in our mind through *abstraction* [Jesseph 1993, p. 10]. Abstraction as basis of mathematics was fully acknowledged in the 17th century. Even mathematicians who were so different in their mathematical preferences like John Barrow and John Wallis – the first a classical geometer, the second delighted with the algebraic treat-

Fig. 10.2.6. Plato (Head of Plato, Roman copy of a Greek sculpture now in the Glyptothek München (Munich) and Aristotle (Bust of Aristotle, marble, Roman copy after the Greek original bronze by Lysippos, about 330 BC now in the National Museum Rom - Palazzo Altemps; Ludovisi Collection [Photo: Jastrow 2006])

ment of the new Analytical Geometry – valued the importance of abstraction very highly.

It is obvious that Berkeley was well aware of the abstractionist philosophy of mathematics; after all he read books by Wallis and Barrow, and the contemporary dictionaries like Joseph Raphson's *Mathematical Dictionary* of 1702 defined mathematics as an abstract science [Jesseph 1993, p. 17]. The idea of abstraction had also penetrated general philosophy; for example in the writings of John Locke. Berkeley on the contrary refused abstraction in general philosophy as in the philosophy of mathematics.

In Berkeley's view abstraction had caused nothing but confusion and chaos in philosophy and the sciences. In the introduction to his *A Treatise concerning the Principles of Human Knowledge* of 1710 he reduced the missing progress in philosophy to false principles [Jesseph 1993, p. 21],

> *'amongst all which there is none, methinks, hath a more wider influence over the thoughts of speculative men, than this of abstract general ideas.'*

Berkeley's argument against the absurdities of abstract ideas is known as 'argument from impossibility'. We do not want to digress too much but only remark that the argument from impossibility is concerned with the idea of impossible objects derived by means of abstraction, which can not be described in a consistent manner. Together with the principle that something impossible can never be understood, this was enough to reject abstraction [Jesseph 1993, p. 21f.]. Contemporary philosophers like Locke made it easy for Berkeley to work with the argument from impossibility; I refer the reader to the inconsistent description of an ideal triangle in Locke's book [Locke 1995, Book IV, Chapter VII, 9, p. 509]. However, Berkeley was otherwise philosophically related to Locke.

But what is then Berkeley's alternative to abstraction? It is the *particular* idea which can be generalised to other problems; Jesseph calls this alternative the 'theory of representative generalisation' [Jesseph 1993, p. 33]. Led by intuition the particular ideas can be thought of being divided into equivalence classes. Every class then contains ideas which are similar with respect to one aspect. For instance, we can draw a particular triangle from the set of all triangles. This triangle is different from all other triangles in its class, but they all share three edges, three angles, and three edge points. If we use only quantities being common to all triangles in the proof of the sum of angles we have proved something for our particular triangle which can be generalised to all other triangles [Jesseph 1993, p. 34].

We do not want to examine the whereabouts of the sources of Berkeley's rejection of the abstract and refer to [Jesseph 1993, p. 38ff.]. That this rejection was inherent in his thinking to a great extent plays also a great role in his rejection of the mathematics of infinitesimals.

10.2.3 Of Infinities

The short essay *Of Infinities* [Berkeley 1985, p. 75-80] was presented to the Dublin Philosophical Society on 19th November 1707 when Berkeley was 22 years old.

There are three important issues on which Berkeley concentrated. Berkeley had taken the difference between 'infinity' and 'infinite' from Locke [Locke 1995, Book II, Chapter XVII, p. 145ff.]. He also took from Locke the semantic argument that one should only use a word if one has an idea corresponding to the meaning of the word. Only on the basis of this argument, talking about infinitely small quantities is enough proof of the impossibility of their existence for Berkeley. He wrote [Jesseph 1993, p. 164]:

> *'For he that, with Mr. Locke, shall duly weigh the distinction there is betwixt infinity of space & space infinitely great or small, & consider that we have an idea of the former, but none at all of the later, will hardly go beyond his notions to talk of parts infinitely small or partes infinitesimæ of finite quantitys, & much less of infinitesimæ infinitesimarum, and so on.'*

The concept: 'we can imagine a smaller quantity to a given one' is completely fine, but: 'we can imagine a quantity being smaller than any given quantity' is nonsense. Infinitesimals were rejected not only because we can not see them, but because we have no idea of them [Jesseph 1993, p. 165f.]. Berkeley goes on [Berkeley 1985, p. 76][1]:

> '*They* [the mathematicians working with infinitesimals] *represent – on paper – quantities of different orders as if they had ideas in their minds which correspond to these words or symbols, or as if it does not contain a contradiction that there may be an infinitely small line and still an even smaller one. It is clear to me that we should not use any symbol without having a corresponding idea, and it is also clear that we do not have an idea of an infinitely small line, on the contrary, ...*'
>
> (Sie [die Infinitesimalmathematiker] stellen – auf dem Papier – infinitesimale Größen verschiedener Ordnungen dar, als hätten sie in ihrem Geist Ideen, die diesen Wörtern oder Zeichen entsprechen, oder als enthielte es keinen Widerspruch, daß es eine unendlich kleine Linie und noch eine unendlich kleinere als sie geben soll. Für mich ist klar, daß wir kein Zeichen ohne eine ihm entsprechende Idee verwenden sollten, und es ist ebenso klar, daß wir keine Idee von einer unendlich kleinen Linie haben, im Gegenteil, ...)

Then Berkeley took up the Nieuwentijt-Leibniz controversy and wrote [Berkeley 1985, p. 77]:

> *'Mr. Nieuwentijt conceded that infinitesimal quantities of first order were genuine quantities, but the differences of differences (differentiae differentiarum) or infinitesimal quantities of the higher orders he eliminated in that he made them to the same number of nils. That is the same as to say that the square, cube, or any other power of a genuinely positive quantity is nil, what apparently is absurd.'*
>
> (Herr Nieuwentijt räumt ein, infinitesimale Größen der ersten Ordnung seien wirkliche Größen, aber die Differenzen von Differenzen (differentiae differentiarum) oder infinitesimale Größen der folgenden Ordnung beseitigt er, indem er sie zu ebensovielen Nullen macht. Das ist dasselbe wie zu sagen, das Quadrat, der Kubus oder eine andere Potenz einer wirklich positiven Größe sei gleich Null, was augenscheinlich absurd ist.)

Not only Nieuwentijt received a put-down, but, of course, Leibniz too. Berkeley is not happy with Leibniz's definition that two quantities are not only equal when their difference is nil, but also if their difference is infinitesimally small. His arguments were quite Aristotelian [Berkeley 1985, p. 78]:

[1] I am translating here Breidert's superb German edition.

> *'»If you«, says he, »add a point of a line to another line, you do not increase its size.« But if lines are infinitely divisible, I ask how there can be any such thing as a point?'*
>
> (»Wenn du«, so sagt er, »einer Linie einen Punkt einer anderen Linie hinzufügst, vermehrst du ihre Größe nicht.« Wenn aber Linien unendlich teilbar sind, frage ich, wie es so etwas wie einen Punkt geben kann.)

10.2.4 The Analyst

The Analyst; or A Discourse addressed to an Infidel Mathematician; Wherein It is examined whether the Object, Principles, and Inferences of the Modern Analysis are more distinctly conceived or more evidently deduced, than Religious Mysteries and Points of Faith was published in 1734 and aimed at two targets, theological and mathematical. Theologically Berkeley wanted to continue his attack on the freethinkers and show that their critique of the revealed religion was unjustified [Jesseph 1993, p. 178]. The same freethinkers who mocked religion on grounds of its mysteries on the other hand had a strong believe in the infinitesimal calculus, which, following Berkeley, was at least as mysterious as religion.

The 'infidel mathematician' appealed to in the title can not be identified with certainty; we can suspect Edmond Halley but that is not important. Berkeley placed religion with its mysteries on one side, mathematics with its rigour on the other, and he wanted to show that the infinitesimal calculus did *not* satisfy the required rigour. He did that in no less than 50 paragraphs.

In paragraph 9, 'Method for finding the Fluxion of a Rectangle of two intermediate Quantities, showed to be illegitimate and false' Berkeley referred to Lemma II in Book II of Newton's *Principia* [Newton 1999, p. 646ff.], [Newton 1963, p. 243f.]. In the translation of [Newton 1999][2] Lemma II reads as:

> *'The moment of a generated quantity is equal to the moment of each of the generating roots multiplied continually by the exponents of the powers of these roots and by their coefficients.'*

The Lemma is not easily understood, but it is concerned with the product rule

$$(f(x)g(x))' = f'(x)g(x) + f(x)g'(x).$$

This becomes clear in Newton's proof [Newton 1999, p. 648]. If AB is a rectangle which is increased in continuous movement, so that the sides $A - \frac{1}{2}a$,

[2]English translation by Cohen and Whitman. A very interesting discussion of terms Newton used and the reasons for the chosen English translation of them can be found there.

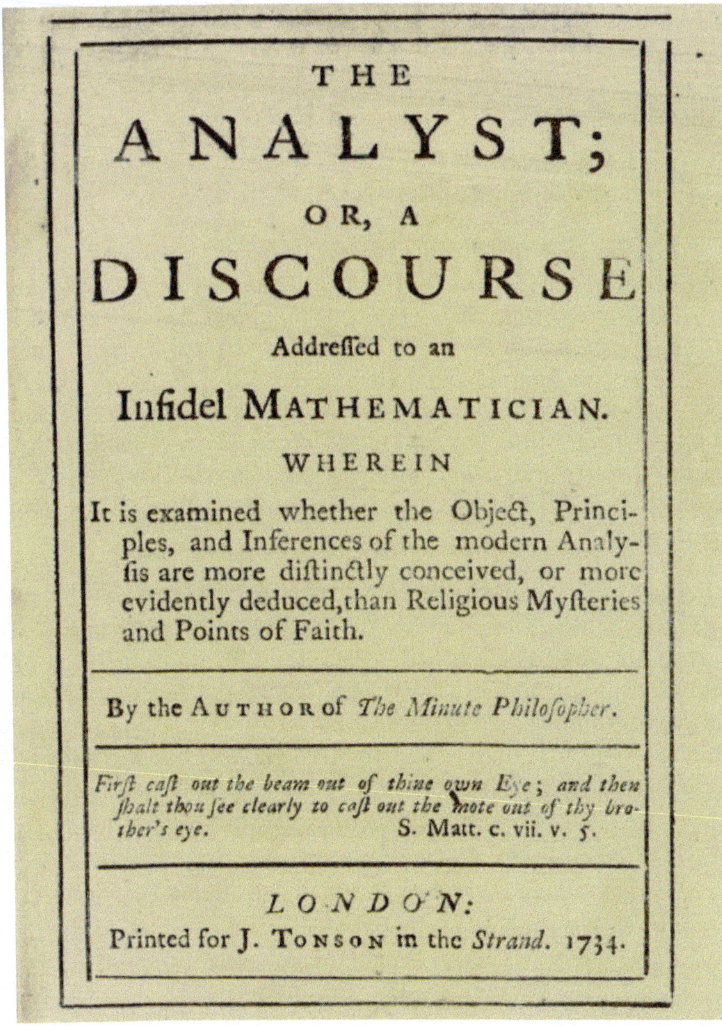

THE
ANALYST;
OR, A
DISCOURSE
Addreſſed to an
Infidel MATHEMATICIAN.
WHEREIN
It is examined whether the Object, Principles, and Inferences of the modern Analyſis are more diſtinctly conceived, or more evidently deduced, than Religious Myſteries and Points of Faith.

By the AUTHOR of *The Minute Philoſopher.*

Firſt caſt out the beam out of thine own Eye; and then ſhalt thou ſee clearly to caſt out the mote out of thy brother's eye. S. Matt. c. vii. v. 5.

LONDON:
Printed for J. TONSON in the *Strand.* 1734.

Fig. 10.2.7. Title page of the *Analyst* (Image Courtesy of the John M. Kelly Library, University of St. Michael's College, Toronto)

$B - \frac{1}{2}b$ become the sides $A + \frac{1}{2}a$, $B + \frac{1}{2}b$, where a and b are moments, i.e. infinitely small quantities, then the initial area is

$$\left(A - \frac{1}{2}a\right) \cdot \left(B - \frac{1}{2}b\right) = AB - \frac{1}{2}Ab - \frac{1}{2}Ba + \frac{1}{4}ab,$$

while after the movement the resulting area is

$$\left(A + \frac{1}{2}a\right) \cdot \left(B + \frac{1}{2}b\right) = AB + \frac{1}{2}Ab + \frac{1}{2}Ba + \frac{1}{4}ab.$$

Apparently, subtraction of these areas yields

$$\left(A+\frac{1}{2}a\right)\cdot\left(B+\frac{1}{2}b\right)-\left(A-\frac{1}{2}a\right)\cdot\left(B-\frac{1}{2}b\right)=Ab+Ba.$$

The individual moments a and b of the sides result in the increment $Ab+Ba$ of the area. Berkeley disagreed [Berkeley 1992, p. 171]:

> *'But it is plain that the direct and true Method to obtain the Moment or Increment of the Rectangle AB, is to take the Sides as increased by their whole Increments, and so multiply them together, A+a with B+b, the Product whereof AB + aB + bA + ab is the augmented Rectangle; whence if we subduct AB, the Remainder aB + bA + ab will be the true Increment of the Rectangle, exceeding that which was obtained by the former illegitimate and indirect Method by the Quantity ab.* [...] *Nor will it avail to say that ab is a Quantity exceeding small: Since we are told that in rebus mathematicis errores quam minimi non sunt contemnendi.'*

Ironically enough, the citation 'in rebus mathematicis errores ...' came from one of Newton's works, meaning literally: 'The most minute errors are not in mathematical matters to be scorned.' [Berkeley 1992, footnote $*$, p. 171]. However, Berkeley was not alone in criticising Newton's Lemma. Charles Hayes (1678–1760), who published the first book on Newton's fluxional calculus in English language in 1704 under the title *Treatise on Fluxions, or an Introduction to Mathematical Philosophy*, was also not satisfied.

Hayes wrote *'There is yet another way to find the Fluxion of any Rectangle* [...]*'* and presented then the derivation we already found in Berkeley's paper. The term ab is denoted $\dot{x}\dot{z}$ by Hayes[3], but he neglected it, because this product is *'infinitely little'* [Jesseph 1993, p. 192].

In paragraph 13 of the *Analyst*, 'The rule for the Fluxions of Powers attained by unfair reasoning', Berkeley presented Newton's derivation of the rule $\frac{d}{dx}x^n = nx^{n-1}$ by means of the binomial theorem [Berkeley 1992, p. 173ff.], [Berkeley 1985, p. 96]. If the quantity x flows uniformly and becomes $x+o$, then x^n becomes $(x+o)^n$, where o is an infinitely small quantity. By the binomial theorem it follows

$$(x+o)^n = x^n + nox^{n-1} + \frac{nn-n}{2}oox^{n-2} + \ldots$$

and the ratio of the increments o and $nox^{n-1} + \frac{nn-n}{2}oox^{n-2} + \ldots$ is the same as 1 to $nx^{n-1} + \frac{nn-n}{2}ox^{n-2} + \ldots$. If we now let the increments o, oo, etc., vanish, then apparently it follows for the ratio

[3] Hayes wrote 'fluxion' for an infinitely small quantity, and thereby assigned a meaning different from Newton's to it. Read 'moment' when Hayes used the word 'fluxion'.

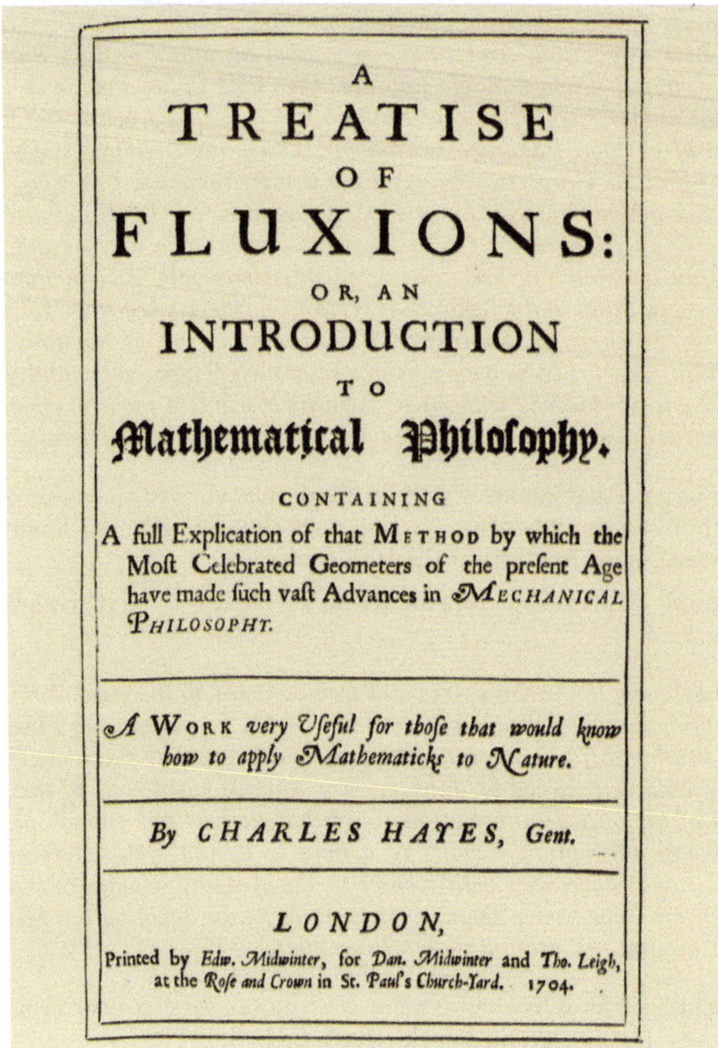

A
TREATISE
OF
FLUXIONS:
OR, AN
INTRODUCTION
TO
Mathematical Philoſophy.
CONTAINING
A full Explication of that METHOD by which the Moſt Celebrated Geometers of the preſent Age have made ſuch vaſt Advances in *MECHANICAL PHILOSOPHY*.

A WORK *very Uſeful for thoſe that would know how to apply Mathematicks to Nature.*

By CHARLES HAYES, Gent.

LONDON,
Printed by *Edw. Midwinter*, for *Dan. Midwinter* and *Tho. Leigh*, at the *Roſe and Crown* in St. *Paul's Church-Yard*. 1704.

Fig. 10.2.8. *A Treatise of Fluxions* [Detail from: Charles Hayes, London 1704]

$$1 : nx^{n-1}.$$

Why, asked Berkeley justly, can an increment o, being assumed a positive quantity at the start of the calculation, be neglected at the end?

In paragraph 14, 'The aforesaid reasoning further unfolded and shew'd to be illogical', Berkeley's critique became merciless [Berkeley 1992, p. 175f.], [Berkeley 1985, p. 98]:

'Hitherto I have supposed that x flows, that x hath a real Increment, that o is something. And I have proceeded all along on that Supposition, without which I should not have been able to have made so much as one single Step. From that Supposition it is that I get at the Increment of x^n, that I am able to compare it with the Increment of x, and that I find the Proportion between the two Increments. I now beg leave to make a new Supposition contrary to the first, i.e. I will suppose that there is no Increment of x, or that o is nothing; which second Supposition destroys my first, and is inconsistent with it, and therefore with every thing that supposeth it. I do nevertheless beg leave to retain nx^{n-1}, which is an Expression obtained in virtue of my first Supposition, which necessarily presupposeth such Supposition, and which could not be obtained without it: All which seems a most inconsistent way of arguing, and such as would not be allowed of in Divinity.'

We have to note that Berkeley's critique painfully showed the weakness of the early infinitesimal mathematics, which only could be silenced finally by the arithmetrisation of Analysis in the 19th century.

In paragraph 17 we read on Newton [Berkeley 1992, p. 178], [Berkeley 1985, p. 101]:

'Considering the various Arts and Devices used by the great Author of the Fluxionary Method: in how many Lights he placeth his Fluxions: and in what different ways he attempts to demonstrate the same Point: one would be inclined to think, he was himself suspicious of the justness of his own demonstrations; and that he was not enough pleased with any one notion steadily to adhere to it. [...] But, whatever the Case might have been with respect to the Author, it appears that his Followers have shewn themselves more eager in applying his Method, than accurate in examining his Principles.'

But Leibniz got away no better than Newton, as we can read in paragraph 18 [Berkeley 1992, S. 179], [Berkeley 1985, S. 102]:

'Leibniz and his Followers in their calculus differentialis making no manner of Scruple, first to suppose, and secondly to reject Quantities infinitely small: [...] As in Fluxions the Point of first importance, and which paves the way to the rest, is to find the Fluxion of a Product of two indeterminate Quantities, so in the calculus differentialis (which Method is supposed to have been borrowed from the former with some small Alterations[4]) the main Point is to obtain the difference of such Product.'

[4]This is to be understood as a cautious remark concerning the priority dispute.

Berkeley arrived at the conclusion that the infinitesimal calculus only worked well because of a compensation of several errors, and he underlined his conclusion by means of some examples [Jesseph 1993, p. 199ff.]. At the end of the *Analyst* Berkeley linked his anti-abstractionist philosophy to his statements concerning the mathematics of infinitesimals.

Under the number 471 Berkeley wrote in his notebook *Philosophial Commentaries* [Berkeley 1871, Vol. IV, p. 426], [Berkeley 1979, p. 61]:

> *'If the wit and industry of the Nihiliarians were employ'd about the usefull & practical mathematiques, what advantage had it brought to mankind!'*

and therewith not only placed practice above theory, but also coined the notion of the 'nihiliarians' depicting mathematicians working with infinitesimals who actually worked with 'nils'. As famous as this neologism is Berkeley's notion of infinitesimals as 'ghosts of departed quantities' in paragraph 35 of the *Analyst* [Berkeley 1992, p. 199], [Berkeley 1985, p. 121]:

> *'And what are these Fluxions? The Velocities of evanescent Increments? And what are these same evanescent Increments? They are neither finite Quantities, nor Quantities infinitely small, nor yet nothing. May we not call them the Ghosts of departed Quantities?'*

10.3 The Reactions to Berkeley's Critique

In contrast to Nieuwentijt's critique of Leibniz's differentials of higher orders Berkeley's *Analyst* initiated a long lasting controversy in English mathematics. Jesseph counts more than a dozen writings concerned with Berkeley in the years 1734 to 1750 only. There was virtually no text on fluxions in which Berkeley's critique was not addressed in one or another way [Jesseph 1993, p. 231]. Incidentally, Berkeley was barely received on the Continent.

The first objection against Berkeley appeared nearly immediately after the *Analyst* was published. It came from James Jurin (1684–1750), a physician and scientist a Guy's Hospital in London.

Jurin had a great satirical talent which he used to defend Newton whose fervent admirer he was. He had studied in Cambridge under the mathematicians Roger Cotes (1682–1716) and William Whiston (1667–1752) and later became Secretary of the Royal Society, when Newton's time of presidency was nearing its end. Already in 1704, the year the *Analyst* was published, Jurin published the pamphlet *Geometry no Friend to Infidelity* under the pseudonym 'Philaletes Cantabrigiensis' (friend of the truth from Cambridge). Berkeley replied 1735 with his paper *Defence of Free-thinking in Mathematics* [Berkeley 1985, p. 142ff.]. There, an appendix on the *Vindication of Sir Isaac Newton's Principles of Fluxions* of one John Walton could be found.

Jurin's riposte was exceptionally polemic since he apparently did not understand Berkeley's justified objections. He was rather concerned with the priority dispute between British and Continental mathematicians that did not concern Berkeley's critique [Breidert 1989, p. 109]. In substance Jurin could not rebut any of Berkeley's arguments. On the contrary, he was taken in by Berkeley in some sense since he unconsciously used the same arguments against which Berkeley was fighting. It was only one view point of Berkeley which Jurin had understood, and that was his anti-abstractionism. No wonder that Jurin also attacked Berkeley's empirist teacher Locke.

John Walton from Dublin also felt qualified to write a pamphlet against Berkeley which appeared under the title *A Vindication of Sir Isaac Newton's Principles of Fluxions, against the objections contained in the Analyst* in 1735. Walton also used arguments which Berkeley had refused and Walton's arguments therefore missed the point. In contrast to his riposte to Jurin Berkeley got very much excited in his *Defence of Free-thinking in Mathematics* against Walton and did not even recoil from personal attacks. Consequentially, Breidert concludes in [Breidert 1989, p. 112] that Berkeley must have known Walton and that Walton probably belonged to the Dublin circle around the Archbishop, who commented disparagingly about Berkeley when the latter had tried to gain the Deanery of Down.

With Berkeley's rebuttal to Jurin and Walton calm did not return. In 1735 their ripostes were published. Jurin published 112 pages under the title *The Minute Mathematician: or, The Free-Thinker no Just-Thinker* .... The title satirised Berkeley's book *Alciphron or the Minute Philosopher* against the

Fig. 10.3.1. James Jurin ([Painting probably by Joseph Highmore, 18th c.], Trinity College, University of Cambridge) and the belfry of Great Court, Trinity College Cambridge; Jurin studied here [Photo: Hans Wolff]

freethinkers, published in 1732. Objectively seen, the contents of the 'Minute Mathematician' was nothing to speak of. Jurin got lost in polemics way below the belt and so it seems consequent that Berkeley did not reply at all. Another issue was Walton's shorter rebuttal *The Catechism of the Author of the Minute Philosopher Fully Answer'd.* At the end of *An Appendix, concerning Mr. Walton's Vindication of Sir Isaac Newtons Principles of Fluxions* Berkeley had advised Walton's pupils to ask their teacher some questions [Breidert 1989, p. 112f.]. Now Walton complied to this advice and answered Berkeley's questions. Walton wrote in a much more sober manner as Jurin and Berkeley payed him a little paper *Reasons for not replying* [Berkeley 1985, p. 192-207][5]. The little work is composed in a satirical tone and with it Berkeley withdrew from the controversy.

10.4 The Impact of Berkley's Critique

After the end of the quarrels between Berkeley, Jurin, and Walton the aftermath of the critique actually began and came from scientists who wanted to put Newton's fluxional calculus on a sounder basis. We have to mention Benjamin Robins (1707–1751), English engineer and mathematician, who is acknowledged as the founder of scientific ballistics and whose work *Principles of Gunnery* of 1742 was translated and massively expanded by none less than Leonhard Euler (1707–1783), who published it under the title *Neue Grundsätze der Artillerie* (New principles in artillery) 1745 in Berlin. Directly encouraged by the controversy about the *Analyst* Robins published the book *A Discourse Concerning the Nature and Certainty of Sir Isaac Newton's Methods of Fluxions, and of Prime and Ultimate Ratios* in which Newton's fluxional calculus was to be defended.

Robins started from Newton's definition of fluxions as being velocities of changing geometrical quantities. The key for understanding fluxions, so Robins wrote, lay in the ratio of fluxions to one another. On page 6 he wrote [Jesseph 1993, p. 260]:

> *'And as different fluents may be understood to be described together in such manner, as constantly to preserve some one known relation to each other; the doctrine of fluxions teaches, how to assign at all times the proportion between the velocities, wherewith homogeneous magnitudes, varying thus together, augment or diminish.'*

Subsequently Robins proved the theorem that the fluxion of x^n is nx^{n-1}. We follow the somehow simplified explanations of Jesseph [Jesseph 1993, p. 260f.],

[5]Complete title: *Reasons For not Replying to Mr. Walton's Full Answer in a Letter to P.T.P. By the Author of the Minute Philosopher.* The identity of P.T.P. remains unknown [Breidert 1989, p. 115, footnote 143].

Fig. 10.4.1. Benjamin Robins received the Copley medal in 1746, the most valuable award of the Royal Society, London (shown here is the medal of 2005 [Photo: Serge Lachinov 2009]). The Copley medal has been awarded to outstanding scientists since 1731. Award winners are (choice): Benjamin Franklin, James Cook, Edward Waring, David Brewster, Antoine César Becquerel, Michael Faradey, Carl Friedrich Gauß, Justus von Liebig, Georg Simon Ohm, John Herschel, Alexander von Humboldt, Léon Foucault, Charles Darwin, Hermann von Hemholtz, Louis Pasteur, James Joseph Sylvester, Arthur Cayley, Thomas Henry Huxley, George Gabriel Stokes, Karl Weierstraß, Felix Klein, Albert Einstein, Max Planck, Niels Bohr, Stephen Hawking, Roger Penrose, and many, many others. In 2015 Peter Higgs was awarded the Copley medal

but also note that Robins's book is available in different formats (photocopies or transcription) on the internet.

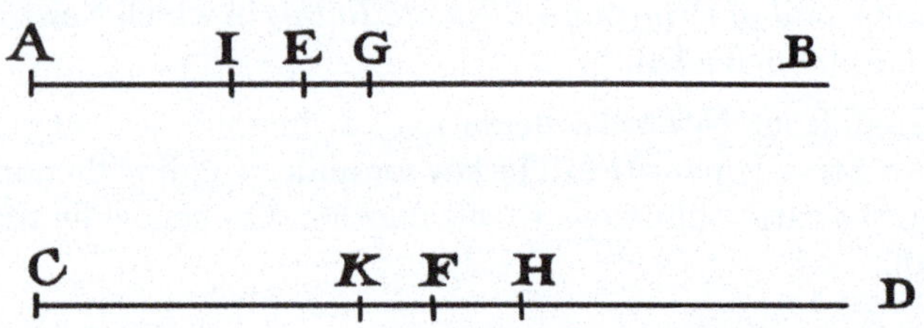

Fig. 10.4.2. Figure from Robins's *A Discourse Concerning the Nature and Certainty of Sir Isaac Newton's Methods of Fluxions, and of Prime and Ultimate Ratios*

Given two line segments AB and CD as in figure 10.4.2 which are thought of being produced by moving points, respectively. The velocities of the two points are thought to be in some ratio. If we denote AE by x, then $CF = x^n$ should be valid. Robins wants to prove that the ratio of the fluxions of CD and AB is everywhere the same. He considers increments EG in AB and FH in CD. Denoting EG by e the assumption may be written as

$$CH = (x + e)^n.$$

Expansion by means of the binomial theorem then yields

$$CH = x^n + nx^{n-1}e + \frac{n(n-1)}{2}x^{n-2}e^2 + \ldots.$$

Subtracting CF from the left, and x^n from the right side (both quantities are identical) leads to

$$CH - CF = FH = nx^{n-1}e + \frac{n(n-1)}{2}x^{n-2}e^2 + \ldots.$$

We assume $n > 1$. On page 9 Robins wrote [Jesseph 1993, p. 261]:

> '... *no line whatever, that shall be greater or less than the line represented by the second term of the foregoing series (viz.* $nx^{n-1}e$*) will bear to the line denoted by* e *the same proportion, as the velocity, wherewith the point moves at* F*, bears to the velocity of the point moving in the line* AB*; but that the velocity at* F *is to that at* E *as* $nx^{n-1}e$ *to* e*, or as* nx^{n-1} *to* 1.'

The proof now proceeds as *reductio ad absurdum* [Sonar 2016, p. 31] with a classical exhaustion argument [Jesseph 1993, p. 262f.].

The proof clearly shows that Robins had fully understood Berkeley's concerns and objections; but it also shows that Newton's calculus of fluxions could in fact be based on classical arguments.

In his essay *A Review of some of the Principal Objections that have been made to the Doctrine of Fluxions and Ultimate Proportions; with some Remarks on the different Methods that have been taken to obviate them* published 1735 in Number 16 of 'The Present State of the Republick of Letters' Robins referred to the futile attempts of Jurin to defuse Berkeley's critique and juxtaposed Jurin's methods with his own. Thereby he came into conflict with Jurin. While it was clear to Robins that last ratios of variables are never actually assumed, Jurin thought otherwise. This new controversy led to excessive quarrels between Robins and Jurin [Jesseph 1993, p. 268]. Nothing showed the importance of Berkeley's critique better than this verbal battle which was fought on several hundred pages in two English journals.

John Colson (1680–1760), Fellow of the Royal Society since 1713 and since 1739 one of Newton's successors on the Lucasian Chair in Cambridge, also tried to give a justification of the calculus of fluxions. He translated Newton's *Methodus Fluxionum et Serierum Infinitarum* as *Method of Fluxions and Infinite Series* into English in 1736 and added an introduction and comments. Unfortunately, Colson was a mediocre mathematician and incapable of answering Berkeley's critique. In his introduction he even confirmed Berkeley's critique more than to reject it [Jesseph 1993, p. 269]. It is characteristic of the state of the British scientific system of the 18th century that Colson came to be considered as Lucasian professor of mathematics at all.

Fig. 10.4.3. left: Thomas Bayes. It is not quite clear whether this actually is a portrait of Thomas Bayes. The picture is taken from *History of Life Insurance* by Terence O'Donnell of the year 1936 where it appeared without information on the source, and right: John Colson [unknown painter, prob. 18th c.]

The cleric Thomas Bayes (1702–1761) is well known in the theory of probability through the Theorem of Bayes allowing the computation of conditional probabilities. It may be less known that Bayes wrote an anonymous reply to Berkeley's *Analyst*.

In 1736 he published his *Introduction to the Doctrin of Fluxions* in which he tried to base the calculus of fluxions on an axiomatic fundament. Jesseph has examined the *Introduction* in some detail and remarks [Jesseph 1993, p. 277]:

> *'We can grant that Bayes's Introduction is a serious work by an important mathematician, but it is hardly a conclusive answer to Berkeley.'*

James Smith also entered the debate with his book *New Treatise of Fluxions* [Jesseph 1993, p. 277ff.] of 1737 which Jesseph calls 'obscure'. Smith could not defuse Berkeley's critique and his book had no influence on the progress of the debate.

Truelly influential was the Scotsman Colin Maclaurin (1698–1746) [Guicciardini 1989, p. 47ff.], [Jesseph 1993, p. 279ff.] with his book *Treatise of Fluxions* which appeared in 1742 in two volumes.

Maclaurin became professor in Aberdeen in 1717 when he was 19 years old. In 1719 he travelled to London where he met Newton and became a member of the Royal Society. On the recommendation of Newton he became professor in Edinburgh in 1726. Maclaurin was without doubt an excellent mathematician in his time, but also a gifted geodesist and geophysicist. While in the past

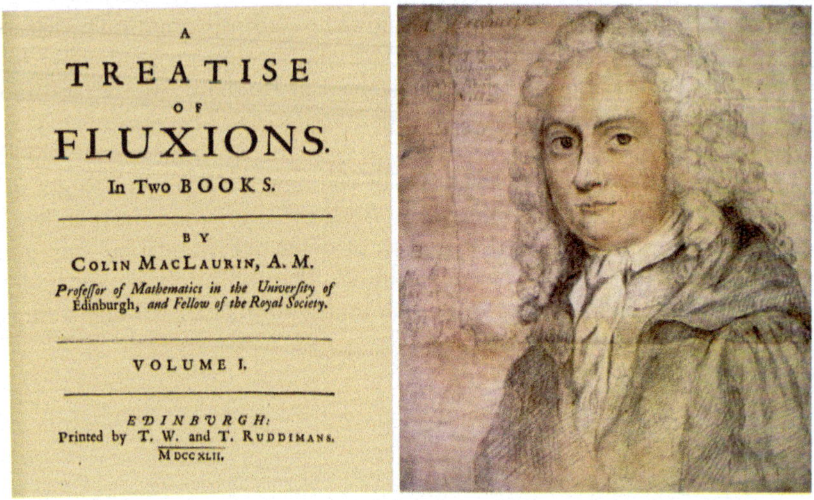

Fig. 10.4.4. Title page of the 1st volume of *A Treatise of Fluxions* (Bayerische Staatsbibliothek München (Munich)), 1742 and Colin MacLaurin [Drawing by the 11th Comte de Buchan, 18th c., after a painting by James Ferguson]

Maclaurin's *Treatise* was thought to be of great influence on the development of the mathematics of infinitesimals in England and Scotland, but not on the Continent, there is now [Grabiner 1997] no doubt that the books were well received also on the Continent and comprised a link between Newton's calculus of fluxions and the analysis on the Continent.

Instead of a simple reply to Berkeley's objections Maclaurin's voluminous *Treatise* not only contained demanding methods but also applied infinitesimal analysis to new problems. Therewith he outshone everything what British mathematics had produced so far. That the work was meant as a reaction to Berkeley's *Analyst* was made clear by Maclaurin already in the preface.

The first volume contains an extensive introduction into the classical method of exhaustion [Sonar 2016, p. 33f.] since Maclaurin wanted to reduce Newton's fluxional calculus to the geometry of the ancient Greeks [Guicciardini 1989, p. 47]. In doing so he argued verbosely against the geometers with their methods of infinitesimals who had entrapped themselves in the 'labyrinths of infinity'. Even a side blow against Descartes's vortex theory was not omitted. However, Guicciardini has pointed out that even Maclaurin had made use of infinitesimal methods in earlier writings, e.g. in his book *Geometria Organica: sive descriptio linearum curvarum universalis* (Organic geometry: universal description of linear curves) of 1720.

In the *Treatise* Maclaurin followed a kinematic definition of fluxions as velocities of a change. A system of axioms allowed him to formulate inequalities between the kinematic quantities [Guicciardini 1989, p. 50]. In the second volume the 'geometry of fluxions' is abandoned in favour of the 'calculus of fluxions', since by the first volume it was clear how the results could be expressed in the language of kinematic quantities. Infinitesimal quantities were banned by this approach; however, the proofs became pronouncedly laborious. The *Treatise* was very positively received in England. Many factors were responsible [Guicciardini 1989, p. 51] for its success. The approach via geometrical methods of kinematics was in the spirit of the time which valued the mathematics of antiquity. One thought of the new analysis as being a generalisation of the methods of Archimedes. Additionally it had been shown that the new calculus did not involve 'ghosts of departed quantities' but quantities which had a clear kinematical meaning. The 'infinitely small' had been banned from the calculus but the British mathematicians still had difficulties to catch up with the Continent.

11 The Aftermath of the War

T. Sonar, *The History of the Priority Dispute between Newton and Leibniz*, https://doi.org/10.1007/978-3-319-72563-5_11

Through all of the 18th century Newton was seen as the spotless champion of the priority dispute in Great Britain, while Leibniz was the devious plagiarist [Rice 2012, p. 89f.]. One of the first authors adjusting the overall picture was the mathematician Augustus De Morgan (1806–1871).

11.1 De Morgan's Work Concerning the Priority Dispute

Augustus De Morgan was born son of a soldier who was stationed in India, but shortly after Augustus's birth the family returned to England. One or two months after birth the baby went blind on one eye. When Augustus was 10-year-old his father died and his mother moved with him to different places in the south-west of England where Augustus attended different schools of which none had a reputation.

His mathematical talent was recognised when he was 14 years old, and hence he became a student at Trinity College, Cambridge at the age of 16. According to the wish of his mother he was supposed to become a cleric. However, he became a nonconformist very early, i.e. he did not accept the doctrines and dogmas of the Church of England. In Cambridge he met the mathematicians William Whewell (1794–1866) and George Peacock (1791–1858). A lifelong friendship developed and De Morgan took on their interest in algebra and logic. His own interests had a wider scope. He loved sciences and acquired great knowledge in numerous other areas than mathematics; he also played the flute in various musical societies which led to his being only Fourth Wrangler in the final exams. He could thus get a Master's degree in Cambridge only after having passed a theological test. This he strictly refused. He thus went to London to work at a court of law. Since the strict religious rules in Oxford

Fig. 11.1.1. Augustus De Morgan [Photo: Sophia Elizabeth De Morgan, 1882]

and Cambridge did not permit catholics, Jews, or nonconformists, some free-thinking men planned the founding of a new university in London, and De Morgan could occupy a professorship of mathematics after some turbulences. He stayed there as a professor for 30 years. For De Morgan the acquisition of knowledge for the sake of knowledge and truth always stood in the foreground. Hence it may not seem surprising that he took some efforts to clarify the circumstances concerning the priority dispute.

11.1.1 In Search of the 'True' Newton

De Morgan wrote on the mood in Great Britain [Rice 2012, p. 96]:

> *'it was in Britain the temper of the age [...] to take for granted that Newton was human perfection.'*

In 1831 a biography *Life of Sir Isaac Newton* of Newton was published. It was written by the Scottish physicist David Brewster (1781–1868) who praised Newton's genius beyond all means, while Leibniz was called a plagiarist in an abrasive tone.

We read [Brewster 1831, p. 199]:

> *'That he* [Leibniz] *was the aggressor is universally allowed. That he first dared to breathe the charge of plagiarism against Newton and that he often referred to it, has been sufficiently apparent; and when arguments failed him he had recourse to threats – declaring that he would publish another Commercium Epistolicum, though he had no appropriate letters to produce.'*

But already four years after Brewster's biography the first publication appeared which scratched the picture of Newton considerably. It was the biography *An account of the Rev. John Flamsteed* of Astronomer Royal Flamsteed by the astronomer Francis Baily (1774–1844), published in London in 1835.

Flamsteed had sent important observational data to Newton with which Newton was able to verify the theories in his *Principia*. Baily had discovered unpublished letters and manuscripts documenting Newton's shabby behaviour towards Flamsteed. Pertinent citations may be found under the title *John Flamsteed, MS material on Newton's character and actions* in [Iliffe 2006, Vol. I, p. 15-22]. Baily's book unsettled the British scientists firmly. Not only had Newton's impeccable behaviour been questioned; there also existed many documents which demonstrated his misconduct.

Besides Brewster the astronomer Stephen Peter Rigaud (1774–1839) in Oxford and William Whewell in Cambridge were particularly alarmed by Baily's biography of Flamsteed. The image of Newton as blameless hero of science

Fig. 11.1.2. David Brewster's book *The Life of Sir Isaac Newton* of 1840 contains a portrait of Sir Isaac Newton. Inner title of *The life of Sir Isaac Newton* by David Brewster; published 1831, here an edition of 1840 (Collection: California Digital Library, americana)

did long since belong to the national self-conception of the ruling class of the British people as it also belonged to the idea of a deeply religious, moral Victorian society. Hence Baily did not only shake up the image of Newton, but also the conventions of society of which the British people were very proud. To defend Newton's reputation Rigaud published the work *Historical essay on the first publication of Sir Isaac Newton's Principia* in 1838 which gave some new information on Edmond Halley. This work did not stand the test of time, but Rigaud's second work to rehabilitate Newton, the *Correspondence of scientific men of the seventeenth century* of 1841 [Rigaud 1965] published by his son Stephen Jordan Rigaud (1816–1859), did. The two volumes contain letters of the years 1606 to 1742 by (amongst others) Newton, Flamsteed, Wallis, Collins, and Halley. Whewell replied to Baily's biography by means of the three volumes of his *History of the inductive sciences* of 1837. In these volumes the moral and intellectual capacities of scientists are discussed and

Fig. 11.1.3. David Brewster in the company of famous scholars. Left to right: Michael Faraday, Thomas H. Huxley, Charles Wheatstone, David Brewster, John Tyndall. David Brewster [unknown photographer, prob. about 1850]

the relation between moral and intellect is illuminated. Following Whewell, Newton was a much grander genius than Flamsteed, hence he deserved much more sympathy than the latter.

De Morgan's first contribution concerning the priority dispute was a biography of Newton of 40 pages, published in 1846. After Baily's biography of Flamsteed the small book by De Morgan was the first in which Newton's weaknesses of character were revealed. He wrote [Rice 2012, p. 101]:

> '[t]*he great fault, or rather misfortune, of Newton's character was one of temperament* [... which] *showed itself in fear of opposition: when he became king of the world of science it made him desire to be an absolute monarch; and never did monarch find more obsequious subjects. His treatment of Leibnitz, of Flamsteed, and (we believe) of Whiston*[1] *is, in each case, a stain upon his memory.'*

The use of the word 'temperament' may seem to indicate that De Morgan wanted to excuse Newton's behaviour. However, we can find a footnote in the *Memoir of Augustus De Morgan* written by De Morgan's wife Sophia Elizabeth (1809–1892) in 1882, saying [De Morgan 1882, p. 257, footnote 2]:

[1] William Whiston (1667–1752) was Newton's successor on the Lucasian Chair.

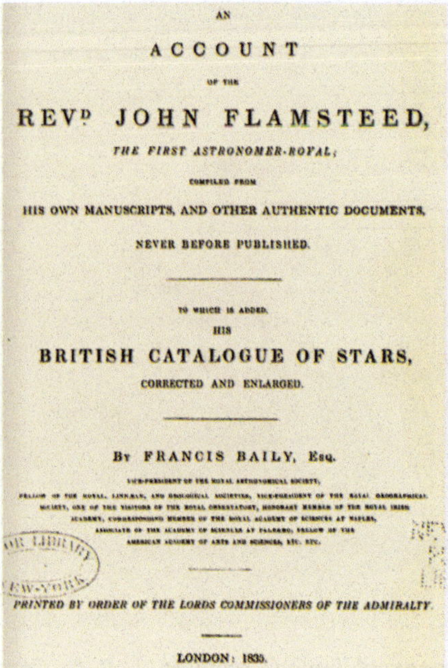
AN
ACCOUNT
OF THE
REVD JOHN FLAMSTEED,
THE FIRST ASTRONOMER-ROYAL;
COMPILED FROM
HIS OWN MANUSCRIPTS, AND OTHER AUTHENTIC DOCUMENTS,
NEVER BEFORE PUBLISHED.

TO WHICH IS ADDED,
HIS
BRITISH CATALOGUE OF STARS,
CORRECTED AND ENLARGED.

BY FRANCIS BAILY, Esq.

PRINTED BY ORDER OF THE LORDS COMMISSIONERS OF THE ADMIRALTY.

LONDON: 1835.

Fig. 11.1.4. Title page of Baily's *Account of the Rev. John Flamsteed* of 1835 (Bayerische Staatsbibliothek München (Munich), Sign: 874072 4 Biogr. 101m, scan S. 7) and a portrait of its author Francis Baily [Photo: before 1844]

> '*My husband always used this word* [i.e. temperament] *for what I should call original character or inborn disposition.*'

Through the work on this biography and through Newton's behaviour towards Flamsteed the interest of De Morgan in the relation between Newton and Leibniz was aroused [Rice 2012, p. 102]. All sources of the time were of the conviction that the commission which had drawn up the *Commercium epistolicum* consisted of six Britons only. In a letter published by Joseph Raphson in his book *History of fluxions* of 1715 Newton wrote that the commission had consisted of men of different nations. De Morgan found in a somehow obscure French source that Abraham de Moivre (1667–1754), a Frenchman living in London, was one of two foreign members of the committee. Hence Newton was right, but De Morgan also found in the source the remark that de Moivre had to give up his always treasured neutrality in order to work in the commission. So De Morgan came to the conclusion, [Rice 2012, p. 102]:

> '*that the Committee in question was thought at the time not to be a judicial body, but one of avowed partizans.*'

De Morgan published his findings as *On a point connected with the dispute between Keill and Leibnitz about the invention of fluxions* in the Philosophical

Fig. 11.1.5. Detail of the title page of the first of the three volumes of *History of the Inductive Sciences* of 1837, right: the author William Whewell

Transactions of the Royal Society in 1846. Although he had expected some resistance his paper was published without delay so that he felt encouraged to write a second paper with the title *A comparison of the first and second editions of the Commercium Epistolicum* in which he clarified a point in favour of Leibniz [Rice 2012, p. 103]. De Morgan found more than 20 changes and additions in the second edition which were not marked, but he did not suspect Newton behind it [Whiteside 1967–81, Vol. VIII, p. 486, footnote 57]. In any case De Morgan's paper was not accepted for publication which greatly distressed him [Rice 2012, p. 104]. Although De Morgan was on good terms with numerous members of the Royal Society he refused all attempts to make him a Fellow.

De Morgan returned to the priority dispute only in 1852, probably because of the correspondence between Newton and Roger Cotes, published by Edlestone in 1850, and the edition of the mathematical papers of Leibniz, edited by Carl Immanuel Gerhardt (1816–1899) and published in seven volumes from 1849 on [Leibniz 2004]. In 1852 De Morgan's paper *A short account of some recent discoveries in England and Germany relative to the controversy on the invention of fluxions* was published in the 'Companion to the almanac for 1852'. Gerhardt's edition of Leibniz's mathematical writings had clearly revealed that Leibniz had found his calculus on his own and independently of Newton. One still believed that the accusation of plagiarism against Leibniz had come only from Newton's henchmen, but not from Newton himself. In a further work of 1852 De Morgan doubted that and suspected that the anonymous author of the review of the *Commercium epistolicum*, *An Account of the Book*

Fig. 11.1.6. Abraham de Moivre ([unknown artist 1736] University of York) and Augustus De Morgan [unknown Photographer, about 1860]

entituled Commercium Epistolicum Collinii & aliorum, ..., was Isaac Newton. De Morgan made him responsible even for the Latin preface, but it took until 1855 until a proof came up. This proof came from a completely unexpected side.

11.1.2 The Conflict with David Brewster

When David Brewster's major work *Memoirs of the life, writings, and discoveries of Sir Isaac Newton* was published in two volumes in 1855, the author had spent some twenty years of work on it [Rice 2012, p. 105]. After the critique of Baily with respect to Newton's treatment of Flamsteed, Brewster thought that a defence was necessary. Brewster had consulted De Morgan on some points which apparently changed some of Brewster's opinions [Brewster 1855, Vol. I, p. xiif.]:

> *'To Professor De Morgan, to whom the public owes a brief but interesting biographical sketch of Newton, and who has carefully investigated various points in the Fluxionary controversy, I have been indebted for much information, and for his kind revision of the sketch I had given of the early history of the Infinitesimal Calculus. On a few questions in the life of Newton, and the history of his discoveries, my opinion differs somewhat from his; but I have been able to confirm, from the documents in my possession, many of his views on important points which he was the first to investigate and to publish.'*

Through the 'documents in my possession' Brewster now became convinced that the author of the introduction to the second edition of the *Commercium epistolicum* as of the review *An Account of the Book entituled Commercium Epistolicum Collinii & aliorum, ...* was Newton himself [Brewster 1855, Vol. II, p. 75]:

> *'In the year 1725, a new edition of the Commercium Epistolicum was published, with notes, a general review of it, and a preface of some length. A question has arisen respecting the authorship of the review and the preface, some ascribing them to Keill, and others to Newton. ..., Professor De Morgan had made it highly probable that both the review and the preface were written by Newton. Of the correctness of this opinion I have found ample evidence in the manuscripts at Hurtsbourne Park[2]; and it is due to historical truth to state, that Newton supplied all the materials for the Commercium Epistolicum, and that, though Keill was its editor, and the committee of the Royal Society the authors of the Report, Newton was virtually responsible for its contents.'*

Notwithstanding this passage Brewster's *Memoirs* remained a document of hero worship for Newton. This is particularly apparent in Brewster's treatment of the priority dispute at the beginning of the second volume. Although he took note of some of De Morgan's discoveries he declined all of them which were in favour of Leibniz. Rice has even suspected [Rice 2012, p. 106f.] that Brewster never read the works of De Morgan, since his judgement of Leibniz reads word for word exactly as in his biography of Newton of 1831.

Shortly before Brewster's new biography was published 'The North British Review' had asked De Morgan to review it. This resulted in a review of 30 pages which can be found on the internet on the pages of the Newton project. Concerning the assessment of Leibniz by Brewster, De Morgan wrote [De Morgan 1855, p. 321f.], [Iliffe 2006, Vol. II, p. 227]:

> *'A passing word on Leibnitz. We shall not stop to investigate the various new forms in which Sir D. Brewster tries to make him out tricking and paltry. We have gone through all the stages which a reader of English works can go through. We were taught, even in boyhood, that the Royal Society had made it clear that Leibnitz stole his method from Newton. By our own unassisted research into original documents we have arrived at the conclusion that he was honest, candid, unsuspecting, and benevolent. His life was passed in law, diplomacy, and public*

[2]After Newton's death his manuscripts came into the possession of John Conduitt who had married Newton's niece Catherine Barton. In 1740 the daughter of the Conduitts, Kitty, married the Viscount Lymington of the Portsmouth family. With her came the manuscripts to Hurstbourne Park, residence of the Plymouth family [Dry 2014, p. 28f.].

business; his leisure was occupied mostly by psychology[3]*, and in a less degree by mathematics. Into this last science he made some incursions, produced one of the greatest of its inventions, almost simultaneously with one of its greatest names, and made himself what Sir D. Brewster calls the 'great rival' of Newton, in Newton's most remarkable mathematical achievement.'*

We know today that De Morgan's assessment of Leibniz's character is as wrong as Brewster's, since also Leibniz's behaviour was not always free of petty jealousy and the pursuit of fame and honour, but certainly De Morgan made conscious use of polemic means. It is due to Augustus De Morgan that the perception of Newton in 19th century England could be adjusted. In any case, in 1925 David Eugene Smith could state [Smith 1958, Vol. II, p. 698]:

'The dispute between the friends of Newton and those of Leibniz as to the priority of discovery was bitter and rather profitless. It was the subject of many articles and of a report by a special committee of the Royal Society. English readers of the 18th century were so filled with the arguments respecting the controversy as set forth in the Commercium Epistolicum (1712) and Raphson's History of Fluxions (1715), that they gave Leibniz little credit for his work. It was not until De Morgan (1846) reviewed the case that they began generally to recognise that they had not shown their usual spirit of fairness. On the other hand, Leibniz was so stung by the accusations of his English critics that he too showed a spirit that cannot always be commended.'

11.2 England's Long Way to Analysis

We have already discussed in Chapter 10 that the infinitely small quantities were eventually driven out of the new analysis in Great Britain in particular by Colin Maclaurin's *Treatise of Fluxions*. In [Guicciardini 1989] Niccoló Guicciardini has thoroughly examined how the understanding and the methods of Newton's calculus of fluxions changed in 18th century Great Britain in the light of different textbooks. His research has shown that British mathematicians – James Stirling, Colin Maclaurin, Brook Taylor, and Roger Cotes have to be mentioned – could not keep up with the mathematicians on the Continent until about 1740. Since 1740 we can detect a decline leading to a crisis resulting finally in a reform. In the first half of the 18th century British mathematicians stood in contact and competition with the mathematicians on the Continent, but they became isolated in the second half and finally could no longer understand mathematical works from the Continent. Concerning analysis the 18th century is characterised by personalities like Leonhard Euler

[3]This is an obvious typing error. It should be 'philosophy' here.

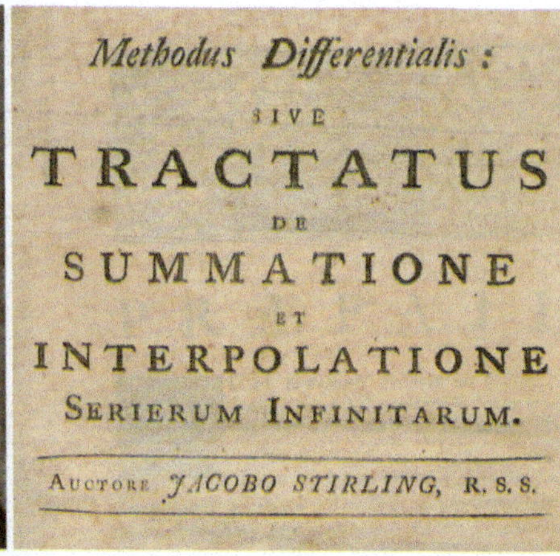

Methodus Differentialis:

SIVE

TRACTATUS

DE

SUMMATIONE

ET

INTERPOLATIONE

SERIERUM INFINITARUM.

AUCTORE *JACOBO STIRLING*, R. S. S.

Fig. 11.2.1. Brook Taylor and a detail of the Italian translation of James Stirling's *Methodus differentialis sive tractatus de summatione et interpolatione...* of 1764 (Bayerische Staatsbibliothek Munich, Sign: 1099397 4 Math.p. 352m)

(1707–1783), Joseph-Louis Lagrange (1736–1813), and Pierre-Simon Laplace (1749–1827). The infinitesimal calculus of functions of several variables made great progress – a development which remained unintelligible to British mathematicians. Euler was called 'analysis incarnate' by his contemporaries, but ontologically he fell short as compared to Leibniz, since he saw the mathematics of infinitely small quantities as an arithmetic with 'nils'. An impetus for a reform of analysis in Great Britain did not come by the works of Euler, but by the works of Laplace.

In 1799 Pierre-Simon Laplace's first volume of the *Traité de Mécanique Céleste* was published which made quite an impression in Great Britain. Here was a Frenchman who had impressively solved problems which were not treated in Newton's *Principia* by means of the new analysis. Reading and understanding Laplace's book became mandatory necessity. Centres of reform of the British analysis were in Scotland, in the military schools of Woolwich and Sandhurst, in Dublin, and in Cambridge [Guicciardini 1989, p. 141f.]. While a group in Dublin was devoted to the applications of analysis the 'Analytical Society' broke up the puristically minded algebraic scene and turned towards the analysis developed by Lagrange. This turn, however, isolated the British again. The 'Analytical Society' was an establishment of some 'young savages', namely Charles Babbage (1791–1871), George Peacock (1791–1858), and John Herschel (1792–1871), who had met during their studies at Cambridge University at the beginning of the 19th century.

The young men wanted to enforce 'pure d-ism against the Dot-age of the University' [Guicciardini 1989, p. 135], hence to replace Newton's dot notation by Leibniz's *d*. The starting point was the experiences Babbage made during a journey to the Continent. In particular he was recommended a French text-

Fig. 11.2.2. Pierre-Simon Laplace ([Painting by Sophie Feytaud] courtesy of the Academic des Sciences Paris, 1842) and Leonhard Euler ([Painting by Jakob Emanuel Handmann 1753] Kunstmuseum Basel)

book *Traité du calcul différentiel et du calcul intégral* by Sylvestre François Lacroix (1765–1843) [Domingues 2008] which Babbage bought for an enormous sum of money in England. The group hence knew Leibniz's calculus and therefore knew more than their tutors in Cambridge which added much to their dissatisfaction. Nevertheless Lagrange's calculus 'won' also in the 'Analytical Society'.

We can only present a short outline here and follow the essay [Grattan-Guinness 2011]. In the second decade of the 19th century three forms of the new analysis could be found in Britain: Newton's calculus of fluxions was

Fig. 11.2.3. John Herschel ([Photo: Julia Margaret Cameron 1867] The Metropolitan Museum of Art), George Peacock [Photo: unknown, 19th c.] and Charles Babbage [Engraving about 1860, The Illustrated London News, 1871]

Fig. 11.2.4. Augustin-Louis Cauchy [Lithography: Gregoire et Deneux, 19th c.] and Joseph-Louis Lagrange

dominant but not further developed for some time, the infinitesimal analysis of Leibniz and Euler, and finally the differential calculus of Lagrange. Between 1799 and 1805 the first four volumes of Laplace's *Mécanique Céleste* had been published and they had a stimulating effect concerning a reform of British analysis [Grattan-Guinness 2011, p. 303f.]. Celestial mechanics eventually occupied a high status in British scientific circles. In case of mathematical applications the differential and integral calculus of Leibniz and Euler was quite popular; the calculus of variations had developed into a more algebraic form to which Lagrange and his δ-operator contributed much. In circles of pure mathematicians reservations had developed against the use of the differential quotient as used by Euler, since it was thought of as a ratio of two infinitely small quantities. Robert Woodhouse in Cambridge and the 'Analytical Society' therefore fancied the version of the calculus presented by Lagrange and which appeared as an essentially algebraic theory [Sonar 2016, p. 472f.]. Lagrange based his calculus on infinite (Taylor) series where he neither proved convergence nor could show their existence for all functions. In order to avoid any reference to infinitely small quantities in formal manipulations Lagrange introduced the notation

$$f' := \frac{dy}{dx}$$

for functions $f = y(x)$.

In fact Lagrange's calculus did not eliminate the problems of the foundations of analysis but displaced them to other areas. Only with the rigorous definition of the limit by Augustin-Louis Cauchy (1789–1857) in his book *Cours d'Analyse* of 1821 came a new, modern understanding of the foundations. Cauchy also gave an example of an arbitrarily smooth function which could not be expanded in a Taylor series, hence Lagrange's algebraic analysis had no chance to survive.

Fig. 11.2.5. Arthur Cayley and George Boole

Out of Lagrange's algebraic analysis two algebraic disciplines developed in Great Britain, functional equations and the operational calculus. If one defines the 'operator'

$$D := \frac{d}{dx},$$

then higher derivatives can be derived by powers of this operator. Since integration is the inversion of differentiation the operation $1/D$ should realise integration. If one writes a differential equation

$$\frac{d^2y}{dx^2} - 2\frac{dy}{dx} - 3y = 0$$

in operator form

$$D^2y - 2Dy - 3y = 0,$$

then factorizing yields

$$(D^2 - 2D - 3)y = (D + 1)(D - 3)y = 0.$$

This factorized equation can be solved if one knows that the function $y_1 = c_1e^{-x}$ solves the equation $(D + 1)y = 0$. As solution of the factorised differential equation one then gets $y = c_1e^{-x} + c_2e^{3x}$. A well known protagonist of such methods was George Boole (1815–1864), who became Professor of Mathematics at Queen's College in Cork, Ireland, in 1848 and who is famous through his works on logic. In his book *Treatise on Differential Equations*, published 1859, the theory of operators can be found in chapters XVI and XVII.

Other important figures keeping the calculus of operators alive were Robert Murphy (1806–1843), Duncan Farquharson Gregory (1813–1844), and not least Arthur Cayley (1821–1895) [Grattan-Guinness 2011, p. 306]. The operational calculus gained new momentum in the 20th century through the intuitive work on the foundations of the theory of electricity by Oliver Heaviside (1850–1925) ('Heaviside calculus') and was placed on firm mathematical ground by the works of Jan Mikusiński (1913–1987).

Together with the interest in operational calculus came a renewed interest in difference equations in the 19th century. Here also George Boole has to be mentioned whose book *A Treatise on the Calculus of Finite Differences* of 1860 is still in print as a mathematical classic [Boole 2003].

The popularity of the operational calculus may be revealed by the title of a contemporary textbook [Grattan-Guinness 2011, p. 307], Robert Daniel Carmichael's *A Treatise on the Calculus of Operations, Designed to Facilitate the Process of the Differential and Integral Calculus and the Calculus of Finite Differences* of 1855. This popularity is easy to explain since the British not only had a preference for all things algebraical, but also had simply not the mathematical skills to follow the modern developments of a Cauchy [Guicciardini 1989, p. 138]:

> *'The main reason for the success of these techniques is that they were easy to learn and offered immense possibilities of dull proliferation. Furthermore, the followers of Babbage, Herschel and Peacock (who, by the way, very soon realised the sterility of their infatuation with Lagrangian analytics), considered the calculus of operators as a 'new continental method', and they were clearly excited to participate in this Renaissance.'*

And further on [Guicciardini 1989, p. 138]:

> *'But the predominance of the algebraic approach to the calculus had its own drawback: it did not allow many British mathematicians influenced by the Analytical Society to appreciate the importance of Cauchy's rigorisation of the calculus, which was motivated by the desire to avoid the 'generalities of algebra'. The shift from the fluxional calculus of the Newtonian tradition, once again left the British isolated.'*

Grattan-Guinness has counted 33 textbooks on the new analysis, 10 on differential equations, and 10 on the calculus of variations published in the 19th century [Grattan-Guinness 2011, p. 309]. For the most part their authors were not outstanding mathematicians. Limits did not occur in modern form and the definitions were mostly wrong, such when William Whewell in his book *The Doctrine of Limits* of 1838 wrote that all which is valid up to the limit would be valid in the limit, too.

The pioneer on the field of limits was Augustus De Morgan. In his book *Elements of Algebra* of 1835 we find a reasonably rigorous definition of a limit [Grattan-Guinness 2011, p. 309]:

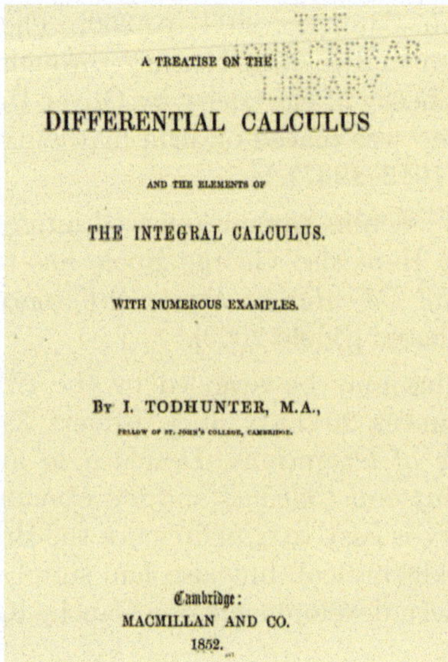

A TREATISE ON THE

DIFFERENTIAL CALCULUS

AND THE ELEMENTS OF

THE INTEGRAL CALCULUS.

WITH NUMEROUS EXAMPLES.

BY I. TODHUNTER, M.A.,

FELLOW OF ST. JOHN'S COLLEGE, CAMBRIDGE.

Cambridge:

MACMILLAN AND CO.

1852.

Fig. 11.2.6. Title page of *A Treatise on the Differential Calculus and the Elements of the Integral Calculus* by Isaac Todhunter of 1852 (University of Chicago, The John Crerar Library) and a portrait of Isaac Todhunter [A. Mcfarlane 1916]

> *'When, under circumstances, or by certain suppositions, we can make A as near as we please to P, ... then P is called the limit of A.'*

In his book *The Differential and Integral Calculus* of 1842 De Morgan already followed Cauchy. The most influential textbook in the Victorian epoch was *A Treatise on the Differential Calculus and the Elements of the Integral Calculus* of 1852, written by Isaac Todhunter (1820–1884), which saw eight editions until 1878 [Grattan-Guinness 2011, p. 310]. However, Todhunter did not present a theory of limits himself but followed De Morgan.

In the following 40 years no British author wrote a book that exceeded the contents of the books by De Morgan and Todhunter or even tried to modernise their books. Grattan-Guinness cites an obituary on the mathematician Andrew Russell Forsyth (1858–1942) of the year 1942 [Grattan-Guinness 2011, p. 311]:

> *'The college lecturers could not read German, and did not read French. One of the most eminent of them in the eighteen-nineties used to speak of the discoverer of the Gamma-function as 'Yewler'* [Euler] *and the founder of the theory of functions as 'Corky'* [Cauchy]*.'*

Only at the end of the 19th century could the British gain a foothold. Andrew Russell Forsyth made great progress with his book *Treatise on Differential Equations* [Forsyth 1996] of 1885 and with the six volumes of *Theory of Differential Equations* which were published between 1890 and 1906. In the preface Forsyth made clear that he was familiar with the mathematical developments on the Continent [Forsyth 1996, p. v]. In contrast to Boole Forsyth dedicated only a few pages to the operational calculus but emphasised works on partial differential equations of first and second order which originated in France between 1780 and 1810 and in Germany by Carl Gustav Jacob Jacobi (1804–1851) in 1830. Today Forsyth's *Treatise* is seen critically since he set the wrong priorities in his selection of material. Leonard Roth wrote on Forsyth's book *Treatise on Differential Equations* [Roth 1971, p. 230]:

> *'I would venture the opinion that this work has done more than anything else to retard the true development of the subject; for over two generations it has continued to put wrong ideas into people's heads concerning the nature and scope of the theory and, thanks to the author's forceful and authoritative style, in this it has been overwhelmingly successful.'*

And concerning the six volumes of *Theory of Differential Equations* [Roth 1971, p. 230f.]:

> *'His major work on the theory of differential equations, a colossal achievement in six volumes, is still today the only treatise in its class which is by a single hand; but a mere glance at the list of contents suffices to reveal that, on the whole, Forsyth looks backward to Lagrange rather than forward to Cauchy.'*

With the appearance of the books by Forsyth other books were published, as for example Horace Lamb's *Elementary Course of Infinitesimal Calculus* of 1897, and George Gibson's *Elementary Treatise on the Calculus with Illustrations from Geometry, Mechanics and Physics* in 1901.

As far as we are concerned a cycle is closing, the construction of which we have started in the first chapter: In 1910 the book *Calculus Made Easy* by Silvanus Phillips Thompson was published, in which Euler's version of the calculus was presented in an easily understandable form. The priority dispute between Leibniz and Newton, fought at the turn of the 17th to the 18th century, had a tragic aftermath for British mathematics which lasted for 200 years. Only in the 20th century could British analysis find the connection to the new mathematics which was initially developed independently by Newton and Leibniz and which nowadays is a major basis of our societies with our high technology industries.

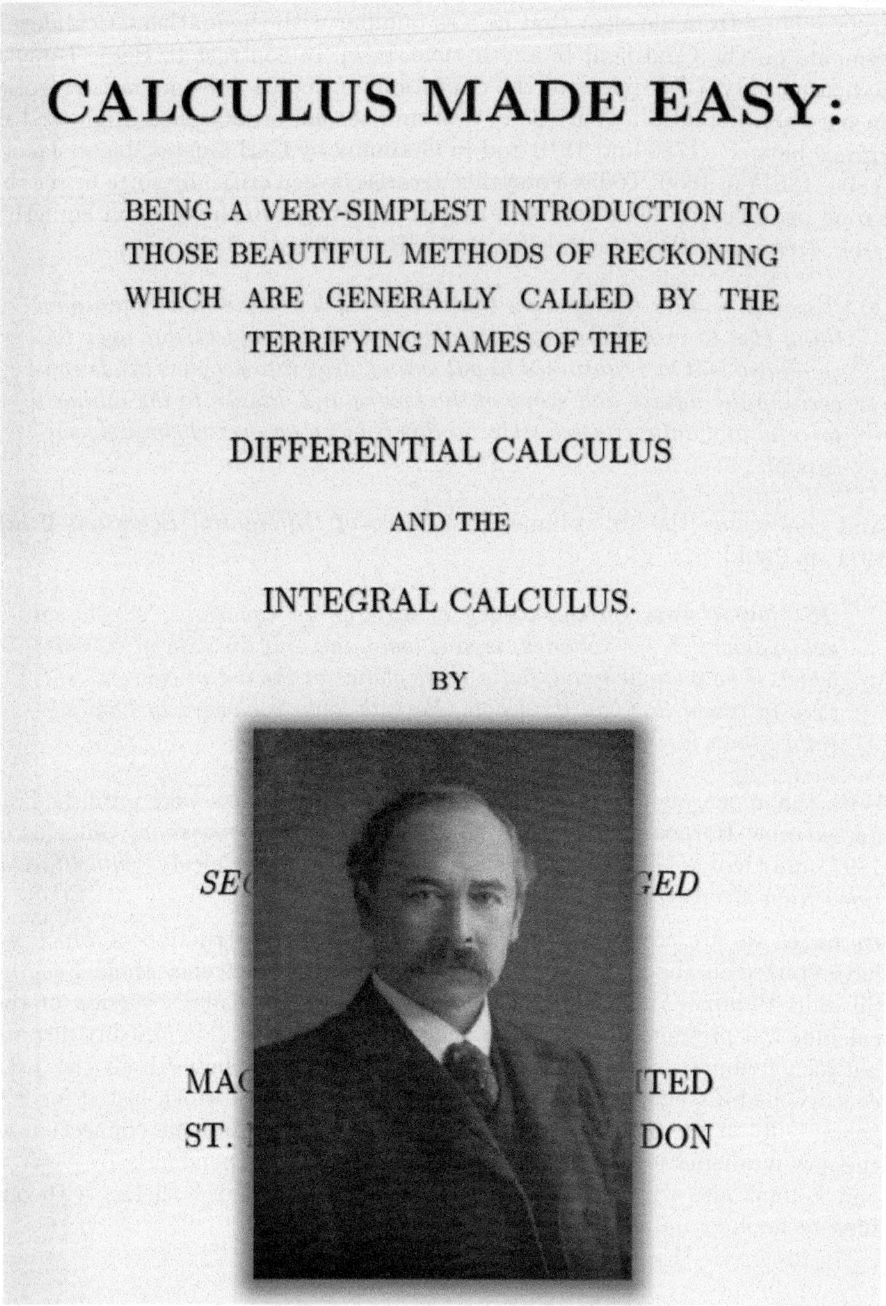

CALCULUS MADE EASY:

BEING A VERY-SIMPLEST INTRODUCTION TO THOSE BEAUTIFUL METHODS OF RECKONING WHICH ARE GENERALLY CALLED BY THE TERRIFYING NAMES OF THE

DIFFERENTIAL CALCULUS

AND THE

INTEGRAL CALCULUS.

BY

SE ED

MA TED

ST. DON

Fig. 11.2.7. Title page of *Calculus made easy* of the second edition of 1914 (First ed. 1910, Projekt Gutenberg) and subsequently inserted the portrait of its author Silvanus P. Thompson from: [Silvanus Thompson, His Life and Letters, Thompson and Thompson, 1920]

Epilogue

T. Sonar, *The History of the Priority Diſpute between Newton and Leibniz*, https://doi.org/10.1007/978-3-319-72563-5

The Messrs. Leibniz and Newton played a certain role early on in my life, starting about 1975, the year in which I got to learn differential calculus. Since I also developed a taste for mechanics shortly afterwards, the names Leibniz and Newton reappeared in this context and already in 1977 I bought the first edition of the truly wonderful book [Szabó 1996] by Istvan Szabó. Shortly before I had ordered Newton's *Principia* in the outdated (which I was not aware of at that time) German translation by Wolfers of 1872, which was reprinted by the 'Wissenschaftliche Buchgesellschaft' (a club for scientific books) from 1963 on. I remember as if it were yesterday that I was keen to learn from the *Principia* how Newton had applied his calculus of fluxions to explain the world, and how disappointed I was when I could not find any calculus of fluxions! At that time there was still no internet; one couldn't just click on Wikipedia or any other wonders of the World Wide Web, but had to visit libraries or to buy books instead. Since I was studying mathematics in Hanover at that time Leibniz lay closer to my heart than Newton and so I started to read everything on Leibniz which I could get my hands on. The biography [Guhrauer 1966] impressed me much and I quickly developed into an adherent of Leibniz and I still am one. I was convinced that Leibniz was the openly operating and innocent polymath who was unjustly attacked by a devious and vengeful Newton. In lectures on the history of mathematics later on I pulled Newton to pieces and when I read Frank Manuels psychoanalytically based biography [Manuel 1968] the reasons that Newton became such a creep seemed obvious to me.

I was certainly not alone in my assessment of Newton. In 2006 I bought Westfall's biography of Newton [Westfall 2006] in the year 1980, in Oxford, which I have read several times since. Westfall attached major importance to the scientific work of Newton and (justly) highlighted Newton's genius. But when it came to the personal side of the genius, his character, Westfall became more and more critical until he was barely off my own assessment. And I was not the only one who felt that way. Rob Iliffe wrote just at the beginning of his short introduction [Iliffe 2007, p. 7]:

> *'In his more orthodox Never at Rest: A Scientific Biography of Isaac Newton of 1980, Richard S. Westfall took Newton's work as the central aspect of his life. Drawing from the full range of Newton's manuscripts that were now available to scholars, his 'scientific biography' engaged with every aspect of Newton's intellectual interests, although his scientific career 'furnishes the central theme'. While he deals ably with Newton's intellectual accomplishments, it is apparent that Westfall's great admiration for this part of Newton's life does not extend to his personal conduct.*
>
> *Ultimately Westfall came to loathe the man whose works he had studied for over 2 decades. He was not the first to feel this way about the Great Man.'*

When I started to work on this book I had not yet changed my consistently negative attitude towards Newton – and my consistently positive towards Leibniz. Therefore I had to be very careful not to worship Leibniz too much and to condemn Newton beforehand. But my view changed when I read Meli's book [Meli 1993] on Leibniz's behaviour concerning the publication of the *Principia.* And that did not remain the only impetus for a change. The yearly book gift of the Gottfried-Wilhelm-Leibniz-Society in 2013 was a book with the ominous title *Komma und Kathedrale* (Comma and cathedral) [Li 2012]. In this book I found the essay *Diplomat in der Gelehrtenrepublik – Leibniz' politische Fähigkeiten im Dienste der Mathematik* (Diplomat in the republic of letters – Leibniz's political skills in service of mathematics) [Wahl 2012] by Charlotte Wahl, and there Leibniz was shown from a different side, namely as an agent on his own account.

Hence, after having finished this book project on the priority dispute, I have come closer to Leibniz's human side. Yes, Newton was indisputably vengeful and devious in his old age. He acted behind the scenes in a cowardly manner when he compiled the *Commercium epistolicum* which accused Leibniz of plagiarism. However, in the development towards the war of final extermination Leibniz also had often not played with open cards. The Newton of before 1700 was certainly an odd fellow and unable to react reasonably to the critique of others concerning his work. The quarrel with Hooke on the theory of colours was highly unwelcome for him, but did he not really know it better than Hooke? That Hooke finally claimed Newton's theory of gravitation must have hit Newton hard. Leibniz enjoyed the advantage of having gathered highly intelligent followers around him after his early publications had appeared – we have mentioned the roles of the brothers Bernoulli and of the Marquis de l'Hospital. Newton was too much of a recluse to gather such a class of followers; the people who fought for him did not belong to the first class of mathematicians. The English warriors fought for national motives and the worship of Newton; not out of a deeper understanding of the new mathematics. Additionally, he came much too late with his publications. The *Epistolae* did not reveal any details of Newton's mathematics but at least Newton wrote very openly if one remembers that he wrote to a person who was a stranger to him.

Much more problematic than Newton and Leibniz was the role of the different 'consultants' on either side who all were more or less warmongers. Fatio did not yet succeed to convince Newton of Leibniz's 'guilt', but finally Keill succeeded. On the Continent it was James Bernoulli who brought Leibniz to think of Newton as being a plagiarist. Leibniz really was an innocent victim – apart from his vanities.

Leibniz's calculus had the clear advantage over Newton's calculus of fluxions and fluents of using a simple and ingenious form of notation, but in principle both men had created the same mathematics which Newton acknowledged in his famous Scholium in the *Principia.* Leibniz's anonymous review of the two appendices of the *Opticks*, in particular *De quadratura*, can in fact be read as a provocation, although I am convinced that it was not meant to be one. It was probably due to the hurry in which Leibniz wrote it. But although Newton wrote in 1711 that he had never read this review, this seems to be unlikely. He probably took a cursory look and did not react at all. Only the overzealous John Keill, who actually declared war, had ill-advised Newton.

Thinking a bit in the direction of counter-factual history in the sense of the German historian Alexander Demandt [Demandt 2011], we can try to imagine what would have happened if Newton and Leibniz had worked in friendship on the new mathematics (and physics). If then men like James Bernoulli and Brook Taylor would have jointly contributed, how fast had we reached a point which only Euler was able to reach in the second half of the 18th century? In this sense the priority dispute has a genuine modern moral and reaches more than 200 years into our own time: Science should be free and advanced by scientists in open and fair collaboration. Secretiveness, prejudices, and jealousy inhibit the progress which we all need urgently in all areas of science.

Thomas Sonar

Eberhard Knobloch's Epilogue

Isaac Newton and the almost three and a half years younger Gottfried Wilhelm Leibniz never met, although Leibniz visited London twice – in 1673 and 1676. Over many decades there was only an indirect correspondence between the two men. Newton sent two letters to the Secretary of the Royal Society, Henry Oldenburg, which were meant for Leibniz: the *epistola prior* and the *epistola posterior*. Leibniz received them in August 1676 and July 1677, respectively. In reverse Oldenburg sent excerpts of letters which he received from Leibniz to Newton. Newton and Leibniz always addressed Oldenburg in their letters and mutually wrote in the third person about each other.

The immediate correspondence between the two men can easily be overlooked. There is just one letter by Leibniz for Newton and vice versa. On 17th March 1693 Leibniz wrote his English associate a friendly and obligatory letter. He had at any possible opportunity publicly acknowledged how much in his view mathematics and physics owed to Newton. Newton had expanded geometry in a wonderful way with his infinite series. Leibniz did not spare his praise concerning Newton's celestial mechanics, although he rejected his unexplained gravitation as a force not based on mechanics.

More than half a year later Newton replied on the 26th October 1693 and emphasized how much he regretted the delay of his answer since he valued the friendship with Leibniz in an extraordinary manner and counted him under the most important geometers of the century. He was concerned that their friendship could be damaged through his reticence because he valued his friends higher than his mathematical inventions. He finished with the affirmation that he wanted to testify with his letter that Leibniz was his immutable friend and that he valued his friendship most highly.

Reading these letters which testify the mutual respect, even friendship, of the two men, one would not suspect that between them one of the most exacerbated controvesy would break out only a few years later. How was this possible?

To give a balanced answer to this question Thomas Sonar takes the background of the history of science and the personal living conditions of both opponents thoroughly as well as completely in view. And in fact: Only the difficult exploration of the historical details allow the reader to understand how it finally could come to the unedifying opposition, characterised by wrong assertions, imputations, and suspicions on either side. A negative snapshot of Newton at the peak of the conflict as was given by Carl Djerassi in his play *Calculus* falls short compared to the historical events, although it was of course not admissible that the English were prosecutors and judges at the same time in the *Commercium epistolicum* of 1712.

Sonar reconstructs the personal experiences and activities of the two main actors of his story. Already the different, if not comparable, childhood expe-

T. Sonar, *The History of the Priority Dispute between Newton and Leibniz*, https://doi.org/10.1007/978-3-319-72563-5

riences made an impact on the two scholars in different ways. The father of the premature born child Isaac Newton had died before the baby was born. Leibniz's father died when the boy Gottfried Wilhelm was six years old. Both men had not, or only barely known their fathers. But the three years old Isaac was separated from his mother by the urging of his stepfather; a traumatic experience for the little boy. Leibniz lost his mother only when he was 17 years old.

Preceding events were interpreted in the light of new actual or supposed knowledge. The 26 years old Leibniz unintentionally revealed his poor knowledge of the mathematical literature at his first visit to London 1673 and he wrote to Oldenburg with far too much pride: 'My calculating machine will completely meet the task and now comes to a completion as it was begun in the time of my absence.' In reality the work on the machine never came to a final conclusion in Leibniz's lifetime. When decades later Leibniz was accused of plagiarism concerning Newton's discoveries these events had not been forgotten and seemed to confirm the bad opinion on Leibniz.

Wrong dates were used accidentally, but also intentionally. The *epistola prior* was sent off to Leibniz on 26th July (after the Julian style used in England); on 5th August in the new style. John Wallis – nationally inclined, if not nationalistic – published the letter in 1699 with Leibniz's knowledge in the third volume of his works where the date of dispatch was given wrongly due to a typing error as '26th June' (6th July following the new style). Newton accepted the wrong date in 1715 and on other occasions against his own knowledge. Following that wrong date Leibniz, who had replied only on 27th August, would have needed many weeks for his reply, since he obviously had to work into the mathematical details. In truth he received the letter only on 24th of August; hence he replied within three days. However, he claimed – wrongly – that he had received the letter the day before.

Similarly confusing is the situation concerning the text and date of dispatch of the *epistola posterior*, the sent copy of which did not even contain all of Newton's changes of the original version.

Which new knowledge could Leibniz have drawn from this letters? His invention of the *calculus* originated in the autumn of 1675. Those who did not know Leibniz's records – and the English side did not – could arrive here at wrong assumptions.

Sonar justly does not hide Leibniz's ineptitudes and provoking remarks. In 1697 Leibniz wrote concerning the solutions of the problem of the brachistochrone which he had received: 'And in fact it seems not unworthy to remark that all those have solved this problem of whom I have assumed that they could do it, namely only those who were able to penetrate the secrets of our differential calculus.' Specifically he mentioned Newton under all those whom he had expected to solve the problem. But how was Newton supposed to understand the sentence 'only those who were able to penetrate the secrets of our differential calculus'?

Leibniz could not have foreseen the injured vanity of Nicolas Fatio de Duillier who was not mentioned when Newton's followers stood openly against Leibniz and thereby provoked the hot phase of the dispute that Newton should be the first inventor of the *calculus*. Almost bluntly he accused Leibniz of plagiarism. He wanted to leave it to the readers of Newton's letters and manuscripts as to whether Leibniz as the second inventor would have borrowed something from Newton.

Hence the disaster ran its course, especially since John Keill blew the same horn. Arguments led to ripostes, so that Leibniz eventually turned the table and accused Newton against his better knowledge. It was all human; all too human.

Thomas Sonar has traced the emergence and the escalation of the dispute, which was additionally fuelled by Leibniz's rejections of Newton's theory of gravitation, in a grand and thrillingly written monograph. With sovereign competence he at the same time explains the mathematical context, so that also non-mathematicians will benefit from this book. Quod erat demonstrandum!

References

[Aiton 1972] Aiton, E. J.: Leibniz on Motion in a Resisting Medium. (Archive for History of Exact Sciences Vol. 9, 1972, S. 257-274).

[Aiton 1972a] Aiton, E. J.: The Vortex Theory of Planetary Motions. London, New York 1972.

[Aiton 1991] Aiton, E. J.: Gottfried Wilhelm Leibniz. Eine Biographie. Aus dem Englischen übertragen von Christiana Goldmann und Christa Krüger. Frankfurt am Main, Leipzig 1991.

[Andersen 1985] Andersen, K.: Cavalieri's Method of Indivisibles. (Archive for History of Exact Sciences, Vol. 31, 1985, No.4, S. 291–367).

[Andriesse 2005] Andriesse, C. D.: Huygens. The Man Behind the Principle. Cambridge, New York, Melbourne, etc. 2005.

[Antognazza 2009] Antognazza, M. R.: Leibniz. An Intellectual Biography. Cambridge, New York, etc. 2009.

[Archimedes 2002] Heath, T.L. (edt.): The Works of Archimedes. Unabridged republication of the edition published by Cambridge University Press in 1897 under the title *The works of Archimedes*, edited in modern notation with introductory chapters, and of the supplement published by Cambridge University Press in 1912 under the title *The method of Archimedes, recently discovered by Heiberg*, a supplement to *The works of Archimedes*. Mineola, New York, 2002.

[Archimedes 2004] Netz, R: The Works of Archimedes. Translated into English, together with Eutocius' commentaries, with commentary, and critical edition of the diagrams. Volume I. Cambridge, New York, etc., 2004.

[Aristoteles 1995] Aristoteles: The Complete Works of Aristotle. The revised Oxford translation. Edited by Jonathan Barnes, 2 vols., sixth printing with corrections, New Jersey 1995.

[Attali 2007] Attali, J.: Blaise Pascal. Biographie eines Genies. Aus dem Französischen von Hans Peter Schmidt. 2. Auflage. Stuttgart 2007.

[Aubrey 1982] Aubrey, J.: Brief Lives. A modern English version. Edited by Richard Barber. Woodbridge 1982.

[Bardi 2006] Bardi, J. S.: The Calculus Wars. Newton, Leibniz, and the Greatest Mathematical Clash of All Time. New York 2006.

[Barrow 1973] Barrow, I.: The Mathematical Works. Edited by W. Whewell. Reprog. Nachdr. der Ausg. Cambridge 1860. 2 Bände in einem Band, Hildesheim, New York 1973.

[Béguin 1998] Béguin, A.: Blaise Pascal. Mit Selbstzeugnissen und Bilddokumenten. 13. Auflage, Reinbek bei Hamburg 1998.

[Bell 1947] Bell, A. E.: Christian Huygens and the Development of Science in the Seventeenth Century. London 1947.

[Berkeley 1871] Berkeley, G.: The Works of George Berkeley D.D.; Formerly Bishop of Cloyne. Including Many of his Writings Hitherto Unpublished. With Prefaces, Annotations, His Life and Letters, and an Account of his Philosophy, by Alexander Campbell Fraser, M.A. Oxford 1871.

[Berkeley 1901] Berkeley, G.: Philosophisches Tagebuch. Philosophical Commentaries. Übers. u. hrsg. von Wolfgang Breidert. Hamburg 1979.

[Berkeley 1979] Berkeley, G.: The Works of George Berkeley D.D.; Formerly Bishop of Cloyne. Including his Posthumous Works. With Prefaces, Annotations, Appen-

T. Sonar, *The History of the Priority Dispute between Newton and Leibniz*, https://doi.org/10.1007/978-3-319-72563-5

dices, an An Account of his Life, by Alexander Campbell Fraser. Vol. I: Philosophical Works. Oxford 1901.

[Berkeley 1985] Berkeley, G.: Schriften über die Grundlagen der Mathematik und Physik. Eingeleitet und übersetzt von Wolfgang Breidert. 1. Auflage 1969. Nachdruck. Frankfurt am Main 1985.

[Berkeley 1992] Berkeley, G.: De Motu and The Analyst. A Modern Edition, with Introductions and Commentary. Edited and translated by Douglas M. Jesseph. The New Synthese Historical Library, Volume 41. Dordrecht 1992.

[Berkeley 1996] Berkeley, G.: Alciphron oder der kleine Philosoph. Übers. von Luise und Friedrich Raab. Mit einer Einleitung versehen und hrsg. von Wolfgang Breidert. Hamburg 1996.

[Blanton 2000] Blanton, J. D. (Übers.): Euler - Foundations of Differential Calculus. New York 2000. [Die Übersetzung beinhaltet den ersten Teil des Werkes *Institutiones calculi differentialis cum eius usu in analysi finitorum ac doctrina serierum*, Eneström Index E212, publiziert 1755.]

[Boas Hall 2002] Boas Hall, M.: Henry Oldenburg. Shaping the Royal Society. Oxford, New York, etc. 2002.

[Boole 2003] Boole, G.: A Treatise on the Calculus of Finite Differences. Reprint of the 2nd ed. London 1872. Mineola, N.Y. 2003.

[Bos 1980] Bos, H. J. M.: Newton, Leibniz and the Leibnizian tradition. in: [Grattan-Guinness 1980], S. 49-93.

[Bourbaki 1971] Bourbaki, N.: Elemente der Mathematikgeschichte. Göttingen 1971.

[Boyer 1959] Boyer, C. B.: The History of the Calculus and its Conceptual Development. Reprint of the first edition 1949. New York 1959.

[Bradley et al. 2015] Bradley, R.E.; Petrilli, S.J.; Sandifer, C.E.: L'Hôpital's Analyse des infiniments petits. An Annotated Translation with Source Material by Johann Bernoulli. Science Networks. Historical Studies 50. Cham, Heidelberg, New York, etc. 2015.

[Breger 1987] Breger, H.: Buchbesprechung. Sudhoffs Archiv, Band 71, 1987. Heft 1, S. 120f.

[Breger/Niewöhner 1999] Breger, H.; Niewöhner, F. (Hrsg.): Leibniz und Niedersachsen. Tagung anlässlich des 350. Geburtstages von G. W. Leibniz, Wolfenbüttel 1996. Stuttgart 1999.

[Breidert 1989] Breidert, W.: George Berkeley 1685–1753. Basel, Boston, Berlin 1989.

[Brewster 1831] Brewster, D.: The Life of Sir Isaac Newton. London 1831.

[Brewster 1855] Brewster, D.: Memoirs of the life, writings, and discoveries of Sir Isaac Newton. 2 Vols. Edinburgh 1855.

[Cajori 1919] Cajori, F.: A History of the Conceptions of Limits and Fluxions in Great Britain from Newton to Woodhouse. Chicago, London 1919.

[Chapman 2004] Chapman, A.: England's Leonardo. Robert Hooke and the Seventeenth-Century Scientific Revolution. London 2004.

[Child 1916] Child, J. M.: Geometrical Lectures of Isaac Barrow. Transl., with notes and proofs, and a discussion on the advance made therein on the work of his predecessors in the infinitesimal calculus by J.M.C., Chicago, London 1916.

[Child 2005] Child, J. M. (Transl.): The Early Mathematical Manuscripts of Leibniz. Transl. and with an introduction by J.M.C. Reprint of the first edition. Mineola, N. Y. 2005.

[Clark/Clark 2000] Clark, D. H., Clark; S. P. H.: Newton's Tyranny. The Suppressed Scientific Discoveries of Stephen Gray and John Flamsteed. New York 2000.

[Clarke 1956] The Leibniz-Clarke correspondence. Together with extracts from Newton's *Principia* and *Opticks*, edited with introduction and notes by H. G. Alexander. Manchester and New York 1956.

[Cohen 1971] Cohen, I. B.: Introduction to Newton's 'Principia'. Harvard 1971.

[Cohen/Smith 2002] Cohen, I. B.; Smith, G. E. (edts.): The Cambridge Companion to Newton. Cambridge, New York, etc. 2002.

[De Morgan 1855] De Morgan, A.: Review of Memoirs of the Life, Writings, and Discoveries of Sir Isaac Newton by Sir David Brewster. The North British Review, 1855, No. 23, S. 307-338. (http://www.newtonproject.sussex.ac.uk/view/texts/normalized/OTHE00094)

[De Morgan 1882] De Morgan, S. E.: Memoir of Augustus De Morgan. London 1882.

[Demandt 2011] Demandt, A.: Ungeschehende Geschichte. Ein Traktat über die Frage: Was wäre geschehen, wenn ...?. Göttingen 2011.

[Descartes 1969] Descartes, R.: Geometrie. Dt. hrsg. von Ludwig Schlesinger. Unveränd. reprogr. Nachdruck der 2., durchges. Aufl., Leipzig 1923. Darmstadt, 1969.

[Descartes 1983] Descartes, R.: Principles of Philosophy. Translated with Explanatory Notes, by Valentine Rodger Miller and Reese P. Miller. Dordrecht, Boston, England 1983.

[Descartes 2005] Descartes, R.: Die Prinzipien der Philosophie. Lateinisch-deutsch. Übers. u. hrsg. von Christian Wohlers. Hamburg 2005.

[Dobbs 1991] Dobbs, B. J. T.: The Janus Faces of Genius. The Role of Alchemy in Newton's Thought. Cambridge 1991.

[Domingues 2008] Domingues, J. C.: Lacroix and the Calculus. Basel, Boston, Berlin 2008.

[Domson 1981] Domson, Ch. A.: Nicolas Fatio de Duillier and the Prophets of London. New York 1981.

[Dry 2014] Dry, S.: The Newton Papers. The Strange & True Odyssey of Isaac Newton's Manuscripts. Oxford, New York, Auckland, etc. 2014.

[DSB 1971] Dictionary of Scientific Biographies, 16 Vols. New York 1971.

[Eberhard/Eckhart 1982] Eberhard, J. A.; Eckhart, J. G.: Leibniz-Biographien. Hildesheim, Zürich, New York 1982.

[Eckhart 1982] Eckhart, J. G.: Lebensbeschreibung des Freyherrn von Leibniz. in: [Eberhard/Eckhart 1982], S. 125-231.

[Edleston 1850] Edleston, J.: Correspondence of Sir Isaac Newton and Professor Cotes, including letters of other eminent men, now first published from the originals in the library of Trinity College, Cambridge; together with an appendix containing other unpublished letters and papers by Newton; with notes, synoptical view of the philosopher's life, and a variety of details illustrative of his history. London und Cambridge 1850.

[Edwards 1979] Edwards Jr., C. H.: The Historical Development of the Calculus. New York, Heidelberg, Berlin 1979.

[Engelsman 1984] Engelsman, S. V.: Families of Curves and the Origins of Partial Differentation. Amsterdam, New York 1984.

[Euchner] Euchner, W.: John Locke zur Einführung. Hamburg 1996.

[Euclid 2013b] Euclid: The Thirteen Books of the Elements. Vol. 2, Books III-IX. Translated with introduction and commentary by Sir Thomas L. Heath. Second Edition. Unabridged, first published in 1956. New York 2013.

[Feingold 1990] Feingold, M. (edt.): Before Newton. The life and times of Isaac Barrow. Cambridge, New York, etc. 1990.

[Feingold 1990a] Feingold, M.: Isaac Barrow – divine, scholar, mathematician. in: [Feingold 1990], S. 1-104.

[Finster/van den Heuvel 1990] Finster, R.; Heuvel, G. van den: Gottfried Wilhelm Leibniz. Mit Selbstzeugnissen und Bilddokumenten. Reinbek bei Hamburg 1990.

[Flasch 2004] Flasch, K.: Nikolaus von Kues in seiner Zeit. Ein Essay. Stuttgart 2004.

[Flasch 2005] Flasch, K.: Nicolaus Cusanus. 2. Auflage. München 2005.

[Flasch 2008] Flasch, K.: Nikolaus von Kues. Geschichte einer Entwicklung. 3. Auflage, Frankfurt/Main 2008.

[Fleckenstein 1956] Fleckenstein, J. O.: Der Prioritätsstreit zwischen Leibniz und Newton. Isaac Newton. 2nd edition Basel 1977.

[Flood/Rice/Wilson 2011] Flood, R.; Rice, A.; Wilson, R. (edts.): Mathematics in Victorian Britain. Oxford, New York, Auckland, etc. 2011.

[Folkerts/Knobloch/Reich 2001] Folkerts, M.; Knobloch, E.; Reich, K.: Maß, Zahl und Gewicht. Mathematik als Schlüssel zu Weltverständnis und Weltbeherrschung. Katalog der Herzog August Bibliothek Wolfenbüttel. 2. Auflage, Wiesbaden 2001.

[Fontenelle 1989] Fontenelle, B. L. B. de: Philosophische Neuigkeiten für Leute von Welt und für Gelehrte. Ausgewählte Schriften. Leipzig 1989.

[Forsyth 1996] Forsyth, A. R.: A Treatise on Differential Equations. Reprint of the 6th edition 1929, Mineola, N. Y. 1996.

[Galilei 1973] Galilei, G.: Unterredungen und mathematische Demonstrationen über zwei neue Wissenszweige, die Mechanik und die Fallgesetze betreffend. Herausgeber Arthur von Oettingen. Sonderausgabe. Nachdruck. Darmstadt 1973.

[Galilei 1982] Galilei, G.: Dialog über die beiden hauptsächlichsten Weltsysteme, das ptolemäische und das kopernikanische. Aus dem Ital. übers. u. erl. v. Emil Strauss. Mit einem Beitrag von Albert Einstein sowie ein Vorw. zur Neuausgabe u. weiteren Erl. von Stillman Drake. Hrsg. von Roman Sexl und Karl von Meyenn. Reprogr. Nachdruck der Ausgabe Leipzig 1891. Sonderausgabe. Nachdruck. Darmstadt 1982.

[Galilei 1989] Galilei, G.: Two New Sciences Including Centers of Gravity and Force of Percussion. Second edition. Translated, with a New Introduction and Notes, by Stillman Drake; and Stillman Drake: History of Free Fall, Aristotle to Galileo, With an Epilogue on π in the Sky. Toronto 1989.

[Galilei 2014] Galilei, G.: Two New Sciences Including Centers of Gravity and Force of Percussion. Second Edition. Translated, with a New Introduction and Notes, by Stillman Drake. Also Including History of Free Fall: Aristotle to Galileo, by Stillman Drake. Third Printing, Toronto 2014.

[Gerhardt 1846] Gerhardt, C. I.: Historia et origo calculi differentialis a G. G. Leibnitio conscripta. Zur 2. Säcularfeier des Leibnizschen Geburtstages aus den Handschriften der Königlichen Bibliothek zu Hannover hrsg. von C. I. G. Hannover 1846.

[Görlich 1987] Görlich, E.: Leibniz als Mensch und Kranker. Dissertation. Medizinische Hochschule. Hannover 1987.

[Goldenbaum/Jesseph 2008] Goldenbaum, U.; Jesseph, D.: Infinitesimal Differences. Controversies between Leibniz and his Contemporaries. Berlin, New York 2008.

[Gowing 1983] Gowing, R.: Ronald Cotes – natural philosopher. Cambridge, London, New York, etc. 1983.

[Grabiner 1997] Grabiner, J. V.: Was Newton's Calculus a Dead End? The Continental Influence of Maclaurin's Treatise of Fluxions. (The American Mathematical Monthly, vol. 104, 1997, S. 393-410).

[Graef 1973] Graef, M. (Hrsg.): 350 Jahre Rechenmaschinen. Vorträge eines Festkolloquiums, veranstaltet vom Zentrum für Datenverarbeitung der Universität Tübingen. München 1973.

[Graeven 1902] Graeven, H.: Leibnizens irdische Überreste. (Hannoversche Geschichtsblätter, Jahrgang 5, 1902, Heft 8, S. 568-571).

[Grattan-Guinness 1980] Grattan-Guinness, I. (edt.): From the Calculus to Set Theory 630-1910. An Introductory History. Princeton and Oxford 1980.

[Grattan-Guinness 2011] Grattan-Guinness, I.: Instruction in the calculus and differential equations in Victorian and Edwardian Britain. in: [Flood/Rice/Wilson 2011] S. 303-319.

[Guhrauer 1966] Guhrauer, G. E.: Gottfried Wilhelm Freiherr von Leibniz – Eine Biographie. 2 Bände. Nachdruck der Originalausgabe von 1846. Hildesheim 1966.

[Guicciardini 1989] Guicciardini, N.: The development of Newtonian calculus in Britain 1700-1800. Cambridge. New York, Port Chester, etc. 1989.

[Guicciardini 1999] Guicciardini, N.: Reading the Principia. Cambridge, New York, etc. 1999.

[Guicciardini 2002] Guicciardini, N.: Analysis and synthesis in Newton's mathematical work. in: [Cohen/Smith 2002] S. 308-328.

[Guicciardini 2009] Guicciardini, N.: Isaac Newton on mathematical certainty and method. Cambridge, Mass., London 2009.

[Haan/Niedhart 2002] Haan, H.; Niedhart, G.: Geschichte Englands vom 16. bis zum 18. Jahrhundert. 2., durchgesehene Auflage. München 2002.

[Hairer/Wanner 2000] Hairer, E.; Wanner, G.: Analysis by Its History. 3. corrected printing, New York, Berlin, Heidelberg, etc. 2000.

[Hall 1980] Hall, A. R.: Philosophers at war - The quarrel between Newton and Leibniz. Cambridge, London, New York, etc. 1980.

[Hall 1993] Hall, A. R.: Newton, his Friends and his Foes. Aldershot, Brookfield 1993.

[Hall 1993a] Hall, A. R.: Horology and Criticism: Robert Hooke. in: [Hall 1993] S. 261-281.

[Hall 1993b] Hall, A. R.: Two Unpublished Lectures of Robert Hooke. in: [Hall 1993] S. 219-230.

[Hall 2002] Hall, A.R.: Newton versus Leibniz. From geometry to metaphysics. in: [Cohen/Smith 2002] S. 431-454.

[Harrison 1978] Harrison, J.: The Library of Isaac Newton. Cambridge, New York, etc. 1978.

[Haupt et al. 2008] Haupt, H. G.; Hinrichs, E.; Martens, S.; Müller, H.; Schneidmüller, B.; Tacke, Ch.: Kleine Geschichte Frankreichs. Hrsg. von Ernst Hinrichs. Ditzingen 2008.

[Hawking 1988] Hawking, S. W.: A Brief History of Time. From the Big Bang to Black Holes. Toronto, New York, London, etc. 1988.

[Hellman 2006] Hellman, H.: Great Feuds in Mathematics. Ten of the Liveliest Disputes Ever. Hoboken, N. J. 2006.

[Hess 1986] Hess, H.-J.: Zur Vorgeschichte der "'Nova Methodus"' (1676-1684). (Studia Leibnitiana, Sonderheft 14: 300 Jahre "'Nova Methodus"' von G. W. Leibniz (1684-1984), Hrsg.: Albert Heinekamp, 1986, S. 64-102).

[Hess 2005] Hess, H.-J.: Leibniz auf dem Höhepunkt seines mathematischen Ruhms. (Studia Leibnitiana, Band 37, 2005, Heft 1, 2005, S. 48-67).

[Heyd 1995] Heyd, M.: "'Be Sober and Reasonable"'. The Critique of Enthusiasm in the Seventeenth and Early Eighteenth Centuries. Leiden, New York, Köln 1995.

[Hill 1991] Hill, Ch.: The World Turned Upside Down. Radical Ideas During the English Revolution. London 1991.

[Hofmann 1939] Hofmann, J. E.: On the Discovery of the Logarithmic Series and Its Development in England up to Cotes. (National Mathematics Magazine, Vol.14, 1939, No.1., S. 37–45).

[Hofmann 1949a] Hofmann, J. E.: Nicolaus Mercator (Kauffman) - sein Leben und Wirken, vorzugsweise als Mathematiker. (Akademie der Wissenschaften und der Literatur zu Mainz. Abhandlungen der Math.-Naturwiss. Klasse, 3. 1950, S. 43-103).

[Hofmann 1966] Hofmann, J. E.: Vom öffentlichen Bekanntwerden der Leibnizschen Infinitesimalmathematik. (Sitzungsberichte der Österreichischen Akademie der Wissenschaften, Math.-naturwiss. Klasse, Abt. 2, Band 175, 1966, S. 209-254).

[Hofmann 1973] Hofmann, J. E.: Leibniz und Wallis. (Studia Leibnitiana, 5. 1973, S. 245-281).

[Hofmann 1974] Hofmann, J. E.: Leibniz in Paris 1672-1676. His Growth to Mathematical Maturity. Cambridge, New York, etc. 1974.

[Hogrebe/Bromand 2004] Hogrebe, W.; Bromand, J. (Hrsg.): Grenzen und Grenzüberschreitungen. XIX. Deutscher Kongress für Philosophie. Vorträge und Kolloquien. Berlin 2004.

[Hooke 2007] Hooke, R.: Micrographia or some physiological descriptions of minute bodies made by magnifying glasses. With observations and inquiries thereupon. New York 2007.

[Hyman 1987] Hyman, A.: Charles Baggage 1791-1871. Philosoph, Mathematiker, Computerpionier. Stuttgart 1987.

[Iliffe 2006] Iliffe, R. (edt.): Early Biographies of Isaac Newton 1660-1885. 2 Vols., London 2006.

[Iliffe 2007] Iliffe, R.: Newton. A Very Short Introduction. Oxford, New York, etc. 2007.

[Inwood 2002] Inwood, S.: The Man Who Knew Too Much. The Strange and Inventive Life of Robert Hooke 1635-1703. London 2002.

[Jahnke 2003] Jahnke, H. N. (edt.): A History of Analysis. Providence, R. I. 2003.

[Jahnke 2003a] Jahnke, H. N.: Algebraic Analysis in the 18th Century. in: [Jahnke 2003] S. 105-136.

[Jardine 2003] Jardine, L.: The Curious Life of Robert Hooke. The Man who Measured London. London 2003.

[Jesseph 1993] Jesseph, D. M.: Berkeley's Philosophy of Mathematics. Chicago, London 1993.

[Jesseph 1999] Jesseph, D. M.: Squaring the Circle. The War between Hobbes and Wallis. Chicago, London 1999.

[Johnson/Wolbarsht 1979] Johnson, L.W.; Wolbarsht, M.L.: Mercury poisoning: A probable cause of Isaac Newton's physical and mental ills. (Notes and Records of the Royal Society of London, 34, 1979, 1, S. 1-9).

[Kant 2005] Kant, I.: Allgemeine Naturgeschichte und Theorie des Himmels oder Versuch von der Verfassung und dem mechanischen Ursprunge des ganzen Weltgebäudes nach Newtonischen Grundsätzen abgehandelt. Nachwort von Jürgen Hamel.

4., erweiterte Auflage. Nachdruck der ersten Auflage Königsberg, Leipzig 1755. Frankfurt am Main, 2005.

[Kerr 2002] Kerr, P.: Dark Matter: The Private Life of Sir Isaac Newton. A Novel. Crown 2002.

[Klein 1987] Klein, J.: Francis Bacon oder die Modernisierung Englands. Hildesheim, Zürich, New York 1987.

[Knobloch 1993] Knobloch, E. (Hrsg.): Gottfried Wilhelm Leibniz – De quadratura arithmetica circuli ellipseos et hyperbolae cujus corollarium est trigonometria sine tabulis. Göttingen 1993.

[Knobloch 2002] Knobloch, E.: Leibniz's rigorous foundation of infinitesimal geometry by means of Riemannian sums. (Synthese, Vol. 133, 2002, S. 59-73).

[Knobloch 2004] Knobloch, E.: Von Nicolaus von Kues über Galilei zu Leibniz. Vom mathematischen Umgang mit dem Unendlichen. in: [Hogrebe/Bromand 2004], S. 490–503.

[Knobloch 2008] Knobloch, E.: Generality and Infinitely Small Quantities in Leibniz's Mathematics. The Case of his Arithmetical Quadrature of Conic Sections and Related Curves. in: [Goldenbaum/Jesseph 2008], S. 171–184.

[Knobloch 2012] Knobloch, E.: Galilei und Leibniz. Hannover 2012.

[Knobloch/Schulenburg 2000] Knobloch, E.; Schulenburg, J.-M. Graf von der (Hrsg.): Gottfried Wilhelm Leibniz. Hauptschriften zur Versicherungs- und Finanzmathematik. Berlin 2000.

[Koyré 1965] Koyré, A.: Newtonian Studies. London 1965.

[Krohn 2006] Krohn, W.: Francis Bacon. Orig.-Ausg. 2., überarbeitete Auflage. München 2006.

[Kulenkampff 1987] Kulenkampff, A.: George Berkeley. Orig.-Ausg. München 1987.

[Lasswitz 1984] Lasswitz, K.: Geschichte der Atomistik vom Mittelalter bis Newton, 2 Bde. Unveränd. fotomech. Nachdr. der Ausgabe Hamburg, Leipzig 1890. Hildesheim 1984.

[Leibniz 1985–1992] Leibniz, G. W.: Philosophische Schriften. 3 Bde. in 5 Bdn. 2., unveränd. Aufl. Darmstadt, 1985. Bd. 4. 1. Aufl. 1992.

[Leibniz 1949] Leibniz, G. W.: Werke. Hrsg. von W. E. Peuckert, Bd. 1: Protogaea. Übers. von Wolf von Engelhardt. Stuttgart 1949.

[Leibniz 2004] Leibniz, G. W.: Mathematische Schriften. Hrsg. v. C. I. Gerhardt. Nachdruck. Faks.-Ausg. 1971-2004. 7 Bde. Hildesheim, Zürich, New York 2004.

[Leibniz 2005] Leibniz, G. W.: Monadologie. Französisch-deutsch. Übers. u. hrsg. von Hartmut Hecht. Stuttgart 2005.

[Leibniz 2008a] Leibniz, G. W.: Sämtliche Schriften und Briefe, Reihe 7: Mathematische Schriften. Mathematische Schriften, Bd. 4: 1670-1673: Infinitesimalmathematik. Berlin 2008.

[Leibniz 2008b] Leibniz, G. W.: Sämtliche Schriften und Briefe, Reihe 7: Mathematische Schriften. Mathematische Schriften, Bd. 5: 1674-1676: Infinitesimalmathematik. Berlin 2008.

[Leibniz 2011] Leibniz, G. W.: Die mathematischen Zeitschriftenartikel. Mit einer CD: Die originalsprachlichen Fassungen. Übersetzt und kommentiert von Heinz-Jürgen Hess und Malte-Ludolf Babin. Hildesheim, Zürich, New York 2011.

[Leibniz/Newton 1998] Leibniz, G. W.; Newton, Sir I.: Über die Analysis des Unendlichen. Abhandlung über die Quadratur der Kurven. Aus dem Lateinischen übers. u. hrsg. von G. Kowalewski. Reprint der Ausg. Leipzig. [Kein Jahr angegeben]. Thun, Frankfurt am Main 1998.

[Leonhardi 1799] Leonhardi, F. G.: Geschichte und Beschreibung der Kreis- und Handelsstadt Leipzig nebst der umliegenden Gegend. Leipzig 1799.

[Levenson 2009] Levenson, T.: Newton and the Counterfeiter. The Unknown Detective Career of the World's Greatest Scientist. Boston, New York 2009.

[Li 2012] Li, W. (Hrsg.): Komma und Kathedrale. Tradition, Bedeutung und Herausforderung der Leibniz-Edition. Berlin 2012.

[Lieb/Hershman 1983] Lieb, J.; Hershman, D.: Isaac Newton: Mercury poisoning or manic depression? (The Lancet, Vol. 322. 1983, Issue 8365. S. 1479-80).

[Locke 1995] Locke, J.: An Essay Concerning Human Understanding. Amherst, N. Y. 1995.

[Loeffel 1987] Loeffel, H.: Blaise Pascal 1623–1662. Basel, Boston, Stuttgart 1987.

[Ludovici 1966] Ludovici, C. G.: Entwurf einer vollständigen Historie der Leibnizschen Philosophie zum Gebrauch seiner Zuhörer herausgegeben, 2 Tle. Reprint der Ausgabe Leipzig 1737, Hildesheim, 1966.

[Mackensen 1973] Mackensen, L. von: Von Pascal zu Hahn. Die Entwicklung der Rechenmaschinen im 17. und 18. Jahrhundert. in: [Graef 1973] S. 21-33.

[Mahoney 1990] Mahoney, M. S.: Barrow's mathematics: between ancients and moderns. in: [Feingold 1990] S. 179ff.

[Malcolm/Stedall 2005] Malcolm, N.; Stedall, J.: John Pell (1611-1685) and his Correspondence with Sir Charles Cavendish. The mental world of an early modern mathematician. Oxford, New York, etc. 2005.

[Mancosu 1996] Mancosu, P.: Philosophy of Mathematics & Mathematical Practice in the Seventeenth Century. New York, Oxford 1996.

[Manuel 1968] Manuel, F. E.: A Portrait of Isaac Newton. Cambridge, Mass. 1968.

[Maurer 2002] Maurer, M.: Kleine Geschichte Englands. Durchges., aktualisierte u. bibliogr. erg. Ausg. Stuttgart 2002.

[McKie 1948] McKie, D.: The Arrest and Imprisonment of Henry Oldenburg. (Notes and Records of the Royal Society of London, Vol. 6, 1948, No. 1, S. 28-47).

[Meli 1993] Meli, D. B.: Equivalence and Priority. Newton versus Leibniz. Including Leibniz's unpubl. manuscripts on the Principia. Oxford, New York, etc. 1993.

[Mercator 1975] Mercator, N.: Logarithmotechnia. Nachdruck der Ausgabe London 1668 u. Rom 1666, Hildesheim, New York 1975.

[Morris 1978] Morris, J. (edt.): The Oxford Book of Oxford. Oxford, London, New York 1978.

[Müller/Krönert 1969] Müller, K.; Krönert, G.: Leben und Werk von Gottfried Wilhelm Leibniz. Eine Chronik. Frankfurt am Main 1969.

[Nagel 2008] Nagel, F.: Nieuwentijt, Leibniz, and Jacob Hermann on Infinitesimals. in: [Goldenbaum/Jesseph 2008] S. 199-214.

[Newton 1963] Newton, I.: Mathematische Prinzipien der Naturlehre. Mit Bemerkungen und Erläuterungen herausgegeben von J. Ph. Wolfers. Unveränd. fotomech. Nachdruck der Ausgabe Berlin 1872. Darmstadt 1963.

[Newton 1979] Newton, Sir I.: Opticks. Or a treatise of the reflections, refractions, inflections and colours of light. With a foreword by Albert Einstein, an introd. by Edmund Whittaker. Unabridged an unaltered republ. of the work orig. publ. in 1931. Based on the 4th ed. London 1730. New York 1979.

[Newton 1995] Newton, I.; The Principia. Translated by Andrew Motte. Nachdruck der Übersetzung der dritten Auflage der *Principia* 1729, Amherst, New York 1995.

[Newton 1999] Newton, I.: The Principia. Mathematical Principles of Natural Philosophy. Translation by I. Bernard Cohen and Anne Whitman. Berkeley, Calif., Los Angeles, London 1999.

[Newton 1999a] Newton, I.: Die mathematischen Prinzipien der Physik. Übersetzt und herausgegeben von Volkmar Schüller. Berlin, New York 1999.

[Nikolaus 1952] Nikolaus von Kues: Schriften, Bd. 11: Die mathematischen Schriften. Übersetzt von Josepha Hofmann. Mit einer Einf. u. Anm. vers. von Joseph Ehrenfried Hofmann. Hamburg 1952.

[Pascal 1997] Pascal, B.: Gedanken über die Religion und einige andere Themen. Hrsg. von Jean-Robert Armogathe. Aus d. Franz. übers. v. Ulrich Kunzmann. Stuttgart 1997.

[Pask 2013] Pask, C.: Magnificient Principia. Exploring Isaac Newton's Masterpiece. Amherst, New York 2013.

[Pepys 2003] Pepys, S.: The Diary of Samuel Pepys: A Selection. Selected and Edited by Robert Latham from *The Diary of Samuel Pepys, a new and complete transcription*, edited by Robert Latham and William Matthews. First published as *The Shorter Pepys*, 1985. London, New York, etc. Reprint under the present title 2003.

[Ploetz 2008] Der große Ploetz. Die Enzyklopädie der Weltgeschichte. Göttingen 2008.

[Purrington 2009] Purrington, R. D.: The First Professional Scientist. Robert Hooke and the Royal Society of London. Basel, Boston, Berlin 2009.

[Rice 2012] Rice, A.: Vindicating Leibniz in the calculus priority dispute: The role of Augustus De Morgan. in: [Wardhaugh 2012] S. 98-114.

[Rigaud 1965] Rigaud, S. J.: Correspondence of Scientific Men of the Seventeenth Century. 2 Vols. Reprint Hildesheim 1965.

[Ritter 1916] Ritter, P.: Bericht eines Augenzeugen über Leibnizens Tod und Begräbnis. (Zeitschrift des Historischen Vereins für Niedersachsen, 81. Jahrgang 1916, Heft 3, S. 247-252).

[Robinson/Adams 1955] Robinson, H. W.; Adams, W.: The Diary of Robert Hooke M.A., M.D., F.R.S, 1672-1680. Transcibed from the orig. in the possession of the Corporation of the City of London. With a foreword by Sir Frederick Gowland Hopkins. London 1935.

[Rosenberger 1987] Rosenberger, F.: Isaac Newton und seine Physikalischen Prinzipien. Ein Hauptstück aus der Entwickelungsgeschichte der modernen Physik. Unveränd. reprogr. Nachdruck der Ausg. Leipzig 1895. Darmstadt 1987.

[Roth 1971] Roth, L.: Old Cambridge Days. (The American Mathematical Monthly, Vol. 78, No. 3. S. 223-236, 1971)

[Rouse Ball 1960] Rouse Ball, W. W.: A Short Account of the History of Mathematics. Reprogr. Nachdruck der 4. Auflage 1908. New York 1960.

[Schmidt-Biggemann 1999] Schmidt-Biggemann, W.: Blaise Pascal. Orig.-Ausg. München 1999.

[Scott 1981] Scott, J. F.: The Mathematical Work of John Wallis, D. D., F. R. S (1616–1703). 2. ed., reprod. of the 1983 ed. With a foreword by E. N. da C. Andrade. New York 1981.

[Scriba 1969] Scriba, Chr. J.: Neue Dokumente zur Entstehungsgeschichte des Prioritätsstreites zwischen Leibniz und Newton um die Erfindung der Infinitesimalrechnung. (Studia Leibnitiana, Supplementa: Akten des internationalen Leibniz-Kongresses Hannover, 14.-19. November 1966, Band II: Mathematik, Naturwissenschaften. 1969).

[Scriba 1970] Scriba, Chr. J.: The Autobiography of John Wallis, F. R. S. (Notes and Records of the Royal Society of London Vol. 25, 1970, S. 17–46).

[Shapiro 1990] Shapiro, A. E.: The *Optical Lectures* and the foundations of the theory of optical imagery. in: [Feingold 1990] S. 250-290.
[Smith 1958] Smith, D. E.: History of Mathematics. 2 vols., Reprint of the first edition Vol. 1: 1923, Vol. 2: 1925. New York 1958.
[Sonar 2008] Sonar, Th.: Der Tod des Gottfried Wilhelm Leibniz. Wahrheit und Legende im Licht der Quellen. in: Abhandlungen der Braunschweigischen Wissenschaftlichen Gesellschaft, Band LIX, S. 203-230, 2008.
[Sonar 2011] Sonar, Th.: 3000 Jahre Analysis. Geschichte, Kulturen, Menschen. Berlin, Heidelberg, New York 2011.
[Sonar 2016] Sonar, Th.: 3000 Jahre Analysis. Geschichte, Kulturen, Menschen. 2te korrigierte Auflage, Berlin, Heidelberg, New York 2016.
[Spargo/Pounds 1979] Spargo, P. E., Pounds, C. A.: Newton's 'Derangement of the Intellect'. New Light on an Old Problem. (Notes and Records of the Royal Society of London, Vol. 34, 1979, Nr. 1, S. 11-32)
[Stäckel 1976] Stäckel, P. (Hrsg.): Variationsrechnung. Abhandlungen von Johann Bernoulli, Jacob Bernoulli, Leonhard Euler, Joseph Louis Lagrange, Adrien Marie Legendre, Carl Gustav Jacobi. Sonderausg. Nachdruck der Ausg. Leipzig 1894. Darmstadt 1976.
[Stedall 2002] Stedall, J. A.: A Discourse Concerning Algebra. English Algebra to 1685. Oxford, New York, etc. 2002.
[Stedall 2004] Stedall, J. A.: The Arithmetic of Infinitesimals: John Wallis 1656. New York: Springer, 2004.
[Szabó 1996] Szabó, I.: Geschichte der mechanischen Prinzipien und ihrer wichtigsten Anwendungen. Hrsg. v. Peter Zimmermann u. Emil A. Fellmann. Korrigierter Nachdruck der 3. korr. u. erw. Aufl. Basel, Boston, Berlin 1996.
[Thiel 1990] Thiel, U.: John Locke. Mit Selbstzeugnissen und Bilddokumenten. Reinbek bei Hamburg 1990.
[Thomas/Smith 1990] Thomas, D. J.; Smith, J. M.: Joseph Raphson, F.R.S. (Notes and Records of the Royal Society of London Vol. 44, 1990, Nr. 2, S. 151-167).
[Thompson 1998a] Thompson, S. P.; Gardner, M.: Calculus Made Easy. Being a very-simplest Introduction to those Beautiful Methods of Reckoning which are Generally Called by the Terrifying Names of the Differential Calculus and the Integral Calculus. Newly Revised, Updated, Expanded, and Annotated for its 1998 edition. New York 1998.
[Thompson 1998b] Thompson, S.P.: Analysis leicht gemacht. Differenzieren und Integrieren verstehen durch Üben. 12. Auflage, unveränd. Nachdruck. Thun, Frankfurt/Main 1998.
[Truesdell 1958] Truesdell, C.: The New Bernoulli Edition. (Isis, Vol. 49, 1958, No. 1, S. 54-62).
[Truesdell 1984] Truesdell, C.: An Idiot's Fugitive Essays on Science. Methods, criticism, training, circumstances. New York, Berlin, Heidelberg, etc. 1984.
[Turnbull 1959–77] Turnbull, H. W. et al. (edts): The Correspondence of Isaac Newton. 7 Vols. Cambridge 1959-1977.
[Verdun 2015] Verdun, A.: Leonhard Eulers Arbeiten zur Himmelsmechanik, Band 1. Berlin, Heidelberg, etc. 2015.
[Vergil 2001] Vergilius Maro, Publius: Bucolica – Hirtengedichte. Lateinisch/-deutsch. Übers., Anm., interpretierender Kommentar u. Nachwort v. Michael von Albrecht. Studienausg. Stuttgart 2001.
[Vermij 1989] Vermij, R. H.: Bernard Nieuwentijt and the Leibnizian Calculus. (Studia Leibnitiana, Band 21, 1989, Heft 1, S. 69-86).

[Volkert 1988] Volkert, K.: Geschichte der Analysis. Mannheim, Wien, Zürich 1988.

[Voltaire 2011] Voltaire: Letters concerning the English Nation. Edited with an Introduction and Notes by Nicholas Cronk. Oxford, New York, etc. 2009.

[Wahl 2012] Wahl, Ch.: Diplomat in der Gelehrtenrepublik – Leibniz' politische Fähigkeiten im Dienste der Mathematik. in [Li 2012] S. 273-291.

[Walsdorf 2014] Walsdorf, A.; Badur, K.; Stein, E.; Kopp, F. O.: Das letzte Original. Die Leibniz-Rechenmaschine der Gottfried Wilhelm Leibniz Bibliothek. Hannover 2014.

[Wardhaugh 2012] Wardhaugh, B.: The History of the History of Mathematics. Case studies for the seventeenth, eighteenth and nineteenth centuries. Oxford, Bern, Berlin, etc. 2012.

[Waschkies 1999] Waschkies, H.-J.: Leibniz' geologische Forschungen im Harz. in [Breger/Niewöhner 1999] S. 187-210.

[Westfall 2006] Westfall, R. S.: Never at Rest. A Biography of Isaac Newton. 18. printing. Cambridge 2006.

[Whiteside 1967–81] Whiteside, D. T. (edt.): The Mathematical Papers of Isaac Newton. 8 Vols. Cambridge 1967–1981.

[Wiegandt/Tille 1988] Wiegandt, P.; Tille, P.: Ein Apfel und Sir Isaak. 2. Auflage. Halle 1988.

[Wielenga 2012] Wielenga, F.: Geschichte der Niederlande. Stuttgart 2012.

[Wußing 1984] Wußing, H.: Isaac Newton. 3., durchges. Aufl. Leipzig 1984.

[Yoder 1988] Yoder, J.G.: Unrolling Time. Christiaan Huygens and the Mathematization of Nature. Cambridge, New York, etc. 1988.

List of Figures

The abbreviation PD in a reference means that these pictures were marked as 'public domain' in the internet at the time of the production of this book, or that they were marked as free of license otherwise. The abbreviation HWK stands for the editor Heiko Wesemüller-Kock.

T. Sonar, *The History of the Priority Dispute between Newton and Leibniz*, https://doi.org/10.1007/978-3-319-72563-5

Index of Names

T. Sonar, *The History of the Priority Dispute between Newton and Leibniz*, https://doi.org/10.1007/978-3-319-72563-5

Subject Index

T. Sonar, *The History of the Priority Dispute between Newton and Leibniz*, https://doi.org/10.1007/978-3-319-72563-5

The manufacturer's authorised representative in the EU is Springer Nature Customer Service Centre GmbH, Europaplatz 3, 69115 Heidelberg, Germany. If you have any concerns regarding our products, please contact ProductSafety@springernature.com

Printed and bound by CPI Group (UK) Ltd, Croydon, CR0 4YY
30/09/2026
02225438-0003